D1577390

Human Motion

Computational Imaging and Vision

This comprehensive book series embraces state-of-the-art expository works and advanced research monographs on any aspect of this interdisciplinary field.

Topics covered by the series fall in the following four main categories:

- Imaging Systems and Image Processing
- Computer Vision and Image Understanding
- Visualization
- Applications of Imaging Technologies

Only monographs or multi-authored books that have a distinct subject area, that is where each chapter has been invited in order to fulfill this purpose, will be considered for the series.

Volume 36

Human Motion
Understanding, Modelling, Capture, and Animation

Edited by

Bodo Rosenhahn

Max-Planck Institute for Computer Science, Saarbrücken, Germany

Reinhard Klette

The University of Auckland, New Zealand

and

Dimitris Metaxas

Rutgers University, USA

 Springer

A C.I.P. Catalogue record for this book is available from the Library of Congress.

ISBN 978-1-4020-6692-4 (HB)
ISBN 978-1-4020-6693-1 (e-book)

Published by Springer,
P.O. Box 17, 3300 AA Dordrecht, The Netherlands.

www.springer.com

Printed on acid-free paper

Contents

Part V Modelling and Animation

Preface

Edward Muybridge (1830–1904) is known as the pioneer in motion capturing with his famous experiments in 1887 called "Animal Locomotion". Since then, the field of animal or human motion analysis has grown in many directions. However, research and results that involve human-like animation and the recovery of motion is still far from being satisfactory.

The modelling, tracking, and understanding of human motion based on video sequences as a research field has increased in importance particularly in the last decade with the emergence of applications in sports sciences, medicine, biomechanics, animation (online games), surveillance, and security. Progress in human motion analysis depends on empirically anchored and grounded research in computer vision, computer graphics, and biomechanics. Though these fields of research are often treated separately, human motion analysis requires the integration of methodologies from computer vision and computer graphics. Furthermore, the understanding and use of biomechanics constraints improves the robustness of such an approach.

This book is based on a June 2006 workshop held in Dagstuhl, Germany. This workshop brought together for the first time researchers from the aforementioned disciplines. Based on their diverse perspectives, these researchers have been developing new methodologies and contributing, through their findings, to the domain of human motion analysis. The interdisciplinary character of the workshop allowed people to present a wide range of approaches that helped stimulate intellectual discussions and the exchange of new ideas.

The goal of the editors of this book is to present an interdisciplinary approach to modelling, visualization, and estimation of human motion, based on the lectures given at the workshop. We invited several authors to contribute chapters in five areas, specifically 2D processing, 3D processing, motion learning, animation and biomechanics, and mathematical foundations of motion modelling and analysis. Approximately five chapters represent each area. Each chapter reflects the current "state of the art" in its respective research area. In addition, many chapters present future challenges. Three experts reviewed each chapter based on the Springer-Verlag guidelines and, if they deemed a chapter acceptable, its author(s) were required to make revisions within a month.

This is the first edited volume by Springer-Verlag on this emerging and increasingly important topic and similar workshops and special meetings have already been scheduled as part of the best computer vision and graphics conferences e.g., CVPR, ICCV, and Eurographics.

The editors would like to thank the authors for their contributions to this volume and we are certain that given the importance of this interdisciplinary domain, many more books will be published and meetings will occur.

Saarbrücken, Auckland, New Jersey   Bodo Rosenhahn, Reinhard Klette, Dimitris Metaxas
July 2007

1

Understanding Human Motion: A Historic Review

Reinhard Klette[1] and Garry Tee[2]

[1] Computer Science Department
The University of Auckland, Auckland, New Zealand
[2] Department of Mathematics
The University of Auckland, Auckland, New Zealand

Summary. Understanding human motion is based on analyzing global motion patterns, rather than on studying local patterns such as hand gestures or facial expressions. This introductory chapter reviews briefly (by selection, not by attempting to cover developments, and with a focus on Western History) people and contributions in science, art, and technology which contributed to the field of human motion understanding. This review basically stops at the time when advanced computing technology became available for performing motion studies based on captured image data or extensive (model-based) calculations or simulations.

1.1 Introduction

Interest in human motion goes back very far in human history, and is motivated by curiosity, needs or methods available at a time. For example, a biomechanical perspective is characterized by the "need for new information on the characteristics of normal and pathological human movement" [35]. It is also possible to outline disciplines of science (e.g., mathematics) or arts (e.g., paintings, sculptures), relevant to human motion, just to indicate briefly the complexity of the subject. Obviously, different disciplines are interested in different aspects of the subject; biomechanics is, for example, focusing on human locomotion, with less interest in muscle models, and when correcting motion (e.g., of disabled children) by surgery, it will be exactly the opposite.

This chapter attempts to inform about a few developments which are of (somehow) joint interest for computer vision, computer graphics, and biomechanics. Those areas collaborate increasingly in research and developments relevant to human motion.

The following chapter could certainly be more detailed about the historic context of developments in the understanding of human motion at various periods of human history. For example, art was definitely a major driving force for many centuries for specifying human motion (see, e.g., comments

1

B. Rosenhahn et al. (eds.), *Human Motion – Understanding, Modelling, Capture, and Animation*, 1–22.
© 2008 *Springer*.

on da Vinci below), or Braune and Fischer were (at the beginning of 20th century; see below) among the first to quantitatively measure human motion, but their work was motivated by improving the efficiency of troop movement.

Many mathematical results had been found in the early civilizations of Mesopotamia and Egypt, but a succession of Greek thinkers (starting with Thales, in the 6th century)[1] developed mathematics as a coherent logically organized structure of ideas. We start our review at this period of time.

1.2 Classical Antiquity

The ancient Greek philosopher Aristotle (-383 to -321) published, besides much other fundamental work, also a (short) text $\Pi EPI \ \Pi OPEIAS \ Z\Omega I\Omega N$ [3] on the gait of animals. He defined locomotion as "the parts which are useful to animals for movement in place". The text is very readable, certainly also due to an excellent translation, and it contains discussions of interesting questions (e.g., "why are man and birds bipeds, but fish footless; and why do man and bird, though both bipeds, have an opposite curvature of the legs"), links to basic knowledge in geometry (e.g., "when ... one leg is advanced it becomes the hypothenuse of a right-angled triangle. Its square then is equal to the square on the other side together with the square on the base. As the legs then are equal, the one at rest must bend ... at the knee ..."), or experiments (e.g., "If a man were to walk parallel to a wall in sunshine, the line described (by the shadow of his head) would be not straight but zigzag..."). This text[2] is the first known document on biomechanics. It already contains, for example, very detailed observations about the motion patterns of humans when involved in some particular activity.

Sculptures, reliefs, or other artwork of classical antiquity demonstrate the advanced level of understanding of human or animal motion, or body poses (often in a historic context).

Classical antiquity already used mathematics for describing human poses or motion, demonstrated in artworks that we have to consider individual poses as well as collective poses (e.g., in Roman arts, a married couple was indicated by showing eye contact between woman and man, possibly enhanced by a pictured handshake), and showed in general that motion and poses need to be understood in context. Motion was only presented by means of static artwork; the first dynamic presentation of motion was by means of moving pictures, and this came nearly 2000 years later, at the end of the 19th century.

A charioteer with horses four-in-hand traditionally had the horses gallop in a race, where gallop is defined as a certain step-sequence by the horses, also

---

[1] We use the astronomical system for numbering years.

[2] In close relation with Aristotle's texts $\Pi EPI \ Z\Omega I\Omega N \ \Gamma ENE\Sigma E\Omega\Sigma$ (*On the Parts of Animals*) and $\Pi EPI \ Z\Omega I\Omega N \ KINH\Sigma E\Omega\Sigma$ (*On the Progression of Animals*).

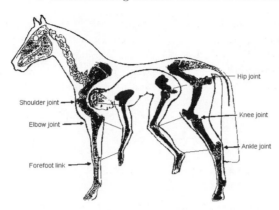

Fig. 1.1. A modification of a drawing published by W. F. Bartels on www. kutschen.com/Gelenke.htm.

including a period of suspension with no hoof touching the ground. However, until the invention of moving pictures, it was an open question whether such a period of suspension does occur.

In classical antiquity, motion patterns of humans were usually studied in close relation to motion patterns of animals. Indeed, those comparative studies have continued to be useful: see Figure 1.1 illustrating evolutionary relations between joints of humans and horses.

Human motion studies today are basically performed by modelling human (dynamic) shape, and by applying perspective geometry (when understanding recorded image sequences or creating animations). Basics of the geometry of three-dimensional (3D) volumes and, to a lesser extent, also of perspective geometry, date back to classical antiquity [13]. Perspective emerged (at first) from geometrical optics (see, e.g., Euclid's[3] *ΟΠΤΙΚΗ* (*Optics*), defining visual rays or visual cones), and it received a major stimulus in art of the European Renaissance (see next section).

1.3 Renaissance

Leonardo da Vinci (1452–1519) stated in his sketchbooks, that "it is indispensable for a painter, to become totally familiar with the anatomy of nerves, bones, muscles, and sinews, such that he understands for their various motions and stresses, which sinews or which muscle causes a particular motion"

---

[3] Euclid worked at the Museion in Alexandria (in about −300), writing on mathematics, optics, astronomy and harmony. His "Elements" gives a very detailed study of plane and solid geometry (together with number theory), and it became one of the most influential books ever written. His treatment of geometrical optics formed a basis for the theory of perspective.

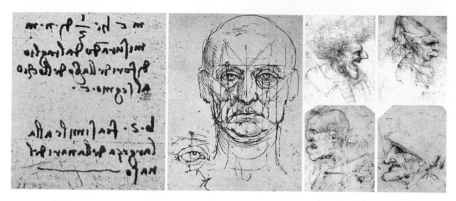

Fig. 1.2. Fragments from da Vinci's sketchbooks (human faces).

of a human.[4] For an example of his modelling of the human anatomy, see Figure 1.2 on the left and in the middle. In his mirror writing he wrote that "$m\ c$ measures $1/3$ of $n\ m$, measured from the outer angle of the eye lid to letter c" and "$b\ s$ corresponds to the width of the nostril". However, a few pages later he showed "funny faces" in his sketchbook (for a few, see right of Figure 1.2), illustrating that a match between model and reality was not always given.

Besides very detailed models of the human anatomy, also characterizing special appearances such as parameters of "a beautiful face" (e.g., in his opinion, in such a face the width of the mouth equals the distance between the middle line of the mouth to the bottom of the chin), da Vinci's sketchbooks also contain quite detailed studies about *kinematic trees*[5] of human motion. For a man going upstairs (see left of Figure 1.3), he writes: "The center of mass of a human who is lifting one foot, is always on top of the center of the sole of foot [on which he is standing]. A human going upstairs shifts weight forward and to the upper foot, creating a counterweight against the lower leg, such that the workout of the lower leg is reduced to moving itself. When going upstairs, a human starts with relieving body weight from that foot which he is going to lift. Furthermore, he dislocates the remaining body mass onto the opposite leg, including the [weight of the] other leg. Then he lifts this other leg and places the foot on the step, which he likes to climb on. Next he dislocates the whole body weight, including that of this leg, onto the upper foot, puts his hand onto his thigh, slides his head forward, and moves towards the tip of the upper foot, quickly lifting the heel of the lower foot. With this push he

---

[4] Quotations of da Vinci are translated from [16]. Today, various books are published on human anatomy specially designed for artists; see, for example, [5].

[5] Today, *kinematic chains* are used for modelling propagations, e.g. of forces, over time along a body part such as an arm or a leg. Da Vinci already considered "divisions" in those propagations (e.g., from the upper body to both the left and the right leg), here indicated by using the name *tree* rather than *chain*.

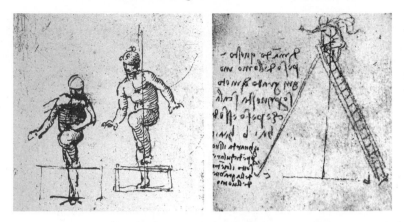

Fig. 1.3. Drawing in da Vinci's sketchbooks (a man going upstairs, or up a ladder).

lifts himself upward, simultaneously he straightens the arm which was resting on the knee. This stretching of the arm pushes body and head upward, and thus also straightens the back which was bended before."

Next to the drawing, shown on the right of Figure 1.3, da Vinci wrote the following: "I ask for the weight [pressure] of this man for every segment of motion when climbing those stairs, and for the weight he places on b and on c. Note the vertical line below the center of mass of this man."

It is certainly impressive to see the level of detail in modelling human shape or motion, given by da Vinci centuries ago. This was illustrated above just by examples, and a comprehensive biomechanical study about his contributions would be a sensible project.

Michelangelo di Lodovico Buonarroti Simoni (1475–1564) is also famous for his realistically portrayed human motion.

Perspective geometry (required for proper modelling of human motion) was established by means of drawing rules by artists of the Renaissance, such as Filippo Di Ser Brunellesco (1377–1446), Piero della Francesca (1420?–1492), Albrecht Dürer (1471–1528), Raphael (1483–1520), and many others. Perspective geometry also became a mathematical theory, pioneered by Girard Desargues (1591–1661) at the beginning of the Baroque era.

1.4 Baroque

The scientist Giovanni Alfonso Borelli (1608–1679) contributed to various disciplines. In his "On the Movement of Animals" [8] (published posthumously in two parts in 1680 and 1681) he applied to biology the analytical and geometrical methods, developed by Galileo Galilei (1564–1642)[6] in the field of

---

[6] Galileo Galilei became a major founder of modern science, applying analytical and geometrical methods in the field of mechanics, combining theory and experiment.

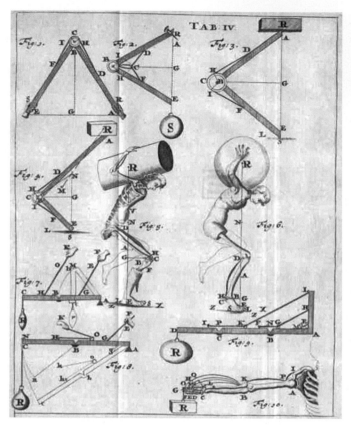

Fig. 1.4. Copper engraving by Borelli in 1680/1681 (Deutsches Museum, Munich).

mechanics.[7] For this reason he is also often called "the father of biomechanics" (with Aristotle as a second alternative, see Section 1), or (one of) the founder(s) of the *Iatrophysic School* (also called *iatromathematic, iatromechanic,* or *physiatric*). A result of basic importance for establishing this school is that the circulation of the blood is comparable to a hydraulic system. This school vanished after some years, but some of the work of Borelli is still worth noting today. He "was the first to understand that the levers of the musculoskeletal system magnify motion rather than force, so that muscles must produce much larger forces than those resisting the motion" [32]. Bones serve as levers and muscles function according to mathematical principles; this became a basic principle for modelling human motion.

Figure 1.4 shows an example of a drawing from [8]. The physiological studies in this text (including muscle analysis and a mathematical discussion of movements, such as running or jumping) are based on solid mechanical

---

[7] Borelli's great work is almost wholly a study of animal statics, since Newton had not yet developed the mathematics of dynamics.

principles. The change from visual (qualitative) observation to quantitative measurements was crucial for the emergence of biomechanics. Borelli also attempted [8] to clarify the reason for muscle fatigue and to explain organ secretion, and he considered the concept of pain.

1.5 Age of Enlightenment

There seem to be not many important contributions to the study of human motion, between the times of Borelli and the latter half of the 19th century, when chronophotography provided a new tool for understanding motion.

Besides studies on human or animal motion in a narrow sense, the foundation of modern dynamics by Isaac Newton (1642–1727),[8] including his three laws of motion, was also a very crucial contribution to the understanding of human motion (these laws are formulated here for this case, slightly modified from [41]):

> *Newton's Law of Inertia.* A human in motion will continue moving in the same direction at the same speed unless some external force (like gravity or friction) acts to change the motion characteristics. (This law was already formulated by Galileo Galilei.)
> *Newton's Law of Acceleration.* $F = ma$. A force F acting on a human motion will cause an acceleration a in the direction of the force and proportional to the strength of the force (m is the mass of the human).[9]
> *Newton's Law of Action-Reaction.* A human's motion against a medium (such as another body) is matched with a reaction force of equal magnitude but opposite direction.

All three laws had been discussed already in some sense by Aristotle when considering the motions of a boat. According to Aristotle, "every movement needs a mover", and his (incorrect !) concept can be expressed as $F = mv$, where v is the velocity [21].

Between the 17th and 19th centuries numerous famous scientists, starting with René Descartes (1596–1650), basically established modern mathematics,

---

[8] Isaac Newton was born on 25 December 1612, in the Julian Calendar which was still used in England. (In the Gregorian calendar the date was 3 January 1613). He is the most influential of all scientists. He made many major advances in mathematics (including differential and integral calculus), and he applied mathematics with immense success to dynamics, astronomy, optics and other branches of science.

[9] Newton's statement of the Second Law of Motion is: "The change of motion is proportional to the motive power impressed; and is made in the direction of the right line in which that force is impressed" (the Motte–Cajori translation). Newton's applications of that law shew that "change of motion" means "rate of change of momentum", and it holds for bodies of varying mass.

including geometrical volumes, analytical geometry, and geometrical algebra. Today's human motion studies benefit from those developments.

We briefly mention Luigi Galvani (1737–1798) and his discovery (1780) of "animal electricity", which was correctly interpreted (in 1800) by Alessandro Volta (1748–1827) as muscles contracting in response to electric current. Hermann von Helmholtz (1821–1894) invented the myograph in 1852, and he used it to study the propagation of electricity along nerves. He was greatly surprised to find that, in a frog nerve, the electrical signal travelled only 27 metres per second [45].

The concepts of energy and of thermodynamics were largely developed between 1820 and 1860 by Nicolas Léonard Sadi Carnot (1796–1832), Rudolf Julius Emmanuel Clausius (1822–1888), William Thomson (later Baron Kelvin, 1824–1907), Herman von Helmholtz (1821–1894), and James Clerk Maxwell (1831–1879).

The mathematician Charles Babbage (1791–1871) had, by 1837, invented all of the basic ideas about computers (and many advanced ideas about them). He attempted to construct various mechanical computers, but he did not succeed in completing any working computer. (In 1991, his Difference Engine No. 2 was completed, following his plans – and it worked.)

Quantitative physiology was founded in 1866 by the chemist Edward Frankland (1825–1899), who demonstrated experimentally that work done by a human matched the energy of the food consumed [11].

In the 19th century, a variety of toys were made which produced moving pictures. In the 1830s, several inventors developed the *Phenakistoscope*, a disk with several radial slots and successive pictures painted between the slots. When the disk was spun with the pictures on the side facing a mirror, a viewer looking towards the mirror from the blank side of the disk would get momentary views (through the slots) of the pictures in cyclic motion. In the 1860s, several inventors improved that to develop the *Zoetrope*, a rotating drum with slits parallel to the axis. A strip of paper could be fitted inside the cylinder, with slits in the paper fitted to the slits in the cylinder, and with successive pictures printed on the paper between the slits [40, pp.16–20].

A major contribution was the work by brothers Ernst Heinrich Weber (1795–1878), Wilhelm Eduard Weber (1804–1891), and Eduard Friedrich

Fig. 1.5. From left to right: Eduard Friedrich Weber, Ernst Heinrich Weber, and Wilhelm Eduard Weber. Right: calculated picture from [46].

Weber (1806–1871); all three collaborated in their research on physics, human anatomy and locomotion. The latter two published the book [46]. Wilhelm Eduard Weber is famous for the invention of the first electromagnetic telegraph in 1833, jointly with Carl Friedrich Gauss (1777–1855).

[46] contains "calculated pictures", see right of Figure 1.5, pioneering today's computer graphics. Subsequent phases of human walking are calculated using differential equations, and visualized by drawings using perspective projection.

[46] analyzed human gait and provided a theory of locomotion, including the prediction of a "walking machine" (with two, four, or more legs, depending on terrain difficulty), moved by steam. In fact, movement control of today's multi-legged robots depends on solutions of PDEs (partial differential equations). The Weber brothers were the first who studied the path of the center of mass during movement.

The mathematician Pafnyutii L'vovich Chebyshev (1821–1894) advanced greatly the theory of mechanical linkages. His inventions include a wheelchair driven by crank handles, and a model of a 4-leg walking machine. Such "walking machines" might have inspired Mrs. Edmund Craster (died 1874) to write her poem *The Centipede*:

> The Centipede was happy quite,
> Until the Toad in fun
> Said 'Pray, which leg goes after which?'
> And worked her mind to such a pitch
> She lay, distracted, in the ditch
> Consid'ring how to run.

1.6 Chronophotography

The French astronomer Pierre Janssen (1824–1907) used on 8 December 1874 a multi-exposure camera (of his own invention) for recording the transit of Venus across the Sun. His "clockwork 'revolver' took forty-eight exposures in seventy-two seconds on a daguerreotype disc. Janssen's work in turn greatly

Fig. 1.6. Left: E.-J. Marey. Right: an 1882 photo by Marey.

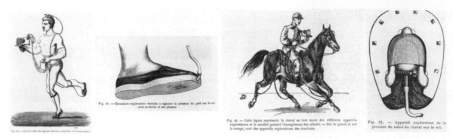

Fig. 1.7. Left: a runner with instruments to record his movements, including a shoe to record duration and phases of ground contact. Right: a trotting horse with instruments to record the horse's leg locomotion, including an instrument for measuring the ground pressure of a hoof [31].

influenced the chronophotographic experiments" [47] of the French scientist Etienne-Jules Marey (1830–1904); see left of Figure 1.6 for a photo of himself. He was interested in locomotion of animals or humans. In his book [31] he reported about motion studies, where data had been collected by various instruments; see Figure 1.7.

His interests in locomotion studies led him later to the design of special cameras allowing a recording of several phases of motion in the same photo. Figure 1.6, right, shows a flying pelican recorded by him around 1882. Marey reported in a 1890 book about locomotion of birds, also using his photographs for illustration and analysis. Later he also used movies (with up to 60 pps in good quality), which was influential pioneering work for the emerging field of cinematography. Figure 1.8 illustrates his work reported in the book [33]. Also see [10] (short-listed in 1994 for Britain's Kraszna-Krausz award).

The British-born Eadweard Muybridge (1830–1904) became a renowned photographer after he emigrated to the USA. Inspired by Marey's recording of motion [29], and by a disputed claim that a galloping horse may have all

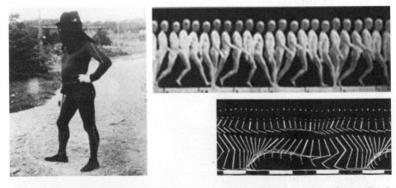

Fig. 1.8. Left: a man in a black dress; limbs are marked by white lines. Right, top: a chronophotograph of a striding man dressed partially in white, and partially in black. Right, bottom: white lines in a chronophotograph of a runner [33].

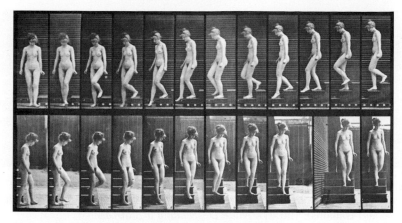

Fig. 1.9. Woman walking downstairs (Muybridge, late 19th century).

four hooves off the ground, in 1878 he set up a series of 12 cameras[10] for recording fast motion alongside a barn, sited on what is now the Stanford University campus. His rapid sequence of photographs of a galloping horse did shew all four hooves off the ground for part of the time [40, p.21] [36]. He invented a machine for displaying the recorded series of images, pioneering motion pictures this way. He applied his technique to movement studies. The human subjects were typically photographed nude or nearly nude, for different categories of locomotion; see Figure 1.9.

Fig. 1.10. Lilienthal's flight, Maihöhe 1893, photographed by O. Anschütz (Otto Lilienthal Museum, Germany).

Muybridge's motion studies, based on multiple images, included walking downstairs, boxing, walking of children, and so forth. They are often cited in the context of the beginning of biomechanics, and they were definitely very influential for the beginning of cinematography at the end of the 19th century. Movies were shot in several countries, shortly after his successful demonstrations of "moving pictures".

A third famous representative of chronophotography was the German inventor Ottomar Anschütz (1846–1907) whose 1884 photographs of gliding

---

[10] In 1879, he increased that to 24 cameras.

Fig. 1.11. From left to right: detail of one of Muybridge's photographic plates, Pablo Picasso's *Les Demoiselles d'Avignon* (1907), Marcel Duchamp's *Nude Descending a Staircase*, and Hananiah Harari's *Nude Descending a Staircase*.

flights of storks inspired Otto Lilienthal's design of experimental gliders. One of Anschütz's inventions is a 1/1000th of a second shutter. Earlier than Muybridge, he invented in 1890 a machine (called *Tachyscop*) for the first moving pictures. It was similar to a *Zoetrope* but used photographs lined up in a cylinder, which could be seen through a slot (e.g., a walking man, a walking woman, the gallop of a horse, and a flying crane). Anschütz also took photos of the first flights of Lilienthal in 1893 and 1894; see Figure 1.10.

Less known, but also a pioneer of the early days of human motion capturing, was Albert Londe (1858–1917) [47]. Londe constructed a camera, fitted with (at the beginning of his studies) 9 lenses arranged in a circle, and used this camera to study the movements of patients (at La Hôpital de la Salpêtrière in Paris) during epileptic fits.

The work by Marey and Muybridge was also of great influence in the arts [15]. Figure 1.11 shows on the left a detail of one of Muybridge's plates, showing a female with a handkerchief. Duchamp points to Marey for origins of his *Nude Descending a Staircase*, and Picasso points to one of Muybridge plates, entitled *Dropping and Lifting of a Handkerchief* (1885), for origins of his *Les Demoiselles d'Avignon.* The English painter Francis Bacon (1909–1992) even compared the importance of Muybridge for his artistic development with that of Michelangelo [15].

1.7 Human Motion Studies in Biomechanics

In the latter half of the 19th century, Christian Wilhelm Braune (1831–1892) and Otto Fischer (1861–1917) started with experimental studies of human gait (e.g., for determining the center of mass), which resulted in the development of prosthesis.

In the 20th century, biomechanics developed into a discipline of science, establishing its own research programs. The French reformer (and 'work physiologist') Jules Amar (1879–1935) published in 1914 the very influential book [2],

Fig. 1.12. Left: A. V. Hill. Right: Testing the acceleration of sprinters by Hill at Cornell University, Ithaca, NY [23]. "The large coils of wire were used to detect a magnet worn by the runner as he sprinted past them. Velocity and acceleration were calculated by knowing the distance between the wire coils." [6].

which soon after defined the standards for human engineering in Europe and the United States. The technology of cinematographic analysis of sprint running allowed a new quality in research (note: the flicker-fusion rate of the human eye is only about 12 Hz); see, for example, papers [17,18] by Wallace O. Fenn (1893–1971), who became the president of the American Physiological Association. Graduate programs in biomechanics developed in the United States in the 1940s. Starting with the 1950s, biomechanics became a worldwide discipline for physical educators, especially in the context of sports. Helmholtz's myograph was developed into the electronic electromyograph, for measuring the electric activity of muscles.

The book [7] by Nicholas Bernstein (1896–1966) pioneered the areas of motor control and coordination. He studied the spatial conception of the degrees-of-freedom problem in the human motor system for walking, running or jumping.

Archibald Vivian Hill (1886–1977) was convinced by F. G. Hopkins (Nobel Prize in Physiology or Medicine, 1929) to "pursue advanced studies in physiology rather than mathematics" [42]. Hill investigated the efficiency and energy cost in human movement (see, e.g., [24]). Based on his solid background in mathematics, he developed mathematic "models describing heat production in muscle, and applied kinetic analysis to explain the time course of oxygen uptake during both exercise and recovery" [42]. His research initiated biophysics [25]. Hill shared the 1922 Nobel Prize in Physiology or Medicine with the German chemist Otto Meyerhof. Hill was honored for his discoveries about the chemical and mechanical events in muscle contraction [22].

Research in computerized gait analysis is today widely supported by marker-based pose tracking systems (see Figure 1.13), which have their origins in the work by G. Johannsson [26] (see Section 1.9). Basically, the camera systems used are fast (e.g., 300 Hz or more), but recorded images are normally restricted to binary information, showing positions of markers only. Computer

Fig. 1.13. Gait laboratory of The University of Auckland (in 2000). Left: walking subject with markers. Middle: walking area (top) and one of the fast cameras (bottom). Right: generated animated 3D stick figure (to) and camera calibration unit (bottom).

vision already helped to create 3D body models for gait analysis (e.g., by using whole-body scanners, based on the principle of structured lighting, or by applying photometric stereo [4]). The increasing availability of high-speed cameras supports the development of marker-less motion tracking systems (e.g., [39]), overcoming the apparent restrictions of marker-based systems.

For a review on past and more recent work in biomechanics, see [43]. Recent textbooks are, for example, [49, 50], or other book publications by Human Kinetics. [35] reviews markerless motion capture for biomechanical applications; see also Chapter 15 in this book. Chapter 16 is about motion variations between clinical gait and daily live, and Chapter 17 on studies to support the optimization of human motion.

1.8 Human Motion Studies in Computer Graphics

Computer animation of human walking is a major area of interest in computer graphics. Compared to biomechanics, the discipline emerged "recently", namely about 50 years ago with the advent of the computer.[11]

Basically, a computer graphics process starts with defining the models used. Figure 1.14 illustrates three options. Tracking markers (see previous

---

[11] The first working computers were built during World War II (with the first COLOSSUS operating at Bletchley Park in December 1943), and now computers have become essential tools in most branches of science, including studies of human motion.

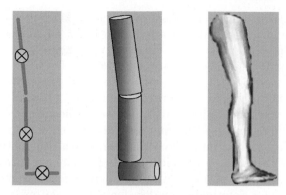

Fig. 1.14. Three options for modelling a leg: stick figure (left), simple geometrical parts (middle), or a (generic) model of the shape of a human leg.

section) allows to generate a stick figure, which may be based on general assumptions into a volumetric model (e.g., defined by cylindric parts, or a generic body model of a human). Then the ways are specified how to present those models, for example rendered with respect to given surface textures and light sources in form of an animation, within a synthesized scene, or just against a monochromatic background (see Figure 1.15 for an example of recent graduate student projects; for more complex animations see commercial products of the movie or game industries, which are major forces pushing for progress in animated human motion).

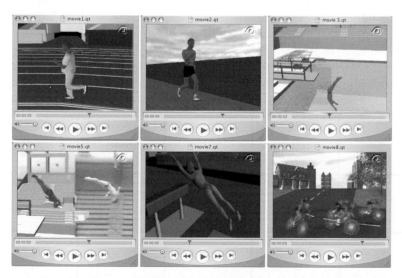

Fig. 1.15. Frames of various animations of human movements [20]. The clip illustrated on the lower left allows to compare synthetic and real human motion.

Static models of moving humans, or dynamic 3D poses, are generated by applying various means. For static whole–body modelling, this can be achieved, efficiently and accurately, by structured lighting or encoded light (e.g., Gray codes) for static bodies (see, e.g., [28]). LEDs, marker-based multi-camera systems (see Figure 1.13), or silhouette-based generic 3D model tracking (e.g., [39]) are options for capturing data about movements of a human.

For recent books, addressing human motion studies in computer graphics (and in computer vision), see [19]. Sun and Metaxas showed in [44] how gait, captured on even ground, can be used to generate realistic motion on uneven terrain. Chapter 22 discusses realistic modelling of human motion. Chapter 24 is about the importance of human motion analysis for character animation. Chapter 23 provides deformable models for a possible option to represent motion.

1.9 Human Motion Studies in Computer Vision

Computer vision exists for about the same time as computer graphics. Instead of only capturing image sequences, to be analyzed by a human observer, now those sequences are digitized, and computer programs are used for an automated analysis of the sequence. As illustrated by Figure 1.8, Marey already used simplifications such as white skeletal curves on a moving human. Braune and Fischer [9] attached light rods to an actor's limbs, which then became known as *Moving Light Displays* (LEDs). Gunnar Johannsson [26,27] pioneered studies on the use of image sequences for a programmed human motion analysis, using LEDs as input (see Figure 1.16). These very limited inputs of information allow an interesting analysis, for example with respect of identifying a particular person.

Motion analysis in computer vision has to solve two main tasks, detecting correspondences between subsequent images, and tracking of an object within a sequence of images. This can be based on different methodologies, such as

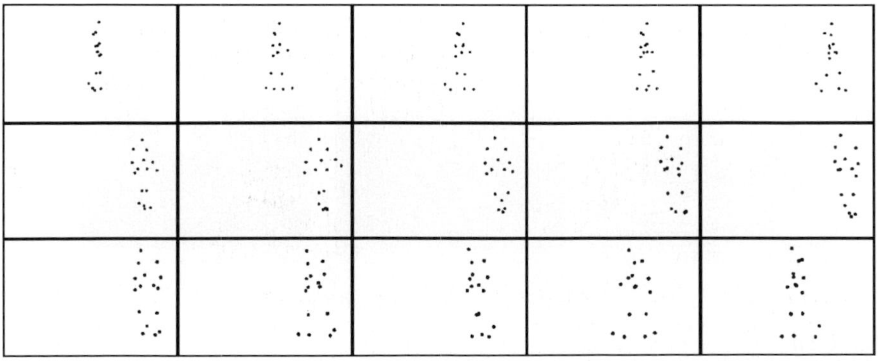

Fig. 1.16. A sequence of 15 LED frames, extracted from an animation on [30].

tracking 2D features at a local (e.g., corners) or global level (e.g., silhouettes, after a "proper" segmentation of images), or tracking based on projecting a generic 3D model (of a human) into the scene (see Figure 1.17).

For reviews on human motion studies in computer vision, see [1, 34]. The use of LEDs in computer vision is reviewed in [14]. For a recent collection of papers, also covering human motion in computer vision, see [37]. Human motion studies in computer vision have been already a subject of an international workshop [48]. As an example of a recent "landmark", we cite [51], discussing in depth the recognition of people based on gait.

This book contains a basic Chapter 7 on various models for human motion and Chapter 18 which reviews human motion analysis. The understanding of human motion from video, and the tracking of moving people, is the subject in Chapters 11, 5, 6, 20, 8. Special issues when tracking clothed people are discussed in Chapter 12. The recognition of human actions is reported in Chapter 3. Chapters 9, 14 are about human motion studies in the context of computer–human interaction. The Chapters 21, 13, 10, 2 provide information about theoretical areas which have proved to be of use for human motion modelling or understanding (Geometrical algebra, simulated annealing, manifold learning). Finally, Chapter 19 discusses the application of human motion studies in a particular application (dummy movements and crash test analysis). The application of motion analysis for cardiac motion studies is reported in Chapter 4.

Fig. 1.17. Human poses are tracked using a generic 3D model of the upper human body; the figure shows the backprojection of the recorded 3D motion into the original 4-camera image sequence, also demonstrating model movements of occluded body parts [39].

1.10 Conclusions

This review certainly proves that studies on human motion were and are interdisciplinary (from the beginning, which was about 2000 years ago). This book aims at contributing to deeper interactions between biomechanics, computer graphics, and computer vision (already existing at advanced levels in some institutes, see, e.g., [12, 38]). Certainly, further areas are of relevance, with biophysics or mathematics as major contributors, or medicine (e.g., rehabilitation technology), robotics (e.g., studies on passive dynamic walking), or sports sciences (e.g., modelling of athlete motion) as important areas of application. However, the book was planned for the time being to remain focused on those three areas, but future development of human motion studies will certainly also benefit from interactions with neurology (e.g., neural control mechanisms for motion), cognitive sciences (e.g., selective attention to understand motion), health exercises (i.e., not just focusing on sports, but also on recreation), and so forth.

Basic concepts for biomechanical studies of human motion were already developed by 1687, when Isaac Newton published his three laws of motion. Basic mathematic tools for human motion studies were already provided in mathematics by the end of the 19th century. The start of photography in the 19th century led to the documentation of human motion, starting at the end of the 19th century. The advent of the computer, and of digital technology in general in the latter half of the 20th century finally provided the tools for analyzing human motion based on digitized image sequences, and for animating or studying human motion using extensive calculations and detailed models of human locomotion.

There is a general demand in more in-depth studies on locomotion (e.g., also on gait disorders). A future major qualitative advance in the area of human motion studies is expected with the widespread use of high-speed, high-resolution, and light-sensitive cameras for recording and analyzing human motion, and at the time when human motion analysis becomes an integrative approach in medical treatments.

There is furthermore a continuous strong progress in computer vision and graphics. Computer graphics, with animations and game applications, are a major force to simplify motion understanding and capturing into integrated vision and graphics systems. Recent applications are already manifold, also including identifications of persons (i.e., by gait patterns), motion advice (e.g., for learning golf), documenting particular performances (e.g., 3D models of dancing), or multimedia presentations (e.g., mapping of recorded motion into a prepared image sequence). Challenging tasks are related to markerless motion (and shape) capture, also outdoors, with partial occlusions, general clothing, real-time (even for highly dynamic motion patterns), and also allowing extreme environments, starting with, for example, those for swimming or rock climbing.

Acknowledgments

The authors thank the anonymous referees for valuable comments.

References

1. Aggarwal, J. K., and Q. Cai: Human motion analysis: a review. *Comput Vis Image Underst*, **73**:428–440, 1999.
2. Amar, J.: *The Human Motor.* Eng. trans. by E. P. Butterworth and G. Wright, Routledge, London, 1920; E. P. Dutton, New York, 1920 (originally published in French in 1914).
3. Aristotle: *ΠΕΡΙ ΠΟΡΕΙΑΣ ΖΩΙΩΝ.* Eng. trans. by A.S.L. Farquharson, eBooks@Adelaide, 2004.
4. Austin, N., Y. Chen, R. Klette, R. Marshall, Y.-S. Tsai, and Y. Zhang: A comparison of feature measurements for kinetic studies on human bodies. In Proc. *Robot Vision*, pages 43–51, LNCS 1998, Springer, Berlin, 2001.
5. Barcsay, J.: *Anatomy for the Artist.* Sterling New York, 2006.
6. Bassett, D. R., Jr.: Scientific contributions of A. V. Hill: exercise physiology pioneer. *J. Appl. Physiol,* **93**:1567–1582, 2002.
7. Bernstein, N.: *The Coordination and Regulation of Movement.* Pergamon Press, Oxford, 1967.
8. Borelli, G. A.: *De Motu Animalium.* Rome, 1680/1681 (*On the Movement of Animals*, translated from Latin to English by P. Maquet, Springer, Berlin, 1989).
9. Braune, W., and O. Fischer: *The Human Gait.* Eng. trans. by P. Maquet and R. Furlong. Springer, 1987 (originally published in German in 1892).
10. Braun, M.: *Picturing Time: the Work of Etienne-Jules Marey (1830-1904).* University of Chicago Press, Chicago, 1992.
11. Brock, W. H.: *Edward Frankland,* in *Dictionary of Scientific Biography* (edited by C. C. Gillispie), Volume 5, pages 124–127, Charles Scribners, New York, 1972.
12. Brown University, entry "Research areas in Computer Vision", www.cs.brown.edu/research/areas/computer_vision.html (visited in Oct. 2006).
13. Calter, P.: Geometry in art and architecture. www.math.dartmouth.edu/~matc/math5.geometry/ (visited in Oct. 2006).
14. Cèdras, C., and M. Shah: A survey of motion analysis from moving light displays. In Proc. *Computer Vision Pattern Recognition*, pages 214–221, 1994.
15. Clinical Gait Analysis, University of Vienna. Link to 'Walking in Art'. http://www.univie.ac.at/cga/ (visited in Oct. 2006).
16. da Vinci, L.: *Skizzenbücher.* Paragon Books, Bath, UK, 2005 (German edition of about 1000 drawings from da Vinci's note books, including translations of his notes); original edition in English: H. A. Suhr (editor): *Leonardo's Notebooks.* Black Dog and Leventhal, New York, 2005.
17. Fenn, W. O.: Mechanical energy expenditure in sprint running as measured in moving pictures. *Am. J. Physiol.*, **90**:343–344, 1929.
18. Fenn, W. O.: A cinematographical study of sprinters. *Sci. Monthly*, **32**:346–354, 1931.

19. Forsyth, D. A., O. Arikan, L. Ikemoto, J. F. O'Brien, and D. Ramanan: *Computational Studies of Human Motion. Foundations and Trends in Computer Graphics and Vision.* Now Publishers, Hanover, Massachusetts, 2006.

20. Georgia Tech, entry 'Simulating Human Motion", www.cc.gatech.edu/gvu/animation/Areas/humanMotion/humanMotion.html (visited in Oct. 2006).

21. Haertel, H., M. Kires, Z. Jeskova, J. Degro, Y. B. Senichenkov, and J.-M. Zamarro: Aristotle still wins over Newton. An evaluation report of a new, simulation-supported approach to teach the concepts of inertia and gravity. In Proc. *EUROCON*, Volume I, pages 7–11, 2003.

22. Hill, A. V.: The mechanism of muscular contraction (Nobel Lecture). http://nobelprize.org/nobel_prizes/medicine/laureates/1922/hill-lecture.html (visited in Oct. 2006).

23. Hill, A. V.: *Muscular Movement in Man: The Factors Governing Speed and Recovery from Fatigue.* McGraw-Hill, New York, 1927.

24. Hill, A. V.: *Muscular Activity.* Herter Lectures. Sixteenth Course. Williams & Wilkins, Baltimore, Maryland, 1926.

25. Hill, A. V.: *Adventures in Biophysics.* University of Pennsylvania Press, Philadelphia, 1931.

26. Johannsson, G.: Visual perception of biological motion and a model for its analysis. *Percept Psychophys,* **14**:201–211, 1973.

27. Johannsson, G.: Visual motion perception. *Sci Am,* 76–88, November 1976.

28. Klette, R., K. Schlüns, and A. Koschan. *Computer Vision: Three-dimensional Data from Digital Images.* Springer, Singapore, 1998.

29. Lefebvre, T.: Marey and chronophotography. http://www.bium.univ-paris5.fr/histmed/medica/marey/marey03a.htm (visited in Oct. 2006).

30. Li, B.: Homepage, entry "Research". www.doc.mmu.ac.uk/STAFF/B.Li/ (visited in Oct. 2006).

31. Marey, E.-J.: *La Machine Animale, Locomotion Terrestre et Aérienne.* Germer Baillière, Paris, 1873.

32. Martin, R. B.: A genealogy of biomechanics. www.asbweb.org/html/about/history/ (visited in Oct. 2006).

33. Marey, E.-J.: *Le Mouvement.* G. Masson, Paris, 1894.

34. Moeslund, T. B., and E. Granum: A survey of computer vision-based human motion capture. *Comput Vis Image Underst,* **81**:231–268, 2001.

35. Mündermann, L., S. Corazza, and T. P. Andriacchi. The evolution of methods for the capture of human movement leading to markerless motion capture for biomechanical applications. *J. Neuroengineering Rehabil.,* **15**:3–6, 2006.

36. Muybridge, E.: *Animals in Motion.* Edited by L. S. Brown, Dover, New York, 1957.

37. Paragios, N., Y. Chen, and O. Faugeras (editors): *The Handbook of Mathematical Models in Computer Vision.* Springer, 2005.

38. Robotics Institute at CMU, entry "Researchers Listed by Interest", http://www.ri.cmu.edu/general/interests.html (visited in Oct. 2006).

39. Rosenhahn, B., U. Kersting, D. Smith, J. Gurney, T. Brox, and R. Klette. A system for marker-less human motion estimation. In Proc. *Pattern Recognition* (DAGM) (W. Kropatsch, R. Sablatnig, and A. Hanbury, editors), LNCS 3663, pages 230-237, Springer, Berlin, 2005 (Main Award of DAGM 2005).

40. Sklar, A.: *A World History of Film.* Harry N. Abrams, New York, 2002.
41. Smith, G. A.: Biomechanics of cross country skiing.
    biomekanikk.nih.no/xchandbook/index.html (visited in Oct. 2006).
42. Sportscience. Entry on A. V. Hill.
    http://www.sportsci.org/ (visited in Oct. 2006).
43. Stergiou, N., and D. L. Blanke: Biomechanics.
    grants.hhp.coe.uh.edu/faculty/Kurz/Biomechanics_Stergiou_Blanke.pdf   (visited in Oct. 2006).
44. Sun, H., and D. Metaxas. Automating gait generation. In Proc. *Siggraph*, pages 261–270, 2001.
45. Turner, R. S.: *Herman von Helmholtz*, in *Dictionary of Scientific Biography* (edited by C. C. Gillispie), Volume 6, pages 241–253, Charles Scribners Sons, New York, 1972.
46. Weber, E. F., and W. E. Weber: *Über die Mechanik der menschlichen Gehwerkzeuge. Eine anatomisch-physiologische Untersuchung* Göttingen, Dieterich, 1836 (Translation by P. Maquet and R. Furlong: *Mechanics of the Human Walking Apparatus.* Springer, Berlin, 1992).
47. Who's who of Victorian cinema.
    www.victorian-cinema.net/news.htm (visited in Oct. 2006).
48. Workshop on "Human Motion" in December 2000 at Austin (Texas),
    www.ece.utexas.edu/projects/cvrc/humo/ (visited in Oct. 2006).
49. Zatsiorsky, V. M.: *Kinematics of Human Motion.* Human Kinetics, Champaign, Illinois, 1997.
50. Zatsiorsky, V. M.: *Kinetics of Human Motion.* Human Kinetics, Champaign, Illinois, 2002.
51. Zhang, R., C. Vogler, and D. Metaxas. Human gait recognition in sagittal plane. *Image Vis Comput*, **25**:321–330, 2007.

Appendix: A Few Definitions

Biomechanics studies living organisms at different levels – from molecular to macroscopic level – as mechanical systems (i.e., internal and external forces acting on the moving body, and effects produced by these forces).

Computer Graphics generates digital pictures or animations for visualizing real or synthetic scenes, objects, or processes.

Computer Vision analyzes 3D scenes, objects, or processes based on captured images, with the purpose of modelling or understanding the pictured scenes, objects, or processes.

Chronophotography ("pictures of time") is applied science (used mainly in the second half of the 19th century), using photography for studies of changes in time.

Dynamics is concerned with the effects of forces on the motion of objects

Locomotion is (in biology) an autonomous motion of an individual (human, bird, fish, and so forth), considered to be defined by a sequence of patterns composed by anatomic parts. Locomotion patterns can be classified into categories such as walking, jumping, swimming, and so forth.

Kinematics is a subfield of biomechanics studying movements with respect to geometrical aspects, without references to its physic causes (i.e., masses or forces).

Kinesiology denoted (in the United States; this name was first applied to the study of muscles and movements in the first decade of the 20th century by Baron Nils Posse and William Skarstrom) the use of mathematical or mechanical principles for studying human motion (note: basically identical to biomechanics, originally the preferred name for this field in Europe, but now widely accepted worldwide).

Kinetics studies what causes a body to move the way it does, by analyzing the actions of forces contributing to the motion of masses. Subfields are statics and dynamics.

Prosthesis is a field in medicine aimed at artificial replacements of body parts.

Statics is concerned with the analysis of loads (force, moment, torque) on physical systems in static equilibrium (i.e., whose relative positions do not vary over time; for example in inertial motion).

Part I

2D Tracking

2

The Role of Manifold Learning
in Human Motion Analysis

Ahmed Elgammal and Chan-Su Lee

Department of Computer Science
Rutgers University, Piscataway, NJ, USA

Summary. The human body is an articulated object with a high number of degrees
of freedom. Despite the high dimensionality of the configuration space, many human
motion activities lie intrinsically on low-dimensional manifolds. Although the intrinsic body configuration manifolds might be very low in dimensionality, the resulting
appearance manifolds are challenging to model given various aspects that affect the
appearance such as the shape and appearance of the person performing the motion,
or variation in the viewpoint, or illumination. Our objective is to learn representations for the shape and the appearance of moving (dynamic) objects that support
tasks such as synthesis, pose recovery, reconstruction, and tracking. We studied various approaches for representing global deformation manifolds that preserve their
geometric structure. Given such representations, we can learn generative models
for dynamic shape and appearance. We also address the fundamental question of
separating style and content on nonlinear manifolds representing dynamic objects.
We learn factorized generative models that explicitly decompose the intrinsic body
configuration (content) as a function of time from the appearance/shape (style factors) of the person performing the action as time-invariant parameters. We show
results on pose recovery, body tracking, gait recognition, as well as facial expression
tracking and recognition.

2.1 Introduction

The human body is an articulated object with high degrees of freedom. The
human body moves through the three-dimensional world and such motion is
constrained by body dynamics and projected by lenses to form the visual
input we capture through our cameras. Therefore, the changes (deformation) in appearance (texture, contours, edges, etc.) in the visual input (image sequences) corresponding to performing certain actions, such as facial
expression or gesturing, are well constrained by the 3D body structure and
the dynamics of the action being performed. Such constraints are explicitly
exploited to recover the body configuration and motion in model-based approaches [1–8] through explicitly specifying articulated models of the body

B. Rosenhahn et al. (eds.), Human Motion – Understanding, Modelling, Capture, and
Animation, 25–56.

parts, joint angles and their kinematics (or dynamics) as well as models for camera geometry and image formation. Recovering body configuration in these approaches involves searching high-dimensional spaces (body configuration and geometric transformation) which is typically formulated deterministically as a nonlinear optimization problem, e.g. [5,9], or probabilistically as a maximum likelihood problem, e.g. [8]. Chapter 13 shows an example of a stochastic global optimization approach for recovering body configurations. Such approaches achieve significant success when the search problem is constrained as in tracking context. However, initialization remains the most challenging problem, which can be partially alleviated by sampling approaches. The dimensionality of the initialization problem increases as we incorporate models for variations between individuals in physical body style, models for variations in action style, or models for clothing, etc. Partial recovery of body configuration can also be achieved through intermediate view-based representations (models) that may or may not be tied to specific body parts [10–19]. In such case constancy of the local appearance of individual body parts is exploited. Alternative paradigms are appearance-based and motion-based approaches where the focus is to track and recognize human activities without full recovery of the 3D body pose [20–28].

Recently, there have been research for recovering body posture directly from the visual input by posing the problem as a learning problem through searching a pre-labeled database of body posture [29–31] or through learning regression models from input to output [32–38]. All these approaches pose the problem as a machine learning problem where the objective is to learn input–output mapping from input–output pairs of training data. Such approaches have great potential for solving the initialization problem for model-based vision. However, these approaches are challenged by the existence of wide range of variability in the input domain.

Role of Manifold:

Despite the high-dimensionality of the configuration space, many human motion activities lie intrinsically on low-dimensional manifolds. This is true if we consider the body kinematics, as well as if we consider the observed motion through image sequences. Let us consider the observed motion. The shape of the human silhouette walking or performing a gesture is an example of a dynamic shape where the shape deforms over time based on the action performed. These deformations are constrained by the physical body constraints and the temporal constraints posed by the action being performed. If we consider these silhouettes through the walking cycle as points in a high-dimensional visual input space, then, given the spatial and the temporal constraints, it is expected that these points will lay on a low-dimensional manifold. Intuitively, the gait is a one-dimensional manifold which is embedded in a high-dimensional visual space. This was also shown in [39]. Such manifold can be twisted, self-intersect in such high-dimensional visual space.

Similarly, the appearance of a face performing facial expressions is an example of a dynamic appearance that lies on a low-dimensional manifold in the visual input space. In fact if we consider certain classes of motion such as gait, or a single gesture, or a single facial expressions and if we factor out all other sources of variability, each of such motions lies on a one-dimensional manifolds, i.e., a trajectory in the visual input space. Such manifolds are nonlinear and non-Euclidean.

Therefore, researchers have tried to exploit the manifold structure as a constraint in tasks, such as tracking and activity recognition, in an implicit way. Learning nonlinear deformation manifolds is typically performed in the visual input space or through intermediate representations. For example, exemplar-based approaches such as [40] implicitly model nonlinear manifolds through points (exemplars) along the manifold. Such exemplars are represented in the visual input space. HMM models provide a probabilistic piecewise linear approximation which can be used to learn nonlinear manifolds as in [41] and in [33].

Although the intrinsic body configuration manifolds might be very low in dimensionality, the resulting appearance manifolds are challenging to model given various aspects that affect the appearance such as the shape and appearance of the person performing the motion, or variation in the viewpoint, or illumination. Such variability makes the task of learning visual manifold very challenging because we are dealing with data points that lies on multiple manifolds on the same time: body configuration manifold, view manifold, shape manifold, illumination manifold, etc.

Linear, Bilinear and Multi-linear Models:

Can we decompose the configuration using linear models? Linear models, such as PCA [42], have been widely used in appearance modelling to discover subspaces for variations. For example, PCA has been used extensively for face recognition such as in [43–46] and to model the appearance and view manifolds for 3D object recognition as in [47]. Such subspace analysis can be further extended to decompose multiple orthogonal factors using bilinear models and multi-linear tensor analysis [48, 49]. The pioneering work of Tenenbaum and Freeman [48] formulated the separation of style and content using a bilinear model framework [50]. In that work, a bilinear model was used to decompose face appearance into two factors: head pose and different people as style and content interchangeably. They presented a computational framework for model fitting using SVD. Bilinear models have been used earlier in other contexts [50, 51]. In [49] multi-linear tensor analysis was used to decompose face images into orthogonal factors controlling the appearance of the face, including geometry (people), expressions, head pose, and illumination. They employed high order singular value decomposition (HOSVD) [52] to fit multi-linear models. Tensor representation of image data was used in [53] for video compression and in [54, 55] for motion analysis and synthesis. N-mode analysis of higher-order tensors was originally proposed and developed in [50, 56, 57]

Fig. 2.1. Twenty sample frames from a walking cycle from a side view. Each row represents half a cycle. Notice the similarity between the two half cycles. The right part shows the similarity matrix: each row and column corresponds to one sample. Darker means closer distance and brighter means larger distances. The two dark lines parallel to the diagonal show the similarity between the two half cycles.

and others. Another extension is algebraic solution for subspace clustering through generalized-PCA [58, 59]

In our case, the object is dynamic. So, can we decompose the configuration from the shape (appearance) using linear embedding? For our case, the shape temporally undergoes deformations and self-occlusion which result in the points lying on a nonlinear, twisted manifold. This can be illustrated if we consider the walking cycle in Figure 2.1. The two shapes in the middle of the two rows correspond to the farthest points in the walking cycle kinematically and are supposedly the farthest points on the manifold in terms of the geodesic distance along the manifold. In the Euclidean visual input space, these two points are very close to each other, as can be noticed from the distance plot on the right of Figure 2.1. Because of such nonlinearity, PCA will not be able to discover the underlying manifold. Simply, linear models will not be able to interpolate intermediate poses. For the same reason, multidimensional scaling (MDS) [60] also fails to recover such manifold.

Nonlinear Dimensionality Reduction and Decomposition of Orthogonal Factors:

Recently some promising frameworks for nonlinear dimensionality reduction have been introduced, e.g. [61–67]. Such approaches can achieve embedding of nonlinear manifolds through changing the metric from the original space to the embedding space based on local structure of the manifold. While there are various such approaches, they mainly fall into two categories: Spectral-embedding approaches and Statistical approaches. Spectral embedding includes approaches such as isometric feature mapping (Isomap) [61], Local linear embedding (LLE) [62], Laplacian eigenmaps [63], and Manifold Charting [64]. Spectral-embedding approaches, in general, construct an affinity matrix between data points using data dependent kernels, which reflect local manifold structure. Embedding is then achieved through solving an eigen-value problem on such matrix. It was shown in [68, 69] that these approaches are all instances of kernel-based learning, in particular kernel principle component analysis KPCA [70]. In [71] an approach for embedding out-of-sample points to complement such approaches. Along the same line, our work [72, 73] introduced a general framework for mapping between input and embedding spaces.

All these nonlinear embedding frameworks were shown to be able to embed nonlinear manifolds into low-dimensional Euclidean spaces for toy examples as well as for real images. Such approaches are able to embed image ensembles nonlinearly into low-dimensional spaces where various orthogonal perceptual aspects can be shown to correspond to certain directions or clusters in the embedding spaces. In this sense, such nonlinear dimensionality reduction frameworks present an alternative solution to the decomposition problems. However, the application of such approaches are limited to embedding of a single manifold.

Biological Motivation:
While the role of manifold representations is still unclear in perception, it is clear that images of the same objects lie on a low-dimensional manifold in the visual space defined by the retinal array. On the other hand, neurophysiologists have found that neural population activity firing is typically a function of a small number of variables, which implies that population activity also lie on low-dimensional manifolds [74].

2.2 Learning a Simple Motion Manifold

2.2.1 Case Study: The Gait Manifold

In order to achieve a low-dimensional embedding of the gait manifold, nonlinear dimensionality reduction techniques such as LLE [62], Isomap [61], and others can be used. Most these techniques result in qualitatively similar manifold embedding. As a result of nonlinear dimensionality reduction we can reach an embedding of the gait manifold in a low-dimension Euclidean space [72]. Figure 2.2 illustrates the resulting embedded manifold for a side view of the walker.[1] Figure 2.3 illustrates the embedded manifolds for five different viewpoints of the walker. For a given view-point, the walking cycle evolves along a closed curve in the embedded space, i.e., only one degree of freedom controls the walking cycle which corresponds to the constrained body pose as a function of the time. Such conclusion is conforming with the intuition that the gait manifold is one dimensional.

One important question is what is the least dimensional embedding space we can use to embed the walking cycle in a way that discriminate different

---

[1] The data used are from the CMU Mobo gait data set which contains 25 people from six different viewpoints. We used data sets of walking people from multiple views. Each data set consists of 300 frames and each containing about 8 to 11 walking cycles of the same person from a certain viewpoints. The walkers were using treadmill which might results in different dynamics from the natural walking.

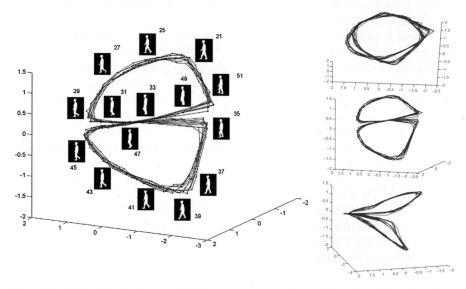

Fig. 2.2. Embedded gait manifold for a side view of the walker. Left: sample frames from a walking cycle along the manifold with the frame numbers shown to indicate the order. Ten walking cycles are shown. Right: three different views of the manifold.

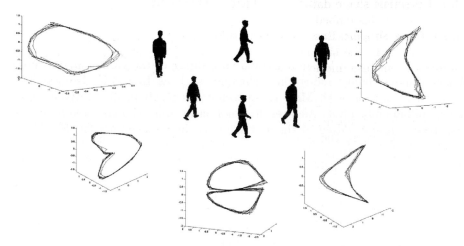

Fig. 2.3. Embedded manifolds for 5 different views of the walkers. Frontal view manifold is the right most one and back view manifold is the leftmost one. We choose the view of the manifold that best illustrates its shape in the 3D embedding space.

poses through the whole cycle. The answer depends on the viewpoint. The manifold twists in the embedding space given the different viewpoints which impose different self-occlusions. The least twisted manifold is the manifold for the back view as this is the least self-occluding view (left most manifold in

Figure 2.3. In this case the manifold can be embedded in a two-dimensional space. For other views the curve starts to twist to be a three-dimensional space curve. This is primarily because of the similarity imposed by the viewpoint which attracts far away points on the manifold closer. The ultimate twist happens in the side view manifold where the curve twists to be a figure eight shape where each cycle of the eight (half eight) lies in a different plane. Each half of the "eight" figure corresponds to half a walking cycle. The cross point represents the body pose where it is totally ambiguous from the side view to determine from the shape of the contour which leg is in front as can be noticed in Figure 2.2. Therefore, in a side view, three-dimensional embedding space is the least we can use to discriminate different poses. Embedding a side view cycle in a two-dimensional embedding space results in an embedding similar to that shown in top left of Figure 2.2 where the two half cycles lies over each other. Different people are expected to have different manifolds. However, such manifolds are all topologically equivalent. This can be noticed in Figure 2.8-c. Such property will be exploited later in the chapter to learn unified representations from multiple manifolds.

2.2.2 Learning the Visual Manifold: Generative Model

Given that we can achieve a low-dimensional embedding of the visual manifold of dynamic shape data, such as the gait data shown above, the question is how to use this embedding to learn representations of moving (dynamic) objects that supports tasks such as synthesis, pose recovery, reconstruction and tracking. In the simplest form, assuming no other source of variability besides the intrinsic motion, we can think of a view-based generative model of the form

$$y_t = T_\alpha \gamma(x_t; a) \tag{2.1}$$

where the shape (appearance), y_t, at time t is an instance driven from a generative model where the function γ is a mapping function that maps body configuration x_t at time t into the image space. The body configuration x_t is constrained to the explicitly modeled motion manifold. i.e., the mapping function γ maps from a representation of the body configuration space into the image space given mapping parameters a that are independent from the configuration. T_α represents a global geometric transformation on the appearance instance.

The manifold in the embedding space can be modeled explicitly in a function form or implicitly by points along the embedded manifold (embedded exemplars). The embedded manifold can be also modeled probabilistically using Hidden Markov Models and EM. Clearly, learning manifold representations in a low-dimensional embedding space is advantageous over learning them in the visual input space. However, our emphasize is on learning the mapping between the embedding space and the visual input space.

Since the objective is to recover body configuration from the input, it might be obvious that we need to learn mapping from the input space to the embedding space, i.e., mapping from $\mathbb{R}^d$ to $\mathbb{R}^e$. However, learning such mapping is not feasible since the visual input is very high-dimensional so learning such mapping will require large number of samples in order to be able to interpolate. Instead, we learn the mapping from the embedding space to the visual input space, i.e., in a generative manner, with a mechanism to directly solve for the inverse mapping. Another fundamental reason to learn the mapping in this direction is the inherent ambiguity in 2D data. Therefore, mapping from visual data to the manifold representation is not necessarily a function. While learning a mapping from the manifold to the visual data is a function.

It is well know that learning a smooth mapping from examples is an ill-posed problem unless the mapping is constrained since the mapping will be undefined in other parts of the space [75]. We argue that, explicit modelling of the visual manifold represents a way to constrain any mapping between the visual input and any other space. Nonlinear embedding of the manifold, as was discussed in the previous section, represents a general framework to achieve this task. Constraining the mapping to the manifold is essential if we consider the existence of outliers (spatial and/or temporal) in the input space. This also facilitates learning mappings that can be used for interpolation between poses as we shall show. In what follows we explain our framework to recover the pose. In order to learn such nonlinear mapping, we use a Radial basis function (RBF) interpolation framework. The use of RBF for image synthesis and analysis has been pioneered by [75, 76] where RBF networks were used to learn nonlinear mappings between image space and a supervised parameter space. In our work we use the RBF interpolation framework in a novel way to learn mapping from unsupervised learned parameter space to the input space. Radial basis function interpolation provides a framework for both implicitly modelling the embedded manifold as well as learning a mapping between the embedding space and the visual input space. In this case, the manifold is represented in the embedding space implicitly by selecting a set of representative points along the manifold as the centers for the basis functions.

Let the set of representative input instances (shape or appearance) be $\mathsf{Y} = \{y_i \in \mathbb{R}^d \ \ i = 1, \cdots, N\}$ and let their corresponding points in the embedding space be $\mathsf{X} = \{x_i \in \mathbb{R}^e, \ \ i = 1, \cdots, N\}$ where e is the dimensionality of the embedding space (e.g., $e = 3$ in the case of gait). We can solve for multiple interpolants $f^k : \mathbb{R}^e \to \mathbb{R}$ where k is k-th dimension (pixel) in the input space and f^k is a radial basis function interpolant, i.e., we learn nonlinear mappings from the embedding space to each individual pixel in the input space. Of particular interest are functions of the form

$$f^k(x) = p^k(x) + \sum_{i=1}^{N} w_i^k \phi(|x - x_i|), \qquad (2.2)$$

where $\phi(\cdot)$ is a real-valued basic function, w_i are real coefficients, $|\cdot|$ is the norm on $\mathbb{R}^e$ (the embedding space). Typical choices for the basis function includes thin-plate spline ($\phi(u) = u^2 log(u)$), the multiquadric ($\phi(u) = \sqrt{(u^2 + c^2)}$), Gaussian ($\phi(u) = e^{-cu^2}$), biharmonic ($\phi(u) = u$) and triharmonic ($\phi(u) = u^3$) splines. p^k is a linear polynomial with coefficients c^k, i.e., $p^k(x) = [1 \quad x^\top] \cdot c^k$. This linear polynomial is essential to achieve approximate solution for the inverse mapping as will be shown.

The whole mapping can be written in a matrix form as

$$f(x) = B \cdot \psi(x), \tag{2.3}$$

where B is a $d \times (N+e+1)$ dimensional matrix with the k-th row $[w_1^k \cdots w_N^k \quad c^{k^\top}]$ and the vector $\psi(x)$ is $[\phi(|x - x_1|) \cdots \phi(|x - x_N|) \quad 1 \quad x^\top]^\top$. The matrix B represents the coefficients for d different nonlinear mappings, each from a low-dimension embedding space into real numbers.

To insure orthogonality and to make the problem well posed, the following additional constraints are imposed

$$\sum_{i=1}^{N} w_i p_j(x_i) = 0, j = 1, \cdots, m \tag{2.4}$$

where p_j's are the linear basis of m-degree polynomial p. Therefore the solution for B can be obtained by directly solving the linear systems

$$\begin{pmatrix} A & P \\ P^\top & 0 \end{pmatrix} B^\top = \begin{pmatrix} Y \\ 0_{(e+1) \times d} \end{pmatrix}, \tag{2.5}$$

where $A_{ij} = \phi(|x_j - x_i|)$, $i, j = 1, \cdots, N$, P is a matrix with i-th row $[1 \quad x_i^\top]$, and Y is $(N \times d)$ matrix containing the representative input images, i.e., $Y = [y_1 \cdots y_N]^\top \in \mathbb{R}^d$ $i = 1, \cdots, N$. Solution for B is guaranteed under certain conditions on the basic functions used. Similarly, a mapping can be learned using arbitrary centers in the embedding space (not necessarily at data points) [72, 75].

Given such mapping, any input is represented by a linear combination of nonlinear functions centered in the embedding space along the manifold. Equivalently, this can be interpreted as a form of basis images (coefficients) that are combined nonlinearly using kernel functions centered along the embedded manifold.

2.2.3 Solving for the Embedding Coordinates

Given a new input $y \in \mathbb{R}^d$, it is required to find the corresponding embedding coordinates $x \in \mathbb{R}^e$ by solving for the inverse mapping. There are two questions that we might need to answer:

1. What are the coordinates of point $x \in \mathbb{R}^e$ in the embedding space corresponding to such input?
2. What is the closest point on the embedded manifold corresponding to such input?

In both cases we need to obtain a solution for

$$x^* = \operatorname*{argmin}_{x} ||y - B\psi(x)|| \tag{2.6}$$

where for the second question the answer is constrained to be on the embedded manifold. In the cases where the manifold is only one-dimensional, (for example in the gait case, as will be shown) only one-dimensional search is sufficient to recover the manifold point closest to the input. However, we show here how to obtain a closed-form solution for x^*.

Each input yields a set of d nonlinear equations in e unknowns (or d nonlinear equations in one e-dimensional unknown). Therefore a solution for x^* can be obtained by least square solution for the over-constrained nonlinear system in 2.6. However, because of the linear polynomial part in the interpolation function, the vector $\psi(x)$ has a special form that facilitates a closed-form least square linear approximation and therefore, avoid solving the nonlinear system. This can be achieved by obtaining the pseudo-inverse of B. Note that B has rank N since N distinctive RBF centers are used. Therefore, the pseudo-inverse can be obtained by decomposing B using SVD such that $B = USV^\top$ and, therefore, vector $\psi(x)$ can be recovered simply as

$$\psi(x) = V\tilde{S}U^T y \tag{2.7}$$

where $\tilde{S}$ is the diagonal matrix obtained by taking the inverse of the nonzero singular values in S as the diagonal terms and setting the rest to zeros. Linear approximation for the embedding coordinate x can be obtained by taking the last e rows in the recovered vector $\psi(x)$. Reconstruction can be achieved by remapping the projected point.

2.2.4 Synthesis, Recovery and Reconstruction

Given the learned model, we can synthesize new shapes along the manifold. Figure 2.4-c shows an example of shape synthesis and interpolation. Given a learned generative model in the form of Equation (2.3), we can synthesize new shapes through the walking cycle. In these examples only 10 samples were used to embed the manifold for half a cycle on a unit circle in 2D and to learn the model. Silhouettes at intermediate body configurations were synthesized (at the middle point between each two centers) using the learned model. The learned model can successfully interpolate shapes at intermediate configurations (never seen in the learning) using only two-dimensional embedding. The figure shows results for three different peoples.

Given a visual input (silhouette), and the learned model, we can recover the intrinsic body configuration, recover the viewpoint, and reconstruct the input and detect any spatial or temporal outliers. In other words, we can simultaneously solve for the pose, viewpoint, and reconstruct the input. A block diagram for recovering 3D pose and viewpoint given learned manifold models are shown in Figure 2.4. The framework [77] is based on learning three components as shown in Figure 2.4-a:

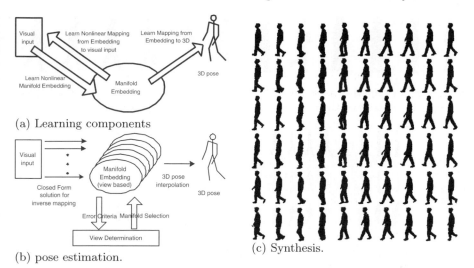

(a) Learning components

(b) pose estimation.

(c) Synthesis.

Fig. 2.4. (a, b) Block diagram for the learning framework and 3D pose estimation. (c) Shape synthesis for three different people. First, third and fifth rows: samples used in learning. Second, fourth, sixth rows: interpolated shapes at intermediate configurations (never seen in the learning).

1. Learning Manifold Representation: using nonlinear dimensionality reduction we achieve an embedding of the global deformation manifold that preserves the geometric structure of the manifold as described in Section 2.2.1. Given such embedding, the following two nonlinear mappings are learned.
2. Manifold-to-input mapping: a nonlinear mapping from the embedding space into visual input space as described in Section 2.2.2.
3. Manifold-to-pose: a nonlinear mapping from the embedding space into the 3D body pose space.

Given an input shape, the embedding coordinate, i.e., the body configuration can be recovered in closed-form as was shown in Section 2.2.3. Therefore, the model can be used for pose recovery as well as reconstruction of noisy inputs. Figure 2.5 shows examples of the reconstruction given corrupted silhouettes as input. In this example, the manifold representation and the mapping were learned from one person data and tested on other people data. Given a corrupted input, after solving for the global geometric transformation, the input is projected to the embedding space using the closed-form inverse mapping approximation in Section 2.2.3. The nearest embedded manifold point represents the intrinsic body configuration. A reconstruction of the input can be achieved by projecting back to the input space using the direct mapping in Equation (2.3). As can be noticed from the figure, the reconstructed silhouettes preserve the correct body pose in each case which shows that solving for the inverse mapping yields correct points on the manifold. Notice that no

Fig. 2.5. Examples of pose-preserving reconstruction results. Six noisy and corrupted silhouettes and their reconstructions next to them.

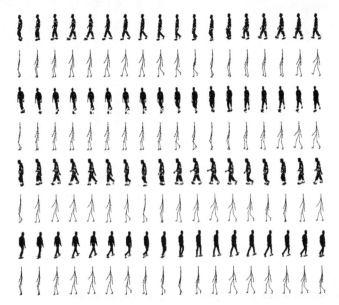

Fig. 2.6. 3D reconstruction for 3 people from different views: (top to bottom): person 70 views 1,2, person 86 views 1 and person 76 view 4.

mapping is learned from the input space to the embedded space. Figure 2.6 shows examples of 3D pose recovery obtained in closed-form for different people from different view. The training has been done using only one subject data from five viewpoints. All the results in Figure 2.6 are for subjects not used in the training. This shows that the model generalized very well.

2.3 Adding More Variability: Factoring out the Style

The generative model introduced in Equation (2.1) generates the visual input as a function of a latent variable representing body configuration constrained to a motion manifold. Obviously body configuration is not the only factor controlling the visual appearance of humans in images. Any input image is a function of many aspects such as person body structure, appearance, viewpoint, illumination, etc. Therefore, it is obvious that the visual manifolds of different people doing the same activity will be different. So, how to handle all these

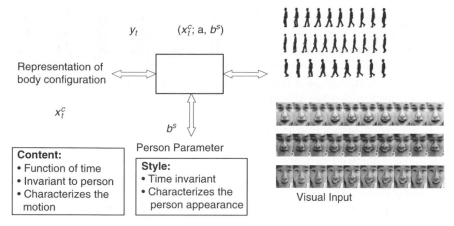

Fig. 2.7. Style and content factors: Content: gait motion or facial expression. Style: different silhouette shapes or face appearance.

variabilities. Let us assume the simple case first, a single viewpoint and we deal with human silhouettes so we do not have any variability due to illumination or appearance. Let the only source of variability be variation in people silhouette shapes. The problem now is how to extend the generative model in Equation (2.1) to include a variable describing people shape variability. For example, given several sequences of walking silhouettes, as in Figure 2.7, with different people walking, how to decompose the intrinsic body configuration through the action from the appearance (or shape) of the person performing the action. we aim to learn a decomposable generative model that explicitly decomposes the following two factors:

- Content (body pose): A representation of the intrinsic body configuration through the motion as a function of time that is invariant to the person, i.e., the content characterizes the motion or the activity.
- Style (people) : Time-invariant person parameters that characterize the person appearance (shape).

On the other hand, given an observation of certain person at a certain body pose and given the learned generative model we aim to be able to solve for both the body configuration representation (content) and the person parameter (style). In our case the content is a continuous domain while style is represented by the discrete style classes which exist in the training data where we can interpolate intermediate styles and/or intermediate contents. This can be formulated as a view-based generative model in the form

$$y_t^s = \gamma(x_t^c; a, b^s) \tag{2.8}$$

where the image, y_t^s, at time t and of style s is an instance driven from a generative model where the function $\gamma(\cdot)$ is a mapping function that maps from a representation of body configuration x_t^c (content) at time t into the image space given mapping parameters a and style dependent parameter b^s that is time invariant.[2] A framework was introduced in [73] to learn a decomposable generative model that explicitly decomposes the intrinsic body configuration (content) as a function of time from the appearance (style) of the person performing the action as time-invariant parameter. The framework is based on decomposing the style parameters in the space of nonlinear functions that maps between a learned unified nonlinear embedding of multiple content manifolds and the visual input space.

Suppose that we can learn a unified, style-invariant, nonlinearly embedded representation of the motion manifold $\mathcal{M}$ in a low-dimensional Euclidean embedding space, $\mathbb{R}^e$, then we can learn a set of style-dependent nonlinear mapping functions from the embedding space into the input space, i.e., functions $\gamma_s(x_t^c) : \mathbb{R}^e \to \mathbb{R}^d$ that maps from embedding space with dimensionality e into the input space (observation) with dimensionality d for style class s. Since we consider nonlinear manifolds and the embedding is nonlinear, the use of nonlinear mapping is necessary. We consider mapping functions in the form

$$y_t^s = \gamma_s(x_t) = C^s \cdot \psi(x_t^c) \tag{2.9}$$

where C^s is a $d \times N$ linear mapping and $\psi(\cdot) : \mathbb{R}^e \to \mathbb{R}^N$ is a nonlinear mapping where N basis functions are used to model the manifold in the embedding space, i.e.,

$$\psi(\cdot) = [\psi_1(\cdot), \cdots, \psi_N(\cdot)]^T$$

Given learned models of the form of Equation (2.9), the style can be decomposed in the linear mapping coefficient space using bilinear model in a way similar to [48, 49]. Therefore, input instance y_t can be written as asymmetric bilinear model in the linear mapping space as

$$y_t = \mathcal{A} \times_3 b^s \times_2 \psi(x_t^c) \tag{2.10}$$

where $\mathcal{A}$ is a third order tensor (3-way array) with dimensionality $d \times N \times J$, b^s is a style vector with dimensionality J, and $\times_n$ denotes mode-n tensor product. Given the role for style and content defined above, the previous equation can be written as

$$y_t = \mathcal{A} \times_3 b^{people} \times_2 \psi(x_t^{pose}) \tag{2.11}$$

Figure 2.8 shows examples for decomposing styles for gait. The learned generative model is used to interpolate walking sequences at new styles as well as to solve for the style parameters and body pose. In this experiment we used five sequences for five different people[3] each containing about 300 frames which are noisy. The learned manifolds are shown in Figure 2.8-c which shows

---

[2] We use the superscript s, c to indicate which variables denote style or content respectively.

[3] The data are from CMU Mobogait database.

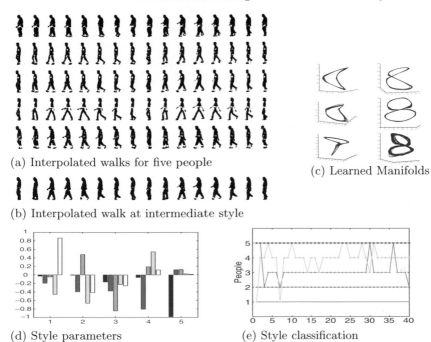

(a) Interpolated walks for five people

(c) Learned Manifolds

(b) Interpolated walk at intermediate style

(d) Style parameters

(e) Style classification

Fig. 2.8. (a) Interpolated walks for five people. (b) Interpolated walk at intermediate style between person 1 and 4. (c) Learned manifolds for the five people and the unified manifold (bottom right). (d) Estimated style parameters given the unified manifold. (e) Style classification for test data of 40 frames for 5 people.

a different manifold for each person. The learned unified manifold is also shown in Figure 2.8-c. Figure 2.8-b shows interpolated walking sequences for the five people generated by the learned model. The figure also shows the learned style vectors. We evaluated style classifications using 40 frames for each person and the result is shown in the figure with correct classification rate of 92%. We also used the learned model to interpolate walks in new styles. The last row in the figure shows interpolation between person 1 and person 4.

2.4 Style Adaptive Tracking: Bayesian Tracking on a Manifold

Given the explicit manifold model and the generative model learned in Section 2.3, we can formulate contour tracking within a Bayesian tracking framework. We can achieve style adaptive contour tracking on cluttered environments where the generative model can be used as an observation model to generate contours of different people shape styles and different poses. The tracking is performed on three conceptually independent spaces: body configuration space, shape style space and geometric transformation space. Therefore, object state combines heterogeneous representations. The manifold provides

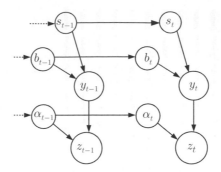

Fig. 2.9. A graphic model for the factorized generative model used for adaptive contour tracking.

a constraint on the motion, which reduces the system dynamics of the global nonlinear deformation into a linear dynamic system. The challenge will be how to represent and handle multiple spaces without falling into exponential increase of the state space dimensionality. Also, how to do tracking in a shape space which can be high dimensional?

Figure 2.9 shows a graphical model illustrating the relation between different variables. The shape at each time step is an instance driven from a generative model. Let $z_t \in \mathbb{R}^d$ be the shape of the object at time instance t represented as a point in a d-dimensional space. This instance of the shape is driven from a model in the form

$$z_t = T_{\alpha_t} \gamma(b_t, s_t; \theta), \tag{2.12}$$

where the $\gamma(\cdot)$ is a nonlinear mapping function that maps from a representation of the body configuration b_t and a representation person shape style s_t, independent from the configuration, into the observation space given the mapping parameter θ. T_{α_t} represents a geometric transformation on the shape instance. Given this generative model, we can fully describe observation instance z_t by state parameters α_t, b_t, and s_t. The mapping $\gamma(b_t, s_t; \theta)$ is a nonlinear mapping from the body configuration state b_t as

$$y_t = \mathcal{A} \times s_t \times \psi(b_t), \tag{2.13}$$

where $\psi(b_t)$ is a kernel induced space, $\mathcal{A}$ is a third order tensor, s_t is a shape style vector o person k and $\times$ is appropriate tensor product. Given this form, the mapping parameter θ is the tensor $\mathcal{A}$.

The tracking problem is then an inference problem where at time t we need to infer the body configuration representation b_t and the person specific parameter s_t and the geometric transformation T_{α_t} given the observation z_t. The Bayesian tracking framework enables a recursive update of the posterior $P(X_t | Z^t)$ over the object state X_t given all observation $Z^t = Z_1, Z_2, .., Z_t$ up to time t:

$$P(X_t|Z^t) \propto P(Z_t|X_t) \int_{X_{t-1}} P(X_t|X_{t-1})P(X_{t-1}|Z^{t-1}) \qquad (2.14)$$

In our generative model, the state X_t is $[\alpha_t, b_t, s_t]$, which uniquely describes the state of the tracking object. Observation Z_t is the captured image instance at time t.

The state X_t is decomposed into three substates α_t, b_t, s_t. These three random variables are conceptually independent since we can combine any body configuration with any person shape style with any geometrical transformation to synthesize a new contour. However, they are dependent given the observation Z_t. It is hard to estimate joint posterior distribution $P(\alpha_t, b_t, s_t|Z_t)$ for its high-dimensionality. The objective of the density estimation is to estimate states α_t, b_t, s_t for a given observation. The decomposable feature of our generative model enables us to estimate each state by a marginal density distribution $P(\alpha_t|Z^t)$, $P(b_t|Z^t)$, and $P(s_t|Z^t)$. We approximate marginal density estimation of one state variable along representative values of the other state variables. For example, in order to estimate marginal density of $P(b_t|Z^t)$, we estimate $P(b_t|\alpha_t^*, s_t^*, Z^t)$, where α_t^*, s_t^* are representative values such as maximum posteriori estimates.

modelling body configuration space: Given a set of training data for multiple people, a unified mean manifold embedding can be obtained as was explained in Section 2.3. The mean manifold can be parameterized by a one-dimensional parameter $\beta_t \in \mathbb{R}$ and a spline fitting function $f : \mathbb{R} \to \mathbb{R}^3$, which satisfies $b_t = f(\beta_t)$, to map from the parameter space into the three-dimensional embedding space.

modelling style shape space: Shape style space is parameterized by a linear combination of basis of the style space. A generative model in the form of Equation (2.13) is fitted to the training data. Ultimately the style parameter s should be independent of the configuration and therefore should be time invariant and can be estimated at initialization. However, we do not know the person style initially, therefore, the style needs to fit to the correct person style gradually during the tracking. So, we formulated style as time variant factor that should stabilize after some frames from initialization. The dimension of the style vector depends on the number of people used for training and can be high dimensional.

We represent new style as a convex linear combination of style classes learned from the training data. The tracking of the high-dimensional style vector s_t itself will be hard as it can fit local minima easily. A new style vector s is represented by linear weighting of each of the style classes s^k, $k = 1, \cdots, K$ using linear weight vector $\lambda = [\lambda^k, \cdots, \lambda^K]^T$:

$$s = \sum_{k=1}^{K} \lambda^k s^k, \qquad \sum_{k=1}^{K} \lambda^k = 1, \qquad (2.15)$$

where K is the number of style classes used to represent new styles. The overall generative model can be expressed as

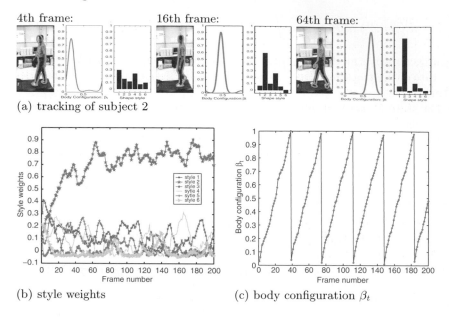

(a) tracking of subject 2

(b) style weights

(c) body configuration β_t

Fig. 2.10. Tracking for a known person.

$$z_t = T_{\alpha_t} \left(\mathcal{A} \times \left[\sum_{k=1}^{K} \lambda_t^k s^k \right] \times \psi(f(\beta_t)) \right). \qquad (2.16)$$

Tracking problem using this generative model is the estimation of the parameters α_t, β_t, and λ_t at each new frame given the observation z_t. Tracking can be done using a particle filter as was shown in [78, 79]. Figures 2.10 and 2.11 show style adaptive tracking results for two subjects. In the first case, the person style is in the training set while in the second case the person was not seen before in the training. In both cases, the style parameter started at the mean style and adapted correctly to the person shape. It is clear that the estimated body configuration shows linear dynamics and the particles are showing a gaussian distribution on the manifold.

2.5 Adding More Variability: A Factorized Generative Model

In Section 2.3 it was shown how to separate a style factor when learning a generative model for data lying on a manifold. Here we generalize this concept to decompose several style factors. For example, consider the walking motion observed from multiple viewpoints (as silhouettes). The resulting data lie on multiple subspaces and/or multiple manifolds. There is the underling motion manifold, which is one dimensional for the gait motion. There is the view

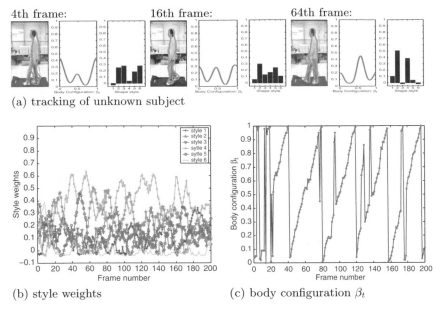

(a) tracking of unknown subject

(b) style weights

(c) body configuration β_t

Fig. 2.11. Tracking for an unknown person.

manifold and the space of different people's shapes. Another example we consider is facial expressions. Consider face data of different people performing different facial dynamic expressions such as sad, smile, surprise, etc. The resulting face data posses several dimensionality of variability: the dynamic motion, the expression type and the person face. So, how to model such data in a generative manner. We follow the same framework of explicitly modelling the underlying motion manifold and over that we decompose various style factors.

We can think of the image appearance (similar argument for shape) of a dynamic object as instances driven from such generative model. Let $y_t \in \mathbb{R}^d$ be the appearance of the object at time instance t represented as a point in a d-dimensional space. This instance of the appearance is driven from a model in the form

$$y_t = T_\alpha \gamma(x_t; a_1, a_2, \cdots, a_n) \tag{2.17}$$

where the appearance, y_t, at time t is an instance driven from a generative model where the function γ is a mapping function that maps body configuration x_t at time t into the image space. i.e., the mapping function γ maps from a representation of the body configuration space into the image space given mapping parameters $a_1, \cdots, a_n$ each representing a set of conceptually orthogonal factors. Such factors are independent of the body configuration and can be time variant or invariant. The general form for the mapping function γ that we use is

$$\gamma(x_t; a_1, a_2, \cdots, a_n) = \mathcal{C} \times_1 a_1 \times \cdots \times_n a_n \cdot \psi(x_t) \tag{2.18}$$

where $\psi(x)$ is a nonlinear kernel map from a representation of the body configuration to a kernel induced space and each a_i is a vector representing a parameterization of orthogonal factor i, $\mathcal{C}$ is a core tensor, $\times_i$ is *mode-i* tensor product as defined in [52, 80].

For example for the gait case, a generative model for walking silhouettes for different people from different viewpoints will be in the form

$$y_t = \gamma(x_t; v, s) = \mathcal{C} \times v \times s \times \psi(x) \tag{2.19}$$

where v is a parameterization of the view, which is independent of the body configuration but can change over time, and s is a parameterization of the shape style of the person performing the walk which is independent of the body configuration and time invariant. The body configuration x_t evolves along a representation of the manifold that is homeomorphic to the actual gait manifold.

Another example is modelling the manifolds of facial expression motions. Given dynamic facial expression such as sad, surprise, happy, etc., where each expression start from neutral and evolve to a peak expression; each of these motions evolves along a one-dimensional manifold. However, the manifold will be different for each person and for each expression. Therefore, we can use a generative model to generate different people faces and different expressions using a model in the form be in the form

$$y_t = \gamma(x_t; e, f) = \mathcal{A} \times e \times f \times \psi(x_t) \tag{2.20}$$

where e is an expression vector (happy, sad, etc.) that is invariant of time and invariant of the person face, i.e., it only describes the expression type. Similarly, f is a face vector describing the person face appearance which is invariant of time and invariant of the expression type. The motion content is described by x which denotes the motion phase of the expression, i.e., starts from neutral and evolves to a peak expression depending on the expression vector, e.

The model in Equation (2.18) is a generalization over the model in Equations (2.1) and (2.8). However, such generalization is not obvious. In Section 2.3 LLE was used to obtain manifold embeddings, and then a mean manifold is computed as a unified representation through nonlinear warping of manifold points. However, since the manifolds twists very differently given each factor (different people or different views, etc.) it is not possible to achieve a unified configuration manifold representation independent of other factors. These limitations motivate the use of a conceptual unified representation of the configuration manifold that is independent of all other factors. Such unified representation would allow the model in Equation (2.18) to generalize to decompose as many factors as desired. In the model in Equation (2.18), the relation between body configuration and the input is nonlinear where other factors are approximated linearly through multilinear analysis. The use of nonlinear mapping is essential since the embedding of the configuration manifold is nonlinearly related to the input.

The question is what conceptual representation of the manifold we can use. For example, for the gait case, since the gait is 1D closed manifold embedded in the input space, it is homeomorphic to a unit circle embedded in 2D. In general, all closed 1D manifold is topologically homeomorphic to unit circles. We can think of it as a circle twisted and stretched in the space based on the shape and the appearance of the person under consideration or based on the view. So we can use such unit circle as a unified representation of all gait cycles for all people for all views. Given that all the manifolds under consideration are homeomorphic to unit circle, the actual data is used to learn nonlinear warping between the conceptual representation and the actual data manifold. Since each manifold will have its own mapping, we need to have a mechanism to parameterize such mappings and decompose all these mappings to parameterize variables for views, different people, etc.

Given an image sequences $y_t^a, t = 1, \cdots, T$ where a denotes a particular class setting for all the factors $a_1, \cdots, a_n$ (e.g., a particular person s and view v) representing a whole motion cycle and given a unit circle embedding of such data as $x_t^a \in \mathbb{R}^2$ we can learn a nonlinear mapping in the form

$$y_t^a = B^a \psi(x_t^a) \qquad (2.21)$$

Given such mapping the decomposition in Equation (2.1) can be achieved using tensor analysis of the coefficient space such that the coefficient B^a are obtained from a multilinear [80] model

$$B^a = \mathcal{C} \times_1 a_1 \times \cdots \times_n a_n$$

Given a training data and a model fitted in the form of Equation (2.18) it is desired to use such model to recover the body configuration and each of the orthogonal factors involved, such as viewpoint and person shape style given a single test image or given a full or a part of a motion cycle. Therefore, we are interested in achieving an efficient solution to a nonlinear optimization problem in which we search for x^*, a_i^* which minimize the error in reconstruction

$$E(x, a_1, \cdots, a_n) = \| y - \mathcal{C} \times_1 a_1 \times \cdots \times_n a_n \times \psi(x) \| \qquad (2.22)$$

or a robust version of the error. In [81] an efficient algorithms were introduced to recover these parameters in the case of a single image input or a sequence of images using deterministic annealing.

2.5.1 Dynamic Shape Example: Decomposing View and Style on the Gait Manifold

In this section we show an example of learning the nonlinear manifold of gait as an example of a dynamic shape. We used CMU Mobo gait data set [82] which contains walking people from multiple synchronized views[4]. For training

---

[4] CMU Mobo gait data set [82] contains 25 people, about 8 to 11 walking cycles each captured from six different viewpoints. The walkers were using a treadmill.

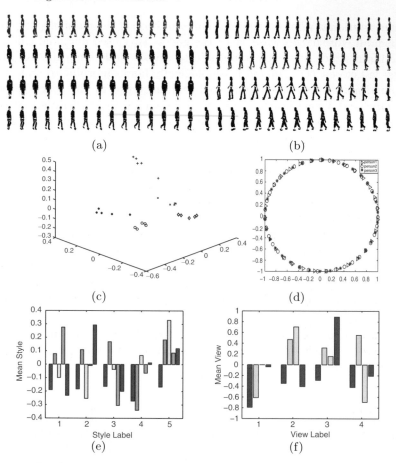

(a)

(b)

(c)

(d)

(e)

(f)

Fig. 2.12. (a, b) Example of training data. Each sequence shows a half cycle only. (a) four different views used for person 1. (b) side views of people $2, 3, 4, 5$. (c) style subspace: each person cycles have the same label. (d) unit circle embedding for three cycles. (e) Mean style vectors for each person cluster. (f) view vectors.

we selected five people, five cycles each from four different views. i.e., total number of cycles for training is $100 = 5$ people $\times$ 5 cycles $\times$ 4 views. Note that cycles of different people and cycles of the same person are not of the same length. Figure 2.12-a,b show examples of the sequences (only half cycles are shown because of limited space).

The data is used to fit the model as described in Equation (2.19). Images are normalized to 60×100, i.e., $d = 6,000$. Each cycle is considered to be a style by itself, i.e., there are 25 styles and 4 views. Figure 2.12-d shows an example of model-based aligned unit circle embedding of three cycles. Figure 2.12-c shows the obtained style subspace where each of the 25 points corresponding to one of the 25 cycles used. Important thing to notice is that the style vectors are clustered in the subspace such that each person style vectors (corresponding to different cycles of the same person) are clustered together which indicate that the model can find the similarity in the shape style between different cycles of the same person. Figure 2.12-e shows the mean style vectors for each of the five clusters. Figure 2.12-f shows the four view vectors.

Figure 2.13 shows example of using the model to recover the pose, view and style. The figure shows samples of a one full cycle and the recovered body configuration at each frame. Notice that despite the subtle differences between the first and second halves of the cycle, the model can exploit such differences to recover the correct pose. The recovery of 3D joint angles is achieved by learning a mapping from the manifold embedding and 3D joint angle from motion captured data using GRBF in a way similar to Equation (2.21). Figure 2.13-b and Figure 2.13-c show the recovered style weights (class probabilities) and view weights respectively for each frame of the cycle which shows correct person and view classification. Figure 2.14 shows examples recovery of the 3D pose and view class for four different people non of them was seen in training.

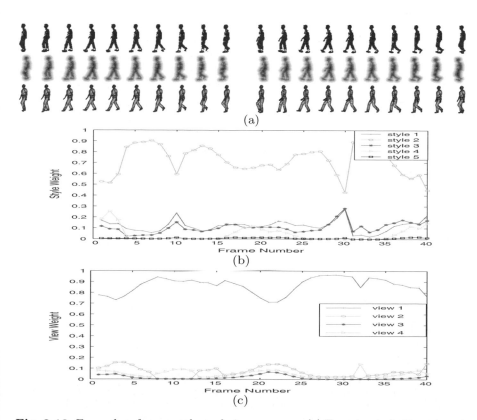

Fig. 2.13. Examples of pose, style, and view recovery. (a) From top to bottom: input shapes, implicit function, and recovered 3D pose. Odd-numbered frames, from 1 to 39, are shown. (b) Recovered style weights at each frame. (c) Recovered view weights at each frame.

Fig. 2.14. Examples of pose and view recovery for four different people from four views.

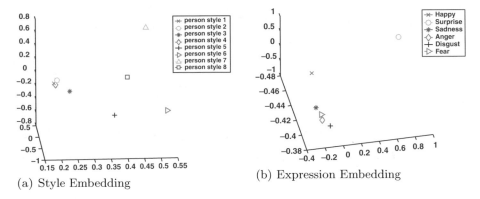

(a) Style Embedding

(b) Expression Embedding

Fig. 2.15. Facial expression analysis for 8 subjects with 6 expressions from Cohn-Kanade Data set. (a) embedding in the style space.(b) embedding in the expression space. First three dimensions are shown.

2.5.2 Dynamic Appearance Example: Facial Expression Analysis

We used the model to learn facial expressions manifolds for different people. We used CMU-AMP facial expression database where each subject has 75 frames of varying facial expressions. We choose four people and three expressions each (smile, anger, surprise) where corresponding frames are manually segmented from the whole sequence for training. The resulting training set contained 12 sequences of different lengths. All sequences are embedded to unit circles and aligned as described in Section 2.5. A model in the form of Equation (2.20) is fitted to the data where we decompose two factors: person facial appearance style factor and expression factor besides the body configuration which is nonlinearly embedded on a unit circle. Figure 2.15 shows the resulting person style vectors and expression vectors.

We used the learned model to recognize facial expression, and person identity at each frame of the whole sequence. Figure 2.16 shows an example of a whole sequence and the different expression probabilities obtained on a frame per frame basis. The figure also shows the final expression recognition after

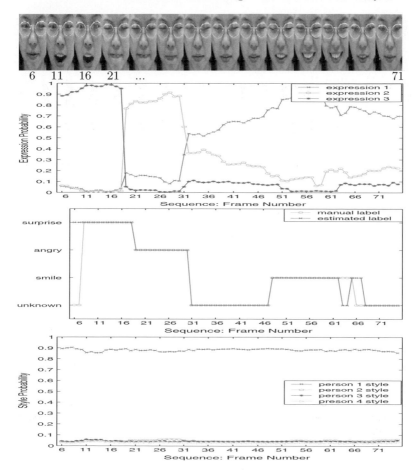

Fig. 2.16. From top to bottom: Samples of the input sequence; Expression probabilities; Expression classification; Style probabilities.

thresholding along manual expression labeling. The learned model was used to recognize facial expressions for sequences of people not used in the training. Figure 2.17 shows an example of a sequence of a person not used in the training. The model can successfully generalizes and recognize the three learned expression for this new subject.

2.6 Conclusion

In this chapter we focused on exploiting the underlying motion manifold for human motion analysis and synthesis. we introduced a framework for learning a landmark-free correspondence-free global representations of dynamic shape and dynamic appearance manifolds. The framework is based on using

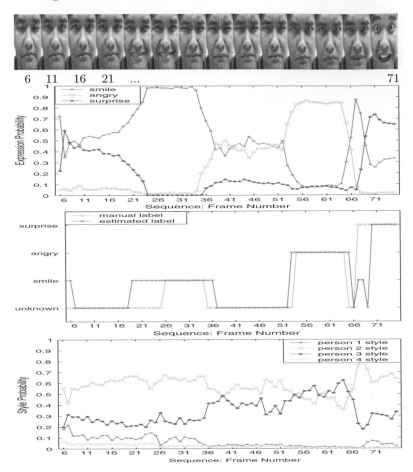

Fig. 2.17. Generalization to new people: expression recognition for a new person. From top to bottom: Samples of the input sequence; Expression probabilities; Expression classification; Style probabilities.

nonlinear dimensionality reduction to achieve an embedding of the global deformation manifold, which preserves the geometric structure of the manifold. Given such embedding, a nonlinear mapping is learned from such embedded space into the visual input space using the RBF interpolation. Given this framework, any visual input is represented by a linear combination of nonlinear bases functions centered along the manifold in the embedded space. In a sense, the approach utilizes the implicit correspondences imposed by the global vector representation which are only valid locally on the manifold through explicit modelling of the manifold and the RBF interpolation where closer points on the manifold will have higher contributions than far away points.

We also showed how approximate solution for the inverse mapping can be obtained in a closed-form, which facilitates the recovery of the intrinsic body configuration. The framework was applied to learn a representation of the gait manifold as an example of a dynamic shape manifold. We showed how the learned representation can be used to interpolate intermediate body poses as well as in recovery and reconstruction of the input. We also learn mappings from the embedded motion manifold to a 3D joint angle representation, which yields an approximate closed-form solution for 3D pose recovery.

We showed how to learn a decomposable generative model that separates appearance variations from the intrinsics underlying dynamics' manifold though introducing a framework for separation of style and content on a nonlinear manifold. The framework is based on decomposing the style parameters in the space of nonlinear functions that maps between a learned unified nonlinear embedding of multiple content manifolds and the visual input space. The framework yields an unsupervised procedure, which handles dynamic nonlinear manifolds. It also improves on past work on nonlinear dimensionality reduction by being able to handle multiple manifolds. The proposed framework was shown to be able to separate style and content on both the gait manifold and simple facial expression manifolds. As mention in [62], an interesting and important question is how to learn a parametric mapping between the observation and a nonlinear embedding space. We partially addressed this question. More details about this topic can be found in [73].

The use of a generative model is necessary since the mapping from the manifold representation to the input space will be well defined in contrast to a discriminative model where the mapping from the visual input to manifold representation is not necessarily a function. We introduced a framework to solve for various factors such as body configuration, view, and shape style. Since the framework is generative, it fits well in a Bayesian tracking framework and it provides separate low-dimensional representations for each of the modeled factors. Moreover, a dynamic model for configuration is well defined since it is constrained to the 1D manifold representation. The framework also provides a way to initialize a tracker by inferring about body configuration, viewpoint, body shape style from a single or a sequence of images.

The framework presented in this chapter was basically applied to one-dimensional motion manifolds such as gait and facial expressions. One-dimensional manifolds can be explicitly modeled in a straight forward way. However, there is no theoretical restriction that prevents the framework from dealing with more complicated manifolds. In this chapter we mainly modeled the motion manifold while all appearance variability are modeled using subspace analysis. Extension to modelling multiple manifolds simultaneously is very challenging. We investigated modelling both the motion and the view manifolds in [83]. The proposed framework has been applied to gait analysis and recognition in [78, 79, 84, 85]. It was also used in analysis and recognition of facial expressions in [86, 87].

Acknowledgments

This research is partially funded by NSF award IIS-0328991

References

1. J.O'Rourke, Badler: Model-based image analysis of human motion using constraint propagation. IEEE PAMI **2**(6) (1980)
2. Hogg, D.: Model-based vision: a program to see a walking person. Image and Vision Computing **1**(1) (1983) 5–20
3. Chen, Z., Lee, H.: Knowledge-guided visual perception of 3-d human gait from single image sequence. IEEE SMC **22**(2) (1992) 336–342
4. Rohr, K.: Towards model-based recognition of human movements in image sequence. CVGIP **59**(1) (1994) 94–115
5. Rehg, J.M., Kanade, T.: Model-based tracking of self-occluding articulated objects. In: ICCV (1995) 612–617
6. Gavrila, D., Davis, L.: 3-d model-based tracking of humans in action: a multiview approach. In: IEEE Conference on Computer Vision and Pattern Recognition. Volume 73–80 (1996)
7. Kakadiaris, I.A., Metaxas, D.: Model-based estimation of 3D human motion with occlusion based on active multi-viewpoint selection. In: Proc. IEEE Conf. Computer Vision and Pattern Recognition, CVPR, Los Alamitos, California, USA, IEEE Computer Society (1996) 81–87
8. Sidenbladh, H., Black, M.J., Fleet, D.J.: Stochastic tracking of 3d human figures using 2d image motion. In: ECCV (2) (2000) 702–718
9. Rehg, J.M., Kanade, T.: Visual tracking of high DOF articulated structures: an application to human hand tracking. In: ECCV (2) (1994) 35–46
10. Darrell, T., Pentland, A.: Space-time gesture. In: Proc IEEE CVPR (1993)
11. Campbell, L.W., Bobick, A.F.: Recognition of human body motion using phase space constraints. In: ICCV (1995) 624–630
12. Wern, C.R., Azarbayejani, A., Darrell, T., Pentland, A.P.: Pfinder: Real-time tracking of human body. IEEE Transaction on Pattern Analysis and Machine Intelligence **19**(7) (1997)
13. Ju, S.X., Black, M.J., Yacoob, Y.: Cardboard people: A parameterized model of articulated motion. In: International Conference on Automatic Face and Gesture Recognition, Killington, Vermont (1996) 38–44
14. Black, M.J., Jepson, A.D.: Eigentracking: Robust matching and tracking of articulated objects using a view-based representation. In: ECCV (1) (1996) 329–342
15. Haritaoglu, I., Harwood, D., Davis, L.S.: W4: Who? when? where? what? a real time system for detecting and tracking people. In: International Conference on Automatic Face and Gesture Recognition (1998) 222–227
16. Yacoob, Y., Black, M.J.: Parameterized modelling and recognition of activities. Computer Vision and Image Understanding: CVIU **73**(2) (1999) 232–247
17. Fablet, R., Black, M.J.: Automatic detection and tracking of human motion with a view-based representation. In: Proc. ECCV 2002, LNCS 2350 (2002) 476–491

18. Sidenbladh, H., Black, M.J., Sigal, L.: Implicit probabilistic models of human motion for synthesis and tracking. In: Proc. ECCV 2002, LNCS 2350 (2002) 784–800

19. Goldenberg, R., Kimmel, R., Rivlin, E., Rudzsky, M.: 'Dynamism of a dog on a leash' or behavior classification by eigen-decomposition of periodic motions. In: Proceedings of the ECCV'02, Copenhagen, Springer, LNCS 2350 (2002) 461–475

20. Polana, R., Nelson, R.C.: Qualitative detection of motion by a moving observer. International Journal of Computer Vision **7**(1) (1991) 33–46

21. Nelson, R.C., Polana, R.: Qualitative recognition of motion using temporal texture. CVGIP Image Understanding **56**(1) (1992) 78–89

22. Polana, R., Nelson, R.: Low level recognition of human motion (or how to get your man without finding his body parts). In: IEEE Workshop on Non-Rigid and Articulated Motion (1994) 77–82

23. Polana, R., Nelson, R.C.: Detecting activities. Journal of Visual Communication and Image Representation (1994)

24. Niyogi, S., Adelson, E.: Analyzing and recognition walking figures in xyt. In: Proc. IEEE CVPR (1994) 469–474

25. Song, Y., Feng, X., Perona, P.: Towards detection of human motion. In: IEEE Computer Society Conference on Computer Vision and Pattern Recognition (CVPR 2000) (2000) 810–817

26. Rittscher, J., Blake, A.: Classification of human body motion. In: IEEE International Conferance on Compute Vision (1999)

27. Bobick, A., Davis, J.: The recognition of human movement using temporal templates. IEEE Transactions on Pattern Analysis and Machine Intelligence **23**(3) (2001) 257–267

28. Cutler, R., Davis, L.: Robust periodic motion and motion symmetry detection. In: Proc. IEEE CVPR (2000)

29. Mori, G., Malik., J.: Estimating human body configurations using shape context matching. In: European Conference on Computer Vision (2002)

30. Kristen Grauman, Gregory Shakhnarovich, T.D.: Inferring 3d structure with a statistical image-based shape model. In: ICCV (2003)

31. Shakhnarovich, G., Viola, P., Darrell, T.: Fast pose estimation with parameter-sensitive hashing. In: ICCV (2003)

32. Howe, Leventon, Freeman, W.: Bayesian reconstruction of 3d human motion from single-camera vidco. In: Proc. NIPS (1999)

33. Brand, M.: Shadow puppetry. In: International Conference on Computer Vision. Volume 2 (1999) 1237

34. Rosales, R., Sclaroff, S.: Inferring body pose without tracking body parts. Technical Report 1999-017 (1999)

35. Rosales, R., Sclaroff, S.: Specialized mappings and the estimation of human body pose from a single image. In: Workshop on Human Motion (2000) 19–24

36. Rosales, R., Athitsos, V., Sclaroff, S.: 3D hand pose reconstruction using specialized mappings. In: Proc. ICCV (2001)

37. Christoudias, C.M., Darrell, T.: On modelling nonlinear shape-and-texture appearance manifolds. In: Proc.of IEEE CVPR. Volume 2 (2005) 1067–1074

38. Rahimi, A., Recht, B., Darrell, T.: Learning appearane manifolds from video. In: Proc.of IEEE CVPR. Volume 1 (2005) 868–875

39. Bowden, R.: Learning statistical models of human motion. In: IEEE Workshop on Human Modelling, Analysis and Synthesis (2000)

40. Toyama, K., Blake, A.: Probabilistic tracking in a metric space. In: ICCV (2001) 50–59
41. Bregler, C., Omohundro, S.M.: Nonlinear manifold learning for visual speech recognition (1995) 494–499
42. Jolliffe, I.T.: Principal Component Analysis. Springer-Verlag (1986)
43. M.Turk, A.Pentland: Eigenfaces for recognition. Journal of Cognitive Neuroscience **3**(1) (1991) 71–86
44. Belhumeur, P.N., Hespanha, J., Kriegman, D.J.: Eigenfaces vs. fisherfaces: Recognition using class specific linear projection. In: ECCV (1) (1996) 45–58
45. Cootes, T.F., Taylor, C.J., Cooper, D.H., Graham, J.: Active shape models: Their training and application. CVIU **61**(1) (1995) 38–59
46. Levin, A., Shashua, A.: Principal component analysis over continuous subspaces and intersection of half-spaces. In: ECCV, Copenhagen, Denmark (2002) 635–650
47. Murase, H., Nayar., S.: Visual learning and recognition of 3d objects from appearance. International Journal of Computer Vision **14** (1995) 5–24
48. Tenenbaum, J., Freeman, W.T.: Separating style and content with bilinear models. Neural Computation **12** (2000) 1247–1283
49. Vasilescu, M.A.O., Terzopoulos, D.: Multilinear analysis of image ensebles: Tensorfaces. In: Proc. of ECCV, Copenhagen, Danmark (2002) 447–460
50. Magnus, J., Neudecker, H.: Matrix Differential Calculus with Applications in Statistics and Econometrics. Wiley, New York (1988)
51. Marimont, D., Wandell, B.: Linear models of surface and illumination spectra. Journal of Optical Society od America **9** (1992) 1905–1913
52. Lathauwer, L.D., de Moor, B., Vandewalle, J.: A multilinear singular value decomposiiton. SIAM Journal On Matrix Analysis and Applications **21**(4) (2000) 1253–1278
53. Shashua, A., Levin, A.: Linear image coding of regression and classification using the tensor rank principle. In: Proc. of IEEE CVPR, Hawai (2001)
54. Vasilescu, M.A.O.: An algorithm for extracting human motion signatures. In: Proc. of IEEE CVPR, Hawai (2001)
55. Wang, H., Ahuja, N.: Rank-r approximation of tensors: Using image-as-matrix representation. (In: Proc IEEE CVPR)
56. Tucker, L.: Some mathematical notes on three-mode factor analysis. Psychometrika **31** (1966) 279–311
57. Kapteyn, A., Neudecker, H., Wansbeek, T.: An approach to n-model component analysis. Psychometrika **51**(2) (1986) 269–275
58. Vidal, R., Ma, Y., Sastry, S.: Generalized principal component analysis (gpca). In: Proceedings of IEEE CVPR. Volume 1 (2003) 621–628
59. Vidal, R., Hartley, R.: Motion segmentation with missing data using powerfactorization and gpca (2004)
60. Cox, T., Cox, M.: Multidimentional scaling. Chapman & Hall (1994)
61. Tenenbaum, J.: Mapping a manifold of perceptual observations. In: Advances in Neural Information Processing. Volume 10 (1998) 682–688
62. Roweis, S., Saul, L.: Nonlinear dimensionality reduction by locally linear embedding. Sciene **290**(5500) (2000) 2323–2326
63. Belkin, M., Niyogi, P.: Laplacian eigenmaps for dimensionality reduction and data representation. Neural Comput. **15**(6) (2003) 1373–1396
64. Brand, M., Huang, K.: A unifying theorem for spectral embedding and clustering. In: Proc. of the Ninth International Workshop on AI and Statistics (2003)

65. Lawrence, N.: Gaussian process latent variable models for visualization of high dimensional data. In: NIPS (2003)
66. Weinberger, K.W., Saul, L.K.: Unsupervised learning of image manifolds by semidefinite programming. In: Proceedings of IEEE CVPR. Volume 2 (2004) 988–995
67. Mordohai, P., Medioni, G.: Unsupervised dimensionality estimation and manifold learning in high-dimensional spaces by tensor voting. In: Proceedings of International Joint Conference on Artificial Intelligence (2005)
68. Bengio, Y., Delalleau, O., Le Roux, N., Paiement, J.F., Vincent, P., Ouimet, M.: Learning eigenfunctions links spectral embedding and kernel pca. Neural Comp. **16**(10) (2004) 2197–2219
69. Ham, J., Lee, D.D., Mika, S., Schölkopf, B.: A kernel view of the dimensionality reduction of manifolds. In: Proceedings of ICML, New York, NY, USA, ACM Press (2004) 47
70. Schölkopf, B., Smola, A.: Learning with Kernels: Support Vector Machines, Regularization, Optimization and Beyond. MIT Press, Cambridge, Massachusetts (2002)
71. Bengio, Y., Paiement, J.F., Vincent, P., Delalleau, O., Roux, N.L., Ouimet, M.: Out-of-sample extensions for lle, isomap, mds, eigenmaps, and spectral clustering. In: NIPS 16 (2004)
72. Elgammal, A.: Nonlinear generative models for dynamic shape and dynamic appearance. In: Proc. of 2nd International Workshop on Generative-Model based vision. GMBV 2004 (2004)
73. Elgammal, A., Lee, C.S.: Separating style and content on a nonlinear manifold. In: Proc. of CVPR (2004) 478–485
74. Seung, H.S., Lee, D.D.: The manifold ways of perception. Science **290**(5500) (2000) 2268–2269
75. Poggio, T., Girosi, F.: Network for approximation and learning. Proc. IEEE **78**(9) (1990) 1481–1497
76. Beymer, D., Poggio, T.: Image representations for visual learning. Science **272**(5250) (1996)
77. Elgammal, A., Lee, C.S.: Inferring 3d body pose from silhouettes using activity manifold learning. In: Proc. IEEE Conference on Computer Vision and Pattern Recognition (2004)
78. Lee, C.S., Elgammal, A.: Style adaptive bayesian tracking using explicit manifold learning. In: Proc BMVC (2005)
79. Lee, C.S., Elgammal, A.: Gait tracking and recognition using person-dependent dynamic shape model. In: International Conference on Automatic Face and Gesture Recognition. Volume 0., IEEE Computer Society (2006) 553–559
80. Vasilescu, M.A.O., Terzopoulos, D.: Multilinear subspace analysis of image ensembles. (2003)
81. Lee, C.S., Elgammal, A.: Homeomorphic manifold analysis: Learning decomposable generative models for human motion analysis. In: Workshop on Dynamical Vision (2005)
82. Gross, R., Shi, J.: The cmu motion of body (mobo) database. Technical Report TR-01-18, Carnegie Mellon University (2001)
83. Lee, C.S., Elgammal, A.M.: Simultaneous inference of view and body pose using torus manifolds. In: ICPR (3) (2006) 489–494

84. Lee, C.S., Elgammal, A.: Gait style and gait content: Bilinear model for gait recogntion using gait re-sampling. In: International Conference on Automatic Face and Gesture Recognition (2004) 147–152
85. Lee, C.S., Elgammal, A.M.: Towards scalable view-invariant gait recognition: Multilinear analysis for gait. In: AVBPA (2005) 395–405
86. Lee, C.S., Elgammal, A.: Facial expression analysis using nonlinear decomposable generative models. In: AMFG (2005) 17–31
87. Lee, C.S., Elgammal, A.M.: Nonlinear shape and appearance models for facial expression analysis and synthesis. In: ICPR (1) (2006) 497–502

3

Recognition of Action as a Bayesian Parameter Estimation Problem over Time

Volker Krüger

Computer Vision and Machine Intelligence Lab
CIT, Aalborg University
Lautrupvang 15
2750 Ballerup
Denmark

Summary. In this chapter we will discuss two problems related to action recognition: The first problem is the one of identifying in a *surveillance* surveillance scenario to determine walk or run gait and approximate direction. The second problem is concerned with the recovery of *action primitives* from observed complex actions. Both problems will be discussed within a *statistical framework*. *Bayesian propagation* over time offers a framework to treat likelihood observations at each time step and the dynamics between the time steps in a unified manner. The first problem will be approached as a pattern recognition and tracking task by a Bayesian propagation of the likelihoods. The latter problem will be approached by explicitly specifying the dynamics while the likelihood measure will estimate how well each dynamical model fits each time step. Extensive experimental results show the applicability of the Bayesian framework for action recognition and round up our discussion.

3.1 Introduction

Understanding how we cognitively interpret the *actions* and activities of other humans is a question that has been extensively studied [17, 43, 44]. This has become interesting to the computer vision scientists in a number of different contexts. In *surveillance* it is of interest to recognize suspicious and unusual actions. In the robotics community, the question regarding how to recognize action is intensively studied in the context of imitation learning and human–humanoid interaction. In *imitation learning*, humans are teaching robots through simple demonstrations of what they are supposed to do. This approach not only minimizes the training time for specific tasks but also enables to teach *humanoid robots* to perform actions in a way that appears familiar to a human observer.

To synthesize and recognize *actions* several techniques are possible. All have in common that they need to detect and track meaningful features in the video data. Usually, detection and tracking of the visual features is considered

B. Rosenhahn et al. (eds.), Human Motion – Understanding, Modelling, Capture, and
Animation, 57–79.

to be a problem of its own and is treated as such. In case of the action recognition from video problem, tracking is treated independently from the actual action recognition task. There are some publications that treat the tracking and the recognition part in conjunction, e.g., Ormoneit et.al. [36]. However, they considers the specific action given a priori to assist the 3D body tracking task.

In this chapter of the book we would like to argue that one should consider both, the tracking and the action recognition, as a joint problem and estimate them at the *same* time.

If we consider the tracking problem from a Bayesian perspective, then the classical *Bayesian propagation* over time for tracking can be formalized as follows:

$$p(\alpha_t|I_1, I_2, \ldots, I_t) \equiv p_t(\alpha_t)$$

$$= \int_{\alpha_{t-1}} p(I_t|\alpha_t)p(\alpha_t|\alpha_{t-1})p_{t-1}(\alpha_{t-1})d\alpha_t \ . \tag{3.1}$$

Here, $\alpha_t = (x_t, y_t, s_t)$. The x, y denote the position of the object to be tracked, s its scale and I_t the images in the video.

Equation 3.1 explains how the density $p(\alpha_t|I_1, I_2, \ldots, I_t)$ changes over time when more image I_t become available. In order to estimate Equation 3.1 variants of *Monte Carlo* methods are usually applied, see [22, 30] for applications in computer vision and [12, 27, 31, 33] for a general treatment.

In their original papers of Isard and Blake [22] and Li and Chellappa [30] only the estimation of the affine tracking parameters is considered.

However, Equation 3.1 offers a general framework for the estimation of density functions over time. Let us consider the separate parts of it: the leftmost part, $p_{t-1}(\alpha_{t-1})$, is the *prior* which summarizes our belief knowledge up to time $t - 1$ of what values the parameters x, y and s should have. The part to the right of the equal sign, $p(I_t|\alpha_t)$, is the *likelihood* measure which computes the likelihood or the "fit" of the new image I_t to our model with model parameters x_t, y_t and s_t. In case of face tracking, the model is a face image, translated and scaled according to the parameters x_t, y_t and s_t. If the parameters are close to the true parameters of the face in the input video, then this results into a large likelihood. If, on the other hand, the parameters are very different from the true ones, then the fit of the new input image with the model is usually small which then results into a small likelihood. A good model is crucial for obtaining large likelihood measures and it is possible to introduce further parameters to increase the model quality. For example, one can use a set of models instead of just one by introducing an additional random variable i as a model identifier. We have done this in the past [50] for the face recognition from video problem where we had introduced a random variable i_t to specify the identity of the person to be tracked. Consider the following refinement of Equation 3.1:

$$p(\alpha_t, i_t | I_1, I_2, \ldots, I_t) \equiv p_t(\alpha_t, i_t)$$

$$= \sum_{i_{t-1}} \int_{\alpha_{t-1}} p(I_t | \alpha_t, i_t) p(\alpha_t, i_t | \alpha_{t-1}, i_{t-1}) p_{t-1}(\alpha_{t-1}, i_{t-1}) \qquad (3.2)$$

where $\alpha_t = (x_t, y_t, s_t)$. For example, if $i_t = k$, then the likelihood function uses the specific face image of individual k in the database for the likelihood measure.

The second last part in Equation 3.1, $p(\alpha_t | \alpha_{t-1})$, is the *propagation* step or *diffusion*-step which represents our belief of how the parameters change at the time step from $t-1$ to t. The propagation step contains a deterministic part, which represents our knowledge of how the parameters *have* to change, as well as a statistical part. In cases where one does not have a good knowledge of the movement, a Gaussian diffusion of the parameters is usually a good choice. On the other hand, a strong and dynamic model can lead to very good model parameter predictions for the likelihood estimation. If the prediction model is wrong, however, it also leads "deterministically" to very bad parameter predictions and thus to very low likelihood estimates. This fact can be used for action recognition.

In that sense, the use of the likelihood measure and of the propagation model is complementary:

- The likelihood measure evaluates a specific observation model on an image, given the set of model parameters. Complex tasks, such as face recognition from video, can be solved with that.
- The prediction model is responsible for choosing the new model parameters.
- The prediction itself can also be a parameterized. Here, the parameters can be estimated in the same way as those of the model used in the likelihood function, with the difference, however, that the prediction parameters are evaluated only indirectly through the likelihood function.

As we will see (Section 3.3), one can do action recognition without the use of an actual action/prediction model. On the other hand, the prediction model itself can be also used for action recognition (Section 3.4). Clearly, for robust action recognition, one would want to use both tools.

In the next section (Section 3.2) we will give an overview of related literature.

In Sections 3.3 and 3.4, we will develop the new approaches and verify them in two large experiments.

In Section 3.3 we will deal with the problem of identifying whether a person is walking or running and, to some extend, identify the direction the person is moving to. We will consider a typical *surveillance* scenario where the camera has a wide field of view and where individuals are far away from the camera with only a small number of pixels in hight.

Each action gives rise to a set of typical silhouettes. By identifying the silhouette in the video at each time step, we become able to identify the action. Technically, our problem of identifying running and walking actions has several subproblems:

1. One needs to identify *where* the person is in the video. This is the tracking problem (Equation 3.1).
2. One needs to identify *what* pose the person has in each particular moment in time. This is similar to the recognition problem in [50].
3. One needs to identify *how* the pose changes over time in order to become able to recognize the action. Clearly, the pose of the person changes according to his/her action. In case of the surveillance scenario, we use this only very rudimentary.

Section 3.3 will be concluded with experimental results (Section 3.3.3).

In Section 3.4 we will consider the problem of action recognition in a different manner by using the propagation model: We will use a parameterized propagation/action model, where the parameter is used to identify a different action model. In order to identify the action, we are left with identifying the right propagation parameter. As an example application, it is assumed that actions are composed out of action primitives. This is similar to speech recognition, where words are composed out of phonemes. In speech recognition, the first step in recognizing a word is to recognize the phonemes. Then, given the phonemes, *Hidden Markov Models* (HMMs) are usually applied to recognize the word. In action recognition, we face the same problem: First, the action primitives need to be recognized. Then, the recognition of a more complex action can follow. However, unlike phonemes which are relatively local in time because of their short duration, action primitives are stretched out over time periods that are often longer than a second in duration. Therefore, techniques from speech recognition are not suitable for this problem. In Section 3.4.3 we will present and experimentally verify a solution where HMMs are used to model the propagation step in Equation 3.2.

3.2 Related Work

A pioneering work in the context our first problem has been presented by Efros et al. [13]. They attempt to recognize simple actions of people whose images in the video are only 30 pixels tall and where the video quality is poor. They use a set of features that are based on blurred optic flow (blurred motion channels). First, the person is tracked so that the image is stabilized in the middle of a tracking window. The blurred motion channels are computed on the residual motion that is due to the motion of the body parts. Spatio-temporal cross-correlation is used for matching with a database. Roh et al. [45] base their action recognition task on curvature scale space templates of a player's silhouette.

A large number of publications work with space-time volumes. One of the main approaches is to use *spatio-temporal XT-slices* from an image volume

XYT [41, 42] where articulated motions of a human can be associated with a typical trajectory pattern. Ricquebourg and Bouthemy [41] demonstrate how XT-slices can facilitate tracking and reconstruction of 2D motion trajectories. The reconstructed trajectory allows a simple classification between pedestrians and vehicles. Ritscher et al. [42] discuss the recognition in more detail by a closer investigation of the XT-slices. Quantifying the braided pattern in the slices of the spatiotemporal cube gives rise to a set of features (one for each slice) and their distribution is used to classify the actions.

Bobick and Davis pioneered the idea of temporal templates [2, 4]. They propose a representation and recognition theory [2, 4] that is based on *motion energy images* (MEI) and *motion history images* (MHI). The MEI is a binary *cumulative motion image*. The MHI is an enhancement of the MEI where the pixel intensities are a function of the motion history at that pixel. Matching temporal templates is based on Hu moments. Bradski et al. [6] pick up the idea of MHI and develop timed MHI (tMHI) for motion segmentation. tMHI allow determination of the normal optical flow. Motion is segmented relative to object boundaries and the motion orientation. Hu moments are applied to the binary silhouette to recognize the pose. A work conceptually related to [4] is by Masound and Papanikolopoulos [34]. Here, motion information for each video frame is represented by a feature image. However, unlike [4], an action is represented by several feature images. PCA is applied for dimensionality reduction and each action is then represented by a manifold in PCA space.

The recovery of phonemes in speech recognition is a closely related to our problem of action primitive recovery (Section 3.4). In speech recognition, acoustic data gets samples and quantized, followed by using the LPC (Linear Predictive Coding) to compute a *cepstral* feature set, or by a PLP (Perceptual Linear Predictive) analysis [18]. In a later step, time slices are analyzed. Gaussians are often used to compute likelihoods of the observations of being a phoneme [20]. An alternative way is to analyze time slices with an Artificial Neural Network [5]. Timeslices seem to work well on phonemes that have a very short duration. In our case, however, the action primitives have usually a much longer duration and one would have a combinatorial problem when considering time slices.

When viewing other agents performing an action, the human visual system seems to relate the visual input to a sequence of *motor primitives*. The neurobiological representation for visually perceived, learned and recognized actions appears to be the same as the one used to drive the motor control of the body [17, 43, 44]. These findings have gained considerable attention from the robotics community [11, 46]. In *imitation learning* the goal is to develop a robot system that is able to relate perceived actions to its own motor control in order to learn and to later recognize and perform the demonstrated actions.

In [24, 25], Jenkins et al. suggest applying a spatiotemporal nonlinear dimension reduction technique on manually segmented human motion capture data. Similar segments are clustered into primitive units which are generalized into parameterized primitives by interpolating between them. In the same manner, they define action units ("behavior units") which can be generalized into actions. In [21] the problem of defining motor primitives is approached from the motor side. They define a set of nonlinear differential equations that form a control policy (CP) and quantify how well different trajectories can be fitted with these CPs. The parameters of a CP for a primitive movement are learned in a training phase. These parameters are also used to compute similarities between movements. In [1, 7, 8] a HMM based approach is used to learn characteristic features of repetitively demonstrated movements. They suggest to use the HMM to synthesize joint trajectories of a robot. For each joint, one HMM is used. In [8] an additional HMM is used to model end-effector movement. In these approaches, the HMM structure is heavily constrained to assure convergence to a model that can be used for synthesizing joint trajectories.

Generally, there is a very large body of literature on action recognition. However, only a small subset is concerned with action primitives and their detection and recognition. In [49], Vecchio and Perona employ techniques from the dynamical systems framework to approach segmentation and classification. System identification techniques are used to derive analytical error analysis and performance estimates. Once, the primitives are detected an iterative approach is used to find the sequence of primitives for a novel action. In [32], Lu et al. also approach the problem from a system theoretic point of view. Their goal is to segment and represent repetitive movements. For this, they model the joint data over time with a second order auto-regressive (AR) model and the segmentation problem is approached by detection significant changes of the dynamical parameters. Then, for each motion segment and for each joint, they model the motion with a damped harmonic model. In order to compare actions, a metric based on the dynamic model parameters is defined. In [24, 25], Jenkins et al. suggest applying a spatiotemporal nonlinear dimension reduction technique on manually segmented human motion capture data. Similar segments are clustered into primitive units which are generalized into parameterized primitives by interpolating between them. In the same manner, they define action units ("behavior units") which can be generalized into actions. While most scientists concentrate on the action representation by circumventing the vision problem, [38] takes a vision-based approach. They propose a view-invariant representation of action based on *dynamic instants* and *intervals*. Dynamic instants are used as primitives of actions which are computed from discontinuities of 2D hand trajectories. An interval represents the time period between two dynamic instants (key poses). A similar approach of using meaningful instants in time is proposed by Reng et al. [39] where key poses are found based on the curvature and covariance of the normalized trajectories. In [10] key poses are found through evaluation of anti-eigenvalues.

3.3 Recognition of Running and Walking Actions

In this section we present an approach that relies on a very simple propagation model but uses instead a flexible set of observation models. The technique to recognize the action can be compared with a dynamic pattern recognition approach, rather than a pattern recognition approach for dynamics. With our approach we will recognize walking and running actions of humans in a *surveillance scenario*. The problem at hand is that the surveillance camera has a wide field of view and the individual has only a small number of pixels in height. Our interest is to detect whether a moving object is a person, whether this person is running or walking, and in which rough direction the person is moving ($0°$, $45°$, or $90°$).

The idea is related to our model-based *background subtraction* system with human walking and running poses at different camera orientations as the model knowledge [28, 29].

The approach consists of two parts: the learning part in which a suitable set of silhouettes has to be captured and structured and the actual recognition part.

Both parts consist of two steps: In a first step we apply a background subtraction method [14] in order to extract the moving image parts. For the training step, in order to capture good silhouettes, we assume a controlled scenario which allows us to extract good *silhouettes*. In the recognition part, we attempt to track and match the silhouettes in the database to the noisy ones in the incoming video. The recognition part will make use of the ideas on Bayesian parameter estimation over time, as outlined above. The statistical integration of the observations over time make the recognition system very robust to noise, occlusion and shadows, as we will show in the experiment section, below.

3.3.1 Learning and Representation of Silhouettes

In order to generate our silhouette model knowledge we apply a classical *background subtraction* method (BGS) [14] to a scenario that is controlled in a manner that facilitates the learning process. In our case, since we want to learn silhouettes of humans, we assure that only humans are visible in the scene during training and that the background variations and shadows are kept as small as possible to minimize distortions. Then, we use this video data directly to capture the proper exemplars. The different classes of silhouettes we have considered are *walking* and *running*, with the different angles of $0°$, $45°$ and $90°$ with respect to the camera.

After the application of a classical BGS, applying mean-shift tracking [9] allows to extract from the BGS output-data a sequence of small image patches containing, centered, the silhouette. This procedure is the same as the one used in [26], however, with the difference that here we do not threshold the BGS output but use *probabilistic silhouettes* (instead of binary ones as in

Fig. 3.1. This image shows examples of the silhouette exemplars as learned in the learning stage. These silhouettes are used as input to Isomap [48].

[26]). These silhouettes still contain for each silhouette pixel the belief of a pixel being part of the foreground. Figure 3.1 shows some example silhouette exemplars.

The set of silhouettes is not a vectorspace. However, for tracking and recognizing the right silhouettes, as we will see later, it is of great advantage to have at least a metric space of the silhouettes so that *similar* silhouettes can be found. We have used Isomap [48] to generate a vector space from the manifold of silhouettes. Each class of silhouettes is treated independently by Isomap. Thus, we end up with six different vector spaces, three for walking and three for running. As distance measure for the Isomap we used the Euclidean distance.[1] The topology of these six spaces resemble to real topology of the silhouettes (see also [15, 16]). The topology of the silhouettes is used as an implicit representation of the allowed pose changes.

3.3.2 Recognizing the Action in the Surveillance Scenario

We consider the problem of recognizing action as a pattern recognition problem to the output of the background subtraction approach. We use the previously learned silhouette exemplars to find the best match with the output $I^p(\mathbf{x})$ of a background subtraction application. Then, the silhouette class from which the silhouettes are derived most of the time identifies the action.

Each pixel in the image $I^p(\mathbf{x})$ contains a value in the range $[0, 1]$, where 1 indicates the highest probability of a pixel being a foreground pixel.

An exemplar is selected and deformed according to a 5D parameter vector

$$\theta = [n, a, s, x, y], \tag{3.3}$$

where x and y denote the position of the silhouette in the image I^p, s its scale, and n is a natural number that refers to a specific silhouette in the action class a.

---

[1] The more natural distance function for our probabilistic silhouettes is arguably the Kullback-Leibler divergence measure. The use of this measure is presently under investigation.

We use normalized correlation to compute the distance between the exemplar silhouette, parameterized according to a deformation vector θ_t and the appropriate region of interest at position x_t and y_t, as specified by θ in the BGS image $I_t^p(\mathbf{x})$.

In order to find at each time-step t the most likely θ_t in the image $I_t^p(\mathbf{x})$, we use, as already mentioned, Bayesian propagation over time

$$p(\alpha_t, a_t | I_1^p, I_2^p, \ldots, I_t^p) \equiv p_t(\alpha_t, a_t)$$
$$= \sum_{a_{t-1}} \int_{\alpha_{t-1}} p(I_t^p | \alpha_t, a_t) p(\alpha_t, a_t | \alpha_{t-1}, a_{t-1}) p_{t-1}(\alpha_{t-1}, a_{t-1}) \quad (3.4)$$

with $\alpha_t = [n, s, x, y]_t$. Here, I_t^p denotes the probability images, we use the variable a to reference the action and n to reference a silhouette of the action a. By considering the joint distribution of the silhouette id n and the action a together with the tracking parameters we are able to view the tracking and the recognition as a single problem. By marginalizing over the geometric parameters $\alpha = (x, y, s, n)$,

$$p(a_t | I_1^p, \ldots, I_t^p) = \int_{\alpha_t} p(\alpha_t, a_t | I_1^p, \ldots, I_t^p) \quad (3.5)$$

we can estimate the likelihood of each action a at any time. In a similar way, we can compute the likelihood of each silhouette n at any time.

As the diffusion density $p(\alpha_t, a_t | \alpha_{t-1}, a_{t-1})$ in Equation 3.2 we use for the tracking parameters, x, y, and s the Brownian motion model due to the absence of a better one. We use the simplifying assumption that the action parameter a is constant since while assuming that the action does not change over time. A new silhouette n is selected also in a Gaussian manner, according to the topological structure computed through isomap.

Once the pdf $p_t(a_t)$ converges to 1 for some action the pdf $p_t(a_t)$ stays constant and new evidence from novel images is ignored. To be able to take into account all incoming data, we employ a voting scheme by counting the number of times the pdf converged to each action and by resetting the prior for the random variable a to the even distribution.

3.3.3 Experiments

In this section we present qualitative and quantitative results obtained from our experiments. In these experiments we wanted to investigate the capabilities of our approach for the recognition of actions and the recognition of the right silhouette in each image of the input video. The results clearly show the potentials for an effective silhouette and action recognition.

As qualitative experiments we have run our approach on a number of test sequences recorded outdoors. The test videos show between one and three individuals walking and running, with partially heavy occlusion and illumination variations. Figure 3.2 shows a scenario, with a pedestrian walking behind

Fig. 3.2. This image shows an example of normal occlusion and shadow. The top image shows the original input image. The bottom left image shows the normal output from the background subtraction function while the bottom right one shows the detected silhouette, parameterized according the the detected parameters.

trees, thereby at times being occluded. Since, in spite of the occlusion and at times heavy noise, e.g., due to trees moving in the wind, the likelihood measures with the occluded input silhouettes are still the largest in the image, this occlusion does not pose any problems to our approach. In addition, due to the structure of the silhouette space, we are able to deduce the right silhouette, even though the visual information coming from the background subtraction method (Figure 3.2, bottom left) would not be sufficient for a unique identification of the right pose. image Figure 3.3 shows the same scenario, in a frame where the pedestrian is heavily occluded. Even in this case, the visual data from the background subtraction function is sufficient to detect the right parameters of the silhouette. Both, action and silhouettes are identified correctly. In addition, even though the visual information coming from the background subtraction (Figure 3.3, middle) would suggest a differ-

Fig. 3.3. This image shows an example of normal occlusion and shadow. The top image shows the original input image. The bottom left image shows the normal output from the background subtraction function while the bottom right one shows the detected silhouette, parameterized according the the detected parameters.

ent pose, the system still detects the correct one. The scenario presented in Figure 3.4, shows two pedestrians walking towards each other. We deal with the case where more than one individual is visible by considering the two individuals as two unrelated recognition problems. For each individual, a separate set of particles is introduced. When a new region of activity is detected in the video, the region is checked whether it is likely to be an individual. At this point we have introduced a conservative threshold value that should allow to distinguish human silhouettes from noise. In a quantitative evaluation we have investigated the correctness of the particle method in matching the correct silhouette. When the background subtraction is started on an input video, the particles are initially evenly distributed. Then, the particle filter usually needed 20–50 frames to find a sufficiently good approximation of the true density. Before convergence, the selected silhouette is still random. Af-

Fig. 3.4. This image shows an example of heavy occlusion. The left image shows the original input image, the middle one the normal output from the background subtraction function, the right image shows the detected silhouette superimposed.

ter 50 frames and without any occlusion, the silhouette with the maximum likelihood is the correct one in ≈98% of the cases. In ≈20% of the cases the ML silhouette was incorrect when e.g., a bush was largely occluding the legs. However, recovery time was within 5 frames. In case of partial occlusion of the entire body through, e.g., small trees, reliability degraded between 1% (slight occlusion) to 10% (considerable occlusion). The silhouette was incorrect in ≈59% of the cases where the legs were fully occluded, e.g., by a car. In the videos the individual was in average 70 px. high. Reliability increased with more pixels on the target. The action was correctly identified in 98% of the cases. However, an interpretation of this result is more complex: The correctness of the detected silhouettes has a direct influence on the recognized action. By definition, an action is correctly identified if the particle filter converges and votes for the correct action most of the time. This was the case in most of our videos where the average occlusion was sufficiently small. On the other hand, in a crowded environment, our approach would break down.

3.4 Recognizing Action Primitives by Combining HMMs with Bayesian Propagation

In this section we want to discuss another approach that extends the Bayesian propagation for action recognition. In the previous section, we have extended the joint density function by an action variable which was used as a prior for the likelihood measurement. In this section, we again extend the joint density function with an action parameter. Here, however, we will use it as a priori for the propagation and diffusion step.

In this section, the scenario concerns the recognition of action primitives from observed complex actions.

There is biological evidence that actions and activities are composed out of action primitives similarly to phonemes being concatenated into words [17, 43, 44].

In this sense, one can define an *action hierarchy* of *action primitives* at the coarsest level, and then *actions* and *activities* as the higher abstract levels where actions are composed out of the action primitives while activities are, in turn, a composition of the set of actions [3, 35].[2]

In order to recognize an action performed by an individual, an action hierarchy makes sense due to the otherwise combinatorial problem. If one follows the approach of using an action hierarchy, then one of the main problems is to recover the *action primitives* from an observed action.

Thus, given a set (or alphabet) of action primitives, we are concerned in this section with the recovery of the sequence of the action primitives from an observed action.

In other words, if we have given an alphabet of action primitives P and if we have given an *action* S which by definition is a sequence $S = a_1 a_2 a_3 \ldots a_T$ of some length T and composed out of these action primitives from P, then it is our interest to recover these primitives and their precise order.

This problem is closely related to speech recognition where the goal is to find the right sequences of phonemes (see Section 3.2). Once we have parsed and detected the sequence of action primitives in the observed sequence, this sequence of action primitives could identify the action. (In speech recognition, the sequence of detected phonemes is used to identify the corresponding word.)

One possibility to recognize then the action from the detected primitives is to define an *action-grammar* for each action, based on the action primitives as the alphabet and to use a parsing approach for recognition, as suggested in [23, 47].

---

[2] In the following, we define the term *action* as a sequence of action primitive of arbitrary length.

In order to take into account possible noise and imperfect data, we base our approach on *Hidden Markov Models* (HMMs) [19,37] and represent our action primitives with HMMs.

Thus, given a set of action primitives where each action primitive is represented by an HMM and given an observed sequence S of these action primitives where

1. The order of the action primitives
2. The duration of each single action primitive and the position of their boundaries

are unknown, we would like to identify the most likely sequence of action primitives in the observation sequence S for subsequent parsing.

According to the biological findings, the representation for action recognition is closely related to the representation for action synthesis (i.e., the motor representation of the action) [17,43,44]. This motivates us to focus our considerations in this section to actions represented in joint space. Thus, our actions are given as sequences of joint settings. A further justification for this approach is that this action representation can then be used, in future work, to bias 3D body trackers as it operates directly on the 3D parameters that are to be estimated by the 3D tracker. The focus of this chapter on joint data is without limiting generality.

In this section, we will recognize the action based on the dynamics. Clearly, one could also employ the approach presented in Section 3.3. In that case, one would evaluate the liklihood of a joint setting given the set of joint settings for a particular action.

3.4.1 Representing and Recognizing Action Primitives Using HMMs

In order to approach the action recognition problem, we model each of the action primitives $P = \{a^1, a^2, \ldots, a^N\}$ with a mixture-HMM where each observation function is a continuous Gaussian mixture with $M \geq 1$ mixtures. The mixture HMMs are trained based on demonstrations of a number of individuals. The Gaussian mixture are able to represent the variability across individuals to allow some degree of invariance across different individuals. The training results into a set of HMMs $\{\lambda_i | i = 1 \ldots N\}$, one for each action primitive.

Once each action primitive is represented with an HMM, the primitives can generally simply be recognized with the classical recognition technique for HMMs by employing a maximum likelihood or a maximum a-posteriori classifier: Given an observation sequence O_t of an action primitive, and a set of HMMs λ_i, the maximum likelihood (ML)

$$\max_i P(O_t | \lambda_i) \tag{3.6}$$

identifies the most likely primitive. An alternative to the ML technique is the maximum a-posteriori (MAP) estimate that allows to take into account the likelihood of observing each action primitive:

$$\max_i P(\lambda_i|O_t) = \max_i P(O_t|\lambda_i)P(\lambda_i) \ , \tag{3.7}$$

where $P(\lambda_i)$ is the likelihood that the action, represented by the HMM λ_i, appears.

3.4.2 Recognition with HMMs

In general, the likelihood of an observation for some HMM λ_i can be computed as

$$P(O|\lambda_i) = \sum_S P(O, S|\lambda_i) \tag{3.8}$$

$$= \sum_S P(O|S, \lambda_i)P(S|\lambda_i) \tag{3.9}$$

$$= \sum_S \prod_{t=0}^{T} P(O_t|S_t, \lambda_i) \prod_{t=0}^{T} P(S_t|S_{t-1}, \lambda_i) \ . \tag{3.10}$$

Here, one marginalizes over all possible state sequences $S = \{S_0, \ldots, S_T\}$ the HMM λ_a can pass through.

To apply this technique to our problem directly is difficult as we would need to know *when* to evaluate, i.e., at what time steps t we should stop and do the maximum-likelihood estimation to find the most likely action primitive that is just now being observed.

Instead of keeping the HMMs distinct, our suggestion is to insert the "action" a of the HMM λ_a as a random variable into Equation (3.10) and to rewrite it as

$$P(O|a) = \sum_S P(S_0, a_0) \prod_{t=1}^{T} P(O_t|S_t, a_t)P(S_t, a_t|S_{t-1}, a_{t-1}) \ . \tag{3.11}$$

In other words, we would like to estimate at each time step the action a and the state S from the previously seen observations, or, respectively, the probability of λ_a being a model of the observed action:

$$P(S_T, a_T|O_{0:T}) = P(S_0, a_0) \prod_{t=1}^{T} P(O_t|S_t, a_t)P(S_t, a_t|S_{t-1}, a_{t-1}) \tag{3.12}$$

The difference in the interpretation becomes more clear when we write Equation (3.12) in a recursive fashion:

$$P(S_{t+1}, a_{t+1}|O_{0:t+1})$$
$$= P(O_{t+1}|S_{t+1}, a_{t+1})P(S_{t+1}, a_{t+1}|S_t, a_t)P(S_t, a_t|O_{0:t}) . \qquad (3.13)$$

This is the classical Bayesian propagation over time. It computes at each time step t the likelihood of observing the action a_t while having observed $O_{0:t}$. If we ignore the action variable a_t, then Equation (3.13) explains the usual efficient implementation of the forward algorithm [19] which allows to compute the likelihood of an observation, given an HMM. Using the random variable a_t, Equation (3.13) defines a pdf across the set of states (where the state vector S_t is the concatenation of state vectors of each individual HMM) and the set of possible actions. The effect of introducing the action a might not be obvious: using the action a, we do not any more estimate the likelihood of an observation, given a HMM λ_a. Instead, we compute *at each time step* the probability mass function (pmf) $P(S_t, a_t|O_{0:t})$ of each state and each identity, given the observations. By marginalizing over the states, we can compute the pmf $P(a_t|O_{0:t})$ for the action at each time step. The likelihood $P(a_t|O_{0:t})$ converges to the most likely action primitive as time progresses and more data becomes available (see Figure 3.5). From Figure 3.5 it is apparent that the pmf $P(a_t|O_{0:t})$ will remain constant after convergence as one action primitive will have the likelihood 1 and all other primitive likelihoods have vanished. Similarly to Section 3.3, we apply a voting scheme that counts the votes after each convergence and then restarts the HMMs. The states are initialized with the present observation likelihoods and then propagated with the transition matrix as usual. Figure 3.6 shows the repeated convergence and the restarting of the HMMs. In the example shown in Figure 3.6 we have used two concatenated action primitives, denoted by the green curve with the "+" and by the blue curve with the "o", respectively. The first action primitive was

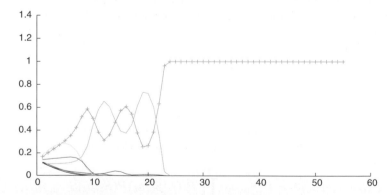

Fig. 3.5. This image shows an example for a typical behavior of the pmf $P(a_t|O_{0:t})$ for each of the actions a as time t progresses. One can see that the likelihood for one particular action (the correct one in this example, marked with "+") converges to 1 while the likelihoods for the others vanish.

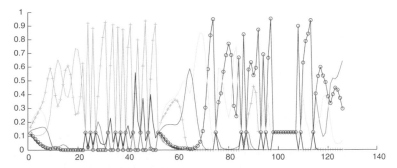

Fig. 3.6. This image shows an example for a typical behavior of the pmf $P(a_t|O_{0:t})$ as time t progresses. The input data consisted of two action primitives: first, action primitive "2", marked with "+", then, action primitive "3", marked with "o". One can see that until $\approx$ sample 52 the system converges to action "2", after sample 70, the system converges to primitive 3. The length of the first sequence is 51 samples, the length of sequence 2 is 71 samples.

in the interval between 0 and 51, while the second action primitive was from sample 52 to the end. One can see that the precise time step when primitive 1 ended and when primitive 2 started cannot be identified. But this does not pose a problem for our recovery of the primitives as for us the order matters but not their precise duration. In Figure 3.5 a typical situation can be seen where the observed data did not give enough evidence for a fast recognition of the true action.

3.4.3 Experiments

For our experiments, we have used our MoPrim [40] database of human one-arm movements. The data was captured using a **FastTrack** Motion capture device with 4 electromagnetic sensors. The sensors are attached to the torso, shoulder, elbow and hand (see Figure 3.7). Each sensor delivers a $6D$ vector, containing $3D$ position and $3D$ orientation thus giving a $24D$ sample vector at each time-step (4 sensors with each $6D$). The MoPrim database consists of 6 individuals, showing 9 different actions, with 20 repetitions for each. The actions in the database are simple actions such as *point forward*, *point up*, *"come here"*, *"stop!"*. Each sequence consists of $\approx$60–70 samples and each one starts with 5 samples of the arm in a resting position where it is simply hanging down.

Instead of using the sensor positions directly, we transform the raw $24D$ sensor data into joint angles: one elbow angle, one shoulder angle between elbow, shoulder and torso and a 3D orientation of the normal of this shoulder–elbow–torso–triangle. The orientation of the normal is given with respect to the normal of this triangle when the arm is in resting position. All angles are given in radians. No further processing of the MoPrim data was done.

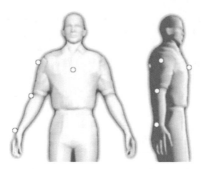

Fig. 3.7. This image shows the positions of the magnetic sensor on the human body.

We have carried out several different experiments:

1. In the first test, we tested for invariance with respect to the performing human. We have trained nine HMM for nine actions. Each of the HMMs was trained on 6 individuals and all the 20 repetitions of the actions. The recognition testing was then carried out on the remaining individual (leave-one-out strategy). The HMMs we use were mixture HMMs with 10 states and 5 mixtures per state.
2. In this test, we tested for invariance with respect to the variations within the repetitions. We have trained nine HMMs for nine actions. Each HMM was trained on all individuals but only on 19 repetitions. The test set consisted of the 20th repetition of the actions.
3. As a base line reference, we have tested how good the HMMs are able to recognize the actions primitives by testing action primitive sequences of length 1. Here, the HMMs were trained as explained under 2 above. This test reflects the recognition performance of the classical maximum-likelihood approach.
4. We have repeated the above three experiments after having added Gaussian noise with zero mean and a standard deviation of $\sigma = 0$, $\sigma = 0.3$ and $\sigma = 1$ to the training and testing data. As all angles are given in radians, thus, this noise is considerable.

To achieve a good statistic we have for each test generated 10.000 test actions of random length ≤ 100. Also, we have systematically left out each individual (action) once and trained on the remaining ones. The results below are averaged across all leave-one-out tests. In each test action, the action primitives were chosen randomly, identically and independently. Clearly, in reality there is a strong statistical dependency between action primitives so that our recognition results can be seen as a lower bound and results are likely to increase considerably when exploiting the temporal correlation by using an action grammar (e.g., another HMM).

The results are summarized in Table 3.1. One can see that the recognition rates of the individual action primitives is close to the general baseline of the

Table 3.1. Summary of the results of our various experiments. In the experiments, the training of the HMMs were done without the test data. We tested for invariance w.r.t. identity and w.r.t. the action. The *baseline* shows the recognition results when the test action was a single action primitives.

Leave-one-Out experiments		
Test	Noise σ	Recognition Result
Identities (Test 1)	0	0.9177
Repetitions (Test 2)	0	0.9097
Baseline (Test 3)	0	0.9417
Identities (Test 1)	0.5	0.8672
Repetitions (Test 2)	0.5	0.8710
Baseline (Test 3)	0.5	0.8649
Identities (Test 1)	1	0.3572
Repetitions (Test 2)	1	0.3395
Baseline (Test 3)	1	0.3548

HMMs. The recognition rates degrade with increasing noise which was to be expected, however, the degradation effect is the same for all three experiments (identities, repetition, baseline).

All actions in the action database start and end in a resting pose. To assure that the resting pose does not effect the recognition results, we have repeated the above experiments on the action primitives where the rest poses were omitted. However, the recognition results did not change notably.

3.5 Conclusions

In this chapter we have discussed two approaches for action recognition that were based on the Bayesian propagation over time. We have used the fact that the Bayesian framework offers a unified framework to combine observation knowledge as well as dynamical knowledge and that this is particularly useful for the action recognition tasks. In the surveillance scenario we have approached the action recognition task as a pattern recognition task with the aim to identify at each time-step the silhouette of the human. Each action gives rise to a set of typical silhouettes. By identifying the silhouette in the video at each time step, we become able to identify the action. The task of identifying the silhouette is similar to the tracking task in order to identify *where* the silhouette is in the video. On the other hand, the identification of the action itself is slightly different as the action is assumed to be constant (an assumption that can be relaxed). Here, we employed a voting scheme which counts the number of times the observed silhouettes give rise to a particular action. We have treated the surveillance scenario strictly as a pattern matching problem as we have not used any dynamic information. The use of dynamic modes was discussed separately in the context of recovering action primitive in observed complex actions. Here, we have modeled the dynamics with Hidden Markov Models which, being Bayesian themselves, fit nicely into the classical Bayesian propagation framework as used usually for tracking. For the recovery, again, a voting scheme was employed which counted the votes for each action primitive.

The two approaches presented are complementing each other within the same framework. Clearly, the action primitives could be estimated in a similar manner as the silhouettes in Section 3.3. Then, we would have exploited the fact that for a particular arm action, only a specific set of arm poses can appear. In Section 3.4, these appearances were encoded implicitly in the mixture models and the actual recognition was carried out based on how these arm poses changed, i.e., on the dynamics.

From the two presented experiments, one gets the impression that a combination of the two approaches, within the common Bayesian framework, will lead to a powerful technique for action recognition.

Acknowledgments

This work was partially funded by PACO-PLUS (IST-FP6-IP-027657).

References

1. A. Billard, Y. Epars, S. Calinon, S. Schaal, and G. Cheng. Discovering Optimal Imitation Strategies. *Robotics and Autonomous Systems*, 47:69–77, 2004.
2. A. Bobick. Movement, Activity, and Action: The Role of Knowledge in the Perception of Motion. *Philosophical Trans. Royal Soc. London*, 352:1257–1265, 1997.
3. A. Bobick. Movements, Activity, and Action: The Role of Knowledge in the Perception of Motion. In *Royal Society Workshop on Knowledge-based Vision in Man and Machine*, London, February 1997.
4. A. Bobick and J. Davis. The Recognition of Human Movement Using Temporal Templates. *IEEE Transactions on Pattern Analysis and Machine Intelligence*, 23(3):257–267, 2001.
5. H. Bourlard and N. Morgan. *Connectionist Speech Recognition: A Hybrid Approach*. Kluwer Press, 1994.
6. G. Bradski and J. Davis. Motion Segmentation and Pose Recognition with Motion History Gradients. *Machine Vision and Applications*, 13, 2002.
7. S. Calinon and A. Billard. Stochastic Gesture Production and Recognition Model for a Humanoid Robot. In *International Conference on Intelligent Robots and Systems*, Alberta, Canada, Aug 2–6, 2005.
8. S. Calinon, F. Guenter, and A. Billard. Goal-Directed Imitation in a Humanoid Robot. In *International Conference on Robotics and Automation*, Barcelona, Spain, April 18–22, 2005.
9. D. Comaniciu, V. Ramesh, and P. Meer. Real-time Tracking of Non-rigid Objects Using mean Shift. In *Computer Vision and Pattern Recognition*, volume 2, pages 142–149, Hilton Head Island, South Carolina, June 13–15, 2000.

10. N. Cuntoor and R. Chellappa. Key Frame-Based Activity Representation Using Antieigenvalues. In *Asian Conference on Computer Vision*, volume 3852 of *LNCS*, Hyderabad, India, Jan, 13–16, 2006.
11. B. Dariush. Human Motion Analysis for Biomechanics and Biomedicine. *Machine Vision and Applications*, 14:202–205, 2003.
12. A. Doucet, S. Godsill, and C. Andrieu. On Sequential Monte Carlo Sampling Methods for Bayesian Filtering. *Statistics and Computing*, 10:197–209, 2000.
13. A. Efros, A. Berg, G. Mori, and J. Malik. Recognizing Action at a Distance. In *Internatinal Conference on Computer Vision*, Nice, France, Oct 13–16, 2003.
14. A. Elgammal and L. Davis. Probabilistic Framework for Segmenting People Under Occlusion. In *ICCV*, ICCV01, 2001.
15. A. Elgammal and C. Lee. Separating Style and Content on a Nonlinear Manifold. In *Computer Vision and Pattern Recognition*, Washington DC, June 2004.
16. A. Elgammal, V. Shet, Y. Yacoob, and L. Davis. Learning Dynamics for Exemplar-based Gesture Recognition. In *Computer Vision and Pattern Recognition*, Madison, Wisconsin, June 16–22, 2003.
17. M. Giese and T. Poggio. Neural Mechanisms for the Recognition of Biological Movements. *Nature Reviews*, 4:179–192, 2003.
18. H. Hermansky. Perceptual Linear Predictive (plp) Analysis of Speech. *Journal of Acoustical Society of America*, 87(4):1738–1725, 1990.
19. X. Huang, Y. Ariki, and M. Jack. *Hidden Markov Models for Speech Recognition*. Edinburgh University Press, 1990.
20. X. Huang and M. Jack. Semi-continous Hidden Markov Models for Speech Signals. *Computer Speech and Language*, 3:239–252, 1989.
21. A. Ijspeert, J. Nakanishi, and S. Schaal. Movement Imitation withNonlinear Dynamical Systems in Humanoid Robots. In *International Conference on Robotics and Automation*, Washington DC, May, 2002.
22. M. Isard and A. Blake. Condensation – Conditional Density Propagation for Visual Tracking. *International Journal of Computer Vision*, 29:5–28, 1998.
23. Y. Ivanov and A. Bobick. Recognition of Visual Activities and Interactions by Stochastic Parsing. *IEEE Transactions on Pattern Analysis and Machine Intelligence*, 22(8):852–872, 2000.
24. O. Jenkins and M. Mataric. Deriving Action and Behavior Primitives from Human Motion Capture Data. In *International Conference on Robotics and Automation*, Washington DC, May 2002.
25. O. Jenkins and M. Mataric. Deriving Action and Behavior Primitives from Human Motion Data. In *International Conference on Intelligent Robots and Systems*, pages 2551–2556, Lausanne, Switzerland, Sept 30–Oct 4, 2002.
26. A. Kale, A. Sundaresan, A. Rjagopalan, N. Cuntoor, A. Chowdhury, V. Krueger, and R. Chellappa. Identification of Humans Using Gait. *IEEE Transactions on Image Processing*, 9:1163–1173, 2004.
27. G. Kitagawa. Monta Carlo Filter and Smoother for Non-gaussian Nonlinear State Space Models. *Journal of Computational and Graphical Statistics*, 5:1–25, 1996.
28. V. Krueger, J. Anderson, and T. Prehn. Probabilistic Model-based Background Subtraction. In *Scandinavian Conference on Image Analysis,*, pages 180–187, June 19-22, Joensuu, Finland, 2005.
29. V. Krueger, J. Anderson, and T. Prehn. Probabilistic Model-based Background Subtraction. In *International Conference on Image Analysis and Processing*, pages 180–187, Sept. 6–8, Cagliari, Italy, 2005.

30. B. Li and R. Chellappa. Simultanious Tracking and Verification via Sequential Posterior Estimation. In *Computer Vision and Pattern Recognition*, Hilton Head Island, South Carolina, June 13–15, 2000.

31. J. Liu and R. Chen. Sequential Monte Carlo for Dynamic Systems. *Journal of the American Statistical Association*, 93:1031–1041, 1998.

32. C. Lu and N. Ferrier. Repetitive Motion Analysis: Segmentation and Event Classification. *IEEE Transactions on Pattern Analysis and Machine Intelligence*, 26(2):258–263, 2004.

33. D. MacKay. In M. Jordan, editor, *Learning in Graphical Models, Introduction to Monte Carlo Methods*, pages 175–204. MIT Press, 1999.

34. O. Masound and N. Papanikolopoulos. A Method for Human Action Recognitoin. *Image and Vision Computing*, 21:729–743, 2003.

35. H.-H. Nagel. From Image Sequences Towards Conceptual Descriptions. *Image and Vision Computing*, 6(2):59–74, 1988.

36. D. Ormoneit, H. Sidenbladh, M. Black, and T. Hastie. Learning and Tracking Cyclic Human Motion. In *Workshop on Human modelling, Analysis and Synthesis at CVPR*, Hilton Head Island, South Carolina, June 13–15 2000.

37. L. R. Rabiner and B. H. Juang. An introduction to hidden Markov models. *IEEE ASSP Magazine*, pages 4–15, January 1986.

38. C. Rao, A. Yilmaz, and M. Shah. View-Invariant Representation and Recognition of Actions. *Journal of Computer Vision*, 50(2), 2002.

39. L. Reng, T. Moeslund, and E. Granum. Finding Motion Primitives in Human Body Gestures. In S. Gibet, N. Courty, and J.-F. Kamps, editors, *GW 2005*, number 3881 in LNAI, pages 133–144. Springer, Berlin Heidelberg, 2006.

40. L. Reng, T. Moeslund, and E. Granum. Finding motion primitives in human body gestures. In S. Gibet, N. Courty, and J.-F. Kamp, editors, *GW 2005*, pages 133–144. Springer, 2006.

41. Y. Ricquebourg and P. Bouthemy. Real-Time Tracking of Moving Persons by Exploiting Spatio-Temporal Image Slices. *Transactions on Pattern Analysis and Machine Intelligence*, 22(8), 2000.

42. J. Rittscher, A. Blake, and S. Roberts. Towards the Automatic Analysis of Complex Human Body Motions. *Image and Vision Computing*, 20, 2002.

43. G. Rizzolatti, L. Fogassi, and V. Gallese. Parietal Cortex: From Sight to Action. *Current Opinion in Neurobiology*, 7:562–567, 1997.

44. G. Rizzolatti, L. Fogassi, and V. Gallese. Neurophysiological Mechanisms Underlying the Understanding and Imitation of Action. *Nature Reviews*, 2:661–670, Sept, 2001.

45. M. Roh, B. Christmas, J. Kittler, and S. Lee. Robust Player Gesture Spotting and Recognition in Low-Resolution Sports Video. In *European Conference on Computer Vision*, Graz, Austria, May 7–13, 2006.

46. S. Schaal. Is Imitation Learning the Route to Humanoid Robots? *Trends in Cognitive Sciences*, 3(6):233–242, 1999.

47. A. Stolcke. An Efficient Probabilistic Context-Free Parsing Algorithm That Computes Prefix Probabilities. *Computational Linguistics*, 21(2):165–201, 1995.

48. J. Tenenbaum, V. de Silva, and J. Langford. A Global Geometric Framework for Nonlinear Dimensionality Reduction. *Science*, 290:2319–2323, 2000.

49. D. Vecchio, R. Murray, and P. Perona. Decomposition of Human Motiom into Dynamics-based Primitives with Application to Drawing Tasks. *Automatica*, 39, 2003.
50. S. Zhou, V. Krueger, and R. Chellappa. Probabilistic Recognition of Human Faces From Video. *IJCV*, 91:214–245, July, 2003.

4

The William Harvey Code: Mathematical Analysis of Optical Flow Computation for Cardiac Motion

Yusuke Kameda[1] and Atsushi Imiya[2]

[1] School of Science and Technology, Chiba University, Japan
[2] Institute of Media and Information Technology, Chiba University, Japan

Summary. For the non-invasive imaging of moving organs, in this chapter, we investigate the generalisation of optical flow in three-dimensional Euclidean space. In computer vision, optical flow is dealt with as a local motion of pixels in a pair of successive images in a sequence of images. In a space, optical flow is defined as the local motion of the voxel of spatial distributions, such as x-ray intensity and proton distributions in living organs. Optical flow is used in motion analysis of beating hearts measured by dynamic cone beam x-ray CT and gated MRI tomography. This generalisation of optical flow defines a class of new constraints for optical-flow computation. We first develop a numerically stable optical-flow computation algorithm. The accuracy of the solution of this algorithm is guaranteed by Lax equivalence theorem which is the basis of the numerical computation of the solution for partial differential equations. Secondly, we examine numerically the effects of the divergence-free condition, which is required from linear approximation of infinitesimal deformation, for the computation of cardiac optical flow from images measured by gated MRI. Furthermore, we investigate the relation between the vector-spline constraint and the thin plate constraint. Moreover, we theoretically examine the validity of the error measure for the evaluation of computed optical flow.

4.1 Introduction

William Harvey (1578–1657) is the first medical doctor who had discovered and had correctly described the mechanism of the circulation of the blood in the cardiac system of animals Furthermore, he had clarified the physical function of the heart [1]. In Harvey's time anatomical surgery was the only way to experiment and observe the function of the beating heart of animals. Nowadays, using mathematics, physics, and computer sciences, we can observe beating hearts in living organs non-invasively. To model the functions of a beating heart from the viewpoint of biomechanics, we are required physical measurements of the motion of each point of the heart wall in living organs.

Optical flow is a non-interactive and non-invasive technique for the detection of motion of an object on a plane and in a space. For medical study and

B. Rosenhahn et al. (eds.), Human Motion – Understanding, Modelling, Capture, and Animation, 81–104.

diagnosis of moving organs in the human body, optical flow of tomographic images provides a fundamental tool [2, 3]. The motion of each point in a sequence of temporal images is computed as optical flow in computer vision. The three-dimensional version of optical flow is used for analysis of motion of each point on a heart.

The non-invasive imaging of moving organs is achieved by MRI, x-ray CT, and ultrasonic CT. Usually the signal-to-noise ratio of non-invasive imaging is low. Therefore, we are required to develop numerically accurate methods of the optical-flow computation for tomographic images.

Variational methods provide a unified framework for image analysis, such as optical flow computation, noise removal, edge detection, and in-painting [4–7] using the variational principle. The fundamental nature of the variational principle governed by the minimisation of the energy functionals for the problems allows us to describe algorithms for the problems of image analysis as the computation of the solution of Euler-Lagrange equations, which are partial differential equations. In this chapter, we investigate the generalisation of optical flow [8–11] in three-dimensional Euclidean space. In computer vision, optical flow is dealt with as the local motion of pixels in a pair of successive images for a sequence of images. Optical flow is used in motion analysis of a three-dimensional beating heart measured by dynamic cone-beam x-ray CT and gated MRI tomography. In a three-dimensional space, optical flow is defined as the local motion of voxels of spatial distribution such as x-ray intensity and proton distributions in living organs. This generalisation of optical flow defines a class of new constraint for optical-flow computation which allows us to detect the motions on the segment-boundary of a three-dimensional deformable object.

We first show some mathematical relations between the Horn-Schunck constraint and Nagel-Enkelmann constraint in three-dimensional Euclidean space. Moreover, we analyse the numerical schemes of optical-flow computation employing Lax equivalence theorem. Lax equivalence theorem guarantees the stability of algorithms and accuracy of solutions computed by numerical schemes for numerical solving partial differential equations, that is, with Lax equivalence theorem, a numerically computed solution converges to the solution of the equation by decreasing the size of grids for numerical computation. Therefore, if an algorithm for optical-flow computation is expressed as a numerical computation of a partial differential equation with the condition which satisfies Lax equivalence theorem, the solution computed by the algorithm is accurate and stable. Secondly, we aim, in this chapter, to examine numerically the effects of the divergence-free condition [14] for the computation of cardiac optical flow from images obtained by gated MRI tomography.

From the viewpoints of elastic deformation of the heart wall, we have an exact *incompressible condition* [13], such that

$$det(\boldsymbol{I} + \nabla \boldsymbol{u}) = 1, \ \nabla \boldsymbol{u} = (\nabla u, \nabla v, \nabla w) \tag{4.1}$$

for a small displacement vector $\boldsymbol{u} = (u, v, w)^\top$ in $\mathbf{R}.^3$ If $|\boldsymbol{u}| \ll 1$, that is, *small displacement gradient approximation* or *small strain and rotation approximation* [13, 14], we have the relation

$$det(\boldsymbol{I} + \nabla \boldsymbol{u}) = 1 + div\boldsymbol{u} + O(|\boldsymbol{u}|^2). \tag{4.2}$$

Therefore, assuming that the deformation is locally infinitesimal, we have the relation

$$\nabla \cdot \boldsymbol{u} = div\boldsymbol{u} \tag{4.3}$$

as a linear approximation of the condition in Equation (4.1), where the vector $\boldsymbol{u} = (u, v, w)^\top$ is optical flow of a voxel in three-dimensional images. This condition is used as an additional constraint for optical-flow computation [13, 14]. Then, we have a convex minimisation problem [8–10, 15],

$$J(\boldsymbol{u}, \alpha, \beta) = J(\boldsymbol{u}, \alpha) + \beta |div\boldsymbol{u}|^2, \tag{4.4}$$

where the first term of the right-hand side of the equation is the usual linear or non-linear optimisation functional combining the data term and the regularisation term with a positive parameter α. A typical and traditional expression[1] of $J(\boldsymbol{u}, \alpha)$ [8–10] is

$$J(\boldsymbol{u}, \alpha) = \int_{\mathbf{R}^3} |\nabla f^\top \boldsymbol{u} + f_t|^2 d\boldsymbol{x} + \alpha N(\boldsymbol{u}, \boldsymbol{u}) d\boldsymbol{x}. \tag{4.5}$$

As we will show in the next section, the constraint $\nabla \cdot \boldsymbol{u} = 0$ in Equation (4.3) and the Horn-Schunck regularisation term,

$$N_{HS}(\boldsymbol{u}, \boldsymbol{u}) = tr(\nabla \boldsymbol{u} \nabla \boldsymbol{u}^\top). \tag{4.6}$$

are dependent regularisation conditions. This analytical property means that the Horn-Schunck condition minimises the condition $div\boldsymbol{u}$ and the other smoothing criteria simultaneously. Furthermore, as an extension of the constraint to the higher order derivatives, we show the mathematical relation between the vector spline constraint, which was introduced for the computation of smooth optical flow, [20, 21] and thin plate constraint, which was introduced for the computation of optical flow on the deformable boundary

---

[1] Recently, for

$$\Psi(s) = \sqrt{s^2 + \varepsilon^2}, \text{ s.t. } 0 < \varepsilon \ll 1,$$

the criterion in the form

$$\int_{\mathbf{R}^3} \Psi(|\nabla f^\top \boldsymbol{u} + f_t|) d\boldsymbol{x} + \alpha \Psi(tr \nabla \boldsymbol{u} \nabla \boldsymbol{u}^\top) d\boldsymbol{x},$$

is proposed [15]. This minimisation problem allows us to detect a smooth optical-flow field. We can deal with the three-dimensional version this minimisation problem.

in three-dimensional Euclidean space. Moreover, we theoretically examine the validity of the error measure for the evaluation of computed optical flow.

We introduce the three-dimensional version of the Nagel-Enkelmann criterion and its dual criterion. Since Nagel-Enkelmann criterion allows us to detect boundary motion, the dual criterion allows us to detect the motion on the boundary of segments. Next we derive a numerical criterion for the convergence of optical-flow computation algorithms with the divergence-zero condition. We also show numerical results for the Horn-Schunck, the Nagel-Enkelmann, and our new criteria.

4.2 Optical Flow in a Space

For a spatio-temporal image $f(\boldsymbol{x},t)$ defined in a spatio-temporal space $\mathbf{R}^3 \times \mathbf{R}_+$, the total derivative with respect to time t is given as

$$\frac{d}{dt}f = \nabla f^\top \boldsymbol{u} + \frac{\partial f}{\partial t}\frac{dt}{dt}, \quad \nabla f = (f_x, f_y, f_z)^\top \tag{4.7}$$

where $\boldsymbol{u} = \dot{\boldsymbol{x}} = \frac{d\boldsymbol{x}}{dt}$ is the motion of each point $\boldsymbol{x}$ in $\mathbf{R}^3$. The motion constraint

$$\frac{d}{dt}f = 0 \tag{4.8}$$

implies that the motion vector $\boldsymbol{u} = (u,v,w)^\top$ of the point $\boldsymbol{x}$ is the solution of the singular equation,

$$\nabla f^\top \boldsymbol{u} + f_t = 0. \tag{4.9}$$

To solve this singular equation, the regularisation

$$J = \int_{\mathbf{R}^3} \left\{ |\nabla f^\top \boldsymbol{u} + f_t|^2 + \alpha N(\boldsymbol{u},\boldsymbol{u}) \right\} d\boldsymbol{x} \tag{4.10}$$

is employed, where $N(x,y)$ is an appropriate positive bilinear form and α is the positive regularisation parameter.

If the regulariser is in the form

$$N(\boldsymbol{u},\boldsymbol{u}) = tr(\nabla \boldsymbol{u} \boldsymbol{N} \nabla \boldsymbol{u}^\top) \tag{4.11}$$

for an appropriate positive definite matrix $\boldsymbol{N}$, where $\nabla \boldsymbol{u}$ is the vector gradient of $\boldsymbol{u}$, that is, $\nabla \boldsymbol{u} = (\nabla u, \nabla v, \nabla w)$, the Euler-Lagrange equation of the energy functional defined in Equation (4.10) is

$$\nabla^\top \boldsymbol{N} \nabla \boldsymbol{u} = \frac{1}{\alpha}(\boldsymbol{S}\boldsymbol{u} + f_t \nabla f), \quad \boldsymbol{S} = \nabla f \nabla f^\top \tag{4.12}$$

where $\boldsymbol{S}$ is the structure tensor of $f(x,y,z,t)$ at time t.

4.3 Generalised Regularisation Term

4.3.1 Orthogonal Decomposition of Regulariser

It is possible to define Nagel-Enkelmann term for the three dimensional problem by

$$
\begin{aligned}
\boldsymbol{N} &= \frac{1}{|tr\boldsymbol{S}| + 3\lambda^2}((tr\boldsymbol{S})\boldsymbol{I} - \boldsymbol{S} + 2\lambda^2 \boldsymbol{I}) \\
&= \frac{1}{|tr\boldsymbol{S}| + 3\lambda^2}(\boldsymbol{T} + 2\lambda^2 \boldsymbol{I}) \\
&= \frac{1}{|\nabla f|^2 + 3\lambda^2}
\begin{pmatrix}
f_y^2 + f_z^2 + 2\lambda^2 & -f_x f_y & -f_x f_z \\
-f_x f_y & f_x^2 + f_z^2 + 2\lambda^2 & -f_y f_z \\
-f_x f_z & -f_y f_z & f_x^2 + f_y^2 + 2\lambda^2
\end{pmatrix} \quad (4.13)
\end{aligned}
$$

where

$$
\boldsymbol{T} = \boldsymbol{S}^2 - \boldsymbol{I} = tr\boldsymbol{S} \times \boldsymbol{I} - \boldsymbol{S}, \qquad (4.14)
$$

such that $\boldsymbol{TS} = \boldsymbol{ST} = 0$, as a generalisation of Nagel-Enkelman criterion for two-dimensional images [9].

Now, we mathematically define the dual constraint matrix $\boldsymbol{N}^\perp$ of the Nagel-Enkelmann regulariser as

$$
\boldsymbol{N} + \boldsymbol{N}^\perp = \boldsymbol{I}, \qquad (4.15)
$$

In three-dimensional Euclidean space the explicit expression of the matrix $\boldsymbol{N}^\perp$ is

$$
\begin{aligned}
\boldsymbol{N}^\perp &= \frac{1}{|tr\boldsymbol{S}| + 3\lambda^2}\left(\boldsymbol{S} + \lambda^2 \boldsymbol{I}\right) \\
&= \frac{1}{|\nabla f|^2 + 3\lambda^2}
\begin{pmatrix}
f_x^2 + \lambda^2 & f_x f_y & f_x f_z \\
f_x f_y & f_y^2 + \lambda^2 & f_y f_z \\
f_x f_z & f_y f_z & f_z^2 + \lambda^2
\end{pmatrix}. \quad (4.16)
\end{aligned}
$$

For matrices $\boldsymbol{N}$ and $\boldsymbol{N}^\perp$, we have the relations

$$
\boldsymbol{N}\nabla f = \frac{2\lambda^2}{|\nabla f|^2 + 3\lambda^2}\nabla f, \quad \boldsymbol{N}\nabla f^\perp = \frac{|\nabla f|^2 + 2\lambda^2}{|\nabla f|^2 + 3\lambda^2}\nabla f^\perp, \qquad (4.17)
$$

$$
\boldsymbol{N}^\perp\nabla f = \frac{|\nabla f|^2 + \lambda^2}{|\nabla f|^2 + n\lambda^2}\nabla f, \quad \boldsymbol{N}^\perp\nabla f^\perp = \frac{\lambda^2}{|\nabla f|^2 + n\lambda^2}\nabla f^\perp. \qquad (4.18)
$$

Therefore, it is possible to derive the eigenvalue decompositions of $\boldsymbol{N}$ and $\boldsymbol{N}^\perp$ as

$$
\boldsymbol{N} = \boldsymbol{R}diag(n_1, n_2, n_3)\boldsymbol{R}^\top, \quad \boldsymbol{N}^\perp = \boldsymbol{R}diag(n_1^\perp, n_2^\perp, n_3^\perp)\boldsymbol{R}^\top \qquad (4.19)
$$

for $\boldsymbol{R} = (\boldsymbol{r}_1, \boldsymbol{r}_2, \boldsymbol{r}_3)^{\top}$, where

$$\boldsymbol{r}_1 = \frac{\nabla f}{|\nabla f|}, \quad \boldsymbol{r}_k^{\top} \boldsymbol{r}_1 = 0, \quad |\boldsymbol{r}_k| = 1 \tag{4.20}$$

for $k = 2, 3$ and

$$n_1 = \frac{2\lambda^2}{|\nabla f|^2 + 3\lambda^2}, \quad n_2 = n_3 = \frac{|\nabla f|^2 + 2\lambda^2}{|\nabla f|^2 + 3\lambda^2}, \tag{4.21}$$

and

$$n_1^{\perp} = \frac{|\nabla f|^2 + \lambda^2}{|\nabla f|^2 + 3\lambda^2}, \quad n_2^{\perp} = n_3^{\perp} = \frac{\lambda^2}{|\nabla f|^2 + 3\lambda^2}. \tag{4.22}$$

These decompositions of the matrix $\boldsymbol{N}$ lead to the conclusion that $0 < n_i < 1$ and $0 < n_i^{\perp} < 1$ for $i = 1, 2, 3$.

4.3.2 Separation of Motion in a Space

An infinitesimal motion in a space is decomposed into a translation and a small-angle rotation around this translation vector [19]. This geometric property of motion in a space implies that surface motion of deformable object in a space is decomposed into motions in the normal direction and on the tangent plane at each point as shown in (a)and (b) of Figure 4.1.

Since the principal minor eigenvector of the matrix $\boldsymbol{N}$ is the gradient ∇f, Nagel-Enkelmann criterion enhances the flow parallel to ∇f and relaxes noise on the plane perpendicular to ∇f. Conversely, the regulariser with matrix $\boldsymbol{N}^{\perp}$ enhances the vector perpendicular to ∇f and relaxes noise in the direction of $\nabla f^{\perp}$. These geometric properties of matrices $\boldsymbol{N}$ and $\boldsymbol{N}^{\perp}$ allow us to decompose flow vectors into two vectors which are parallel to ∇f and perpendicular to ∇f. In a three-dimensional space, the dimension of tangent space spanned by vectors perpendicular to ∇f is 2. Furthermore, the zero-cross set of the gradient map $|\nabla f|$, that is,

$$E(\boldsymbol{x}) = \{\boldsymbol{x} \,|\, |\nabla f| = 0\},$$

is the the edge-curve set of segments. Therefore, the portion of the optical-flow vector on the tangent plane at each point is approximately the motion on the surface boundary of a deformable object. This geometrical relation of the region boundary implies that regularisers $tr(\nabla \boldsymbol{u} \boldsymbol{N} \nabla \boldsymbol{u}^{\top})$ and $tr(\nabla \boldsymbol{u} \boldsymbol{N}^{\perp} \nabla \boldsymbol{u}^{\top})$ allow us to detect the normal and tangent motions of the deformation of an object surface.

4.3.3 Vector Regularisation Term

As described in section 1, for the three-dimensional optical-flow computation problem, the additional assumption such

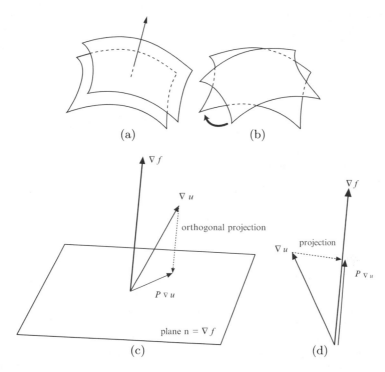

Fig. 4.1. Motion of boundary in a space. Surface motion of deformable object in a space is decomposed into motions in the normal direction (a) and on the tangent plane (b) at each point (c) and (d) show the geometrical relations among the gradient of gray-values, and the vectors $\boldsymbol{N}\boldsymbol{u}$ and $\boldsymbol{N}^{\perp}\boldsymbol{u}$ in a three-dimensional space.

$$div\boldsymbol{u} = 0 \tag{4.23}$$

is considered. Therefore, our regularisation problem becomes

$$\int_{\mathbf{R}^3} \left\{ |\nabla f^{\top}\boldsymbol{u} + f_t|^2 d + \alpha tr(\nabla\boldsymbol{u}\boldsymbol{N}\nabla\boldsymbol{u}^{\top}) + \beta|div\boldsymbol{u}|^2 \right\} d\boldsymbol{x}. \tag{4.24}$$

The Euler-Lagrange Equation of the energy functional defined in Equation (4.24) is

$$\nabla^{\top}\boldsymbol{N}\nabla\boldsymbol{u} = \frac{1}{\alpha}(\nabla_t f)^{\top}\boldsymbol{v}\nabla f + \frac{\beta}{\alpha}\nabla\nabla^{\top}\boldsymbol{u} \tag{4.25}$$

for $\boldsymbol{v} = (\boldsymbol{u}^{\top}, 1)^{\top}$. Therefore, the embedding of the Euler-Lagrange equation to the parabolic PDE is

$$\frac{\partial}{\partial\tau}\boldsymbol{u} = \nabla^{\top}\boldsymbol{N}\nabla\boldsymbol{u} - \frac{1}{\alpha}(\nabla_t f^{\top}\boldsymbol{v})\nabla f - \frac{\beta}{\alpha}\nabla\nabla^{\top}\boldsymbol{u}. \tag{4.26}$$

Next, we analyse mathematical property of an additional regularisation term. Setting

$$shu = \sqrt{\sum_{i=1}^{3}\left(\frac{\partial}{\partial x_i}u_i - \frac{\partial}{\partial x_{i+1}}u_{(i+1)}\right)^2 + \left(\frac{\partial}{\partial x_i}u_i + \frac{\partial}{\partial x_{i+1}}u_{(i+1)}\right)^2}, \quad (4.27)$$

where $u_4 = u_1$ and $x_4 = x_1$ for $\boldsymbol{u} = (u, v, w)^\top = (u_1, u_2, u_3)^\top$ and $\boldsymbol{u} = (x, y, z)^\top = (x_1, x_2, x_3)^\top$, respectively, we have the relation

$$tr(\nabla\boldsymbol{u}\nabla\boldsymbol{u}^\top) = \frac{1}{2}(|divu|^2 + |rotu|^2 + |shu|^2). \quad (4.28)$$

Analytically, Equation (4.27) implies that the optical-flow vector $\boldsymbol{u}$ is a continuous function at all points with respect to all variables. The constraints

$$divu = 0, \; rotu = \boldsymbol{0}, \; shu = 0, \quad (4.29)$$

derive the regulariser

$$J_{GA} = \beta_1(divu)^2 + \beta_2|rotu|^2 + \beta_3 sh^2u = \boldsymbol{w}^\top\boldsymbol{Q}\boldsymbol{w} \quad (4.30)$$

for a block symmetry matrix $\boldsymbol{Q}$ and the vector

$$\boldsymbol{w} = \text{vec}(\nabla\boldsymbol{u}) = \begin{pmatrix} \nabla u \\ \nabla v \\ \nabla w \end{pmatrix}.$$

This relation implies that the Horn-Schunck regulariser $tr(\nabla\boldsymbol{u}\nabla\boldsymbol{u}^\top)$ minimises all of $divu$, $rotu$, and shu selection $\beta_1 = \beta_2 = \beta_3$ in Equation (4.30). Furthermore, we have the relation

$$\alpha tr(\nabla\boldsymbol{u}\nabla\boldsymbol{u}^\top) + \beta divu = \left(\frac{1}{2}\alpha + \beta\right)divu + \frac{1}{2}\alpha|rotu|^2 + \frac{1}{2}\alpha sh^2u. \quad (4.31)$$

This relation implies that the term $tr(\nabla\boldsymbol{u}\nabla\boldsymbol{u}^\top)$ and $divu$ are mathematically dependent terms.[2]

4.4 Numerical Scheme

4.4.1 Convergence Analysis

For the discretisation, we adopt Forward-Time Centred-Space, FTCS in the abbreviated form, such as

---

[2] On a two-dimensional plane, for planar optical flow $\boldsymbol{u} = (u, v)^\top$.

$$rotu = \frac{\partial}{\partial x}v - \frac{\partial}{\partial y}u$$

and

$$sh^2u = \left(\frac{\partial}{\partial x}u - \frac{\partial}{\partial y}v\right) + \left(\frac{\partial}{\partial x}v - \frac{\partial}{\partial y}u\right).$$

Therefore, shu geometrically expresses the discontinuity of the vector function $\boldsymbol{u}(x, y) = (u(x, y), v(x, y))^\top$ on each point.

$$\frac{\partial u}{\partial x} = \frac{u^n(i+h) - u^n(i-h)}{2h}, \tag{4.32}$$

$$\frac{\partial^2 u}{\partial x^2} = \frac{u^n(i+h) - 2u^n(i) + u^n(i-h)}{h^2}, \tag{4.33}$$

$$\frac{\partial^2 u}{\partial x \partial y} = \frac{1}{4h^2}\{u^n(i+h, j+h) - u^n(i-h, j+h)$$
$$- u^n(i+h, j-h) + u^n(i-h, j-h)\} \tag{4.34}$$

$$\frac{\partial u}{\partial \tau} = \frac{u^{n+1}(i) - u^n(i)}{\Delta \tau}, \tag{4.35}$$

where n, h, an $\Delta \tau$ are the iteration number, the unit length of the spatial grid, and unit length of the time grid, respectively.

We analyse the convergence property for the case of $N = I$. Setting L to be the discrete Laplacian with an appropriate boundary condition, the discrete version of Equation (4.12) becomes,

$$Lu_{ijk} = \frac{1}{\alpha}S_{ijk}u_{ijk} + \frac{1}{\alpha}f_t s_{ijk}, \quad S = s_{ijk}s_{ijk}^\top, \quad s_{ijk} = \nabla f(i, j, k). \tag{4.36}$$

For optical flow vectors

$$u_{ijk} = (u(i, j, k), v(i, j, k), w(i, j, k))^\top, \tag{4.37}$$

in a space, we define the permutation matrix P as

$$P\text{vec}(u_{111}, u_{112}, \cdots, u_{MmM}) = \text{vec}\begin{pmatrix} u_{111}^\top \\ u_{112}^\top \\ \vdots \\ u_{MMM}^\top \end{pmatrix} \tag{4.38}$$

Equation (4.36) is a point-wise equation. Therefore, for vector functions x, setting

$$x := \begin{pmatrix} x_{111} \\ x_{112} \\ \vdots \\ x_{MMM} \end{pmatrix} \tag{4.39}$$

we have the matrix equation

$$Lu = \frac{1}{\alpha}Su + \frac{1}{\alpha}f_t s \tag{4.40}$$

for

$$L := I_3 \otimes P^\top(D_2 \otimes I \otimes I + I \otimes D_2 \otimes I + I \otimes I \otimes D_2)P, \tag{4.41}$$

$$S = Diag(S_{111}, S_{112}, \cdots, S_{MMM}). \tag{4.42}$$

P is the permutation matrix of Equation (4.38).

The matrix D_2 is tridiagonal matrix [16–18] such that

$$
D_2 = \begin{pmatrix}
-2 & 1 & 0 & 0 & \cdots & 0 & 0 \\
1 & -2 & 1 & 0 & \cdots & 0 & 0 \\
0 & 1 & -2 & 1 & \cdots & 0 & 0 \\
\vdots & \vdots & \vdots & \vdots & \ddots & \vdots & \vdots \\
0 & 0 & 0 & \cdots & 0 & 1 & -2
\end{pmatrix},
$$

for Dirichlet boundary condition, [17] such that

$$
D_2 = \begin{pmatrix}
-1 & 1 & 0 & 0 & \cdots & 0 & 0 \\
1 & -2 & 1 & 0 & \cdots & 0 & 0 \\
0 & 1 & -2 & 1 & \cdots & 0 & 0 \\
\vdots & \vdots & \vdots & \vdots & \ddots & \vdots & \vdots \\
0 & 0 & 0 & \cdots & 0 & 1 & -1
\end{pmatrix},
$$

for von Neumann boundary condition, and

$$
D_2 = \left(\alpha \begin{pmatrix}
-1 & 1 & 0 & 0 & \cdots & 0 & 0 \\
1 & -2 & 1 & 0 & \cdots & 0 & 0 \\
0 & 1 & -2 & 1 & \cdots & 0 & 0 \\
\vdots & \vdots & \vdots & \vdots & \ddots & \vdots & \vdots \\
0 & 0 & 0 & \cdots & 0 & 1 & -1
\end{pmatrix} + \beta \begin{pmatrix}
-2 & 1 & 0 & 0 & \cdots & 0 & 0 \\
1 & -2 & 1 & 0 & \cdots & 0 & 0 \\
0 & 1 & -2 & 1 & \cdots & 0 & 0 \\
\vdots & \vdots & \vdots & \vdots & \ddots & \vdots & \vdots \\
0 & 0 & 0 & \cdots & 0 & 1 & -2
\end{pmatrix} \right)
$$

for the third kind boundary condition.

Furthermore, the matrix D_1 is also tridiagonal matrix such that

$$
D_1 = \frac{1}{2} \begin{pmatrix}
0 & -1 & 0 & \cdots & 0 & 0 \\
1 & 0 & -1 & 0 & \cdots & 0 & 0 \\
0 & 1 & 0 & \cdots & 0 & 0 \\
0 & 0 & 1 & \cdots & 0 & 0 \\
\vdots & \vdots & \vdots & \vdots & \ddots & \vdots & \vdots \\
0 & 0 & 0 & \cdots & 1 & 0 & -1 \\
0 & 0 & 0 & \cdots & 0 & 1 & 0
\end{pmatrix}.
$$

The operation $\frac{\partial}{\partial \tau}$ is expressed as

$$
\frac{\partial}{\partial \tau} = \frac{1}{\Delta \tau}(I u^{n+1} - I u^n)
$$

in the matrix form.

For a positive constant $\Delta \tau$, setting

$$
A = \left(I + \frac{\Delta \tau}{\alpha} S\right), \quad B = (I + \Delta \tau L), \quad c = -\frac{\Delta \tau}{\alpha} f_t s, \tag{4.43}
$$

Equation (4.36) becomes

$$\boldsymbol{Au} = \boldsymbol{Bu} + \boldsymbol{c}. \tag{4.44}$$

where $\boldsymbol{A}^\top = \boldsymbol{A}$ and $\boldsymbol{B}^\top = \boldsymbol{B}$. Setting $\rho(\boldsymbol{M})$ to be the spectrum of the Matrix $\boldsymbol{M}$, we have the relation $\rho(\boldsymbol{A}) > 1$. Furthermore, we have the next theorem.

Theorem 1. *We have the relation $\rho(\boldsymbol{A}) > 1$. If $\rho(\boldsymbol{B}) \leq 1$, the iteration*

$$\boldsymbol{Au}^{(m+1)} = \boldsymbol{Bu}^{(m)} + \boldsymbol{c} \tag{4.45}$$

converges to the solution of Equation (4.40) [18], that is,

$$\lim_{m \to \infty} \boldsymbol{u}^{(m)} = arg\left\{\boldsymbol{u} \,|\, \boldsymbol{Lu} = \frac{1}{\alpha}(\boldsymbol{Su} + f_t \boldsymbol{s})\right\}. \tag{4.46}$$

4.4.2 Convergence Condition

Setting $\boldsymbol{U}$ and $\boldsymbol{\Lambda}$ to be the discrete cosine transform matrix and the diagonal matrix respectively, we have the relation

$$\boldsymbol{L} = \boldsymbol{U\Lambda U}^\top, \tag{4.47}$$

Substituting this relation to

$$\rho\left(\boldsymbol{I} + \Delta\tau\boldsymbol{L}\right) < 1 \tag{4.48}$$

we have the relation

$$\rho\left(\boldsymbol{U}\left(\boldsymbol{I} + \Delta\tau\boldsymbol{\Lambda}\right)\boldsymbol{U}^\top\right) < 1 \tag{4.49}$$

Then, finally we have the convergence condition

$$\frac{\Delta\tau}{h^2} < \frac{1}{6}, \tag{4.50}$$

since

$$\rho\left(\boldsymbol{I} + \Delta\tau\boldsymbol{\Lambda}\right) = \max_i |1 + \Delta\tau\lambda_i| < 1 \tag{4.51}$$

and $-\frac{4\cdot3}{h^2} \leq \lambda_i \leq 0$. For

$$\overline{\boldsymbol{D_1}} = \begin{pmatrix} \boldsymbol{D}_1 \otimes \boldsymbol{I} \otimes \boldsymbol{I} \\ \boldsymbol{I} \otimes \boldsymbol{D}_1 \otimes \boldsymbol{I} \\ \boldsymbol{I} \otimes \boldsymbol{I} \otimes \boldsymbol{D}_1 \end{pmatrix}, \quad \overline{\boldsymbol{D_1}}^* = \left(\boldsymbol{D}_1 \otimes \boldsymbol{I} \otimes \boldsymbol{I}, \boldsymbol{I} \otimes \boldsymbol{D}_1 \otimes \boldsymbol{I}, \boldsymbol{I} \otimes \boldsymbol{I} \otimes \boldsymbol{D}_1\right) \tag{4.52}$$

and

$$\overline{\boldsymbol{N}} = Diag(\boldsymbol{N}_{111}, \boldsymbol{N}_{112}, \cdots, \boldsymbol{N}_{MMM}), \quad \overline{\boldsymbol{N}^\perp} = Diag(\boldsymbol{N}_{111}^\perp, \boldsymbol{N}_{112}^\perp, \cdots, \boldsymbol{N}_{MMM}^\perp), \tag{4.53}$$

setting

$$\overline{\boldsymbol{M}} = \boldsymbol{P}\overline{\boldsymbol{N}}\boldsymbol{P}, \quad \overline{\boldsymbol{M}^\perp} = \boldsymbol{P}\overline{\boldsymbol{N}^\perp}\boldsymbol{P}, \tag{4.54}$$

we have the relation

$$\boldsymbol{M} = \boldsymbol{I}_3 \otimes \overline{\boldsymbol{D}_1}^{*} \overline{\boldsymbol{M} \boldsymbol{D}_1}, \;\; \boldsymbol{M}^{\perp} = \boldsymbol{I}_3 \otimes \overline{\boldsymbol{D}_1}^{*} \overline{\boldsymbol{M}^{\perp} \boldsymbol{D}_1}. \tag{4.55}$$

Using diagonal matrices $\overline{\boldsymbol{\Lambda}}$ and $\overline{\boldsymbol{\Lambda}_{\perp}}$, and the orthogonal matrix $\overline{\boldsymbol{R}}$ such that

$$\overline{\boldsymbol{N}} = \overline{\boldsymbol{R} \boldsymbol{\Lambda} \boldsymbol{R}^{\top}}, \;\; \overline{\boldsymbol{N}^{\perp}} = \overline{\boldsymbol{R} \boldsymbol{\Lambda}_{\perp} \boldsymbol{R}^{\top}} \tag{4.56}$$

we define

$$\hat{\boldsymbol{L}} = \hat{\boldsymbol{D}}_1^{\top} \hat{\boldsymbol{D}}_1, \;\; \hat{\boldsymbol{L}}_{\perp} = \hat{\boldsymbol{D}}_{1\perp}^{\top} \hat{\boldsymbol{D}}_{1\perp} \tag{4.57}$$

where

$$\hat{\boldsymbol{D}}_1 = (\overline{\boldsymbol{\Lambda}})^{\frac{1}{2}} \overline{\boldsymbol{R}}, \;\; \hat{\boldsymbol{D}}_1 = (\overline{\boldsymbol{\Lambda}_{\perp}})^{\frac{1}{2}} \overline{\boldsymbol{R}}, \tag{4.58}$$

Then, we have the relation

$$\boldsymbol{B} = \boldsymbol{I} + \Delta\tau \hat{\boldsymbol{L}}, \;\; \boldsymbol{B} = \boldsymbol{I} + \Delta\tau \hat{\boldsymbol{L}}_{\perp} \tag{4.59}$$

for the Nagel-Enkelmann and orthogonal Nagel-Enkelmann constraints, respectively. For symmetry matrices $\boldsymbol{N}$ and $\boldsymbol{N}^{\perp}$, eigenvalues are positive and less than one, that is, $0 < n_i < 1$ and $0 < n_i^{\perp} < 1$, for $\lambda \neq 0$, we have the inequalities $\rho(\overline{\boldsymbol{N}}) < 1$ and $\rho(\overline{\boldsymbol{N}^{\perp}})$. These algebraic properties yield that inequalities

$$\rho(\boldsymbol{I} + \Delta\tau \hat{\boldsymbol{L}}) < 1, \;\; \rho(\boldsymbol{I} + \Delta\tau \hat{\boldsymbol{L}}_{\perp}) < 1. \tag{4.60}$$

if

$$\rho(\boldsymbol{I} + \Delta\tau \boldsymbol{L}) < 1. \tag{4.61}$$

Therefore, our numerical schemes for the Horn-Schunck, Nagel-Enkelmann, and the orthogonal Nagel-Enkelmann regularisers converge to unique solutions.

4.4.3 Parameter Tuning

Tuning of the Regularisation Parameter

For the matrix

$$\boldsymbol{A}_{ijk} = \boldsymbol{I} + \frac{1}{\alpha} \boldsymbol{S}_{ijk}$$

we have the relation

$$\boldsymbol{A}_{ijk}^{-1} = \frac{1}{1 + \frac{1}{\alpha} tr \boldsymbol{S}_{ijk}} (\boldsymbol{I} + \frac{1}{\alpha} \boldsymbol{T}_{ijk}), \tag{4.62}$$

where

$$\boldsymbol{T}_{ijk} = tr \boldsymbol{S}_{ijk} \times \boldsymbol{I} - \boldsymbol{S}_{ijk}.$$

For $\boldsymbol{A}_{ijk}$ and $\boldsymbol{A}_{ijk}^{-1}$, we have the relation

$$\boldsymbol{A}_{ijk} \nabla f_{ijk} = (1 + \frac{1}{\alpha} tr \boldsymbol{S}_{ijk}) \nabla f_{ijk}, \;\; \boldsymbol{A}_{ijk} \nabla f_{ijk}^{\perp} = \nabla f_{ijk}^{\perp}, \tag{4.63}$$

and

$$A_{ijk}^{-1}\nabla f_{ijk} = \frac{1}{1 + \frac{1}{\alpha}trS_{ijk}}\nabla f_{ijk}, \quad A_{ijk}^{-1}\nabla f_{ijk}^{\perp} = \nabla f_{ijk}^{\perp}, \tag{4.64}$$

These relations imply that for the stable computation of A^{-1}, we should select α as

$$O(\alpha) = O(\rho(S)) = O(|\nabla f|_{\max}^2). \tag{4.65}$$

This algebraic property leads to the next assertion.

Assertion 1 *The order of α is the same order as that of $|\nabla f|^2$ to achieve accurate computation of optical flow.*

The embedding of the Euler-Lagrange equation in the parabolic PDE is

$$\frac{\partial}{\partial \tau}u = \nabla^{\top}N\nabla u - \frac{1}{\alpha}Fv, \quad v = \begin{pmatrix} u \\ 1 \end{pmatrix}, \tag{4.66}$$

where

$$F = \nabla f \nabla_t f = (S, f_t \nabla f), \tag{4.67}$$

for $\nabla_t f = (f_x, f_y, f_z, f_t)^{\top}$. The trace of the principal maximum square matrix of F is $trS = |\nabla f|^2$. Therefore, if the order of α is the same as that of $|\nabla f|^2$, the second term affects the computation.

Tuning of the Relaxation Parameter

Since the rank of matrix $S = \nabla f \nabla f^{\top}$ is one, we deal with the average of S,

$$\Sigma = \frac{1}{|\Omega(x)|}\int_{\Omega(x)} S dx, \tag{4.68}$$

in the neighbourhood of the point x instead of S. Then, the averaged Nagel-Enkelmann criterion becomes

$$N = \frac{1}{tr\Sigma + 3\lambda^2}((tr\Sigma)I - T + 2\lambda^2 I). \tag{4.69}$$

This matrix is numerically stable compared with the original Nagel-Enkelmann criterion, since Σ is a smoothed version of S. Furthermore, the orthogonal smoothed Nagel-Enkelmann regulariser becomes

$$N^{\perp} = \frac{1}{tr\Sigma + 3\lambda^2}\left(\Sigma + \lambda^2 I\right). \tag{4.70}$$

Using this the smoothed Nagel-Enkelmann regulariser, we analyse geometric properties of this matrix. For

$$\Sigma u_i = t_i^2 u_i, \quad i = 1, 2, 3, \tag{4.71}$$

where $t_1^2 \geq t_2^2 \geq t_3^2 \geq 0$, setting

$$\boldsymbol{\Lambda} = Diag(t_1^2, t_2^2, t_3^2), \quad \boldsymbol{U} = (\boldsymbol{u}_1, \boldsymbol{u}_2, \boldsymbol{u}_3), \qquad (4.72)$$

the Nagel-Enkelmann criterion is expressed as

$$\boldsymbol{N} = \boldsymbol{U}\boldsymbol{D}\boldsymbol{U}^\top, \quad \boldsymbol{D} = Diag(f(t_1), f(t_2), f(t_3)), \qquad (4.73)$$

where

$$f(x) = 1 - \frac{x^2 + \lambda^2}{tr\boldsymbol{\Sigma} + 3\lambda^2}. \qquad (4.74)$$

If

$$f(x) = \frac{x^2 + \lambda^2}{tr\boldsymbol{\Sigma} + n\lambda^2}, \qquad (4.75)$$

we have the orthogonal Nagel-Enkelmann criterion.

From Equation (4.74), we have the relations

$$\lim_{\lambda \to 0} f(x) = 1 - \frac{t_i^2}{tr\boldsymbol{\Sigma}}, \quad \lim_{\lambda \to \infty} f(x) = 1 - \frac{1}{3}. \qquad (4.76)$$

These relations show that for large λ, the Nagel-Enkelmann criterion is unaffected by the structure tensor $\boldsymbol{\Sigma}$ and converges to a general the Horn-Schunck criterion. Conversely, for small λ, the Nagel-Enkelmann criterion is uniformly affected by the structure tensor $\boldsymbol{S}$. The second relation of Equation (4.76) shows that the Horn-Schunck criterion with the dimension factor is

$$V_{HS}(\nabla \boldsymbol{u}) = \frac{2}{3} tr\nabla \boldsymbol{u}\nabla \boldsymbol{u}^\top \qquad (4.77)$$

Furthermore, the orthogonal the Horn-Schunck criterion is linear to the original criterion, that is,

$$V_{HS^\top} = \frac{1}{3} tr\nabla \boldsymbol{u}\nabla \boldsymbol{u}^\top = \frac{1}{3} V_{HS}. \qquad (4.78)$$

Moreover, the first relation of Equation (4.76) shows that for large λ^2, Nagel-Enkelmann constraint becomes

$$V_{NE\infty} = \frac{3}{2} \left(V_{HS} - \frac{1}{tr\boldsymbol{\Sigma}} V_{HS} \right). \qquad (4.79)$$

This analytical property means that if $tr\boldsymbol{T}$ is large, the criterion is linear to the Horn-Schunck criterion. This means that for a high-contrast image the Nagel-Enkelmann criterion with large parameter λ is approximately equivalent to the Horn-Schunck criterion. Therefore, for low contrast images with a moderate value of parameter λ, we can obtain sufficient results using the Nagel-Enkelmann criterion. From Equation (4.74), for the preservation of the positive semidefinite condition on the matrix $\boldsymbol{N}$, the relation

$$1 > \frac{t_i^2 + \lambda^2}{tr\, \Sigma + 3\lambda^2}, \; i = 1, 2, 3 \tag{4.80}$$

is derived. This inequality implies the relation

$$\frac{4}{2} t_1^2 < \lambda^2, \tag{4.81}$$

since $tr\, \Sigma \le 3 \max t_i^2$.

4.5 Numerical Examples

Figures 4.2, 4.3, and 4.4 show examples of optical flow fields computed by the Horn-Schunck, the Nagel-Enkelmann, and the orthogonal Nagel-Enckelmann regularisers, respectively. For each regulariser, we show results for $\alpha = 10^5$, and $\alpha = 5 \times 10^5$. These values were determined to balance the orders of the diffusion term and the reaction term of diffusion-reaction equation. Roughly speaking, the parameter α is of the same order as that of $|\nabla f|^2$, since the parameter α is the denominator of the reaction term.

As expected, results in Figure 4.3 show motion on the boundary of the heart. Furthermore, as the parameter α increases the results becomes

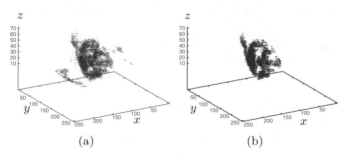

(a)   (b)

Fig. 4.2. Optical flow computed by Horn-Schunck regulariser for $\frac{\Delta \tau}{h^2} = 0.166$ From top to bottom, results for $\alpha = 10^5$, and $\alpha = 5 \times 10^5$, respectively.

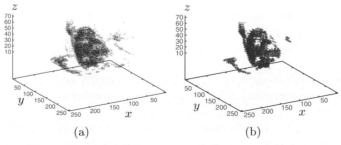

(a)   (b)

Fig. 4.3. Optical flow computed by Nagel-Enkelmann regulariser for $\frac{\Delta \tau}{h^2} = 0.166$ and $\lambda = 1$. From top to bottom, results for $\alpha = 10^5$, and $\alpha = 5 \times 10^5$, respectively.

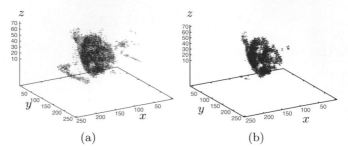

Fig. 4.4. Optical flow computed by the orthogonal Nagel-Enkelmann regulariser for $\frac{\Delta\tau}{h^2} = 0.166$ and $\lambda = 1$. From top to bottom, results for $\alpha = 10^5$, and $\alpha = 5 \times 10^5$, respectively.

smoother. Therefore, the parameter α acts as a scale in linear scale analysis. Specially for $\alpha \to \infty$, the reaction term disappears and a smoothed vector field is computers as optical flow.

With the divergence-free condition, the matrix $\boldsymbol{D}_2$ of Equation (4.41) is replaced by

$$\boldsymbol{E} := \boldsymbol{D}_2 + \frac{\beta}{\alpha}\boldsymbol{K}, \quad \boldsymbol{K} = \begin{pmatrix} \frac{\partial^2}{\partial x^2} & \frac{\partial^2}{\partial xy} & \frac{\partial^2}{\partial xz} \\ \frac{\partial^2}{\partial xy} & \frac{\partial^2}{\partial y^2} & \frac{\partial^2}{\partial yz} \\ \frac{\partial^2}{\partial xz} & \frac{\partial^2}{\partial yz} & \frac{\partial^2}{\partial z^2} \end{pmatrix}. \tag{4.82}$$

For convergence analysis of the optical-flow computation with the divergence-zero condition, we examine the relation between the ratio β/α and the grid condition $\Delta\tau/h^2$, assuming that $h = \Delta x = \Delta y = \Delta z$. For the numerical evaluation, we used the image sequence provided from Professor John Barron and Roberts Research Institute at the University of Western Ontario [12]. This is an image sequence of 20 frames. An image in this sequence is $1.25\,\text{mm}^3$ in spatial resolution. $256 \times 256 \times 75$ in size.

We have evaluated the norm $|\boldsymbol{u}^{n+1} - \boldsymbol{u}^n|$ for each point in 3D image. For the measure

$$E^n = \frac{\max_{i,j,k} |\boldsymbol{u}^{n+1}_{i,j,k} - \boldsymbol{u}^n_{i,j,k}|^2}{|\boldsymbol{u}_{i,j,k}|^{n+1}}, \tag{4.83}$$

if $E_n < 10^{-3}$, we concluded the computation converges, and if

$$E^n > 2\min_{i<n} E^i, \tag{4.84}$$

the computation does not converge. We used frames 1 and 2 of MRI 5 phase images. Specks of the computer used in our numerical evaluation are listed in Table 4.1.

If we assume that non-diagonal values of the matrix $\boldsymbol{K}$ are small, that is, $u_{\alpha\beta}$, $v_{\alpha\beta}$, and $w_{\alpha\beta}$, for $\alpha, \beta \in \{x, y, z\}$ are sufficiently small, we can approximate the matrix $\boldsymbol{K}$ of Equation (4.82) as

Table 4.1. Specks of PC.

CPU	Intel Xeon 3.06GHz
Main Memory Size	4 GB
OS	turbolinux 10 Desktop
HDD Size	100 GB
Program Language	C++
Compiler	GCC 3.3.1
Option	-O3 -march=pentium4 -mfpmath=sse

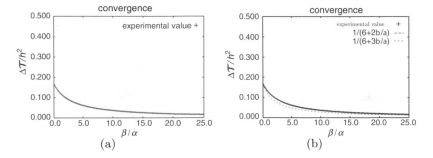

Fig. 4.5. Relation between β/α and $\Delta\tau/h^2$. (a) Curve derived by numerical computation. (b) Curve derived by the small-coupling-term assumption.

$$
\boldsymbol{K} := \begin{pmatrix} \frac{\partial^2}{\partial x^2} & & O \\ & \frac{\partial^2}{\partial y^2} & \\ O & & \frac{\partial^2}{\partial z^2} \end{pmatrix}.
\tag{4.85}
$$

This assumption is valid if the effects of coupling terms of the system of diffusion-reaction equations for the optical-flow computation are small. The condition that the spectral radius of this matrix is smaller than 1 is

$$
3\frac{\Delta\tau}{h^2} + \frac{\beta}{\alpha}\frac{\Delta\tau}{h^2} < \frac{1}{2}
\tag{4.86}
$$

In Figure 4.5 (b), we show both curves defined by (4.86) and derived by numerical computation. These results show that the effects of the coupling terms are sufficiently small if $\Delta\tau/h^2$ satisfies the convergence property for the numerical equation.

Figure 4.5 shows the relation between β/α and $\Delta\tau/h^2$. (a) shows curve derived by numerical computation. (b) shows curve derived by the small-coupling-term assumption. Figure 4.6 shows the curves for Nagel-Enkelmann and the orthogonal Nagel-Enklemann criteria in (a) and (b), with the same assumptions, respectively. These results for the Nagel-Enkelmann and the orthogonal Nagel-Enklemann criteria show that the effects of the coupling terms are small if the grid condition $\Delta\tau/h^2$ satisfies the convergence property for the numerical equation.

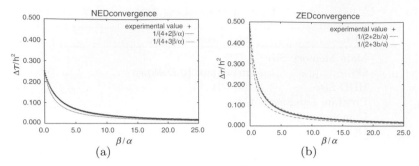

Fig. 4.6. Relation between β/α and $\Delta\tau/h^2$. (a) Curve for Nagel-Enkelman constraint and the divergence-zero condition. (b) Curve for orthogonal Nagel-Enkelmann constraint and the divergence-zero condition.

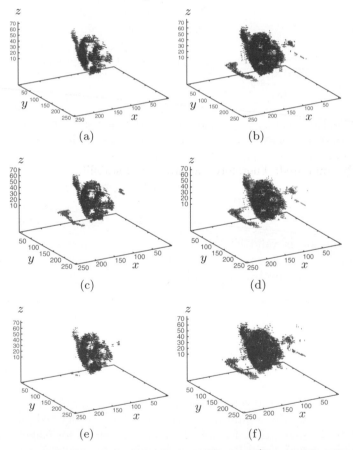

Fig. 4.7. Optical flow computed for $\beta = 0$. We set $\frac{\Delta\tau}{h^2} = 0.166$. From top to bottom in the left, results for Horn-Schunck, Nagel-Enkelmann, and the orthogonal Nagel-Enkelmann criteria, We set $\alpha = 5 \times 10^5$, $\lambda = 1$, and $\frac{\Delta\tau}{h^2} = 0.133$. With the Divergence-zero term for $\beta/\alpha = 0.5$. From top to bottom in the light, results for Horn-Schunck, Nagel-Enkelmann, and the orthogonal Nagel-Enkelmann criteria, We set $\alpha = 5 \times 10^5$ and $\lambda = 1$.

Figure 4.7 left shows optical flow computed for $\beta = 0$ with $\Delta\tau/h^2 = 0.133$. From top to bottom, results for the Horn-Schunck, Nagel-Enkelmann, and the orthogonal Nagel-Enkelmann criteria, are shown. We set $\alpha = 5 \times 10^5$ and $\lambda = 1$.

Figure 4.7 right shows optical flow computed for $\beta/\alpha = 0.5$, with $\Delta\tau/h^2 = 0.133$. From top to bottom, results for the Horn-Schunck, Nagel-Enkelmann, and the orthogonal Nagel-Enkelmann criteria, are shown. We set $\alpha = 5 \times 10^5$ and $\lambda = 1$.

These results show that, without the divergence-zero term, it is possible to compute stable optical flow. Furthermore, optical flow computed by the orthogonal Nagel-Enkelmann constraint allows to detect motion on the surface of a beating heart.

4.6 Validity of Error Analysis

As described in the previous section, the condition defined in Equation (4.50) satisfies Lax equivalence theorem. Lax equivalence theorem guarantees the stability and accuracy of the numerical scheme for solving numerically partial differential equations. Therefore, starting from the zero vector, the iteration form of Equation (4.45) converges to the solution of the Euler-Lagrange equation, which minimises the Hamiltonian of the optical flow. This numerical property implies that computational results in the examples are accurately and stably computed. Therefore, accuracy and stability of the solution only depend on the numerical schemes. In the other words, numerical schemes determine the accuracy and stability of the computed solution. This property means that optical flow is reliable if an computation algorithm is carefully elected and implemented. As analysed in the previous section, our algorithm derives accurate and stable solution.

In traditional researches [10, 11], the accuracy and stability of algorithms were evaluated by statistical analysis for the difference between the computed result and the ground truth of flow vector, which was pre-computed from a synthetic image sequence. A traditional measure for the evaluation of the optical-flow computation algorithms is the angle between the computed result and the ground truth at each point in the images. In this section, we mathematically analyse the validity of this measure and develop a new measure which is computable without the ground truth pre-computed from synthetic image.

For an accurate algorithm, we can assume that $|\boldsymbol{u}| \cong |\boldsymbol{u}^*|$, where $\boldsymbol{u}$ and $\boldsymbol{u}^*$ are the ground truth and the computed flow by an algorithm. Setting $|\boldsymbol{u}| = a$, we have the relation

$$|\boldsymbol{u} - \boldsymbol{u}^*|^2 \cong 2(a^2 - \boldsymbol{u}^\top \boldsymbol{u}^*). \tag{4.87}$$

Since

$$\frac{\boldsymbol{u}^\top \boldsymbol{u}^*}{a^2} = \cos \angle[\boldsymbol{u}, \boldsymbol{u}^*], \tag{4.88}$$

where $\angle[\boldsymbol{u}, \boldsymbol{u}^*]$ is the angle between vectors $\boldsymbol{u}$ and $\boldsymbol{u}^*$, and

$$\cos\theta \cong 1 - \frac{1}{2}\theta^2$$

for a small θ, we have an approximate relation

$$\cos^{-1}\angle[\boldsymbol{u}, \boldsymbol{u}^*] \cong \frac{|\boldsymbol{u} - \boldsymbol{u}^*|}{|\boldsymbol{u}|}. \tag{4.89}$$

The relation shows that the angle between the ground truth and the computed one is an approximation of the absolute error of computed optical flow.

For the computation of Equation (4.89), we are required to prepare an appropriate synthetic image. Therefore, we derive a criterion without the ground truth. Setting $\boldsymbol{v} = (\boldsymbol{u}^\top, 1)^\top$ and $\boldsymbol{v}^* = \boldsymbol{v} + \boldsymbol{\varepsilon}$ for $\boldsymbol{\varepsilon} = (\boldsymbol{\delta}, 0)^\top$, where $\boldsymbol{\delta} = (\boldsymbol{u}^* - \boldsymbol{u})$, we have

$$\epsilon = |(\boldsymbol{m}^\top \boldsymbol{v}^*)\boldsymbol{n}| = \frac{|\nabla f^\top \boldsymbol{u}^* + f_t|}{\sqrt{(|\nabla f|^2 + f_t^2)}}, \tag{4.90}$$

where $\boldsymbol{m} = \frac{\nabla_t f}{|\nabla_t f|}$. Since the solution of the data term of the optical-flow computation is $\nabla_t f^\top \boldsymbol{v} = 0$, the vector $((\boldsymbol{n}^\top \boldsymbol{v}^*)\boldsymbol{n})$ is the error in the direction of the vector $\nabla_t f$. Furthermore, if $\boldsymbol{u} \cong \boldsymbol{u}^*$ and $|\boldsymbol{\delta}| \ll 1$, we have the relations $\epsilon \cong |\boldsymbol{u} - \boldsymbol{u}^*|$ and $\theta \cong \frac{\epsilon}{|\boldsymbol{u}|}$. However, for the computation of Equation (4.90), we are not required to prepare any ground truth computed from a synthetic image. This mathematical property implies that the error measure defined by Equation (4.90) is suitable for the comparison of the results computed from real images. These criteria are suitable for the evaluation of two-dimensional optical-flow computation algorithms, since the difference in the directions characterised by the angles between two vectors.

For the evaluation of the three-dimensional optical-flow computation algorithms, we are required to compute the covariance of the errors in a three-dimensional space, since flow vectors are three-dimensional vectors. For this purpose, we are required to compute

$$\sigma = \frac{1}{n}\sum \boldsymbol{\delta}\boldsymbol{\delta}^\top, \quad \boldsymbol{\delta} = \boldsymbol{u}^* - \boldsymbol{u} \tag{4.91}$$

for a large number of synthetic image sequences.

4.7 Second Order Constraint Problem

In this section, we discuss some perspectives for optical-flow computation. Optical flow of images defines the vector field. For the interpolation of the vector field, the solution of the minimiser,

$$J_v(\boldsymbol{u}) = \int_{\mathbf{R}^3} \gamma_1 |\nabla div u|^2 + \gamma_2 |\nabla rot u|^2 dx \tag{4.92}$$

for $u(x_i) = u_i$, where (x_i, u_i) are samples of the vector field, derives vector spline. In a three-dimensional space, the gradient operation in the second term of Equation (4.92) is computed as the vector gradient, that is, for vector $rotu = (v_1, v_2, v_3)^\top$, $\nabla rotu = (\nabla v_1, \nabla v_2, \nabla v_3)$.

Using this criterion, the criterion

$$E_v(u) = E_o(u) + J_v(u) \tag{4.93}$$

is introduced [13,14,20,21]. The Euler-Lagrange equation of this minimisation problem is PDE with fourth-order spatial derivation.

Setting Hu to be the Hessian matrix of the vector field $u = (u, v, w)^\top$, which is defined as

$$Hu = diag(H_u, H_v, H_w) \tag{4.94}$$

we define the second order constrain as[3]

$$J_2(u) = \int_{R^3} (tr Hu H_u^\top) dx. \tag{4.95}$$

For $tr Hu H_u^\top$, we have the relation

$$tr Hu H_u^\top = |\nabla divu|^2 + |\nabla rotu|^2 + 2h^2(u) + 2k^2(u) \tag{4.96}$$

for

$$h^2(u) = \begin{vmatrix} v_{xy} & u_{xy} \\ \Delta_{xy}v & \Delta_{xy}u \end{vmatrix} + \begin{vmatrix} w_{xy} & v_{xy} \\ \Delta_{yz}w & \Delta_{yz}v \end{vmatrix} + \begin{vmatrix} u_{xy} & w_{xy} \\ \Delta_{zx} & \Delta_{zx}w \end{vmatrix} \tag{4.97}$$

$$k^2(u) = div s, \quad s = \left(\begin{vmatrix} v_y & w_y \\ v_z & w_z \end{vmatrix}, \begin{vmatrix} w_z & u_z \\ w_x & u_x \end{vmatrix}, \begin{vmatrix} u_x & v_x \\ u_y & v_y \end{vmatrix} \right)^\top \tag{4.98}$$

where

$$\Delta_{\alpha\beta} = \frac{\partial^2}{\partial\alpha^2} + \frac{\partial^2}{\partial\beta^2}, \quad \overline{\Delta}_{\alpha\beta} = \frac{\partial^2}{\partial\alpha^2} - \frac{\partial^2}{\partial\beta^2} \tag{4.99}$$

for $\alpha, \beta \in \{xy, yz, zy\}$. These relations show that $tr Hu H_u^\top$, and $|\nabla divu|^2$ and $|\nabla rotu|^2$ are dependent terms. Therefore, considering the regularisation

---

[3] Setting $D^2 f$ to be Whitney array of the order 2 generated by f, that is,

$$D^2 f = (f_{xx}, f_{yx}, f_{zx}, f_{xy}, f_{yy}, f_{zy}, f_{xz}, f_{yz}, f_{zz})^\top$$

we have the relation

$$tr HH^\top = |D^2 f|^2.$$

Therefore, Equation (4.95) is expressed in the vector form as

$$J_2(u) = \int_{R^3} \{|D^2 u|^2 + |D^2 v|^2 + |D^2 w|^2\} dx.$$

term $tr\mathbf{H_u H_u^\top}$ is equivalent to solve vector spline minimisation for optical flow computation.

The variation of $J_2(\mathbf{u})$ with respect to $\mathbf{u}$ derives the biharmonic equation

$$\Delta^2 \mathbf{u} = 0. \tag{4.100}$$

This analytical property concludes that the Euler-Lagrange equation of the variational problem

$$J_{\alpha\beta}(\mathbf{u}) = \int_{\mathbf{R}^3} \left\{ |\nabla f^\top \mathbf{u} + f_t|^2 d + \alpha tr \nabla \mathbf{u} \nabla \mathbf{u}^\top + \beta tr \mathbf{H_u H_u^\top} \right\} d\mathbf{x}. \tag{4.101}$$

is

$$\frac{\beta}{\alpha} \Delta^2 \mathbf{u} - \Delta \mathbf{u} + \frac{1}{\alpha}(\nabla f^\top \mathbf{u} + f_t)\nabla f = 0, \tag{4.102}$$

and its embedding into evolution equation is

$$\frac{\partial}{\partial \tau} \mathbf{u} = -\frac{\beta}{\alpha} \Delta^2 \mathbf{u} + \Delta \mathbf{u} - \frac{1}{\alpha}(\nabla f^\top \mathbf{u} + f_t)\nabla f. \tag{4.103}$$

Setting

$$\frac{\mathbf{u}^{n+1} - \mathbf{u}^n}{\Delta \tau} = -\frac{\beta}{\alpha} \Delta^2 \mathbf{u}^n + \Delta \mathbf{u}^n - \frac{1}{\alpha}(\mathbf{S}\mathbf{u}^n + f_t \nabla f), \tag{4.104}$$

we have the matrix iteration equation in the form

$$\mathbf{A}\mathbf{u}^{n+1} = \mathbf{B}\mathbf{u}^n + \mathbf{c} \tag{4.105}$$

where

$$\mathbf{A} = \mathbf{I} + \frac{\Delta \tau}{\alpha}\mathbf{S}, \quad \mathbf{B} = (\mathbf{I} + \Delta \tau \mathbf{L} - \Delta \tau \frac{\beta}{\alpha}\mathbf{L}^2), \quad \mathbf{c} = -\frac{\Delta \tau}{\alpha} f_t \nabla f. \tag{4.106}$$

This equation computed in the same manner with the iteration form derived from the diffusion-reaction equation for optical-flow computation. Since[4]

$$\mathbf{L}^2 = \mathbf{U}\boldsymbol{\Lambda}^2\mathbf{U}^\top, \tag{4.107}$$

we have the relation

$$\rho\left(\mathbf{I} + \Delta \tau \mathbf{L} - \frac{\beta \Delta \tau}{\alpha}\mathbf{L}^2\right) = \max_i |1 + \Delta \tau \lambda_i - \Delta \tau \frac{\beta}{\alpha}\lambda_i^2|. \tag{4.108}$$

where $-\frac{4 \cdot 3}{h^2} \le \lambda_i \le 0$, and $0 \le \lambda_i^2 \le \left(\frac{4 \cdot 3}{h^2}\right)^2$ Therefore, if

$$\max_i \left| 1 - \Delta \tau \frac{4 \cdot 3}{h^2} - \Delta \tau \frac{\beta}{\alpha}\left(\frac{4 \cdot 3}{h^2}\right)^2 \right| < 1, \tag{4.109}$$

the iteration form of Equation (4.105) converges to the solution.

---

[4] From Equation (4.41), we have the equation

$$\mathbf{L} := \mathbf{I}_3 \otimes \mathbf{P}^\top(\mathbf{D}_2^2 \otimes \mathbf{I} \otimes \mathbf{I} + \mathbf{I} \otimes \mathbf{D}_2^2 \otimes \mathbf{I} + \mathbf{I} \otimes \mathbf{I} \otimes \mathbf{D}_2^2$$
$$+ 2\mathbf{D}_2 \otimes \mathbf{D}_2 \otimes \mathbf{I} + 2\mathbf{D}_2 \otimes \mathbf{I} \otimes \mathbf{D}_2 + 2\mathbf{I} \otimes \mathbf{D}_2 \otimes \mathbf{D}_2)\mathbf{P}.$$

4.8 Concluding Remarks

We investigated the generalisation of optical flow in three-dimensional Euclidean space. First, we analysed the effects of the divergence-free condition on the computation of optical flow for three-dimensional distributions in three-dimensional Euclidean space. With this condition, we obtained a system of reaction-diffusion equations with coupling terms. We numerically evaluated the convergence condition for this condition. The accuracy of the solution computed by our algorithm is guaranteed by Lax equivalence theorem which is the basis of numerical computation of the solution for partial differential equations.

We derived the relation between vector regularisation and thin plate constraint for optical-flow computation, that is, we clarified that thin plate constraint is decomposed into four terms, the gradient of divergence, the vector gradient of rotation, and two other terms which express the continuity of vector functions in a space. Thin plate constraint derives diffusion with the biharmonic term. This equation is numerically solved using the same scheme with the usual diffusion-reaction equation for optical-flow computation.

Furthermore, we analysed the validity of the evaluation measure for optical-flow computation and derived mathematical the background for the established error measure.

Acknowledgements

Images for our experiment were provided from Roberts Research Institute at the University of Western Ontario through Professor John Barron. [12] We express our thanks to Professor John Barron for allowing us to use his data set.

References

1. Harvey, W., *Exceritation anatomica de motu cordis et sanguinis animalibus* Francofurti, Sumptibus G. Fitzel 1628, (Japanese Translation, Iwanami 1961).
2. Sachse, B. F., *Computational Cardiology, Modelling of Anatomy, Electrophysiology, and Mechanics*, LNCS 2966 Tutorial, 2004.
3. Ayache, N., ed. *Handbook of Numerical Analysis Vol XII, Special Volume: Computational Models for the Human Body*, Elsevier, 2004.
4. Morel, J.-M., Solimini, S., *Variational Methods in Image Segmentation*, Rirkhaäuser, 1995.
5. Aubert, G., Kornprobst, P., *Mathematical Problems in Image Processing:Partial Differential Equations and the Calculus of Variations*, Springer, 2002.
6. Sapiro, G., *Geometric Partial Differential Equations and Image Analysis*, Cambridge University Press, 2001.

7. Osher, S., Paragios, N., eds., *Geometric Level Set Methods in Imaging, Vision, and Graphics*, Springer, 2003.
8. Horn, B. K. P., Schunck, B. G., Determining optical flow, Artificial Intelligence, **17**, 185–204, 1981.
9. Nagel, H.-H., On the estimation of optical flow: Relations between different approaches and some new results. Artificial Intelligence, **33**, 299–324, 1987.
10. Barron, J. L., Fleet, D. J., Beauchemin, S. S., Performance of optical flow techniques, International Journal of Computer Vision, **12**, 43–77, 1994.
11. Beauchemin, S. S., Barron, J. L., The computation of optical flow, ACM Computer Surveys **26**, 433–467, 1995.
12. Moore, J., Drangova, M., Wierzbicki, M., Barron, J., Peters, T., A High resolution dynamic heart model based on averaged MRI data, LNCS, **2878**, 15–18, 2003.
13. Zhou, Z., Synolakis, C. E., Leahy, R. M., Song, S. M., Calculation of 3D internal displacement fields from 3D X-ray computer tomographic images, in Proceedings of Royal Society: Mathematical and Physical Sciences, **449**, 537–554, 1995.
14. Song, S. M., Leahy, R. M., Computation of 3-D velocity fields from 3-D cine images of a human heart, IEEE Transactions on Medical Imaging, **10**, 295–306, 1991.
15. Weickert, J., Schnörr, Ch., Variational optic flow computation with a spatio-temporal smoothness constraint, Journal of Mathematical Imaging and Vision **14**, 245–255, 2001.
16. Demmel, J.W., *Applied Numerical Linear Algebra*, SIAM, 1997.
17. Grossmann, Ch., Roos, H.-G., *Numerik partieller Differentialgleichungen*, Trubner, 1994.
18. Varga, R.S., *Matrix Iteration Analysis*, 2nd Edn., Springer, 2000.
19. Selig, J. M., *Geometrical Method in Robotics*, Springer, 1996.
20. Suter, D., Motion estimation and vector spline, Proceedings of CVPR'94, 939–942, 1994.
21. Suter, D., Chen F., Left ventricular motion reconstruction based on elastic vector splines, IEEE Trans. Medical Imaging, 295–305, 2000.

5

Detection and Tracking of Humans in Single View Sequences Using 2D Articulated Model

Filip Korč[1] and Václav Hlaváč[2]

[1] University of Bonn, Department of Photogrammetry
   Nussallee 15, Bonn, 53115, Germany

[2] Czech Technical University in Prague, Center for Machine Perception
   Karlovo náměstí 13, Prague 2, 121 35, Czech Republic

Summary. This work contributes to detection and tracking of walking or running humans in surveillance video sequences. We propose a 2D model-based approach to the whole body tracking in a video sequence captured from a single camera view. An extended six-link biped human model is employed. We assume that a static camera observes the scene horizontally or obliquely. Persons can be seen from a continuum of views ranging from a lateral to a frontal one. We do not expect humans to be the only moving objects in the scene and to appear at the same scale at different image locations.

5.1 Introduction and Problem Formulation

Detection of walking or running humans in surveillance video sequences and tracking them is an appealing computer vision problem. The problem has been approached by many researchers. A satisfactory solution, however, has not been presented yet.

This work suggests a model-based approach to tracking in a video sequence captured from a single camera view. Matching of a human model to video stream is performed in two dimensions (2D) on the image frame level. 2D tracking aims at following the image projections of humans. The 3D displacement of a human is perspectively projected into a planar displacement that can be modeled as a 2D transformation. An adaptive model is required to handle appearance changes due to perspective effects or due to the change of the body parts position relative to each other. Two different postures in 3D may result in identical 2D projection. This ambiguity in a 2D model matching approach makes the tracking task rather difficult. The 2D–3D pose estimation problem is further discussed in Chapter 12.

It has been previously argued by Gavrila [1] that the choice of a solution is, to a great extent, application-driven. Our goal is to develop a method which

B. Rosenhahn et al. (eds.), Human Motion – Understanding, Modelling, Capture, and Animation, 105–130.

could track humans in surveillance sequences, i.e., the main focus of our work is
in tracking humans when precise pose recovery is not critical. We assume that
a single static camera observes the scene horizontally or obliquely. We do not
consider a top view. Model-based approach could in longer perspective help to
overcome the assumption of a static observing camera. The walking/running
persons can be seen from a continuum of views ranging from a lateral to a
frontal one. We do not expect humans to be the only moving objects in the
scene and to appear at the same scale at different image locations. The method
copes with a slowly changing background. It is assumed that the whole human
body can be seen in the sequence. A short occlusion of a tracked human by
other objects in the scene is admissible too.

5.2 Our Approach Informally

Our approach is based on a similar philosophy as the seminal paper by
Hogg [2]. We also employ an articulated 2D model of a human consisting
of several rectangles modelling a simplified 2D view of a human body.

The biped model of a human is composed of six rectangles which are fit
to corresponding parts of a human body: head, torso, left/right thigh and
left/right calf, see Figure 5.1a. Our extended model also takes into account
data in the model neighborhood.

Our detection and tracking algorithm is designed as a loop consisting of
three steps: (a) detecting human candidates, (b) validating model of a human,
(c) tracking of the model in consequent frames.

Input of our algorithm is the outcome of pixel-based motion detector (background subtraction). When detecting human candidates in step (a), a coarse
model is used to direct attention to places in the motion image where humans
could be. The outcome is a region of interest (ROI). Detector of ROIs has to
find almost no false negatives.

In the validation step (b) a slightly more sophisticated extended biped
human model, recall Figure 5.1a, is employed within ROI to verify the
appearance of a human.

The model tracking process (c) is launched after successful model initialization. If the model fails to explain image data satisfactorily in the tracking
process then the algorithm is stopped and restarted in the next frame.

The coarse detection of ROI could be viewed as employing a simple model
allowing the search of human candidates in a computationally efficient manner. After the ROI is found, a model with a more refined structure better
explaining the observation could take over.

Such methodology opens a future way to employing models with a varying
refinement of model structure appropriate to available image data resolution
at which a human in the scene is present. This would lead to tracking a
person at a proper level of detail, attempting to track body parts only when

possible, for example. Another potential is in adapting computational demand to changing degree of certainty in the tracked human. This work reports results in which only ROI detection and the extended biped human model are used.

5.3 Related Work

There has been a number of works adopting a model-based approach and an articulated biped model. A survey of publications in human motion analysis can be found in [1]. This survey identifies a number of applications and provides an overview of developments in the domain of visual analysis of human movement preceding the year 1999. Many of the recent works, however, were devoted to more precise body pose recovery and subsequent action recognition in high resolution videos. We feel that there is a number of issues as self-occlusion which need to be further addressed to merely achieve better robustness in tracking of articulated objects in low resolution surveillance videos.

One of the early monocular model-based approaches to human detection and tracking may be found in Rohr [3]. They matched gray value edge lines to volume model contours and employed Kalman filter to estimate the model parameters in the consecutive frames. A person moving parallel to the image plane was assumed to reduce the complexity of the recognition task.

As far as the model is concerned, our work is close to Ju et al. [4]. They adopted a method for tracking articulated motion of human limbs in a sequence using parameterized models of optical flow. They made the assumption that a person can be represented by a set of connected planar patches: "cardboard person model". Constraints introduced between the patches enforce articulated motion. Their approach is limited to constant viewpoint.

Our model is also similar to Zhang et al. [5]. They apply a 2D five-link biped model to the problem of gait recognition using human body movements. Their work is constrained to a human body observed laterally. We took over their model and extended it significantly.

The model posture evaluation and the idea of a rough calibration of a scene to estimate the size of the model in Beleznai et al. [6] was an inspiration for us. Their method adopts a simple human model described by three rectangles to detect individual humans within groups and to verify their hypothesized configuration. The approach is capable of real time operation, and handles multiple humans and their occlusions Beleznai et al. [7].

The model and the focus of our work relates to Lan [8]. They use a pictorial structure model and aim at handling both self-occlusions and changes in the viewpoint. This way of exploiting constraints provided by walking may be considered as a possible improvement in our initialization step. Lan [9] extended the approach by taking limbs coordination into account. Both publications work with silhouette data captured from a single camera viewpoint.

The idea of our model is also related to Lim [10]. They present a multi-cue tracking procedure. Their shape model is comprised of several silhouettes learned off-line from training sequences. The shape model is used along with an appearance model learned online for a given individual. The shape model is learned from a built collection of normalized foreground probability maps of humans. These probability maps are clustered into sets using K-means clustering algorithm. A mean image is then built for each set creating a representation of a pose. Tracking is then formulated as finding a set of warp parameters which map a foreground blob in a frame onto one of the silhouettes in the shape model. Their appearance model could be an inspiration for a further enhancement of our approach.

Another source of inspiration may be found in Howe [11]. The author aims at applying silhouette lookup to monocular 3D pose tracking. A knowledge base of silhouettes associated with known poses is populated. A silhouette extracted from an input frame identifies a set of silhouettes in the knowledge base. Subsequent Markov chaining exploits the temporal dependency of human motion to eliminate unlikely pose sequences. This solution is further smoothed and optimized. The idea of maximizing the per-frame match to the observations and the temporal similarity between successive frames simultaneously may also be applied to further enhance our solution. Adding a term to a cost function employed in our method rewarding a solution close to the one found in the preceding frame would only represent a minor alteration of our algorithm.

Our approach also relates to Collins [12]. They adopted object tracking approach based on color histogram appearance models. In addition, they consider both object and background information to better distinguish an object from its surroundings. Their mean-shift tracking system models object and background color distributions and tracks targets through partial occlusions and pose variations. Their work is restricted to tracking of rigid objects. As opposed to their approach we use mean-shift to fit an articulated model to image data. Our method is based on the motion segmentation computed using the adaptive background model.

Using the methodology introduced in Section 5.2, the model introduced in [6] could be adopted as a model with the least refined structure suitable for tracking persons at a low image resolution. Our model could be viewed as a more refined model, half-way to a model representing all body parts. More complicated models accounting for arm motion and aspiring to higher accuracy seem to be impractical in our context as the computational demand appears to be too high.

Substantial part of this work was done during diploma thesis project of the first author defended in February 2006 at the Czech Technical University in Prague.

5.4 Extended Six-link Biped Model

A human body may project to the image in a variety of forms. The articulated structure of a person makes a vision-based tracking difficult even if a constrained type of motion as walking or running is assumed. The idea behind the model-based approach is exploiting explicit a priori knowledge about the human body appearance in 2D.

A desired human model for surveillance should be simple for computational speed reasons and should enable capturing appearance of a variety of individuals. On the contrary, the model should possess enough structure allowing the expression of distinguishing appearance of a human.

5.4.1 Body Model

We adopted the 2D five-link biped articulated model of [5] and extended it into a six-link model. Our model has added an articulated head and it can also cope with a frontal view of a person. The biped human model consists of six rectangles representing individual body parts. The rectangles are connected by the joints. The model is shown in Figure 5.1a.

Arms are not included in the model as a reliable recovery of exact arm positions of distant pedestrians is often difficult. The biped model is also intentionally kept simple having in mind that low computational demand is desired.

The model is parameterized by eight parameters. Finding a posture T using the biped model M_b means determining its eight parameters, $T = \{C = \{x, y\}, \Theta = \{\theta_1, \ldots, \theta_6\}\}$, where C is the model center and Θ is the inclination vector consisting of angles between the axes of the body part and the vertical axis y as shown in Figure 5.1a.

The biped human model is designed to cope with changing scale which is common in surveillance sequences due to perspective projection from

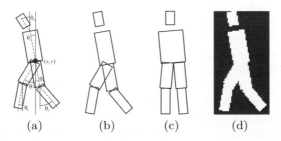

(a)   (b)   (c)   (d)

Fig. 5.1. Six-link biped human model (a) Lateral (b) and frontal (c) model view. Silhouette mask (d) used for evaluation of the extended model.

$3D \rightarrow 2D$. The humans further from the observer look smaller. A scale parameter of the model deals with the foreshortening of a projected human.

Each rectangle of the model is defined by its height l and length of its base b. Each body part is described by $\{\alpha, l\}$, where $\alpha = b/l$ is a base-to-height ratio. Body part heights are normalized with respect to the height of the torso l_1 and the shape model is thus scale-invariant. A shape S of a human is parameterized as follows: $S = \{K, R\}$, where $K = \{\alpha_1, \ldots, \alpha_6\}$ is the base-to-height ratio vector, $R = \{r_1, \ldots, r_6\}$ the relative height vector, $r_i = l_i/l_1$, $i = 1, \ldots, 6$. The six-link biped model is composed as $M_b = \{T, S\}$. The frontal and lateral view is coped with this model. The difference is in parameter α_1 (base-to-height ratio of the torso) and positions of joints connecting the thighs and the torso. The proportions can be seen in Figure 5.1b,c.

5.4.2 Extended Body Model

We extended the biped model to consider both the region explained by the model and the region in the neighborhood of the model which is not represented by the model. A similar idea has been previously applied to a selection of a configuration of models explaining observed clutter of humans [6].

The extended model M is formed by adding a rectangular neighborhood to the biped model M_b. This neighborhood has the height of the biped model M_b. The width of the rectangular neighborhood used is chosen so that it is always possible that the whole walking body fits in the region. The width is calculated as $0.53 \cdot$ height. The rectangular neighborhood and the biped model M_b share a common center.

5.4.3 Model Evaluation

The motion image I is the result of a pixel-based motion detector and constitutes the input to the human detection/tracking system. I is formed as a binary image the values of which tell if the pixel moves/is stationary with respect to the local background, see Figure 5.8a, b, c for examples of the outcome of motion segmentation. The commonly used motion detection algorithm is based on a finite mixture of Gaussians, see Stauffer [13]. This approach can deal with a slowly changing background. The implementation of the motion detection algorithm by Fexa [14] is used.

To evaluate the model in current frame, a silhouette mask is created according to current model view and posture, see Figure 5.1d. This mask is used to compute the amount of missing measurements in the region RB explained by the biped model (white region) and the unexplained observations in the model neighborhood RB (black region). Region RE represents a complement of the region RB within the area of the extended model.

The cost of the current posture of the model M is calculated from the motion image I as

$$C(I \mid M) = \exp\left(-a\left[1 - \frac{1}{A_{RB}\sum_{x,y\in RB}}I(x,y)\right]\right.$$

$$\left.-b\left[\frac{1}{A_{RE}\setminus A_{RB}}\sum_{x,y\in RE\setminus RB}I(x,y)\right]\right) \tag{5.1}$$

where A_{RB} and A_{RE} are the areas of the regions RB and RM, respectively. Scaling parameters a and b were determined experimentally.

5.5 Matching Extended Model Against Data

A model M^* of a human which fits the motion image I best is sought. Using the cost function in Equation (5.1), the problem can be formulated as

$$M^* = \operatorname*{argmax}_{M} C(I \mid M) \tag{5.2}$$

5.5.1 Mean Shift Optimization

Finding globally optimal parameters of the model M^* in Equation (5.2) is computationally complex. We engage the mean shift algorithm, an iterative procedure that shifts a data point to the average of data points in its neighborhood, to find locally optimal solution.

The mean shift idea was introduced by Fukunaga [15]. It was shown that the mean shift vector pointing to a sample mean of local samples generally points in the direction of higher density and thus provides a gradient estimate. The mean shift algorithm was proposed in Fukunaga [16]. The method was then generalized and presented as a mode-seeking process by Cheng [17]. At last, the convergence of the iterated mean shift procedure on discrete data was proved in Comaniciu [18].

We start from some initial estimate of the current stance $\mathbf{v} = (v_1, \ldots, v_8)^T = (x, y, \theta_1, \ldots, \theta_6)^T$. The local mean shift vector represents an offset to a posture vector $\mathbf{v}'$, which is a translation towards the nearest mode according to the cost function in Equation (5.1). The local mean shift vector is computed within the local neighborhood $\{\mathbf{v}_i\}_{i=1\ldots n}$ of the point $\mathbf{v}$. The point $\mathbf{v}_i$ is assigned a weight $C(I \mid M(\mathbf{v}_i))$ according to Equation (5.1), where $M(\mathbf{v}_i) = \{T(\mathbf{v}_i), S\} = \{\{\{v_1, v_2\}, \{v_3, \ldots, v_8\}\}, S\}$.

The new posture vector $\mathbf{v}'$ is computed employing the uniform kernel as

$$\mathbf{v}' = \frac{\sum_{i=1}^{n} \mathbf{v}_i C(I \mid M(\mathbf{v}_i))}{\sum_{i=1}^{n} C(I \mid M(\mathbf{v}_i))} \tag{5.3}$$

Starting from the initial estimate of the current stance, the new posture vector is repeatedly computed until it converges to the sought stance.

5.5.2 Decreasing Computational Complexity

The model M has eight degrees of freedom and so we are faced with a multidimensional optimization problem. An iterative optimization procedure was suggested in Section 5.5.1 to solve the task. However, seeking for an optimal posture in eight dimensional space, i.e., computing an eight dimensional mean shift would still be too time consuming. We decreased the computational demand by both reducing the size and dimension of the considered model parameter space.

The size of the mean shift search space may be reduced in several ways. First, impossible orientations of the individual body parts are considered. Second, we exploit the fact that an observation of walking/running humans is assumed. We thus expect the torso to stay upright and further constrain the ranges of possible orientations of the limbs according to the type of motion assumed. Last, location constraints are considered when initializing the model. This last simplification will be further addressed in Section 5.6.2.

The complexity of the problem is further decreased by reducing the dimension of the considered search space. This is achieved by not optimizing all the parameters at once, but always only a subset of those. Satisfactory results were achieved by splitting the problem into several 2D and one 1D optimization problems instead of considering all eight dimensions. This point will be discussed in more detail in Section 5.6.2.

If a subset of parameters is optimized then only body parts involved are used to evaluate the model, i.e., to create a model silhouette mask. All body parts are used only if the current body posture is evaluated to decide whether the model provides sufficient explanation of seen data and hence if tracking can be started or continued.

5.6 Detection and Tracking

In our algorithm, we first use simple means to find a human candidate, i.e., we focus attention to a ROI. Subsequently, the model is validated at the candidate location. At last, the tracker is started to follow the target in the subsequent frames.

5.6.1 Human Candidate Detection

The attention is focused to the ROI in the motion image I where human candidates could be as illustrated in Figure 5.8a. This step significantly reduces the amount of processed data and contributes to computational efficacy. ROI is detected by correlation of a uniform rectangular mask with the motion image I. The local maxima of the correlation point to moving entities of human size in the image. These maxima are used as centers of ROI.

The local maxima of the correlation refer to moving entities of human size in the motion image. At present, only the strongest correlation maximum is used as the current implementation tracks only a single person.

The height and width of the correlation mask has to be set. We work with expected scaling of humans at a given vertical image location. This information is obtained after a simple calibration performed manually for a given image capturing setup. It is assumed that the scale of a human is linearly dependent on the vertical image location. Such consideration is based on the assumption that a camera provides an oblique view of a planar scene. The more distant humans appear smaller and located higher in the 2D projection.

This step is inspired by [6]. We added the least-squares estimate of the scale function which is based on the height of multiple humans present in the scene at different vertical image positions.

In present implementation, it is also assumed the humans are roughly of the same size. Even though minor height variations should be tolerated by the model at considered scale, taking toddlers into account would require a scale adjustment in the initialization phase.

The width of the mask used in correlation is chosen so that it is always possible that the whole walking body fits in the region. The width is calculated as $0.53 \cdot$ height.

5.6.2 Model Validation

Having detected human candidate, we try to validate our model, i.e., to see whether the model explains data satisfactorily. We first initialize our model at the candidate location. The initialized posture is then further refined in attempt to achieve better fit of the model to data. The resulting pose is evaluated according to Equation (5.1).

Model initialization: After a ROI is detected, a 2D 6-link biped human model is initialized at the candidate location in the motion image I, see Figure 5.8b for illustration. This means that the position of the model and orientation of individual body parts in the image are determined. The pose explaining the data optimally is searched using the mean shift algorithm introduced in Section 5.5.1.

It is convenient to initialize the model incrementally. The torso is initialized first, followed by limbs and head. Finding the initial body posture as a whole would be computationally demanding. Starting from torso leads to insensitivity to unimportant phenomena in the image such as shadows (to be described later).

A subregion corresponding to the torso within the ROI is sought first. This subregion has the width of the ROI. The height and the position of the subregion correspond to the trunk of a person at this scale. The trunk region is initialized in the middle of the selected subregion. The optimal horizontal position and orientation of the trunk is found according to Equation (5.3) within this subregion. The horizontal position of the model is updated.

Detected ROI yields a possible body center of a human. We expect, however, this information to be only a rough estimate of the true position. We try to improve our estimate by handling the data on a finer resolution. For this reason, we do not start by setting the model center to the middle point of the detected human silhouette, as opposed to [5]. We assume that some parts of the segmented region may not belong to the human body. A shadow underneath the body is very common. Torso tends to be less affected by such shadows. The body center update based on torso initialization usually does not take such spurious subregions into account when the body posture is estimated.

After positioning the torso, we proceed by placing the limbs. Orientation of both thighs is determined at once, followed by positioning of both calves. Treating thighs/calves simultaneously yields a better fit. Finally, the head orientation is found.

Initializing the body parts incrementally yields three 2D and one 1D optimization problems instead of one huge 8D problem. The remaining one degree of freedom (vertical position of the whole model) was already initialized when ROI was set. This reduces significantly the problem complexity.

Both frontal and lateral model views are initialized in this step. The model view better explaining the motion image according to Equation (5.1) is chosen.

Posture refinement: The model initialized in the previous step, keeps the vertical position of the previously detected ROI. Its horizontal position is allowed only within the ROI. The posture refinement does not take this limitation into account any longer. We aim to optimize the previously initialized human stance. The possible wrong vertical position of the model estimated in the previous step is refined in this phase.

Individual model parts are again handled separately with the goal of decreasing computational demands. There is a tradeoff between the model/image data fit and the degree of model separation in optimizations.

Satisfactory results are achieved when following iterations are run until convergence. First, the horizontal position of the model is optimized. Second, the vertical position of the model together with the trunk orientation is calculated. Third, both thighs are fit simultaneously. Fourth, both calves are adjusted to the data at the same time. Finally, the orientation of the head is estimated. See Figure 5.8c, d for outcome illustration. The found model is superimposed on both the original data (Figure 5.8d) and the motion image (Figure 5.8c).

The model fails to explain data satisfactorily if the cost, recall Equation (5.1), of the current model posture does not exceed an experimentally estimated threshold. In this case, the ROI location is not considered any more in the current frame.

5.6.3 Tracking

When the initialization of the model is successful then the mean shift tracking is launched in the next frame.

The pose of the model is available from the previous frame. By considering dynamics of human motion, we take the advantage of temporal constraints provided by walking and running and modify the pose. Dynamics are implemented as half a dozen hand coded rules which take periodic motion into account. For instance if the limit position of limbs is reached then the movement in the opposite direction is anticipated. Another rule is meant to anticipate the swing of the calf when both legs are occluded. The speed of this body part is greatest in this phase of a gait cycle and the tracker looses the body part in case the number of frames per second is insufficient. This step aims at anticipating the human pose in the current frame based on the model pose found in the preceding frame. Introducing dynamics helps to overcome the local character of the optimization procedure.

Tracking uses the same iterative procedure described in Section 5.6.2 which starts from the anticipated pose and is repeated until convergence.

It is assumed that the frame rate is sufficiently high so that the person tracked remains in the scope of the mean shift kernel. In case the tracked person is lost the algorithm is restarted.

5.7 Experiments and Lessons Learned from Them

The method was tested on several real sequences, see the CMP Demo Page [19]. The motion detection algorithm [13] provides the sequence of binary images telling which pixels are in motion. This sequence is the input to our implementation. Our model-based tracking is implemented in MATLAB.

5.7.1 Body Model

We choose an articulated biped model without arms. The choice of the employed model can be justified by observing the found model postures superimposed on the silhouette images of tracked subjects.

The extension of the model by the head improves matching the model against image data. The head together with shoulders is a significant characteristics for a frontal human appearance especially if limbs cannot be distinguished in the outcome of the motion segmentation. This feature helps to position the model horizontally, see Figure 5.6b. The improvement of employing the articulated head is noticeable when the person is seen laterally and at the same time the head is not coaxial with the torso. A better fit in vertical direction is found in this case, see Figure 5.3b.

It can be seen that there is very little information in the motion images at considered scale for recovering the position of arms. This task would require using additional features other than segmented motion and would lead to slowing the system down. The position of arms is not crucial in our context. Hence, we do not consider arms in our model. The legs and the head, on the contrary, appear to be distinguishing features. For this reason, we include these parts in the employed model. By observing the silhouette images, it may be verified that the used model corresponds to the level of detail provided in data at the considered scale. See the images in Figure 5.8c for examples.

The frontal model view and the lateral model view are used to fit a person seen from a corresponding perspective. The cost function in Equation (5.1) provides a criterion for choosing a view when facing the task of fitting the model to data on individual frame level. Figure 5.8a, b, c shows examples of initializing the model to motion data. These images depict several poses and both the frontal and the lateral view. In Figure 5.8d, the resulting posture and the chosen model view are superimposed on the original image data. Figure 5.2 illustrates that both the frontal and the lateral model view may be employed to fit a person observed from a range of diagonal views. In Figure 5.2a, b fitting of the frontal model view to a person viewed diagonally may be observed. Figure 5.2c, d depicts how the lateral model view is matched against a human seen diagonally. The motion images are displayed on the left of the original data.

We tested the robustness of the model by fitting it to images of persons walking under different conditions changing slightly their appearance in the motion data. Figure 5.3 illustrates outcome of the experiment. The tested conditions included walking with a backpack (Figure 5.3a, b) and walking with a plant in hand (Figure 5.3c, d). Again, the motion images are displayed on the left of the original data in Figure 5.3. The model was matched properly in both cases. It appears that slight variations in the appearance of the silhouette image are handled properly by the model.

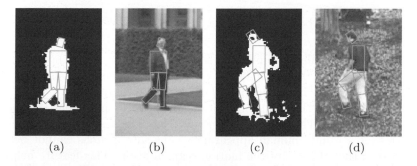

(a)	(b)	(c)	(d)

Fig. 5.2. Fitting the frontal model view (a, b) and the lateral model view (c, d) to a person seen diagonally. The detected model is superimposed on the motion data (a, c) and the original images (b, d). The images of the segmented motion appear on the left of the corresponding original image.

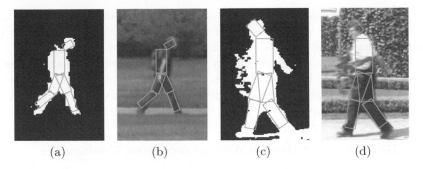

| (a) | (b) | (c) | (d) |

Fig. 5.3. Matching the model against silhouette images of persons walking under different conditions: walking with a backpack (a, b) and walking with a plant in hand (c, d). The detected model is superimposed on both the motion data (a, c) and the original images (b, d). The images of the segmented motion appear on the left of the corresponding original image.

| (a) | (b) | (c) |

Fig. 5.4. Improving accuracy by employing articulated model (a,b). A simple rectangular model employed to find ROI is drawn in cyan. Our articulated human model is drawn in green. Discriminating human appearance from a moving nonhuman object (c).

Our experiments also illustrate that a model capable of accounting for a human articulated structure allows not only to detect a desired object but also provides better accuracy as opposed to one that is less refined. Model center found by the simple rectangular model employed to find the ROI is typically improved by employing articulated model. Figure 5.4a, b shows two detailed views from a sequence with a person walking in a garden (WalkingPersonKinskyGarden.avi, [19]). The detected ROIs and the corresponding model centers are drawn in cyan. Final locations of the models are drawn in green. Frame in Figure 5.4a presents a person seen laterally. The position was improved in the direction of both x (6 pixels) and y (4 pixels) image axes. The height of the model at the ROI location and at the final location was 89 and 88 pixels respectively. Figure 5.4b displays a person seen frontally. The center location was improved in both x and y image axis by 4 and 6 pixels respectively. The height of the model at the ROI location and at the final location was 83 and 82 pixels respectively. Variable scale of the human present in the scene is caused by the perspective foreshortening.

All individuals in our experiments were matched using the same model. It is only the scaling factor that varies. Experimental results illustrate in the presented images that the chosen model may be fit to a variety of individuals at different scales.

5.7.2 Distinguishing Human Appearance from Nonhumans

The ability to discriminate humans from other moving objects present in the scene is illustrated in a sequence which contains a walking person and a small agricultural tractor moving in the background (MovingPersonAnd-TractorWallensteinGarden.avi, [19]). A frame where both the figure and the moving object are far apart is shown in Figure 5.4c. Both moving objects represent two separate regions in corresponding motion image. Current version of the proposed algorithm tracks one individual. However, all detected ROIs were processed by our algorithm in this particular frame. Human appearance has been correctly detected by the model and the moving tractor remained unnoticed. However, it is only the small agricultural tractor that was present in the sequence. This experiment provided thus a relatively easy case to test. Hence, the ability to distinguish humans from nonhumans needs to be further tested on subjects closer to the appearance of persons.

Figure 5.5 shows two different detailed views from the sequence mentioned above. Both the motion segmentation (Figure 5.5a, c) and the original data (Figure 5.5b, d) are provided. The walking person occludes the moving object in the first view (Figure 5.5b). As a result, regions in the motion image (Figure 5.5a) corresponding to the person and the moving object merge into

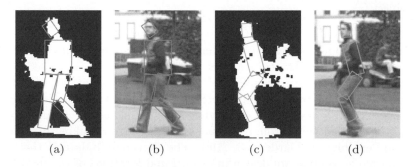

(a)   (b)   (c)   (d)

Fig. 5.5. Detecting a walking person occluding other nonhuman moving object. A rejected model is superimposed on the motion data (a) and on the original image (b). A successfully validated model is superimposed on the motion data (c) and the original image (d). The images of the segmented motion appear on the left of the corresponding image of the original data.

a single blob of foreground pixels. A ROI was found at this location and both model views were initialized. The frontal model view was classified as a view better explaining the observed data. The detected human yields a relatively good fit as far as the region explained by the biped model is concerned. However, there are too many foreground pixels in the biped model neighborhood in this case. As a result, the data is classified as unexplained and the model is rejected. In the second view (Figure 5.5d), the person still partially occludes the moving object. Again, a relatively good fit is found with regard to the region explained by the biped model. However, there are less unexplained observations in the region outside the biped model in this example (Figure 5.5c). The data is thus classified as sufficiently explained and the human appearance is validated. It can be observed that a correct view has been identified in this case.

The pose found by the proposed algorithm may yield a correct pose even in the case the data is classified as unexplained at last. Regardless of the incorrectly chosen view, the pose found in Figure 5.5b may be described as a correct one.

A rejected model may represent a correct pose provided characteristic features are present in the input data. It is the head and the legs that help the algorithm to place the model in this particular case. However, the uncertainty that a human appearance was found in this particular case was considered too high and the detected pose was rejected.

A found posture is rejected whenever the characteristic appearance of the human silhouette is lost in a large foreground blob. Motion data do not allow to discriminate the desired object any longer in this case and the process needs to be supported using other information. An appearance model [10] would provide additional cues on the individual frame level. These two approaches have complementary strengths, and may support each other. See Elgammal 2 for more details on learning representations for the shape and the appearance of moving objects. Introducing a model of human dynamics would help to tackle the problem using temporal information. The dynamical model could further be extended by integrating prior knowledge via an a priori pose distribution. Gall 13 discusses how such pose distribution may be learned from training samples. In the current work, however, only tracking employing the proposed extended human model is implemented. A multi-cue tracking is considered in the future work.

The motion detection algorithm is capable of handling slowly moving background. However, background clutter is still very common in the outdoor motion images. Vegetation such as grass, trees or bushes swaying in the wind is often a cause of such background clutter. It is thus crucial to distinguish the desired object from such spurious artifacts. The ROI detector uses a simple rectangular model to find human-like rectangular regions. A threshold is responsible for accepting the ROI. Such region possesses a portion of foreground pixels which may constitute a human silhouette. A good fit of the model is then possible provided a human is really present. The rectangular model considers the data in the region as a whole. The extended model, on the contrary, provides means to account for the structure of the data. The configuration of the foreground pixels has to be such that most of the pixels appear in the subregion explained by the biped model and minimum elsewhere in the considered region. A human-like silhouette is formed in this case. Our articulated model together with basic kinematic constraints allows to validate variety of human silhouettes. Our experiments suggest that the model is sufficiently robust in the case of background clutter.

5.7.3 Employing Extended Model to Fit Data

The principal idea of fitting the proposed extended model to a motion image is to minimize the amount of missing measurement in the region explained by the model and at the same time the amount of unexplained observations in the model neighborhood. To illustrate how the extended model improves matching the model against data as compared to the biped model without the extension we performed two types of experiments. The model was initialized in a region with a human silhouette. First, we initialized the biped model without the extension. This is achieved by simply neglecting the second term in the cost function in Equation (5.1), i.e, by setting the parameter $b = 0$. The unexplained observations in the model neighborhood have no influence on the value of the cost function in this case. In the second experiment, we initialized the proposed extended model. These two experiments show that employing the extended model brings two main benefits.

First benefit, the model placement is improved. A shadow underneath the human body is common and often causes a significant artifact in the motion image. The head on the other hand is usually less pronounced. A strong shadow and a less pronounced head sometimes lead to a false model placement. This happens in the case when we only consider the region explained by the model. It would result in fitting the model to the strongly pronounced but unwanted region. The second term in Equation (5.1) is responsible for taking the unexplained observation into account. This helps to tackle the problem by lowering the cost of the posture when a shadow is present, helping thus to drive the model away and identifying the head at the same time. Figure 5.6a, b shows a detailed view from the sequence with a person walking in a garden (WalkingPersonKinskyGarden.avi, [19]). Only motion segmentation data are displayed. The images present a rather noisy silhouette of a person seen frontally. A well pronounced shadow region is present underneath the body in the motion image. The model superimposed on the motion data in Figure 5.6a was initialized without the extension. On the contrary, the model displayed in Figure 5.6b was initialized employing the proposed model extension. The position of the extended model was improved by two pixels in vertical direction in this particular case.

Second benefit, the extension of the model improves limbs placement. A fitting procedure is desired which yields a proper position when (a) one leg

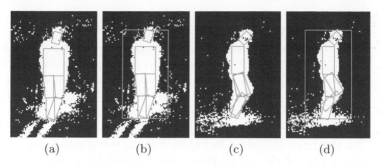

(a)   (b)   (c)   (d)

Fig. 5.6. A slightly improved model placement and limbs localization. Matching model against motion data without (a, c) and with extension (b, d).

occludes the other and also (b) it rewards placing of limbs astride if possible, as illustrated in Figure 5.6c, d. A procedure minimizing only the missing observation in the region explained by the model would lead to a proper position in the case (a), however, it would often result in placing occluded legs also in the case (b). The first term in Equation (5.1) forces the body parts to cover the motion region while minimizing the missing measurement in the region explained by the model and solving thus the case (a). The second term in Equation (5.1) should be insignificant in the case (a) when little unexplained observation is present in the model neighborhood. In the case (b), however, the cost function should force the body parts to cover the region maximally thus lowering the value of the second term. Figure 5.6c, d shows a detailed view from the sequence mentioned in the preceding paragraph. Again, motion segmentation data are displayed. This time, a person seen frontally is present in the images. In this experiment, we aim to illustrate the criterion presented in the Equation (5.1). However, the outcome of both experiments may be difficult to evaluate if the optimization employed in our algorithm is used. If two local optima are found in both experiments, it is difficult to judge what leads to such outcome. The resulting dissimilarity of the results may be caused by the criterion. However, it may also be caused by the local optimization which detected two local optima which are close to each other or far apart. Hence, the model position was fixed and the pose was found by the exhaustive search in this case. This enables to illustrate the fitting criterion and to neglect the influence of the optimization procedure. Figure 5.6c illustrates how the model was initialized without the extension. The model displayed in the Figure 5.6d was initialized with the use of the proposed model extension. Improved placement of the legs may be observed in the case of the proposed extended model. Figure 5.6c, d illustrates that the cost function in Equation (5.1) rewards placing the limbs in a way that covers the foreground region maximally. This helps to find the correct pose on the individual frame level and leads to a better fit in case (b) mentioned above.

5.7.4 Detection and Tracking

Detecting human candidates: The first column of images in Figure 5.8 illustrate the outcome of the ROI detector. It may be observed that a human silhouette is found approximately. This step is clearly influenced by the shadow underneath the body which shifts the detected model center under its true location. Hence, the ROI detector fails to include the head in these cases. In addition, these results illustrate that the horizontal position also needs to be refined. The third row of images in Figure 5.8 has two parallel traces present in the bottom part of the view. These were caused by a sudden change of illumination. Resulting shape of these artifacts was caused by surrounding trees. Such artifacts further influence the task of ROI detection.

Figure 5.7 shows the outcome of the motion segmentation of two frames from the sequence with a person walking in a garden (WalkingPersonKinskyGarden.avi, [19]). The outcome of ROI detection is superimposed on the motion data. The case presented in Figure 5.7a shows a person with legs being apart and the body appearing approximately symmetrical with regard to the vertical image axis. A relatively good estimate of the model center is yielded by the ROI detector in this case. In addition, the shadow underneath the body does not seem to influence the detection in this particular case and most of the region representing the head is included in the ROI. In the second frame displayed in Figure 5.7b, however, the person does not appear symmetrical with respect to the vertical image axis any more. The estimated model center is clearly shifted to the left regarding the true position. Additionally, due to the pronounced shadow underneath the body, part of the head is excluded from the detected region, as shown in Figure 5.7b.

This step provides only a rough estimate of the true position of human candidate.

Model initialization: The ROI detector yields a location where a human could be. The extended human model is then initialized in the detected region.

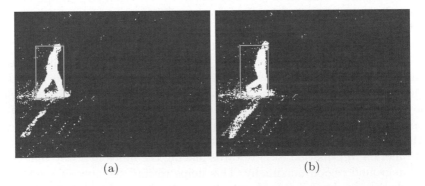

(a)   (b)

Fig. 5.7. Detecting the region of interest (ROI). The found ROI is superimposed on the motion data.

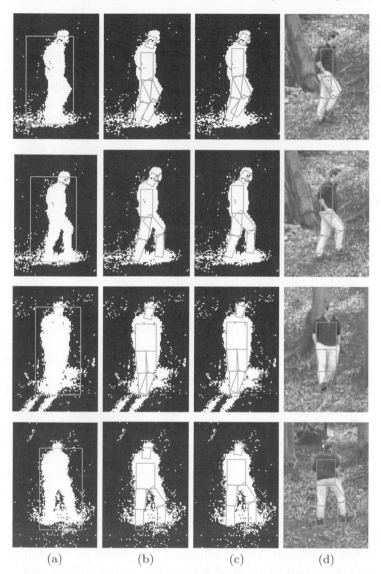

(a)   (b)   (c)   (d)

Fig. 5.8. ROI detection (a), model initialization (b), posture refinement (c) and outcome (d). Images of segmented motion are displayed as binary black and white images. Original data is displayed in the column (d). The found model is superimposed on both types of data. Courtesy A. Fexa for the sequence.

The model initialization step is crucial for several reasons. First, it has to be verified whether a person is present at the candidate location. Second, provided a human is really present in the ROI, initializing the model to data should result in a fit that explains observed data sufficiently. If the above conditions are fulfilled then the tracker is started. Third, the correctly initialized view and posture lead to proper tracking. Initialization procedure that would lead to frequent erroneous rejection of a human candidate would cause the high false negative rate. That is undesirable in the considered application area. For these reasons, it is vital to initialize the model correctly for successive tracking. We tested the functionality of the proposed initialization procedure in three experiments. The outcomes were evaluated by observation.

In the first two experiments we considered models detected in every frame. We looked at several issues. These issues included: portion of frames with a detected person (regardless of the fact whether the model was successfully initialized or rejected), portion of frames with a successfully initialized model, portion of frames with a properly identified model view, portion of frames which have a model with all the parts positioned correctly and at last, portion of frames which contain a model with maximally one misplaced part. A person is considered as detected in case the region representing the human appears mostly in the region of the extended model. A part is classified as correctly matched provided that a greater part of the region representing the part is explained by the corresponding model part.

The process is influenced by a number of aspects with regard to the considered scene and camera setup. Hence, for the first and the second experiment we chose two sequences from our data set with relatively steady conditions that allowed evaluation of the chosen approach. On the other hand, both experiments presented persons carrying an object. We applied the ROI detection and subsequent model initialization to every frame of the test sequence. No information from the preceding frames was propagated to subsequent ones.

Sequence 1 had 137 frames of 360 × 288 pixels each, 25 frames per second and presented a relatively fast walking pedestrian carrying a backpack (WalkingPersonWithBackpackCharlesSquare.avi, [19]). This made the model view choice more challenging. The person walked from the left to the right and was fully visible in 108 frames. The person was observed laterally at a constant scale. The height of the person in the images was 75 pixels. A ROI is initialized when an object is believed to appear fully in the image. For this reason, we only used these frames for a test assessment, i.e., 108 frames. All the detected models were considered in the evaluation including the rejected ones. See Figure 5.3b for detailed view of the tracked person and Table 5.1 for results.

Sequence 2 had 116 frames of 360 × 288 pixels each, 25 frames per second and showed a walking gardener with a plant in hand (WalkingPersonWith-PlantWallensteinGarden.avi, [19]). Again, this should make the model view choice and the fit itself more challenging. First ROI was detected in a frame number 26. Subsequently, ROI was provided in the next 60 frames. The person walked from the right to the left. The person was observed laterally at a constant scale. The height of the person in the image was 150 pixels. Again, only those frames for a test assessment were used in which a ROI was available, here

Table 5.1. Human candidate detection and model initialization. Summary of the experiment where ROI detection and subsequent model initialization are applied to every frame separately. Table shows number of frames, portion of frames where ROI has been detected, portion of frames where model has also been success- fully initialized, portion of frames with correctly identified model view, portion of frames where model with all its parts is positioned correctly and at last, portion of frames which contain a model with maximally one misplaced part. A person is present in all considered frames.

Test sequence	Frames	Detected	Initialized	Correct view	All parts	Max. 1 part
Sequence 1	108	100%	100%	99%	96%	100%
Sequence 2	60	100%	100%	98%	87%	100%

60 frames. All the detected models were used in evaluation. Figure 5.3d shows detailed view of the tracked person and Table 5.1 results of the experiment.

The purpose of the third experiment was to learn how the individual steps of our algorithm preceding the tracking itself contribute to a resulting initial fit. Another goal was to assess the overall functionality of the initialization step under more varying conditions. The sequence which has a person walking in a garden (WalkingPersonKinskyGarden.avi, [19]) was found to be appropriate for the task. The tested conditions included variable scale, changing viewpoint, present shadows, background clutter and noisy data yielded by the motion detector. The scale of the person in these experiments varied from 82 to 88 pixels. An overview of the experiment is given in Figure 5.8.

The first column of images in Figure 5.8 shows the outcome of the ROI detector. These views illustrate how the detection was influenced by shadows, background clutter and noise in the data. Results of ROI detection shown in the first column of images in Figure 5.8 suggest that the model center needs to be refined in the direction of both image axes.

Views in the second column in Figure 5.8 depict how the model was initial- ized in the ROI. Typically, the horizontal position of the model is improved and the first rough fit of the model is found in this step. However, the model vertical position is fixed in this phase and keeps the values yielded by the ROI detector. The model view was correctly chosen in these examples.

As previously explained, the vertical position yielded by the ROI detector needs to be corrected in the presented examples. The third column of images in Figure 5.8 shows the final fit of the model after the refinement step. The final model posture superimposed on the original data may be seen in the last column. Comparison of these images with the ones illustrating the rough initialization reveals that it is the vertical location in the first place that was refined in this step. Further inspection shows that orientations of the remaining body parts have also been slightly improved. See the second and the third image in the first row in Figure 5.8 for example. The vertical location was clearly improved in this case. This instance further illustrates the role of the model extended by the head when dealing with unwanted phenomena such as shadows. The model was driven away from a well pronounced region to a region less significant in the image. However, a location was found which better corresponds to the employed model and the head was thus identified. These two views also illustrate how, for instance, the orientation of the front calf was improved in the refinement step.

Tracking: Our data set contains videos of six individuals, who are walking in the outdoor environment under varying conditions: walking with a backpack, walking with a plant in hand, walking with a jacket over the shoulder and walking with a small agricultural tractor moving in the background. The types of motion tested included slow walk, fast walk, running, standing still shortly and their transitions. Both tracking at the constant and slightly variable scale was tested. Our tracking algorithm was presented with persons seen in continuum of views ranging from frontal to lateral. In addition, view transitions were tested. At last, the experiments illustrate how partial and full occlusions are handled. The sequences were processed by the proposed tracking algorithm, see CMP Demo Page [19], all using the same parameter setting (the view calibration was different for each scene, of course).

Let us further demonstrate the results on two sequences in more detail. It has been already said that a model part is classified as correctly matched provided that a greater part of the region representing the projected body part is explained by the corresponding model part. Results were again evaluated by observation. Only those frames were used for a test assessment, where a ROI was available.

In the first experiment, the proposed tracking algorithm was tested on the sequence which has a relatively fast walking pedestrian carrying a backpack (WalkingPersonWithBackpackCharlesSquare.avi, [19]). This sequence was previously introduced as Sequence 1.

The person has been correctly detected and the model initialized in the first considered frame. The algorithm tracked the person properly throughout the sequence. No reinitialization of the tracking procedure was needed. All the model parts were matched correctly in 98% of the frames. In the remaining 2% of the frames, only one body part was classified as being not positioned properly. However, these cases were still found to be close to the correct position.

In the second experiment, the tracker was tested on the sequence which showed a walking gardener with a plant in hand (WalkingPersonWithPlant-WallensteinGarden.avi, [19]). This sequence was previously introduced as Sequence 2.

The person has been correctly detected and the model initialized in the first frame. The algorithm tracked the person properly in the next 53 frames. The tracker reinitialized twice at the end of the sequence. The first reinitialization happened due to the wrong position of a leg. The second reinitialization was due to many unexplained observations in the motion data caused by the carried object. The tracking process continued afterwards. All the model parts

Table 5.2. Tracking. Summary of the experiment where ROI detection, model initialization and subsequent tracking are applied. Table shows number of frames, number of times tracking has been reinitialized, portion of frames where model with all its parts is positioned correctly and at last, portion of frames which contain a model with maximally one misplaced part. A person is present in all considered frames.

Test sequence	Frames	# of reinitializations	All parts	Max 1 Part
Sequence 1	108	0	98%	100%
Sequence 2	60	2	91%	100%

were matched correctly in 91% of the frames. Only one body part was classified being positioned improperly in the remaining 9% of the frames. It appears that an incorrect scale of the person determined during the calibration step is the cause of the misplaced parts. It was mostly the head that appeared above its true position. Smaller model may yield a better fit in these cases. Results of the the two experiments are summarized in Table 5.2. Processed sequences can be viewed at the CMP Demo Page [19].

The sequence which has a person walking in a garden (WalkingPersonKinskyGarden.avi, [19]) was found appropriate for testing slightly variable scale, changing viewpoint while walking or turning, present shadows, background clutter and noisy data yielded by the motion detector. The scale of the person in the experiment varied from 82 to 88 pixels. Variety of conditions mentioned above were satisfactorily handled by the proposed algorithm. Results may be observed at the CMP Demo Page [19].

5.8 Conclusion

This work contributes to the 2D model-based whole body tracking in motion images. The extended biped model of humans was proposed. It has been shown that the model is general enough to capture the appearance of a variety of individuals. Our experiments also suggest that the model possesses enough structure to express the distinguishing characteristics of humans observed in the motion images. Model parametrization is scale independent and allows tracking of a person at a variable scale. The frontal and the lateral views are treated in a single framework. As a result, a person tracked by our method may be seen from continuum of viewpoints, as opposed to [3–5].

The model to data fit criterion considers both motion image data in the region explained by the biped model and in its neighborhood. Both the number of missing measurements in the foreground and the number of unexplained observations in the background are minimized simultaneously. This provides an inherent mechanism which allows it to cope with limbs placement in the case of self-occlusion. As opposed to [12], where the use of background is also made, we do not restrict ourselves to tracking of rigid objects. When tracking, both the foreground and the local background need to be explained at the same time. This provides means for tracking assessment and dropping the tracking in case the target is occluded by a large still or moving object. Such fit criterion also copes with spurious artifacts such as shadows.

Besides tracking, our solution addresses both target detection and tracking initialization. We employ a computationaly unexpensive method first to focus attention to a region of interest before more refined model is used to validate human appearance. This allows to formulate tracking as a local mean shift optimization and decrease computational complexity of articulated tracking. As opposed to [10], our approach is inherently model-based and does not require training sequences and offline training.

Our algorithm aims at tracking of humans in low resolution videos. Processed sequences, see [19], illustrate the capability to track humans under realistic outdoor scene conditions. A person is tracked under variable illumination and with shadows present. Oclusions, slightly variable scale are handled, basic discrimination between a human and a nonhuman object has been illustrated. As opposed to [6,7], we do not assume humans to be the only moving objects in the scene and provide means for discrimination between moving human and nonhuman. A person walking, stopping, turning around is successfully tracked. Last, our experiments illustrate that the approach is robust to typical conditions changing slightly human appearance.

As the future work, we propose to test the approach more extensively, improve the dynamics by treating both abrupt changes in motion as well as smooth transitions. Biped model pose initialization can be improved by taking coordination between limbs [9] into account. This could suppress the ambiguity caused by projection of 3D world into 2D images. Blake [22] suggests how a dynamical model can be learned.

Basic multi-person tracking may already be achieved within current framework. This would mean using our algorithm to initialize multiple ROIs and tracking these models in the consecutive frames as described in the text. The tracker does not try to fit the model to a large foreground blob that can not be explained using single model. As a result, tracking would automatically be dropped in case of occlusions and started after the tracked person appears unoccluded again. This is a desired behavior when using the motion information only as the segmentation does not provide means to distinguish human appearance in this case. Occlusions would thus be handled in a naive yet sensible way.

The tracking through occlusions can be supported by adopting an appearance model described in [10, 12] and by utilizing the temporal information according to [11]. Initializing multiple models to a foreground blob representing a clutter of persons is another possible extension of the multi-person tracking. For instance, the approach to detection of humans within groups and verification of their hypothesized configuration described in [6] could be extended using articulated models.

In future work, we also intend to test the approach more extensively on data-sets that allow comparison with different tracking algorithms. Hence, we consider testing the proposed method on both the USF Gait Challenge data-set [20] and the CMU MoBo data set [21].

The time performance has not been tested in the current work. Further experiments are needed to evaluate the potential of the chosen approach for real time applications. Our current implementation is in MATLAB. We consider re-implementation in C++ and time performance testing. We would like to learn in what degree our more complicated model as compared to [6] slows the system down and, on the other hand, brings better accuracy and robustness.

Acknowledgments

We would like to thank to Aleš Fexa for motion detection code and one of the image sequences. The first author is grateful to Dr. Csaba Beleznai from Advanced Computer Vision, GmbH., Vienna, Austria who supervised his summer project in person detection in summer 2005. Both authors were supported by the EC projects FP6-IST-004176 COSPAL, INTAS 04-77-7347 PRINCESS, MRTN-CT-2004-005439 VISIONTRAIN and by the Czech Ministry of Education under project 1M0567. Any opinions expressed in this chapter do not necessarily reflect the views of the European Community. The Community is not liable for any use that may be made of the information contained herein.

References

1. Gavrila, D. M.: The Visual Analysis of Human Movement: A Survey. In *Computer Vision and Image Understanding*, **73**:82–98, 1999.
2. Hogg, D. C.: Model-based Vision: A Program to See a Walking Person. In *Image and Vision Computing*, **1**:5–20, 1983.
3. Rohr, K.: Incremental Recognition of Pedestrians from Image Sequences. In Proc. *IEEE Conference on Computer Vision and Pattern Recognition*, 8–13, 1993.
4. Ju, S. X., M. J. Black and Y. Yacoob: Cardboard People: A Parameterized Model of Articulated Image Motion. In Proc. *IEEE International Conference on Automatic Face and Gesture Recognition*, 38–44, 1996.
5. Zhang, R., C. Vogler and D. Metaxas: Human Gait Recognition. In Proc. *IEEE Conference on Computer Vision and Pattern Recognition Workshop*, **1**: 18, 2004.
6. Beleznai, C., B. Frühstück and H. Bischof: Human Detection in Groups Using a Fast Mean Shift Procedure. In Proc. *IEEE International Conference on Image Processing*, 349–352, 2004.
7. Beleznai, C., B. Frühstück and H. Bischof: Tracking Multiple Humans Using Fast Mean Shift Mode Seeking. In Proc. *IEEE International Workshop on Visual Surveillance and Performance Evaluation of Tracking and Surveillance*, 25–32, 2005.

8. Lan, X. and D. P. Huttenlocher: A Unified Spatio-temporal Articulated Model for Tracking. In Proc. *IEEE Conference on Computer Vision and Pattern Recognition*, **1**:722–729, 2004.

9. Lan, X. and D. P. Huttenlocher: Beyond Trees: Common-Factor Models for 2D Human Pose Recovery. In Proc. *IEEE International Conference on Computer Vision*, **1**:470–477, 2005.

10. Lim, J. and D. Kriegman: Tracking Humans Using Prior and Learned Representations of Shape and Appearance. In Proc. *IEEE International Conference on Automatic Face and Gesture Recognition*, 869–874, 2004.

11. Howe, N. R: Silhouette Lookup for Automatic Pose Tracking. In Proc. *IEEE Conference on Computer Vision and Pattern Recognition Workshop*, **1**:15–22, 2004.

12. Collins, R., Y. Liu and M. Leordeanu: On-Line Selection of Discriminative Tracking Features. In Proc. *IEEE Transaction on Pattern Analysis and Machine Intelligence*, **27**:1631–1643, 2005.

13. Stauffer, C. and E. Grimson: Adaptive Background Mixture Models for Real-time Tracking. In Proc. *IEEE Conference on Computer Vision and Pattern Recognition*, **2**:246–252, 1999.

14. Fexa, A.: Separation of individual persons in a crowd from a videosequence. Diploma Thesis, Faculty of Mathematics and Physics, Charles University in Prague, July 2004.

15. Fukunaga, K.: *Introduction to Statistical Pattern Recognition*. Academic Press, New York, 1972.

16. Fukunaga, K. and L. Hostetler: The Estimation of the Gradient of a Density Function, with Applications in Pattern Recognition. In Proc. *IEEE Transactions on Information Theory*, **21**:21–40, 1975.

17. Cheng, Y.: Mean Shift, Mode Seeking, and Clustering. In Proc. *IEEE Transaction on Pattern Analysis and Machine Intelligence*, **17**:790–799, 1995.

18. Comaniciu, D., V. Ramesh and P. Meer: Real-Time Tracking of Non-rigid Objects Using Mean Shift. In Proc. *IEEE Conference on Computer Vision and Pattern Recognition*, **2**:142–149, 2000.

19. CMP Demo Page: http://cmp.felk.cvut.cz/demos/Tracking/TrackHumansKorc/. Center for Machine Perception, Czech Technical University in Prague, 2006.

20. Phillips, P. J., S. Sarkar, I. Robledo, P. Grother, K. W. Bowyer: The Gait Identification Challenge Problem: Data Sets and Baseline Algorithm. In Proc. *IEEE Conference on Computer Vision and Pattern Recognition*, **1**:385–388, 2002.

21. Gross, R. and J. Shi: The CMU Motion of Body (MoBo) database. Technical Report CMU-RI-TR-01-18, Robotics Institute, Carnegie Mellon University, June 2001.

22. Blake, A., B. North and M. Isard: Learning Multi-class Dynamics. In Proc. *Conference on Advances in Neural Information Processing Systems*, 389–395, 1999.

Part II

Learning

6

Combining Discrete and Continuous 3D Trackers

Gabriel Tsechpenakis[1], Dimitris Metaxas[1], and Carol Neidle[2]

[1] Center for Computational Biomedicine, Imaging and modelling (CBIM)
   Computer Science Department, Rutgers University
[2] Linguistics Program
   Department of Modern Foreign Languages and Literatures, Boston University

Summary. We present a data-driven dynamic coupling between discrete and continuous methods for tracking objects of high *dofs*, which overcomes the limitations of previous techniques. In our approach, two trackers work in parallel, and the coupling between them is based on the tracking error. We use a model-based continuous method to achieve accurate results and, in cases of failure, we reinitialize the model using our discrete tracker. This method maintains the accuracy of a more tightly coupled system, while increasing its efficiency. At any given frame, our discrete tracker uses the current and several previous frames to search into a database for the best matching solution. For improved robustness, object configuration sequences, rather than single configurations, are stored in the database. We apply our framework to the problem of 3D hand tracking from image sequences and the discrimination between fingerspelling and continuous signs in American Sign Language.

6.1 Introduction

The 3D shape estimation and tracking of deformable and articulated objects from monocular images or sequences is a challenging problem with many applications in Computer Vision. In this chapter we present a novel framework for 3D tracking, which uses a learning-based coupling between a continuous and a discrete tracker. To demonstrate its performance, we apply our generic method to the case of 3D hand tracking, since hands are a difficult case of articulated objects with high *dofs*, with frequent self-occlusions and complex motions (strong rotations etc.).

There are generally two major approaches in deformable and articulated object tracking: (i) *continuous* (or *temporal*) methods that use both temporal and static information from the input sequence, and (ii) *discrete* methods, which handle each frame separately, using only static information and some kind of prior knowledge.

For the case of 3D hand tracking, several techniques treat the hand configuration estimation as a continuous 3D hand tracking problem [9,10,12,27].

133

B. Rosenhahn et al. (eds.), *Human Motion – Understanding, Modelling, Capture, and Animation*, 133–158.
© 2008 *Springer*.

A possible drawback of some of them is that they introduce error accumulation over time, leading to the loss of track, and when this occurs, they usually cannot recover. This is the reason why several shape estimation techniques have been developed in the past few years [4, 13], treating each frame independently from the previous ones, although these techniques usually require higher computational times.

Both continuous and discrete methods for 3D hand tracking can be divided into two main classes: (a) model-based [8, 10, 16], where 3D hand models are constructed and a matching takes place between the input image features and the respective features of the model projection onto the image plane, and (b) appearance-based approaches [14, 19], which involve mapping of the image feature space to the hand configuration space.

In the past few years, some approaches that use hand configuration databases have been proposed [3, 17, 20] in which the 3D hand pose estimation problem is converted into a database indexing or learning problem. The main problem that arises in these methods, apart from the computational complexity, is that multiple matches between the input hand image and the database samples may occur. In our proposed discrete method, the database consists of hand configuration sequences, instead of single configurations. In this way, we impose a temporal continuity constraint: we take into account not only the current hand configuration but also the recent past, i.e., which sequence of configurations this hand shape resulted from.

Another problem tackled by some methods [2, 4, 19, 21], is the background complexity, i.e., discrimination between the hand and the background edges, when using edges as the visual cues for hand configuration estimation. The method of [21] is based on Elastic Graph Matching (EGM), using a combination of features, such as edges and texture. In [4, 19] matching between the input edge map and shape templates, derived from a database, is used to find the best matching hand edges. In [2], the Hidden State Shape Models (HSSMs), similar to the Hidden Markov Models (HMMs) are introduced, also using prior knowledge (shape templates) for training purposes. In our work, we apply a skin classification method, namely a Support Vector Machine (SVM) to model the skin color. In this way, we manage to maintain the skin regions and the hand edges, rejecting most of the background edges, as described in Section 6.2.4, without the time consuming template matching. In cases where the tracked hand is close to or occludes another skin region, e.g., the face, which actually happens often in ASL videos, we also rely on the temporal continuity constraints of either the model-based continuous tracker [10] or our discrete tracker.

In this chapter, we present a new framework for robust 3D tracking, to achieve high accuracy and robustness, combining the aforementioned advantages of the continuous and discrete approaches. Our approach consists of a data-driven dynamic coupling between a continuous tracker and a novel discrete shape estimation method. Our discrete tracker utilizes a database, which contains object shape sequences, instead of single shape samples, introducing a temporal continuity constraint. The two trackers work in parallel, giving solutions for each frame separately. While a tightly coupled system would require high computational complexity, our framework chooses instantly the best solution from the two trackers, based on an error. This is the *actual* 3D error, i.e., the difference between the expected 3D shape and the estimated one. When tracking objects with high *dofs* and abrupt motions, it is difficult to obtain such 3D information, since there is no ground-truth shape available. In our framework, we learn off-line the 3D shape error, based on the 2D appearance error, i.e., the difference between the tracked object's edges and the edges of the utilized model's projection on the image plane. For this learning we use Support Vector Regression (SVR).

We apply our framework to 3D hand tracking, with particular attention to the case of American Sign Language (ASL) analysis. So far, only 2D tracking schemes have been applied to ASL and the recognition is based on estimated 2D features, which serve as observations for learning schemes such as HMMs [24,25]. In our work, we use the obtained 2D and 3D information for the discrimination between fingerspelling and non-fingerspelled signs in ASL, which is a crucial problem in the recognition task. This is because the internal linguistic structure of these two types of signs differs significantly, and thus the strategies required for recognition of these signs must differ accordingly. Also, we chose to use this application for our approach because of the large fingerspelling segmentation ground-truth available. (The video data and annotations are available to the research community from http://www.bu.edu/asllrp/cslgr/.).

This chapter is organized as follows. In the next subsection, we give a brief description of previous work on 3D hand tracking. In Section 6.2 we describe our framework and its individual parts: (a) the existing continuous tracking we use, in 6.2.1, (b) our discrete tracking method with its individual parts, in 6.2.2, namely the database, the 2D single-frame features, the multi-frame features, and the matching between the database samples and the input frames, (c) the dynamic coupling in subsection 6.2.3, and (d) the feature extraction in cluttered backgrounds in 6.2.4. In Section 6.3 we present our results on the 3D hand tracking, focusing on the case of ASL, including fingerspelling discrimination, in 6.3.1. Finally, in Section 6.4 we describe our conclusions and our future work.

6.2 Our Approach

The parts that our method consists of are: (i) a continuous tracker, (for the case of hand tracking we use the model-based tracker described in subsection 6.2.1), (ii) our novel discrete tracker, described in subsection 6.2.2, and (iii) the error-based coupling between the two trackers, described in subsection 6.2.3.

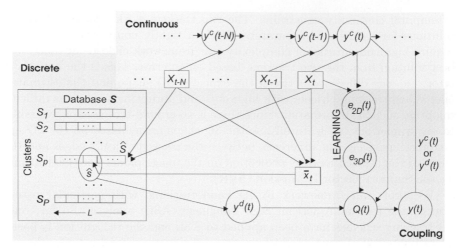

Fig. 6.1. Graphic representation of our overall scheme: (i) model-based continuous tracker giving solution $y^c(t)$ at an instance t, from the previous solution $y^c(t-1)$ and the input observation X_t, (ii) our discrete tracker that utilizes a clustered database **S**, consisting of shape sequences s, and uses the $N+1$ most recently received observations, and (iii) the data-driven coupling of the two trackers, based on the 3D error of the continuous tracker.

Our framework is shown in Figure 6.1. A continuous tracker gives solution $y^c(t)$ at any time t for the tracked object's 3D shape, from the previously estimated shape $y^c(t-1)$ and the input 2D image cues X_t. In parallel, a discrete tracker gives solution $y^d(t)$ at each time t independently. For this module, a database **S** is constructed, consisting of short shape sequences of the object's template. For search efficiency purposes, the database is clustered manually (off-line) into P clusters of L sample sequences s. Each cluster is represented as S_p, where $p = 1, \ldots, P$. At each time t, we first find the best matching cluster $\hat{S}$, where the tracker's solution is located. According to the properties of the tracked object we can use different geometric features from the recent $N+1$ frames for the cluster selection . In our application of the 3D hand tracking, the current observation X_t and the observation at time $t-N$ are used to find the most appropriate database cluster $\hat{S}$, as explained in Sections 6.2.2 and 6.2.3. Then, the $N+1$ most recent observations $(X_{t-N}, \ldots, X_t)$ are used to describe the shape changes over the $N+1$ most recent input frames. These observations are integrated into a quantity $\bar{x}_t$, after dimensionality reduction. As will be explained below, dimensionality reduction is used for matching efficiency, without significant loss of information. This quantity is used to search into the located cluster $\hat{S}$ for the best matching shape sequence $\hat{s}$. After $\hat{s}$ is found, the configuration of the last shape of this sample sequence is derived as the solution $y^d(t)$ of the discrete tracker at time t.

For each time t, two independent solutions are being derived from a continuous and a discrete tracker. The third part of our framework is the selection of the best solution, according to an error indication; that is, based on an indication, when the continuous tracking fails, we use the discrete method for model re-initialization. The continuous tracker's solution (3D model) is projected onto the image plane, and the difference between the tracked object's 2D shape and the 2D model projection is calculated. This is what we call 2D tracking error $e_{2D}(t)$ and it will be defined below. As we show in Section 6.2.3, this 2D error is not a sufficient indication of how well a 3D tracker performs, since a small 2D error may correspond to two completely different 3D shapes. To obtain the actual 3D shape error, we apply off-line learning (Support Vector Regression) for the correspondence between 2D and 3D errors, exploiting information from the database used for our discrete tracker. Having the 3D error of the continuous tracker, we calculate the *switching* parameter $Q(t) = \{-1, 1\}$ (switching between trackers at time t). According to the value of $Q(t)$, we either use the continuous $(Q(t) = 1)$ or the discrete tracker $(Q(t) = -1)$.

The main advantage of having two independent solutions, and choosing one of them as final result, is the lower complexity, compared to a more tightly coupled framework; in a tightly coupled system, the solution of one tracker would depend on the performance of the other, or the final solution would be a fusion of the two independent solutions. Both cases would need explicit description of which part of the object is tracked well with one or the other method. Also, learning the correspondence between the 2D and 3D errors, with prior knowledge about the continuous tracker's failures, is a more robust approach than just using a threshold for the directly calculated 2D tracking error.

Finally we should note that our generic framework could be used for different types of trackers. In our application we used our existing model-based continuous tracker of [10], and our novel discrete tracker that is explicitly described in this chapter. We developed the latter such that it can work with efficiency in the specific framework. One could modify this scheme using a different continuous method or using another discrete tracker such as the methods of [3, 17].

6.2.1 Continuous Tracking

In the model-based continuous tracking of [10] that we use in our framework, 2D edge-driven forces, optical flow, and shading are computed. They are converted into 3D features using a perspective camera model, and the results are used to calculate velocity, acceleration, and the new position of the hand. A Lagrangian second order dynamic hand model is used to predict finger motion between the previous and the current frame. A model shape refinement process is also used, based on the error from the cue constraints, to improve the fitting of the 3D hand model onto the input data.

At each frame visibility checking is performed in order to match correctly image and model points. The computation of the relative motion to the palm of occluded fingers, is based on the rigid motion of the hand. When the relative motion is not too large, the finger edges are picked up when they reappear. This method will fail when the fingers undergo significant relative motions when occluded, and when the hand undergoes strong rotations. Since this method actually belongs to the existing literature, we will not further describe its details; the reader is therefore referred to the literature [10].

6.2.2 Discrete Tracking

For each input frame at time t of the examined sequence, we extract 2D features X_t (what we have called *observations* so far), which will be used to describe the current frame. From now on, we will call these features *single-frame features*. Also, these features will be integrated with the respective features of a number of past frames, $X_{t-1}, \ldots, X_{t-N}$, to serve as multi-frame descriptors $\bar{x}_t$, what we will call *multi-frame features*.

According to our discrete method, instead of matching between single images as, e.g., in [3, 17], we perform multi-frame matching between the most recent input frames and the samples from our synthetic hand database, described in the next paragraph. The database search is efficient, when the discrete tracking is used in our integrated framework, as illustrated in Fig. 6.1. We search our clustered database in two steps: (i) according to a subset of the single-frame features of X_t and X_{t-N}, namely the *projection information* and the *finger counting* result (as described in 6.2.2), we find the most appropriate database cluster $\hat{S}$, and (ii) using the multi-frame features $\bar{x}_t$, we search for the best matching sample sequence $\hat{s}$ inside the chosen cluster $\hat{S}$. The last hand configuration is chosen as the solution $y^d(t)$ of the discrete tracker for the input frame at time t. In this way, we are able to exclude fast a large number of samples from our database and locate the solution in a small subset (cluster). Also, taking into account the most recently estimated hand configurations, we impose a temporal continuity constraint, which is not as strict as in the case of the continuous tracker; therefore, we avoid multiple matches from the database (characteristic of most of the existing discrete methods) without any additional computational load.

The Database

Our synthetic hand model has 20 *dofs*, as shown in Figure 6.2, and its advantage is the good skin texture, which can be used for hand edge extraction.

Similar to all of the discrete tracking approaches that utilize configuration databases, the tracker's accuracy and efficiency depend on the size of the database, i.e., how many hand configuration sequences are included. To demonstrate our framework, we have created 200 configuration sequences, under 29 views, and each sequence has $N + 1$ frames. The choice of the sample length $N + 1$ depends on the temporal information we want to take under consideration. As explained below, we need N to be as small as possible for

Fig. 6.2. Synthetic hand model with 20 *dofs* describing all possible articulations.

computational efficiency, but if N is too small we cannot obtain sufficient temporal information. In our experiments we used $N + 1 = 5$ for $30fps$ videos, which turned out to be optimal for both slow and fast hand movements/articulations.

For each sample s we have stored the $N + 1$ joint angle sets, each set containing 20 angles, corresponding to its successive hand configurations. We have also extracted and stored (i) single-frame and (ii) multi-frame features of each configuration sequence, as described below.

The database $\mathbf{S}$ is organized according to which side of the hand is visible (projection information) in the last frame, and how many fingers are visible in the first and last frame of each sample sequence. Thus, we have divided our database into $P = 108$ clusters S_p ($p = 1, \dots, P$), containing $L = 54$ samples each, on average.

Figure 6.3 illustrates an example of three samples of the virtual hand database. Each sample is a hand configuration sequence of length $N + 1 = 5$, and all of these samples ((a)–(c)) correspond to the same hand gesture (under different camera views). As mentioned above, the database is clustered manually according to how many fingers are visible in the first ($\sharp 1$) and last ($\sharp 5$) frame, and which view of the palm is visible in the last ($\sharp 5$) frame. Thus, these three samples are assigned to different clusters, although they correspond to the same hand gesture. Specifically, as will be explained in the following paragraph, the last two samples ((b) and (c)) are assigned to the same cluster, different from the cluster that the first sample (a) belongs to. We consider that in both (b) and (c) samples there are all the fingers visible in the first frame, only two fingers are visible in the last frame, and the knuckles view is visible in the last frame. On the other hand, for the first sample, only three fingers are visible in the first frame, one finger is visible in the last frame and the palm is in its side view in the last frame. We calculate the number of fingers that are visible using the 2D hand shape as described below. The palm view is assigned manually.

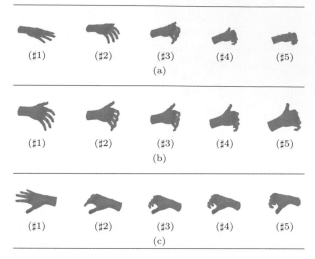

Fig. 6.3. Three different sample configuration sequences of the virtual hand database: (b) and (c) belong to the same cluster, while (a) belongs to a different cluster from the other two.

2D Single-frame Features

From every input frame, we extract both boundary and region-based features of the captured hand.

1. *Boundary-based:*

Let $C(i) = (x(i), y(i))$ be the extracted hand contour on the image plane, where $(x(i), y(i))$ are the cartesian coordinates of each contour point, with $i = 1, \ldots, I$. The curvature function of the contour is

$$K(i) = \frac{\dot{x}(i)\ddot{y}(i) - \ddot{x}(i)\dot{y}(i)}{[\dot{x}^2(i) + \dot{y}^2(i)]^{3/2}}, \tag{6.1}$$

where $\dot{x}(i)$, $\dot{y}(i)$ and $\ddot{x}(i)$, $\ddot{y}(i)$ are the first and second derivatives at location i respectively. We locate the contour zero-crossings

$$z = \{i \in [1, I-1] : K(i) \cdot K(i+1) < 0\}, \tag{6.2}$$

i.e., the points where the curvature changes sign. We evolve the contour using gaussian smoothing in different scales σ:

$$x'(i) = x(i) * g(i, \sigma), \ \ y'(i) = y(i) * g(i, \sigma), \tag{6.3}$$

where $g(i, \sigma) = \frac{1}{\sigma \cdot \sqrt{2\pi}} \cdot e^{-\frac{i^2}{2\sigma^2}}$, and for each scale we locate the new zero-crossings. The evolution stops when no zero-crossings are found. We extract the Curvature Scale Space (CSS) map [1, 11], and its peaks (most important maxima) indicate the most important zero-crossings of the contour, i.e., the most preserved zero-crossings in the different scales. In this way curvature is an efficient shape descriptor and the fact that it is not affine invariant allows us to view shape changes under complex rotations and scaling.

2. *Region-based:*

Let E be the edge map of an input frame, extracted using the canny edge detector. For all edge points (x_e, y_e) of the hand region, we calculate the orientations $\vartheta_i \in [0, \ldots, 180)$ and we extract the histogram H_ϑ, as in [4, 19], using a fixed number B of bins. In our application we used $B = 50$. The orientation histogram can provide us with information about the edges in the interior of the hand, especially when we have sequences of high resolution. We should note that the number of edges in each orientation is normalized, based on the total number of edges detected; in this way, we avoid the effects of both missing edges and scaling that results to different edge resolution.

3. *Projection information:*

From the currently estimated hand configuration of the input video, we obtain the pose information for the next frame, i.e., which side of the hand is visible (palm, side or knuckles view), assuming that the hands' global (palm) pose does not change significantly in two successive frames.

4. *Finger counting:*

For each input frame we count the clearly visible fingers F, by calculating the most important zero-crossings, as described in (i).

In Section 6.2.4 we discuss how we extract the hand edges and contour in a cluttered background, for the calculation of the edge orientation histograms and the CSS maps.

$2D$ Multi-frame Features

In our database, we store hand configuration sequences, and instead of matching between single frames, we match short sequences. In this way, we take into account the $N + 1$ most recent frames of the input sequence, imposing a temporal continuity constraint, less strict than the constraints of most continuous tracking methods.

The first step of our database search is to find the most appropriate cluster. For this step we use the projection information and the result of the finger counting, in the same way as we use them to cluster our database. For the projection information, we use the palm pose of the last estimated input hand configuration.

The next step is to extract the hand temporal information from the most recent $N + 1$ input frames, as hand movement (or hand shape changes) signature. This information will be used for matching between the input frames and the database samples inside the chosen cluster. From the single-frame features, we use the hand curvatures and the edge orientation histograms to describe the hand movements. To reduce the computational complexity of the matching between the input frames and the database samples, we integrate the successive curvatures and edge orientation histograms into two vectors.

Instead of matching $I \times (N+1)$ matrices, for $N+1$ CSS functions with I points, and $B \times (N+1)$ matrices, for $N+1$ edge orientation histograms with B bins, we follow dimensionality reduction, that can be expressed as follows.

For I points of an object contour and its CSS function (B bins of the edge orientation histogram), over $N+1$ successive frames of the input video segment, we assume that we have I points (B points) in an $(N+1)$-dimensional space, and we can obtain I points (B points) in the 1D *space with dimensionality reduction.*

Thus, the problem is transformed into a dimensionality reduction task. For the hand tracking application, we used the nonlinear local Isomap embeddings proposed by Tenenbaum et al. [18], keeping $(I, B) >> N+1$, to keep the residual (mapping) error low. We chose to use Isomap, instead of using a linear embedding, e.g., PCA (Principal Component Analysis), because we have nonlinear degrees of freedom, and we are interested in a global *hand movement signature*, i.e., a globally optimal solution [15].

Thus, if $\overline{K} = [K_n|n = 0, \ldots, N]$ and $\overline{H}_\vartheta = [H_{\vartheta,n}|n = 0, \ldots, N]$ are the sets of $N+1$ CSS functions K_n and edge orientation histograms $H_{\vartheta,n}$, extracted over $N+1$ frames, the embedded 1D multi-frame features are respectively,

$$\tilde{k} = M^{(N+1,1)}(\overline{K}), \quad and \quad \tilde{h}_\vartheta = M^{(N+1,1)}(\overline{H}_\vartheta), \tag{6.4}$$

where $M^{(N+1,1)}$ represents the Isomap embedding from the $N+1$ to the 1 dimensional space.

Matching Between Input Frames and Samples

As matching criterion between the input frames and the database samples, we use the undirected chamfer distance. In general, given two point sets A and B, their undirected chamfer distance $d(A, B)$ is defined by the forward $d^f(A, B)$ and backward $d^b(A, B)$ distances:

$$d^f(A, B) = \frac{1}{\|A\|} \cdot \sum_{a^i \in A} \min_{b^j \in B} \|a^i - b^j\|, \tag{6.5}$$

$$d^b(A, B) = \frac{1}{\|B\|} \cdot \sum_{b^i \in B} \min_{a^j \in A} \|a^j - b^i\|, \tag{6.6}$$

$$d(A, B) = d^f(A, B) + d^b(A, B) \tag{6.7}$$

Replacing A and B with the multi-frame features $\tilde{k}(u) = [\tilde{k}^i(u)|i = 1, \ldots, I]$ and $\tilde{k}(s) = [\tilde{k}^i(s)|i = 1, \ldots, I]$, where I is the number of contour points, we obtain the chamfer distance $d_k(u, s)$ between the embedded CSS functions of the successive input frames u and the database configuration sequence s, respectively.

Similarly, replacing A and B with $\tilde{h}_\theta(u) = [\tilde{h}_\theta^i(u)|i = 1, \ldots, B]$ and $\tilde{h}_\theta(s) = [\tilde{h}_\theta^i(s)|i = 1, \ldots, B]$, where B is the number of bins, we obtain the respective chamfer distance $d_h(u, s)$ between the embedded edge orientation histograms.

For a chosen database cluster $\hat{S} = S_p \in \mathbf{S}$, where $\mathbf{S}$ is the set of all clusters, i.e., the entire database, $p = 1, \ldots, P$, and a set of input frames u, the best matching sample $\hat{s} \in \hat{S}$ ($\in \mathbf{S}$) is given by,

$$\hat{s} = \arg \min_{s \in \hat{S}} \sqrt{d_k(u, s)^2 + d_h(u, s)^2}, \tag{6.8}$$

6.2.3 Coupling Two Trackers

The main idea of our coupling is to obtain model reinitialization from the discrete scheme when continuous tracking fails. This framework provides computational efficiency and allows us to use different trackers as separate procedures. The decision to choose as final solution the results of one or the other tracker, at each time instance t, is obtained using the 2D error described below.

If $E_b^f(t)$ is the boundary edge set of the input frame and $E_b^m(t)$ is the boundary edge set of the model projection onto the image plane, at time t, then the $2D$ tracking error is defined as,

$$e_{2D}(t) = d(E_b^f(t), E_b^m(t)), \tag{6.9}$$

where $d(\cdot)$ represents the chamfer distance given in Equations (6.5)–(6.7). The tracking error defined above can serve only as an indication but does not give the *actual* tracking error, i.e., the 3D configuration error. Sometimes the 2D error is small but the estimated configuration is very different from the actual one, especially for the fingers' joint angles. Thus, we need to know the relationship between the 2D error, which is directly estimated from the current frame of the input sequence and the corresponding estimated configuration, and the actual 3D error, which is defined as,

$$e_{3D}(t) = \sum_{i=1}^{\varphi} \|y_i(t) - y_i^c(t)\|^2 \tag{6.10}$$

where $y(t) = [y_i(t)|i = 1, \ldots, \varphi]$ and $y^c(t) = [y_i^c(t)|i = 1, \ldots, \varphi]$ are the actual (ground-truth) and estimated (using the continuous tracking) configurations respectively, and φ is the number of the *dofs*. The 3D error cannot be extracted directly from the input sequence since there is no ground-truth configuration $y(t)$ available for the tracked hand.

Let $Q(t)$ be the decision parameter at time t, of choosing either the continuous or the discrete tracking results, $y^c(t)$ and $y^d(t)$ respectively. It is

$Q(t) = \{1, -1\}$, where 1 stands for continuous and -1 stands for discrete, and the decision is taken according to,

$$p(Q|e_{2D}) = \int_{e_{3D}} p(Q|e_{3D}) \cdot p(e_{3D}|e_{2D})de_{3D} \qquad (6.11)$$

where $p(Q|e_{3D})$ is the probability density of the decision Q, given the 3D error e_{3D}, and $p(e_{3D}|e_{2D})$ is the probability density of having the 3D tracking error e_{3D} given the 2D error e_{2D}.

To estimate the probability densities, we apply the continuous tracker to a set of M samples of our database, compute the 2D and 3D errors from Equations (6.9), (6.10), and mark the cases where tracking fails.

Thus, from the marked cases of the continuous tracking failure ($Q = -1$), we have the probability at time t,

$$P(Q(t) \ |e_{3D}(t)) = \sum_{i=1}^{M} p(Q(i) \ |e_{3D}(i)) \qquad (6.12)$$

For the estimation of the probability $P(e_{3D}(t)|e_{2D}(t))$, we need to determine the correspondence between the errors e_{3D} and e_{2D}, i.e., we need to estimate a distribution f, such that

$$e_{3D} = f(e_{2D}) \qquad (6.13)$$

To estimate the distribution f, we apply off-line learning to the subset M of our database, using Support Vector Regression [22] with gaussian kernel, and Equation (6.13) can be rewritten as,

$$e_{3D} = f(e_{2D}, \sigma_{svr}, \mathbf{a}, b), \qquad (6.14)$$

where σ_{svr} is the variance used for the kernel, $\mathbf{a}$ is the estimated set of Lagrange multipliers and b is obtained using the Karush–Kuhn–Tucker (KKT) conditions [7]. In the next section we describe the training and testing sets used for the SVR learning, and we give the result for the estimated parameters and the testing set.

6.2.4 2D Feature Extraction in Cluttered Background

In this section we describe how we obtain the 2D features used in the discrete tracking approach, in cases where the background is cluttered, i.e., contains strong edges close to the object boundaries.

The approach we follow consists of the following steps:

1. *Skin color modelling:* We collect different skin samples and we train a classifier to model the skin color. In our case we used an SVM. An alternative way for the hand extraction is the method proposed by Viola et al. [23].
2. *Edge detection:* For every input frame, we extract all the edges using the canny edge detector.

3. *Skin pixels classification:* We classify the pixels of the input frame, into skin and non-skin pixels, based on the SVM results.
4. *Hand boundary edges extraction:* Using the input frame's edge map and the skin classification results, we extract the boundary edges: each edge on the hand boundaries must separate two regions, a skin and a non-skin region.
5. *Internal hand edges extraction:* Similarly to the boundary edges, we extract the internal hand edges: every edge in the interior of the hand area must separate two similar skin regions.
6. *Hand's convex hull and contour estimation:* From the extracted hand edges, we estimate the hand's convex hull and then its contour, using an active contour. In our application we use a snake model.

The procedure described above can be seen in two examples. For the first example, Figures 6.4 and 6.5 illustrate the hand edges and contour extraction of a hand moving in cluttered background. Specifically, in Figure 6.4 each

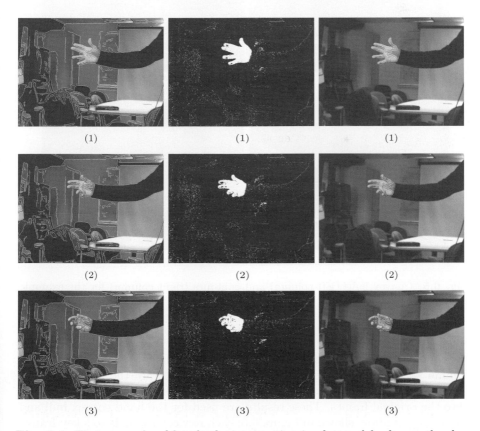

Fig. 6.4. First example of hand edges extraction in cluttered background: edge detection (left column), skin pixels classification (middle column), and the extracted hand edges (right column).

146   G. Tsechpenakis et al.

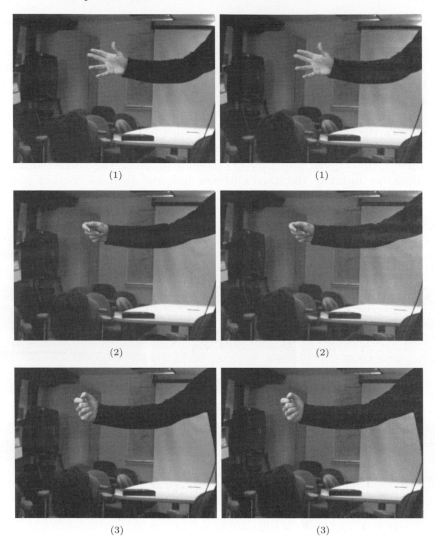

(1)   (1)

(2)   (2)

(3)   (3)

Fig. 6.5. Convex hulls (left column) and contours (right column) for the example of Figure 6.4.

row corresponds to a key-frame; the first column illustrates the results of the canny edge detector (superimposed onto the original frames), the second column shows the classification results (white pixels are the estimated skin pixels), and the third column shows the extracted hand edges superimposed onto the original frames. These results are used for the hand convex hull and contour estimation, shown in Figure 6.5: each row corresponds to a key-frame, the first column shows the convex hulls, and the second column illustrates the final hand contours.

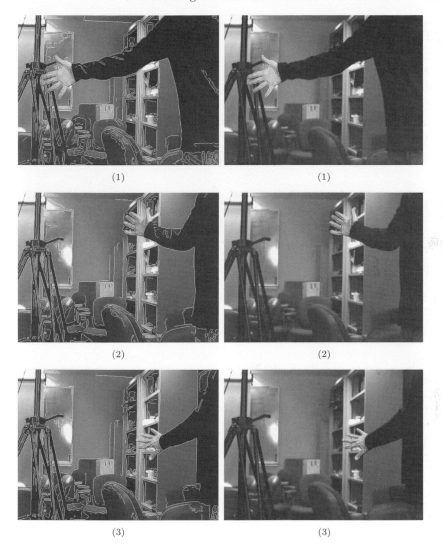

(1)   (1)

(2)   (2)

(3)   (3)

Fig. 6.6. Second example of hand edges extraction in cluttered background: edge detection (left column) and the extracted hand edges (right column).

Figures 6.6 and 6.7 represent the second example of the hand edges and contour extraction in cluttered background. In Figure 6.6 each row corresponds to a key-frame, the first column shows the edge detection results, and the second column shows the extracted hand edges. Figure 6.7 illustrates the corresponding convex hulls (first column) and the final hand contours (second column).

From these examples one can see that the hand edges and contours can be extracted successfully and fast, in cases of cluttered background, where no

148   G. Tsechpenakis et al.

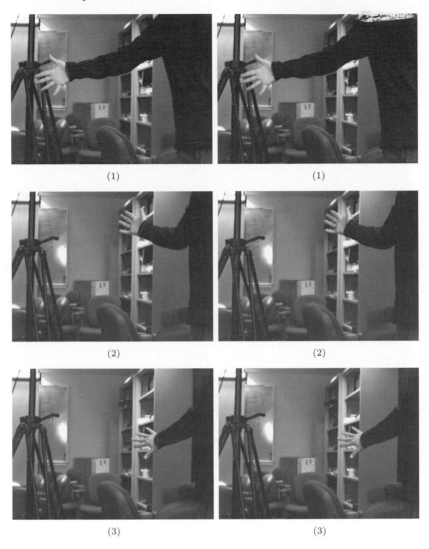

Fig. 6.7. Convex hulls (left column) and contours (right column) for the example of Figure 6.6.

other skin regions exist. In cases where more skin regions exist, apart from the tracked hand, we also rely on the trackers' temporal continuity constraints to choose the edges that correspond to the hand. If the tracked hand is moving away from these undesired skin regions, the hand position is used to exclude the undesired skin edges. In cases where the hand is moving close to or occluding the undesired skin regions, the hand model fits to these edges that best satisfy the continuity assumptions.

6.3 Experimental Results

In our experiments we used a Pentium-4, 3.06 GHz machine. We constructed
the virtual hand database described in Section 6.2.2 using the hand model
shown in Figure 6.2, with 20 *dofs*. As explained above, we created configuration sequences, instead of single configurations, of length $N + 1 = 5$. In
practice, $N + 1$ represents the number of input frames we use for matching
in the database, i.e., the number of input frames we utilize for estimating
the current configuration in our discrete tracking. We choose $N + 1 = 5$, for
$30 fps$ input videos, as the minimum number of input frames that can provide
sufficient temporal information. For 200 gestures under 29 camera views, we
stored 5, 800 sample sequences, or 29, 000 configurations.

Along with the $N + 1$ joint angle sets, for each sample, we stored the
corresponding 2D single-frame features described in Section 6.2.2: (i) projection information, (ii) number of fingers visible, and the multi-frame features
described in Section 6.2.2: (i) embedded CSS functions and (ii) embedded
edge orientation histograms. It should be noted that for the multi-frame features, larger values of N would result to higher residual error during the
Isomap dimensionality reduction. Also, we used fixed number of curvature
points $I = 150$ and orientation histogram bins $B = 50$.

We have clustered the database manually according to the stored single-
frame features, and we obtained $P = 108$ clusters of $L = 54$ sample sequences
each. While the average computational time for the continuous tracker is
$5 fps$, the chosen number of samples in our database does not decrease our
system's efficiency. Including more samples in the database would effect the
computational complexity of our framework.

In the SVR-based off-line learning, for the estimation of the distribution
f of Equation (6.14), we used $M = 500$ sample sequences of our database
as training set, and 100 samples for testing. All samples in the training set
include tracking failures. Figure 6.8 illustrates the SVR results $(e_{3D} = f(e_{2D}))$,
using $\sigma_{svr} = 1.5$: x and y axes represent the 2D and 3D errors respectively.
For the training set we have $\{\bar{e}_{3D} = 0.3383, \ \sigma^2_{e_{3D}} = 0.0744\}$, and $\{\bar{e}_{2D} = 5.9717, \ \sigma^2_{e_{2D}} = 18.4998\}$, mean values and variances for the 3D and 2D errors

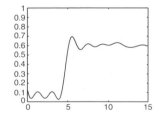

Fig. 6.8. Support vector regression with $\sigma_{svr} = 1.5$: $e_{3D} = f(e_{2D})$.

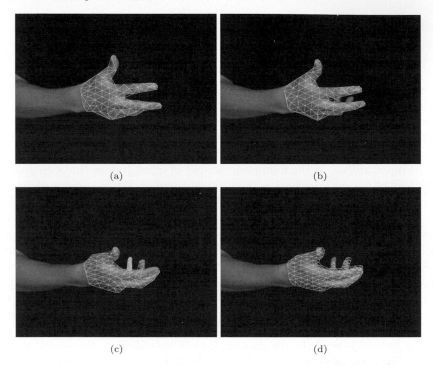

(a)   (b)

(c)   (d)

Fig. 6.9. Continuous tracking failure due to fingers' occlusions ((a)–(c)), and model reinitialization (d) from the discrete tracker.

respectively; note that the 2D error is measured in pixels whereas the 3D error is measured in radians. In this figure one can also see the case where small 2D errors do not necessarily indicate small 3D (actual) errors (steep part of the distribution). This explains why the 2D projection error is not sufficient to indicate tracking failures and justifies experimentally the reason why we chose to use the 3D configuration error as tracking error.

A representative example of continuous tracking failure is illustrated in Figure 6.9. The tracking results are shown superimposed onto the original input frame in a simplified cyan grid. As mentioned in subsection 6.2.1, continuous tracking fails when fingers undergo significant movements when occluded. In key-frame (a), the ring and the little finger are occluded while the hand is rotating and the fingers are articulating. After a few frames (key-frame (b)), the ring finger appears but the continuous tracker does not fit the hand model to the right ring finger position. After a few frames (key-frame (c)), both the ring and the little finger appear, but the hand model fits in the wrong configuration; the hand model also misfits onto the index finger. In Figure 6.9(d) we show the model re-initialization; in our database we included a sample with successive hand configurations similar to the hand movement in the last 5 input frames (including the key-frame (c)). Although the final configuration does not match exactly with the 3D configuration of the input hand, it serves as a good model reinitialization for the continuous tracker. In this way we avoid error accumulation that would occur using the continuous tracker after the key-frame (c).

Figures 6.10 and 6.11 illustrate our framework results for two sequences. Each column shows the original frame (left) and the final result (right). Specifically, Fig. 6.10 shows the case of strong hand rotation without any significant finger articulations, in 9 key-frames. The results are shown with the virtual hand, even when continuous tracking is applied. Figure 6.11 shows a more difficult case, where the hand undergoes not only complicated rotations but also finger articulations. In this case, the time intervals where continuous tracking was successfully applied are smaller, i.e., in the coupling framework the discrete method initializes the hand model in more frames, based on the corresponding 3D error.

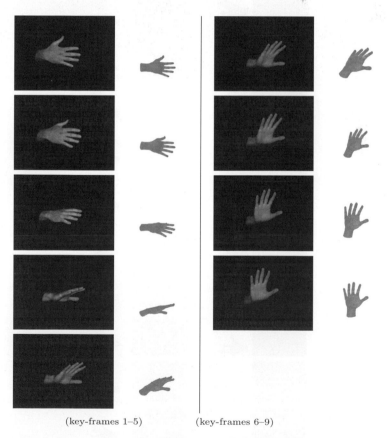

(key-frames 1–5)   (key-frames 6–9)

Fig. 6.10. Tracking results for strong hand rotation: in each column, the left images are the original frames and the right images show the final result.

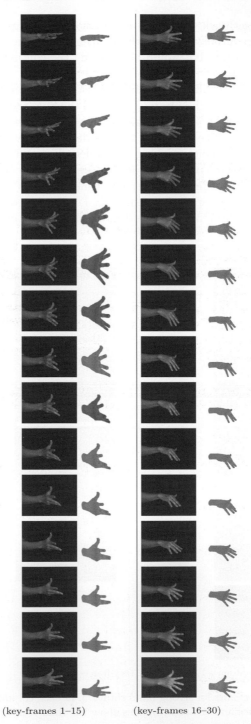

(key-frames 1–15)   (key-frames 16–30)

Fig. 6.11. Tracking results for strong hand rotation and finger articulations: in each column, the left images are the original frames and the right images show the final result.

6.3.1 American Sign Language

Signs in American Sign Language (ASL) and other signed languages are articulated through the use of particular handshapes, orientations, locations of articulation relative to the body, and movements. Lexical signs are either inherently one- or two-handed; that is, some signs are articulated by only one hand, while others are necessarily articulated with both hands. (In addition, there are some instances where a one-handed sign is mirrored on the other hand and where a two-handed sign drops the nondominant hand.) During the articulation of a sign, changes in handshape sometimes occur; however, the number of allowable changes is limited (at most 3 handshapes are allowed within a given sign).

However, finger-spelled signs in ASL – generally corresponding to proper names and other borrowings from spoken language – are instead produced by concatenation of handshapes that correspond to the alphabet of the dominant spoken language (for ASL, this is English) [26]. In ASL, fingerspelled words are articulated using one hand (usually the dominant one) in a specific area of the signing space (in front of and slightly above the signer's shoulder). Unlike lexical signs, the number of handshapes in fingerspelled signs can be greater than three. In addition, the handshape changes are much more rapid than those found in the articulation of lexical signs, and there is greater independence of the movements of individual fingers than is found in non-finger spelled signs.

Clearly, recognition strategies for finger-spelling [5] must be fundamentally different from those used to identify other types of signs. However, automatic identification of finger-spelling portions within a fluent stream of signing is nontrivial, as many of the same handshapes that are used as letters are also used in the formation of other types of signs. Furthermore, there is a high degree of co-articulation in finger-spelling [6]; so finger-spelling frequently contains hand poses that deviate from the canonical handshapes of the letters themselves.

The framework described in this chapter exploited certain properties of finger-spelling to facilitate discrimination of finger-spelled and non-finger-spelled signs: e.g., the fact that finger-spelling tends to occur in neutral space, with relatively smaller displacements of the hands (although frequently a slight gradual movement from left to right) and more rapid movements of individual fingers than are typically found in production of other types of signs. For our experiments, we used videos of native ASL signers collected and annotated at Boston University as part of the American Sign Language Linguistic Research Project (http://www.bu.edu/asllrp/), in conjunction with the National Center for Sign Language and Gesture Resources. Transcription was performed using *SIgnStream* (http://www.bu.edu/asllrp/SignStream/) and these annotations served as ground truth for testing our algorithms. The video data and annotations associated with this project are available to the research community from http://www.bu.edu/asllrp/cslgr/.

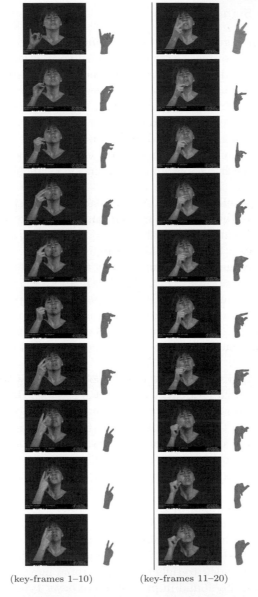

(key-frames 1–10)   (key-frames 11–20)

Fig. 6.12. Tracking results for ASL: in each column, the left images are the original frames and the right images show the final result.

In the experiment of Figure 6.12, we track a hand articulating sequentially both fingerspelled letters and continuous signs, as the person is asking the question, "Who did John see yesterday?" which in sign language is expressed by a fingerspelled word, J−O−H−N followed by three signs corresponding roughly to the English words meaning SEE WHO YESTERDAY.

Our coupling method performs well even in this case, where there are lighting changes and complicated background, which are handled by the continuous tracking method of [10]. In each column, the left images represent the input frames and the right images show the final result.

For the discrimination between finger-spelling and continuous signs in ASL, we use (i) the 2D hand displacements obtained from the contours extracted from the input frames, and (ii) the 3D information provided by our framework. For the classification, we used an SVM. Our training set consists of 200 short ASL sequences, 100 for fingerspelling and 100 for continuous signs. The testing set consists of 100 short ASL sequences.

More specifically, from the extracted contours used to estimate the curvature functions, we obtain the successive hand positions x_p on the image plane. Figure 6.13 illustrates the SVM classification results using the 2D hand displacements dx_p/dt ($x2$-axis), and the estimated configuration changes ($x1$-axis),

$$D_{\tilde{y}} = \frac{1}{\varphi} \sum_{i=1}^{\varphi} d\tilde{y}_i/dt, \qquad (6.15)$$

where $\tilde{y} = [\tilde{y}_i|i = 1,\ldots,\varphi]$ are the estimated configurations and φ is the number of the joint angles (dofs).

Table 6.1 shows the finger-spelling segmentation results for eight ASL videos. The videos we used are annotated so that we know what is said (signed) and we have as ground-truth the actual frames where finger-spelling is performed. The first column shows the video number, the second column shows the actual number of the finger-spelling segments in the sequence, the third column represents the ground-truth finger-spelling frame windows, and

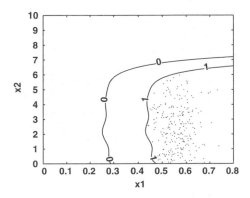

Fig. 6.13. SVM classification results for finger-spelling segmentation: $x1$-axis denotes the estimated configuration changes, and $x2$-axis corresponds to the hand displacements.

Table 6.1. Finger-spelling segmentation results for eight ASL videos.

Video	Segments	Ground-truth	Segmented finger spelling frames
(1)	1	(43–55)	(36–57)
(2)	1	(151–181)	(146–185)
(3)	1	(43 – 65)	(45–69)
(4)	1	(71–81)	(69–85)
(5)	2	(53–67, 87–101)	(55–71, 85–101)
(6)	2	(51–69, 83–101)	(46–73, 81–101)
(7)	2	(25–47, 145–173)	(21–45, 143–175)
(8)	2	(21–29, 125–159)	(19–31, 121–163)

the fourth column shows our segmentation results using the 2D hand displacement and the 3D hand configuration changes. The main reason for the difference between the actual and estimated boundaries is that before and after the actual finger-spelling, there is increased finger articulation, which does not correspond to the finger-spelling phase, but it is just a transition to and from this phase.

6.4 Summary and Conclusions

We have presented a dynamic data-driven framework, coupling continuous tracking with a novel discrete method for 3D object tracking in monocular sequences. We have applied our generic framework to the case of 3D hand tracking, since it is a very challenging task with a wide variety of applications. We have shown how our approach handles articulations, rotations, abrupt movements and cluttered background. We have shown how our framework is applied in the case of American Sign Language and how we discriminate between fingerspelling and non-finger-spelled signs. Our aim is to further evolve our framework to be used for tracking of a much larger type of articulated and nonrigid motions. We plan to extend our work on ASL recognition using both 2D and 3D information. We are currently working on constructing a more realistic hand database for ASL recognition, replacing the artificially created samples with samples obtained by combining Immersion Corporation's CyberGlove, Polhemus' Fastrack and Alias' Mocap. Finally, we should note that although a robust real-time hand tracking and ASL recognition system is a very challenging aim and requires a lot of research effort to be solved in its general form, our framework gives us promising results in terms of robustness under complex and fast hand movements.

References

1. Abbasi S., Mokhtarian F., and Kittler J., Curvature Scale Space Image in Shape Similarity Retrieval, *Multimedia Systems*, 7(6):467–476, 1999.

2. Athitsos V., Wang J., Sclaroff S., and Betke M., Detecting Instances of Shape Classes That Exhibit Variable Structure, *Boston University Computer Science Tech. Report*, No. 2005–021, June 13, 2005.
3. Athitsos V. and Sclaroff S., Database Indexing Methods for 3D Hand Pose Estimation, *Gesture Workshop*, Genova, Italy, April 2003.
4. Athitsos V. and Sclaroff S., Estimating 3D Hand Pose from a Cluttered Image, *IEEE Conference on Computer Vision and Pattern Recognition*, Wisconsin, June 2003.
5. Grobel K. and Hienz H., Video-based recognition of fingerspelling in real time, *Workshops Bildverarbeitung fuer die Medizin*, Aachen, 1996.
6. Jerde T. E., Soechting J. F., and Flanders M., Coarticulation in Fluent Finger-spelling, *The Journal of Neuroscience*, 23(6):2383, March 2003.
7. Kuhn H.W. and Tucker A.W., Nonlinear Programming, *2nd Berkeley Symposium on Mathematical Statistics and Probabilistics*, pp. 481–492, University of California Press, 1951.
8. Lee J. and Kunii T., Model-based Analysis of Hand Posture, *IEEE Computer Graphics and Applications*, 15, pp. 77–86, 1995.
9. Lin J., Wu Y., and Huang T.S., modelling the Constraints of Human Hand Motion, *5th Annual Federated Laboratory Symposium (ARL2001)*, Maryland, 2001.
10. Lu S., Metaxas D., Samaras D., and Oliensis J., Using Multiple Cues for Hand Tracking and Model Refinement, *IEEE Conference on Computer Vision and Pattern Recognition*, Wisconsin, June 2003.
11. Mokhtarian F. and Mackworth A., A Theory of Multiscale, Curvature-based Shape Representation for Planar Curves, *Pattern Analysis and Machine Intelligence*, 14(8), pp. 789–805, 1992.
12. Rehg J. and Kanade T., Model-based Tracking of Self-occluding Articulated Objects, *IEEE International Conference on Computer Vision*, Cambridge, MA, June 1995.
13. Rosales R., Athitsos V., Sigal L., and Sclaroff S., 3D Hand Pose Reconstruction Using Specialized Mappings, *IEEE International Conference on Computer Vision*, Vancouver, Canada, July 2001.
14. Shimada N., Kimura K., and Shirai Y., Real-time 3D Hand Posture Estimation based on 2D Appearance Retrieval Using Monocular Camera, *IEEE ICCV Workshop on Recognition, Analysis and Tracking of Faces and Gestures in Real-time Systems*, Vancouver, Canada, July 2001.
15. de Silva V. and Tenenbaum J.B., Global versus Local Methods in Nonlinear Dimensionality Reduction, *Advances in Neural Information Processing Systems 15*, (eds.) M.S. Baker, S. Thrun, and K. Obermayer, Cambridge, MIT Press, pp. 705–712, 2002.
16. Stenger B., Mendonca P.R.S., and Cipolla R., Model-based 3D Tracking of an Articulated Hand, *IEEE Conference on Computer Vision and Pattern Recognition*, Kauai, December 2001.
17. Stenger B., Thayananthan A., Torr P.H.S., and Cipolla R., Hand Pose Estimation Using Hierarchical Detection, *International Workshop on Human-Computer Interaction*, Prague, Czech Republic, May 2004.
18. Tenenbaum J.B., de Silva V., and Langford J.C., A Global Geometric Framework for Nonlinear Dimensionality Reduction, *Science Magazine*, 290, pp. 2319–2323, 2000.

19. Thayananthan A., Stenger B., Torr P. H. S., and Cipolla R., Shape Context and Chamfer Matching in Cluttered Scenes, *IEEE Conference on Computer Vision and Pattern Recognition*, Madison, June 2003.
20. Tomasi C., Petrov S., and Sastry A., "3D Tracking = Classification + Interpolation," *IEEE International Conference on Computer Vision*, Nice, France, October 2003.
21. Triesch J. and von der Malsburg C., A System for Person-Independent Hand Posture Recognition against Complex Backgrounds, *Pattern Analysis and Machine Intelligence*, 23(12), December, 1999.
22. Vapnik V., *Statistical Learning Theory*, Wiley, New York, 1998.
23. Viola P. and Jones M., Rapid Object Detection using a Boosted Cascade of Simple Features, *IEEE Conference on Computer Vision and Pattern Recognition*, Hawaii, December, 2001.
24. Vogler C. and Metaxas D., A Framework for Recognizing the Simultaneous Aspects of American Sign Language, *Computer Vision and Image Understanding*, 81, pp. 358–384, 2001.
25. Vogler C., Sun H., and Metaxas D., A Framework for Motion Recognition with Applications to American Sign Language and Gait Recognition, *Workshop on Human Motion*, Austin, TX, December 2000.
26. Wilcox S., The Phonetics of Fingerspelling, *Studies in Speech Pathology and Clinical Linguistics*, 4, Amsterdam and Philadelphia, John Benjamins, 1992.
27. Zhou H. and Huang T.S., Tracking Articulated Hand Motion with Eigen Dynamics Analysis, *IEEE International Conference on Computer Vision*, Nice, France, October 2003.

7

Graphical Models for Human Motion Modelling

Kooksang Moon and Vladimir Pavlović

Rutgers University, Department of Computer Science
Piscataway, NJ 08854, USA

Summary. The human figure exhibits complex and rich dynamic behavior that is both nonlinear and time-varying. To automate the process of motion modelling we consider a class of learned dynamic models cast in the framework of dynamic Bayesian networks (DBNs) applied to analysis and tracking of the human figure. While direct learning of DBN parameters is possible, Bayesian learning formalism suggests that hyperparametric model description that considers all possible model dynamics may be preferred. Such integration over all possible models results in a subspace embedding of the original motion measurements. To this end, we propose a new family of Marginal Auto-Regressive (MAR) graphical models that describe the space of all stable auto-regressive sequences, regardless of their specific dynamics. We show that the use of dynamics and MAR models may lead to better estimates of sequence subspaces than the ones obtained by traditional non-sequential methods. We then propose a learning method for estimating general nonlinear dynamic system models that utilizes the new MAR models. The utility of the proposed methods is tested on the task of tracking 3D articulated figures in monocular image sequences. We demonstrate that the use of MAR can result in efficient and accurate tracking of the human figure from ambiguous visual inputs.

7.1 Introduction

modelling the dynamics of human figure motion is essential to many applications such as realistic motion synthesis in animation and human activity classification. Because the human motion is a sequence of human poses, many statistical techniques for sequence modelling have been applied to this problem. A dynamic Bayesian newtwork (DBN) is one method for motion modelling that facilitates easy interpretation and learning. We are interested in learning dynamic models from motion capture data using DBN formalisms and dimensionality reduction methods.

Dimensionality reduction / subspace embedding methods such as Principal Components Analysis (PCA), Multidimensional Scaling (MDS) [11], Gaussian Process Latent Variable Models (GPLVM) [9] and others, play an important

B. Rosenhahn et al. (eds.), Human Motion – Understanding, Modelling, Capture, and
Animation, 159–183.

role in many data modelling tasks by selecting and inferring those features that lead to an intrinsic representation of the data. As such, they have attracted significant attention in computer vision where they have been used to represent intrinsic spaces of shape, appearance, and motion. However, it is common that subspace projection methods applied in different contexts do not leverage inherent properties of those contexts. For instance, subspace projection methods used in human figure tracking [5, 16, 20, 22] often do not fully exploit the dynamic nature of the data. As a result, the selected subspaces sometimes do not exhibit temporal smoothness or periodic characteristics of the motion they model. Even if the dynamics are used, the methods employed are sometimes not theoretically sound and are disjoint from the subspace selection phase.

In this chapter we present a graphical model formalism for human motion modelling using dimensionality reduction. A new approach to subspace embedding of sequential data is proposed, which explicitly accounts for their dynamic nature. We first model the space of sequences using a novel Marginal Auto-Regressive (MAR) formalism. A MAR model describes the space of sequences generated from all possible AR models. In the limit case MAR describes all *stable* AR models. As such, the MAR model is weakly-parametric and can be used as a prior for an arbitrary sequence, without knowing the typical AR parameters such as the state transition matrix. The embedding model is then defined using a probabilistic GPLVM framework [9] with MAR as its prior. A GPLVM framework is particularly well suited for this task because of its probabilistic generative interpretation. The new hybrid GPLVM and MAR framework results in a general model of the space of all *nonlinear dynamic systems* (NDS). Because of this it has the potential to theoretically soundly model nonlinear embeddings of a large family of sequences.

This chapter is organized as follows. We first justify modelling human motion with subspace embedding. Next we define the family of MAR models and study some properties of the space of sequences modeled by MAR. We also show that MAR and GPLVM result in a model of the space of all NDS sequences and discuss its properties. The utility of the new framework is examined through a set of experiments with synthetic and real data. In particular, we apply the new framework to modelling and tracking of the 3D human figure motion from a sequence of monocular images.

7.2 Related Work

Manifold learning approaches to motion modelling have attracted significant interest in the last several years. Brand proposed nonlinear manifold learning that maps sequences of the input to paths of the learned manifold [4]. Rosales and Sclaroff [13] proposed the Specialized Mapping Architecture (SMA) that utilizes forward mapping for the pose estimation task. Agarwal and Triggs [1] directly learned a mapping from image measurement to 3D pose using Relevance Vector Machine (RVM).

However, with high-dimensional data, it is often advantageous to consider a subspace of, e.g., the joint angles space that contains a compact representation of the actual figure motion. Principal Component Analysis (PCA) [8] is the most well-known linear dimensionality reduction technique. Although PCA has been applied to human tracking and other vision application [12, 15, 21], it is insufficient to handle the nonlinear behavior inherent to human motion. Nonlinear manifold embedding of the training data in low dimensional spaces using isometric feature mapping (Isomap), Local linear (LLE) and spectral embedding [3, 14, 19, 24], have shown success in recent approaches [5, 16]. While these techniques provide point-based embeddings implicitly modelling the nonlinear manifold through exemplars, they lack a fully probabilistic interpretation of the embedding process.

The GPLVM, a Gaussian Processes [25] model, produces a continuous mapping between the latent space and the high-dimensional data in a probabilistic manner [9]. Grochow et al. [7] use a Scaled GPLVM (SGPLVM) to model inverse kinematics for interactive computer animation. Tian et al. [20] use a GPLVM to estimate the 2D upper body pose from the 2D silhouette features. Recently, Urtasun et al. [22] exploit the SGPLVM for 3D people tracking. However these approaches utilize simple temporal constraints in the pose space that often introduce "dimensionality curse" to nonlinear tracking methods such as particle filters. Moreover, such methods fail to explicitly consider motion dynamics during the embedding process. Our work addresses both of these issues through the use of a novel marginal NDS model. Wang et al. [23] introduced Gaussian Process Dynamical Models (GPDM) that utilize the dynamic priors for embedding. Our work extends the idea to tracking and investigates the impact of dynamics in the embedded space on tracking in real sequences.

7.3 Dynamics modelling and Subspace Embedding

Because the human pose is typically represented by more than 30 parameters (e.g., 59 joint angles in the marker-based motion capture system), modelling human motion is a complex task. Suppose $\mathbf{y}_t$ is a M-dimensional vector consisting of joint angles at time t. modelling human motion can be formulated as learning a dynamic system:

$$\mathbf{y}_t = h(\mathbf{y}_0, \mathbf{y}_1, ..., \mathbf{y}_{t-1}) + \mathbf{u}_t$$

where $\mathbf{u}_t$ is a (Gaussian) noise process.

A common approach to modelling linear motion dynamics would be to assume a T-th order linear auto-regressive (AR) model:

$$\mathbf{y}_t = \sum_{i=1}^{T} \mathbf{A}_i \mathbf{y}_{t-i} + \mathbf{u}_t \tag{7.1}$$

where $\mathbf{A}_i$ is the auto-regression coefficients and $\mathbf{u}_t$ is a noise process. For instance, 2nd order AR models are sufficient for modelling of periodic motion and higher order models lead to more complex motion dynamics. However, as the order of the model increases the number of parameters grows as $M^2 \cdot T + M^2$ (transition and covariance parameters). Learning this set of parameters may require large training sets and can be prone to overfitting.

Armed with the intuition that correlation between the limbs such as arms and legs always exists for a certain motion, many researchers have exploited the dynamics in the lower projection space rather than learned the dynamics in the high-dimensional pose space for human motion modelling. By inducing a hidden state $\mathbf{x}_t$ of dimension N ($M \gg N$) satisfying the first-order Marko-vian condition, modelling human motion is cast in the framework of dynamic Bayesian networks (DBNs) depicted in Figure 7.1:

$$\mathbf{x}_t = f(\mathbf{x}_{t-1}) + \mathbf{w}_t$$
$$\mathbf{y}_t = g(\mathbf{x}_t) + \mathbf{v}_t$$

where $f(\cdot)$ is a transition function, $g(\cdot)$ represents any dimensional reduction operation, and $\mathbf{w}_t$ and $\mathbf{v}_t$ are (Gaussian) noise processes.

The above DBN formalism implies that predicting future observation $\mathbf{y}_{t+1}$ based on the past observation data $\mathcal{Y}_0^t = \{\mathbf{y}_0, \ldots, \mathbf{y}_t\}$ can be stated as the following inference problem:

$$P(\mathbf{y}_{t+1}|\mathcal{Y}_0^t) = \frac{P(\mathcal{Y}_0^{t+1})}{P(\mathcal{Y}_0^t)} = \frac{\sum_{\mathbf{x}_{t+1}} \cdots \sum_{\mathbf{x}_0} P(\mathbf{x}_0) \prod_{i=0}^{t} P(\mathbf{x}_{i+1}|\mathbf{x}_i) \prod_{i=0}^{t+1} P(\mathbf{y}_i|\mathbf{x}_i)}{\sum_{\mathbf{x}_t} \cdots \sum_{\mathbf{x}_0} P(\mathbf{x}_0) \prod_{i=0}^{t-1} P(\mathbf{x}_{i+1}|\mathbf{x}_i) \prod_{i=0}^{t} P(\mathbf{y}_i|\mathbf{x}_i)}.$$

This suggests that the dynamics of the observation (pose) sequence $\mathbf{Y}$ posses a more complicated form. Namely, the pose $\mathbf{y}_t$ at time t becomes dependent on all previous poses $\mathbf{y}_{t-1}, \mathbf{y}_{t-2}, \ldots$ effectively resulting in an infinite order AR model. However, such model can use a smaller set of parameters than the AR model of Equation (7.1) in the pose space. Assuming a 1st-order linear dynamic system (LDS) $\mathbf{x}_t = \mathbf{F}\mathbf{x}_{t-1} + \mathbf{w}$ and the linear dimensionality reduction process $\mathbf{y}_t = \mathbf{G}\mathbf{x}_t + \mathbf{v}$ where $\mathbf{F}$ is the transition matrix and $\mathbf{G}$ is the inverse of the dimensionality reduction matrix, the number of parameters to be learned is $N^2 + N^2 + N \cdot M + M^2 = 2N^2 + M \cdot (N+M)$ (N^2 in F, NM in G and $N^2 + M^2$ in the two noise covariance matrices for $\mathbf{w}$ and $\mathbf{v}$). When $N \ll M$

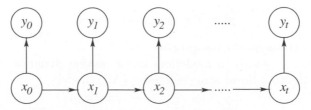

Fig. 7.1. A graphical model for human motion modelling with the subspace modelling.

the number of parameters of the LDS representation becomes significantly smaller than that of the "equivalent" AR model. That is, by learning both the dynamics in the embedded space and the subspace embedding model, we can effectively estimate $\mathbf{y}_t$ given all $\mathcal{Y}_0^{t-1}$ at any time t using a small set of parameters.

To illustrate the benefit of using the dynamics in the embedded space for human motion modelling, we take 12 walking sequences of one subject from CMU Graphics Lab Motion Capture Database [27] where the pose is represented by 59 joint angles. The poses are projected into a 3D subspace. Assume that the dynamics in the pose space and in the embedded space are modeled using the 2nd-order linear dynamics. We perform leave-one-out cross-validation for these 12 sequences – 11 sequences are selected as a training set and the 1 remaining sequence is reserved for a testing set. Let M_{pose} be the AR model in the pose space learned from this training set and M_{embed} be the LDS model in the latent space. Figure 7.2 shows the summary statistics of the two negative log-likelihoods of $P(Y_n|M_{pose})$ and $P(Y_n|M_{embed})$, where Y_n is a sequence reserved for testing.

The experiment indicates that with the same training data, the learned dynamics in the embedded space models the unseen sequences better than the dynamic model in the pose space. The large variance of $P(Y_n|M_{pose})$ for

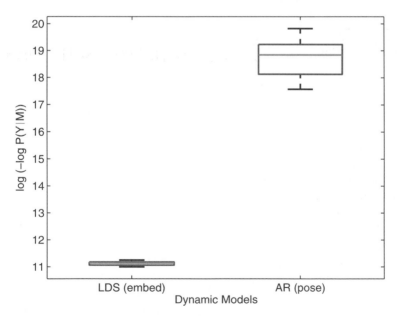

Fig. 7.2. Comparison of generalization abilities of AR ("pose") and LDS ("embed") models. Shown are the medians, upper and lower quartiles (boxes) of the negative log likelihoods (in log space) under the two models. The whiskers depict the total range of the values. Note that lower values suggest better generalization properties (fit to test data) of a model.

different training sets also indicates the overfitting problem that is generally observed in a statistical model that has too many parameters.

As shown in Figure 7.1, there are two processes in modelling human motion using a subspace embedding. One is learning the embedding model $P(\mathbf{y}_t|\mathbf{x}_t)$ and the other is learning the dynamic model $P(\mathbf{x}_{t+1}|\mathbf{x}_t)$. The problem of the previous approaches using the dimensionality reduction in human motion modelling is that these two precesses are decoupled into two separate stages in learning. However, coupling the two learning processes results in the better embedded space that preserves the dynamic nature of original pose space. For example, if the prediction by the dynamics suggests that the next state will be near a certain point we can learn the projection that retains the temporal information better than naive projection disregarding this prior knowledge. Our proposed framework formulates this coupling of the two learning processes in a probabilistic manner.

7.4 Marginal Auto-Regressive Model

We develop a framework incorporating dynamics into the process of learning low-dimensional representations of sequences. In this section, a novel marginal dynamic model describing the space of all stable auto-regressive sequences is proposed to model the dynamics of unknown subspace.

7.4.1 Definition

Consider sequence $\mathbf{X}$ of length T of N-dimensional real-valued vectors $\mathbf{x}_t = [\mathbf{x}_{t,0}\mathbf{x}_{t,1}...\mathbf{x}_{t,N-1}] \in \Re^{1 \times N}$. Suppose sequence $\mathbf{X}$ is generated by the 1st-order AR model $AR(\mathbf{A})$:

$$\mathbf{x}_t = \mathbf{x}_{t-1}\mathbf{A} + \mathbf{w}_t, \ t = 0, ..., T-1 \qquad (7.2)$$

where $\mathbf{A}$ is a specific $N \times N$ state transition matrix and $\mathbf{w}_t$ is a white iid Gaussian noise with precision, α: $\mathbf{w}_t \sim \mathcal{N}(0, \alpha^{-1}\mathbf{I})$. Assume that, without loss of generality, the initial condition $\mathbf{x}_{-1}$ has normal multivariate distribution with zero mean and unit precision: $\mathbf{x}_{-1} \sim \mathcal{N}(0, \mathbf{I})$.

We adopt a convenient representation of sequence $\mathbf{X}$ as a $T \times N$ matrix $\mathbf{X} = [\mathbf{x}_0'\mathbf{x}_1'...\mathbf{x}_{T-1}']'$ whose rows are the vector samples from the sequence. Using this notation Equation (7.2) can be written as

$$\mathbf{X} = \mathbf{X}_\Delta \mathbf{A} + \mathbf{W}$$

where $\mathbf{W} = [\mathbf{w}_0'\mathbf{w}_1'...\mathbf{w}_{T-1}']'$ and $\mathbf{X}_\Delta$ is a *shifted/delayed* version of $\mathbf{X}$, $\mathbf{X}_\Delta = [\mathbf{x}_{-1}'\mathbf{x}_0'...\mathbf{x}_{T-2}']'$. Given the state transition matrix $\mathbf{A}$ and the initial condition, the AR sequence samples have the joint density function

$$P(\mathbf{X}|\mathbf{A}, \mathbf{x}_{-1}) = (2\pi)^{-\frac{NT}{2}} \exp\left\{-\frac{1}{2}tr\left\{(\mathbf{X} - \mathbf{X}_\Delta\mathbf{A})(\mathbf{X} - \mathbf{X}_\Delta\mathbf{A})'\right\}\right\}. \qquad (7.3)$$

The density in Equation (7.3) describes the distribution of samples in a T-long sequence for a particular instance of the state transition matrix $\mathbf{A}$. However, we are interested in the distribution of all AR sequences, regardless of the value of $\mathbf{A}$. In other words, we are interested in the marginal distribution of AR sequences, over all possible parameters $\mathbf{A}$.

Assume that all elements a_{ij} of $\mathbf{A}$ are iid Gaussian with zero mean and unit precision, $a_{ij} \sim \mathcal{N}(0,1)$. Under this assumption, it can be shown [10] that the *marginal* distribution of the AR model becomes

$$P(\mathbf{X}|\mathbf{x}_{-1}, \alpha) = \int_{\mathbf{A}} P(\mathbf{X}|\mathbf{A}, \mathbf{x}_{-1}) P(\mathbf{A}|\alpha) d\mathbf{A}$$

$$= (2\pi)^{-\frac{NT}{2}} |\mathbf{K}_{xx}(\mathbf{X}, \mathbf{X})|^{-\frac{N}{2}} exp\left\{ -\frac{1}{2} tr\{\mathbf{K}_{xx}(\mathbf{X}, \mathbf{X})^{-1}\mathbf{X}\mathbf{X}'\}\right\}$$

$$(7.4)$$

where

$$\mathbf{K}_{xx}(\mathbf{X}, \mathbf{X}) = \mathbf{X}_{\Delta}\mathbf{X}'_{\Delta} + \alpha^{-1}\mathbf{I}. \qquad (7.5)$$

We call this density the *Marginal AR* or MAR density. α is the hyperparameter of this class of models, $MAR(\alpha)$. Intuitively, Equation (7.4) favors those samples in $\mathcal{X}$ that do not change significantly from t to $t+1$ and $t-1$. The graphical representation of MAR model is depicted in Figure 7.3. Different treatments to the nodes are represented by different shades.

MAR density models the distribution of all (AR) sequences of length T in the space $\mathcal{X} = \Re^{T \times N}$. Note that while the error process of an AR model has Gaussian distribution, the MAR density is not Gaussian. We illustrate this in Figure 7.4. The figure shows joint pdf values for four different densities: MAR, periodic MAR (see Section 7.4.2), AR(2), and a circular Gaussian, in the space of length-2 scalar-valued sequences $[\mathbf{x}_0\mathbf{x}_1]'$. In all four cases we assume zero-mean, unit precision Gaussian distribution of the initial condition. All models have the mode at $(0,0)$. The distribution of the AR model is multivariate Gaussian with the principal variance direction determined by the state transition matrix $\mathbf{A}$. However, the MAR models define non-Gaussian

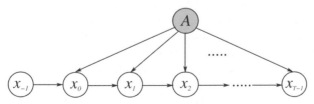

Fig. 7.3. Graphical representation of MAR model. White shaded nodes are optimized while gray shaded node is marginalized.

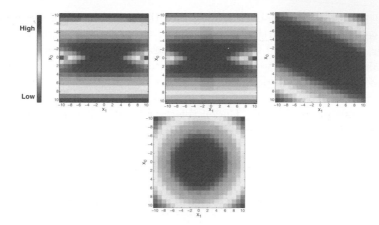

Fig. 7.4. Distribution of length-2 sequences of 1D samples under MAR, periodic MAR, AR, and independent Gaussian models.

distributions with no circular symmetry and with directional bias. This property of MAR densities is important when viewed in the context of sequence subspace embeddings, which we discuss in Section 7.5.

7.4.2 Higher-Order Dynamics

The above definition of MAR models can be easily extended to families of arbitrary D-th order AR sequences. In that case the state transition matrix $\mathbf{A}$ is replaced by an $ND \times N$ matrix $\mathbf{A} = [\mathbf{A}_1' \mathbf{A}_2'...\mathbf{A}_D']'$ and $\mathbf{X}_\Delta$ by $[\mathbf{X}_\Delta \mathbf{X}_{1\Delta}...\mathbf{X}_{D\Delta}]$. Hence, a $MAR(\alpha, D)$ model describes a general space of all D-th order AR sequences. Using this formulation one can also model specific classes of dynamic models. For instance, a class of all periodic models can be formed by setting $\mathbf{A} = [\mathbf{A}_1' \ -\mathbf{I}]'$, where $\mathbf{I}$ is an identity matrix.

7.4.3 Nonlinear Dynamics

In Equation (7.2) and Equation (7.4) we assumed linear families of dynamic systems. One can generalize this approach to nonlinear dynamics of the form $\mathbf{x}_t = g(\mathbf{x}_{t-1}|\zeta)\mathbf{A}$, where $g(\cdot|\zeta)$ is a nonlinear mapping to an L-dimensional subspace and $\mathbf{A}$ is a $L \times N$ linear mapping. In that case $\mathbf{K}_{xx}$ becomes a nonlinear kernel using justification similar to, e.g. [9]. While nonlinear kernels often have potential benefits, such as robustness, they also preclude closed-form solutions of linear models. In our preliminary experiments we have not observed significant differences between MAR and nonlinear MAR.

7.4.4 Justification of MAR Models

The choice of the prior distribution of AR model's state transition matrix leads to the MAR density in Equation (7.4). One may wonder, however, if the choice of iid $\mathcal{N}(0,1)$ results in a physically meaningful space of sequences. We suggest that, indeed, such choice may be justified.

Namely, Girko's circular law [6] states that if $\frac{1}{N}\mathbf{A}$ is a random $N \times N$ matrix with $\mathcal{N}(0,1)$ iid entries, then in the limit case of large $N(> 20)$ all real and complex eigenvalues of $\mathbf{A}$ are *uniformly distributed on the unit disk*. For small N, the distribution shows a concentration along the real line. Consequently, the resulting space of sequences described by the MAR model is that of *all stable AR systems*.

7.5 Nonlinear Dynamic System Models

In this section we develop a Nonlinear Dynamic System view of the sequence subspace reconstruction problem that relies on the MAR representation of the previous section. In particular, we use the MAR model to describe the structure of the subspace of sequences to which the extrinsic representation will be mapped using a GPLVM framework of [9].

7.5.1 Definition

Let $\mathbf{Y}$ be an extrinsic or measurement sequence of duration T of M-dimensional samples. Define $\mathbf{Y}$ as the $T \times M$ matrix representation of this sequence, similar to the definition in Section 7.4, $\mathbf{Y} = [\mathbf{y}_0'\mathbf{y}_1'...\mathbf{y}_{T-1}']'$. We assume that $\mathbf{Y}$ is a result of the process $\mathbf{X}$ in a lower-dimensional MAR subspace $\mathcal{X}$, defined by a nonlinear generative or forward mapping

$$\mathbf{Y} = f(\mathbf{X}|\theta)\mathbf{C} + \mathbf{V}.$$

$f(\cdot)$ is a nonlinear $\Re^N \to \Re^L$ mapping, $\mathbf{C}$ is a linear $L \times M$ mapping, and $\mathbf{V}$ is a Gaussian noise with zero-mean and precision β.

To recover the intrinsic sequence $\mathbf{X}$ in the embedded space from sequence $\mathbf{Y}$ it is convenient not to focus, at first, on the recovery of the specific mapping $\mathbf{C}$. Hence, we consider the family of mappings where $\mathbf{C}$ is a stochastic matrix whose elements are iid $c_{ij} \sim \mathcal{N}(0,1)$. Marginalizing over all possible mappings $\mathbf{C}$ yields a marginal Gaussian Process [25] mapping:

$$P(\mathbf{Y}|\mathbf{X},\beta,\theta) = \int_{\mathbf{C}} P(\mathbf{Y}|\mathbf{X},\mathbf{C},\theta)P(\mathbf{C}|\beta)d\mathbf{C}$$

$$= (2\pi)^{-\frac{MT}{2}}|\mathbf{K}_{yx}(\mathbf{X},\mathbf{X})|^{-\frac{M}{2}}exp\left\{-\frac{1}{2}tr\{\mathbf{K}_{yx}(\mathbf{X},\mathbf{X})^{-1}\mathbf{Y}\mathbf{Y}'\}\right\}$$

where

$$\mathbf{K}_{yx}(\mathbf{X},\mathbf{X}) = f(\mathbf{X}|\theta)f(\mathbf{X}|\theta)' + \beta^{-1}\mathbf{I}.$$

Notice that in this formulation the $\mathbf{X} \to \mathbf{Y}$ mapping depends on the inner product $\langle f(\mathbf{X}), f(\mathbf{X}) \rangle$. The knowledge on the actual mapping f is not necessary; a mapping is uniquely defined by specifying a positive-definite kernel

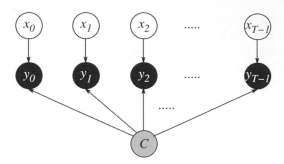

Fig. 7.5. Graphical model of NDS. White shaded nodes are optimized while gray-shaded node is marginalized and black-shaded nodes are observed variables.

$\mathbf{K}_{yx}(\mathbf{X}, \mathbf{X}|\theta)$ with entries $\mathbf{K}_{yx}(i,j) = k(\mathbf{x}_i, \mathbf{x}_j)$ parameterized by the hyperparameter θ. A variety of linear and non-linear kernels (RBF, square exponential, various robust kernels) can be used as $\mathbf{K}_{yx}$. Hence, our likelihood model is a nonlinear Gaussian process model, as suggested by Lawrence [9]. Figure 7.5 shows the graphical model of NDS.

By joining the MAR model and the NDS model, we have constructed a Marginal Nonlinear Dynamic System (MNDS) model that describes the joint distribution of all measurement and all intrinsic sequences in a $\mathcal{Y} \times \mathcal{X}$ space:

$$P(\mathbf{X}, \mathbf{Y}|\alpha, \beta, \theta) = P(\mathbf{X}|\alpha)P(\mathbf{Y}|\mathbf{X}, \beta, \theta). \tag{7.6}$$

The MNDS model has a MAR prior $P(\mathbf{X}|\alpha)$, and a Gaussian process likelihood $P(\mathbf{Y}|\mathbf{X}, \beta, \theta)$. Thus it places the intrinsic sequences $\mathbf{X}$ in the space of all AR sequences. Given an intrinsic sequence $\mathbf{X}$, the measurement sequence $\mathbf{Y}$ is zero-mean normally distributed with the variance determined by the nonlinear kernel $\mathbf{K}_{yx}$ and $\mathbf{X}$.

7.5.2 Inference

Given a sequence of measurements $\mathbf{Y}$ one would like to infer its subspace representation $\mathbf{X}$ in the MAR space, without needing to first determine a particular family of AR models $AR(\mathbf{A})$, nor the mapping $\mathbf{C}$. Equation (7.6) shows that this task can be, in principle, achieved using the Bayes rule $P(\mathbf{X}|\mathbf{Y}, \alpha, \beta, \theta) \propto P(\mathbf{X}|\alpha)P(\mathbf{Y}|X, \beta, \theta)$.

However, this posterior is non-Gaussian because of the nonlinear mapping f and the MAR prior. One can instead attempt to estimate the mode $\mathbf{X}^*$

$$\mathbf{X}^* = \arg\max_{\mathbf{X}} \{\log P(\mathbf{X}|\alpha) + \log P(\mathbf{Y}|\mathbf{X}, \beta, \theta)\}$$

using nonlinear optimization such as the Scaled Conjugate Gradient in [9].

To effectively use a gradient-based approach, one needs to obtain expressions for gradients of the log-likelihood and the log-MAR prior. Note that the expressions for MAR gradients are more complex than those of, e.g., GP due to a linear dependency between $\mathbf{X}$ and $\mathbf{X}_\Delta$ (see Appendix).

7.5.3 Learning

MNDS space of sequences is parameterized using a set of hyperparameters (α, β, θ) and the choice of the nonlinear kernel $\mathbf{K}_{yx}$. Given a set of sequences $\{\mathbf{Y}^{(i)}\}, i = 1, .., S$ the learning task can be formulated as the ML/MAP estimation problem

$$(\alpha^*, \beta^*, \theta^*)|_{\mathbf{K}_{yx}} = \arg\max_{\alpha, \beta, \theta} \prod_{i=1}^{S} P(\mathbf{Y}^{(i)}|\alpha, \beta, \theta).$$

One can use a generalized EM algorithm to obtained the ML parameter estimates recursively from two fixed-point equations:

E-step $\mathbf{X}^{(i)*} = \arg\max_{\mathbf{X}} P(\mathbf{Y}, \mathbf{X}^{(i)}|\alpha^*, \beta^*, \theta^*)$
M-step $(\alpha^*, \beta^*, \theta^*) = \arg\max_{(\beta, \alpha, \theta)} \prod_{i=1}^{S} P(\mathbf{Y}^{(i)}, \mathbf{X}^{(i)*}|\alpha, \beta, \theta)$

7.5.4 Learning of Explicit NDS Model

Inference and learning in MNDS models result in the embedding of the measurement sequence $\mathbf{Y}$ into the space of all NDS/AR models. Given $\mathbf{Y}$, the embedded sequences $\mathbf{X}$ estimated in Section 7.5.3 and MNDS parameters α, β, θ, the explicit AR model can be easily reconstructed using the ML estimation on sequence $\mathbf{X}$, e.g.:

$$\mathbf{A}^* = (\mathbf{X}'_\Delta \mathbf{X}_\Delta)^{-1} \mathbf{X}'_\Delta \mathbf{X}.$$

Because the embedding was defined as a GP, the likelihood function $P(\mathbf{y}_t|\mathbf{x}_t, \beta, \theta)$ follows a well-known result from GP theory: $\mathbf{y}_t|\mathbf{x}_t \sim \mathcal{N}(\mu, \sigma^2 \mathbf{I})$

$$\mu = \mathbf{Y}' \mathbf{K}_{yx}(\mathbf{X}, \mathbf{X})^{-1} \mathbf{K}_{yx}(\mathbf{X}, \mathbf{x}_t) \tag{7.7}$$

$$\sigma^2 = \mathbf{K}_{yx}(\mathbf{x}_t, \mathbf{x}_t) - \mathbf{K}_{yx}(\mathbf{X}, \mathbf{x}_t)' \mathbf{K}_{yx}(\mathbf{X}, \mathbf{X})^{-1} \mathbf{K}_{yx}(\mathbf{X}, \mathbf{x}_t). \tag{7.8}$$

The two components fully define the explicit NDS.

In summary, a complete sequence modelling algorithm consists of the following set of steps.

Algorithm 1 NDS learning

INPUT Measurement sequence $\mathbf{Y}$ and kernel family $\mathbf{K}_{yx}$
OUTPUT $NDS(\mathbf{A}, \beta, \theta)$

1. Learn subspace embedding $MNDS(\alpha, \beta, \theta)$ model of training sequences $\mathbf{Y}$ as described in Section 7.5.3.
2. Learn explicit subspace and projection model $NDS(\mathbf{A}, \beta, \theta)$ of $\mathbf{Y}$ as described in Section 7.5.4.

7.5.5 Inference in Explicit NDS Model

The choice of the nonlinear kernel $\mathbf{K}_{yx}$ results in a nonlinear dynamic system model of training sequences $\mathbf{Y}$. The learned model can then be used to infer subspace projections of a new sequence from the same family. Because of the nonlinearity of the embedding, one cannot apply the linear forward–backward or Kalman filtering/smoothing inference. Rather, it is necessary to use nonlinear inference methods such as (I)EKF or particle filtering/smoothing.

It is interesting to note that one can often use a relatively simple sequential nonlinear optimization in place of the above two inference methods:

$$\mathbf{x}_t^* = \arg\max_{\mathbf{x}_t} P(\mathbf{y}_t|\mathbf{x}_t, \beta^*, \theta^*) P(\mathbf{x}_t|\mathbf{x}_{t-1}^*, \mathbf{A}^*).$$

Such sequential optimization yields local modes of the true posterior $P(\mathbf{X}|\mathbf{Y})$. While one would expect such approximation to be valid in situations with few ambiguities in the measurement space and models learned from representative training data, our experiments show the method to be robust across a set of situations. However, dynamics seem to play a crucial role in the inference process.

7.5.6 Example

We illustrate the concept of MNDS on a simple synthetic example. Consider the AR model $AR(2)$ from Section 7.4. Sequence $\mathbf{X}$ generated by the model is projected to the space $\mathcal{Y} = \Re^{2\times3}$ using a linear conditional Gaussian model $\mathcal{N}(\mathbf{XC}, \mathbf{I})$. Figure 7.6 shows negative likelihood over the space $\mathcal{X}$ of the MNDS,

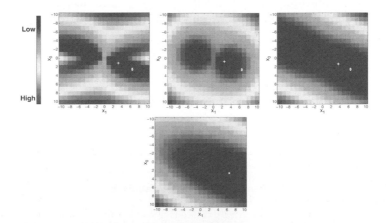

Fig. 7.6. Negative log-likelihood of length-2 sequences of 1D samples under MNDS, GP with independent Gaussian priors, GP with exact AR prior and LDS with the true process parameters. "o" mark represents the optimal estimate $\mathbf{X}^*$ inferred from the true LDS model. "+" shows optimal estimates derived using the three marginal models.

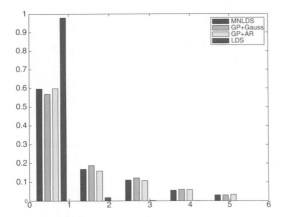

Fig. 7.7. Normalized histogram of optimal negative log-likelihood scores for MNDS, a GP model with a Gaussian prior, a GP model with exact AR prior and LDS with the true parameters.

a marginal model (GP) with independent Gaussian priors, a GP with the exact $AR(2)$ prior, and a full LDS with exact parameters. All likelihoods are computed for the fixed $\mathbf{Y}$. Note that the GP with Gaussian prior assumes no temporal structure in the data. This example shows that, as expected, the maximum likelihood subspace estimates of the MNDS model fall closer to the "true" LDS estimates than those of the nonsequential model. This property holds in general. Figure 7.7 shows the distribution of optimal negative log likelihood scores, computed at corresponding $\mathbf{X}^*$, of the four models over a 10000 sample of $\mathbf{Y}$ sequences generated from the true LDS model. Again, one notices that MNDS has a lower mean and mode than the nonsequential model, GP+Gauss, indicating MNDS's better fit to the data. This suggests that MNDS may result in better subspace embeddings than the traditional GP model with independent Gaussian priors.

7.6 Human Motion Modelling Using MNDS

In this section we consider an application of MNDS to modelling of the human motion from sequences of video images. Specifically, we assume that one wants to recover two important aspects of human motion: (1) 3D posture of the human figure in each image and (2) an intrinsic representation of the motion.

We propose two models in this context. The first model (Model 1) is a fully generative model that considers the natural way of image formation. Given the 3D pose space represented by joint angles $\mathbf{y}_t$, the mapping into a sequence of features $\mathbf{z}_t$ computed from monocular images (such as the silhouette-based alt moments, orientation histograms etc.) is given by a Gaussian process model $P(\mathbf{Z}|\mathbf{Y}, \theta_{zy})$. An NDS is used to model the space $\mathcal{Y} \times \mathcal{X}$ of poses and intrinsic motions $P(\mathbf{X}, \mathbf{Y}|\mathbf{A}, \beta, \theta_{yx})$. The second model (Model 2) relies on the

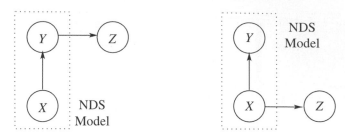

Fig. 7.8. Graphical Models for human motion modelling using MNDS. Left: model 1. Right: model 2.

premise that the correlation between the pose and the image features can be modeled using a latent-variable model. In this model, the mapping into the image feature space from the intrinsic space is given by a Gaussian process model $P(\mathbf{Z}|\mathbf{X}, \theta_{zx})$. The model for the space of poses and intrinsic motions is defined in the same way as the first model. Similar models are often used in other domains such as computational language modelling to correlate complex processes. The graphical models of these two approaches are shown in Figure 7.8.

7.6.1 Model 1

When the dimension of image feature vector $\mathbf{z}_t$ is much smaller than the dimension of pose vector $\mathbf{y}_t$ (e.g., 10-dimensional vector of alt Moments vs. 59-dimensional joint angle vector of motion capture data), estimating the pose given the feature becomes the problem of predicting a higher-dimensional projection in the model $P(\mathbf{Z}|\mathbf{Y}, \theta_{zy})$. It is an undetermined problem. In this case, we can utilize the practical approximation by modelling $P(\mathbf{Y}|\mathbf{Z})$ rather than $P(\mathbf{Z}|\mathbf{Y})$ – it yielded better results and still allowed a fully GP-based framework. That is to say, the mapping into the 3D pose space from the feature space is given by a Gaussian process model $P(\mathbf{Y}|\mathbf{Z}, \theta_{yz})$ with a parametric kernel $\mathbf{K}_{yz}(\mathbf{z}_t, \mathbf{z}_t|\theta_{yz})$.

   As a result, the joint *conditional* model of the pose sequence $\mathbf{Y}$ and intrinsic motion $\mathbf{X}$, given the sequence of image features $\mathbf{Z}$ is approximated by

$$P(\mathbf{X}, \mathbf{Y}|\mathbf{Z}, \mathbf{A}, \beta, \theta_{yz}, \theta_{yx}) \approx P(\mathbf{Y}|\mathbf{Z}, \theta_{yz})P(\mathbf{X}|\mathbf{A})P(\mathbf{Y}|\mathbf{X}, \beta, \theta_{yx}).$$

Learning

In the training phase, both the image features $\mathbf{Z}$ and the corresponding poses $\mathbf{Y}$ are known. Hence, the learning of GP and NDS models becomes decoupled and can be accomplished using the NDS learning formalism presented in the previous section and a standard GP learning approach [25].

Algorithm 2 Human motion model learning

INPUT Image sequence $\mathbf{Z}$ and joint angle sequence $\mathbf{Y}$
OUTPUT Human motion model

1. Learn Gaussian Process model $P(\mathbf{Y}|\mathbf{Z}, \theta_{yz})$ using e.g. [25].
2. Learn NDS model $P(\mathbf{X}, \mathbf{Y}|\mathbf{A}, \beta, \theta_{yx})$ as described in Section 7.5.

Inference and Tracking

Once the models are learned they can be used for tracking of the human figure in video. Because both NDS and GP are nonlinear mappings, estimating current pose $\mathbf{y}_t$ given a previous pose and intrinsic motion space estimates $P(\mathbf{x}_{t-1}, \mathbf{y}_{t-1}|\mathbf{Z}_{0..t})$ will involve nonlinear optimization or linearizion, as suggested in Section 7.5.5. In particular, optimal point estimates $\mathbf{x}_t^*$ and $\mathbf{y}_t^*$ are the result of the following nonlinear optimization problem:

$$(\mathbf{x}_t^*, \mathbf{y}_t^*) = \arg\max_{\mathbf{x}_t, \mathbf{y}_t} P(\mathbf{x}_t|\mathbf{x}_{t-1}, \mathbf{A})P(\mathbf{y}_t|\mathbf{x}_t, \beta, \theta_{yx})P(\mathbf{y}_t|\mathbf{z}_t, \theta_{yz}). \qquad (7.9)$$

The point estimation approach is particularly suited for a particle-based tracker. Unlike some traditional approaches that only consider the pose space representation, tracking in the low-dimensional intrinsic space has the potential to avoid problems associated with sampling in high-dimensional spaces.

A sketch of the human motion tracking algorithm using particle filter with N_P particles and weights $(w^{(i)}, i = 1, ..., N_P)$ is shown below. We apply this algorithm to a set of tracking problems described in Section 7.7.2.

Algorithm 3 Particel filter in human motion tracking

INPUT Image $\mathbf{z}_t$, Human motion model (GP+NDS) and prior point estimates
$(w_{t-1}^{(i)}, \mathbf{x}_{t-1}^{(i)}, \mathbf{y}_{t-1}^{(i)})|\mathbf{Z}_{0..t-1}, i = 1, ..., N_P$
OUTPUT Current pose/intrinsic state estimates $(w_t^{(i)}, \mathbf{x}_t^{(i)}, \mathbf{y}_t^{(i)})|\mathbf{Z}_{0..t}, i = 1, ..., N_P$

1. Draw the initial estimates $\mathbf{x}_t^{(i)} \sim p(\mathbf{x}_t|\mathbf{x}_{t-1}^{(i)}, \mathbf{A})$.
2. Compute the initial poses $\mathbf{y}_t^{(i)}$ from the initial $\mathbf{x}_t^{(i)}$ and NDS model.
3. Find optimal estimates $(\mathbf{x}_t^{(i)}, \mathbf{y}_t^{(i)})$ using nonlinear optimization in Equation (7.9).
4. Find point weights $w_t^{(i)} \sim P(\mathbf{x}_t^{(i)}|\mathbf{x}_{t-1}, \mathbf{A})P(\mathbf{y}_t^{(i)}|\mathbf{x}_t^{(i)}, \beta, \theta_{yx})P(\mathbf{y}_t^{(i)}|\mathbf{z}_t, \theta_{yz})$.

7.6.2 Model 2

Given the sequence of image feature $\mathbf{Z}$, the joint *conditional* model of the pose sequence $\mathbf{Y}$ and the corresponding embedded sequence $\mathbf{X}$ is expressed as

$$P(\mathbf{X}, \mathbf{Y}|\mathbf{Z}, \mathbf{A}, \beta, \theta_{zx}, \theta_{yx}) = P(\mathbf{Z}|\mathbf{X}, \theta_{zx})P(\mathbf{X}|\mathbf{A})P(\mathbf{Y}|\mathbf{X}, \beta, \theta_{yx}).$$

Comparing with Model 1, this model needs no approximation related to the dimension of the image features.

Learning

Given both the sequence of poses and corresponding image features, a generalized EM algorithm as in Section 7.5.3 can be used to learn a set of model parameters $(\mathbf{A}, \beta, \theta_{zx}, \theta_{yx})$.

Algorithm 4 Human motion model learning with EM

INPUT Image sequence $\mathbf{Z}$ and joint angle sequence $\mathbf{Y}$
OUTPUT Human motion model

E-step $\mathbf{X}^* = \arg\max_{\mathbf{X}} P(\mathbf{X}, \mathbf{Y}|\mathbf{Z}, \mathbf{A}^*, \beta^*, \theta_{zx}^*, \theta_{yx}^*)$
M-step $(\mathbf{A}^*, \beta^*, \theta_{zx}^*, \theta_{yx}^*) = \arg\max_{(\mathbf{A}, \beta, \theta_{zx}, \theta_{yx})} P(\mathbf{X}^*, \mathbf{Y}|\mathbf{Z}, \mathbf{A}, \beta, \theta_{zx}, \theta_{yx})$

Inference and Tracking

Once the model is learned the joint probability of the pose $\mathbf{Y}$ and the image features $\mathbf{Z}$ can be approximated by

$$P(\mathbf{Y}, \mathbf{Z}) \approx P(\mathbf{Y}, \mathbf{Z}, \mathbf{X}^*) = P(\mathbf{Y}|\mathbf{X}^*)P(\mathbf{Z}|\mathbf{X}^*)P(\mathbf{X}^*)$$

where
$$\mathbf{X}^* = \arg\max_{\mathbf{X}} P(\mathbf{X}|\mathbf{Y}, \mathbf{Z}).$$

Because we have two conditionally independent GPs, estimating current pose (distribution) $\mathbf{y}_t$ and estimating current point $\mathbf{x}_t$ in the embedded space can be separated. First the optimal point estimate $\mathbf{x}_t^*$ is the result of the following optimization problem:

$$\mathbf{x}_t^* = \arg\max_{\mathbf{x}_t} P(\mathbf{x}_t|\mathbf{x}_{t-1}, \mathbf{A})P(\mathbf{z}_t|\mathbf{x}_t, \beta, \theta_{zx}). \tag{7.10}$$

A particle-based tracker is utilized to estimate the optimal $\mathbf{x}_t$ as tracking with the previous Model 1. A sketch of this procedure using particle filter with N_P particles and weights $(w^{(i)}, i = 1, ..., N_P)$ is shown below.

After estimating the optimal $\mathbf{x}_t^*$, we can easily compute the distribution of the pose $\mathbf{y}_t$ by using the same result from GP theory as Equation 7.7 and Eqaution 7.8 in Section 7.5.4. Note that Model 2 does explicitly provide the estimates of pose variance, which Model 1 does not. The mode can be selected as the final pose estimate:

$$\mathbf{y}_t^* = \mathbf{Y}'\mathbf{K}_{yx}(\mathbf{X}, \mathbf{X})^{-1}\mathbf{K}_{yx}(\mathbf{X}, \mathbf{x}_t^*).$$

Algorithm 5 Particel filter in human motion tracking

INPUT Image $\mathbf{z}_t$, Human motion model (NDS) and prior point estimates
$(w_{t-1}^{(i)}, \mathbf{x}_{t-1}^{(i)}, \mathbf{y}_{t-1}^{(i)})|\mathbf{Z}_{0..t-1}, i = 1, ..., N_P$

OUTPUT Current intrinsic state estimates $(w_t^{(i)}, \mathbf{x}_t^{(i)})|\mathbf{Z}_{0..t}, i = 1, ..., N_P$

1. Draw the initial estimates $\mathbf{x}_t^{(i)} \sim p(\mathbf{x}_t|\mathbf{x}_{t-1}^{(i)}, \mathbf{A})$.
2. Find optimal estimates $\mathbf{x}_t^{(i)}$ using nonlinear optimization in Equation (7.10).
3. Find point weights $w_t^{(i)} \sim P(\mathbf{x}_t^{(i)}|\mathbf{x}_{t-1}, \mathbf{A})P(\mathbf{z}_t^{(i)}|\mathbf{x}_t^{(i)}, \beta, \theta_{zx})$.

7.7 Experiments

7.7.1 Synthetic Data

In our first experiment we examine the utility of MAR priors in a subspace selection problem. A 2nd order AR model is used to generate sequences in a $\Re^{T \times 2}$ space; the sequences are then mapped to a higher dimensional nonlinear measurement space. An example of the measurement sequence, a periodic curve on the Swiss-roll surface, is depicted in Figure 7.9.

We apply two different methods to recover the intrinsic sequence subspace: MNDS with an RBF kernel and a GPLVM with the same kernel and independent Gaussian priors. Estimated embedded sequences are shown in Figure 7.10. The intrinsic motion sequence inferred by the MNDS model more closely resembles the "true" sequence in Figure 7.9. Note that one dimension (blue/dark) is reflected about the horizontal axis, because the embeddings are unique up to an arbitrary rotation. These results confirm that proper dynamic priors may have crucial role in learning of embedded sequence subspaces. We study the role of dynamics in tracking in the following section.

7.7.2 Human Motion Data

We conducted experiments using a database of motion capture data for a 59 d.o.f body model from CMU Graphics Lab Motion Capture Database [27]. Figure 7.11 shows the latent space resulting from the original GPLVM and our MNDS model. Note that there are breaks in the intrinsic sequence of the original GPLVM. On the other hand, the trajectory in the embedded space of MNDS model is smoother, without sudden breaks. Note that the precisions for the points corresponding to the training poses are also higher in our MNDS model.

For the experiments on human motion tracking, we utilize synthetic images as our training data similar to [1, 20]. Our database consists of seven walking sequences of around 2000 frames total. The data was generated using the software (3D human model and Maya binaries) generously provided by the authors of [17,18]. We train our GP and NDS models with one sequence of 250 frames and test on the remaining sequences. In our experiments, we exclude

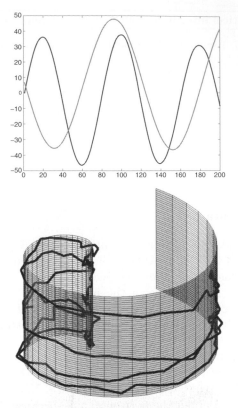

Fig. 7.9. A periodic sequence in the intrinsic subspace and the measured sequence on the Swiss-roll surface.

15 joint angles that exhibit small movement during walking (e.g., clavicle and figures joint) and use the remaining 44 joints. Our choice of image features are the silhouette-based Alt moments used in [13, 20]. The scale and translational invariance of Alt moments makes them suitable to a motion modelling task with little or no image-plane rotation.

In the model learning phase we utilize the approach proposed in Section 7.5. Once the model is learned, we apply the two tracking/inference approaches in Section 7.6 to infer motion states and poses from sequences of silhouette images. The pose estimation results with the two different models show no much difference. The big difference between two models is the speed, which we discuss in the following Section 7.7.3.

Figure 7.12 depicts a sequence of estimated poses. The initial estimates for gradient search are determined by the nearest neighborhood matching in the Alt moments space alone. To evaluate our NDS model, we estimate the same input sequence with the original GPLVM tracking in [20]. Although the silhouette features are informative for human pose estimation, they are also prone to ambiguities such as the left/right side changes. Without proper

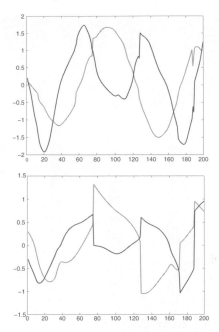

Fig. 7.10. Recovered embedded sequences. Left: MNDS. Right: GPLVM with iid Gaussian priors.

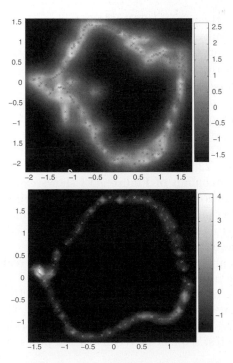

Fig. 7.11. Latent space with the grayscale map of log precision. Left: pure GPLVM. Right: MNDS.

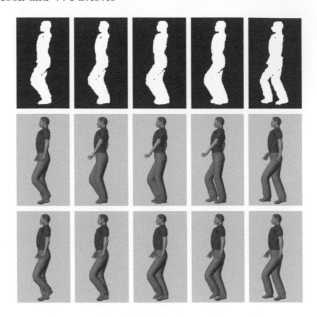

Fig. 7.12. Firs row: Input image silhouettes. Remaining rows show reconstructed poses. Second row: GPLVM model. Third row: NDS model.

dynamics modelling, the original GPLVM fails to estimate the correct poses because of this ambiguity.

The accuracy of our tracking method is evaluated using the mean RMS error between the true and the estimated joint angles [1], $D(\mathbf{y}, \mathbf{y}') = \frac{1}{44} \sum_{i=1}^{44} |(\mathbf{y}_i - \mathbf{y}'_i) mod \pm 180°|$. The first column of Figure 7.13 displays the mean RMS errors over the 44 joint angles, estimated using three different models. The testing sequence consists of 320 frames. The mean error for NDS model is in range $3° \sim 6°$. The inversion of right and left legs causes significant errors in the original GPLVM model. Introduction of simple dynamics in the pose space similar to [22] was not sufficient to rectify the "static" GPLVM problem. The second column of Figure 7.13 shows examples of trajectories in the embedded space corresponding to the pose estimates with the three different models. The points inferred from our NDS model follow the path defined by the MAR model, making them temporally consistent. The other two methods produced less-than-smooth embeddings.

We applied the algorithm to tracking of various real monocular image sequences. The data used in these experiments was the sideview sequence in CMU mobo database made publicly available under the HumanID project [26]. Figure 7.14 shows one example of our tracking result. This testing sequence consists of 340 frames. Because a slight mismatch in motion dynamics between the training and the test sequences, reconstructed poses are not geometrically

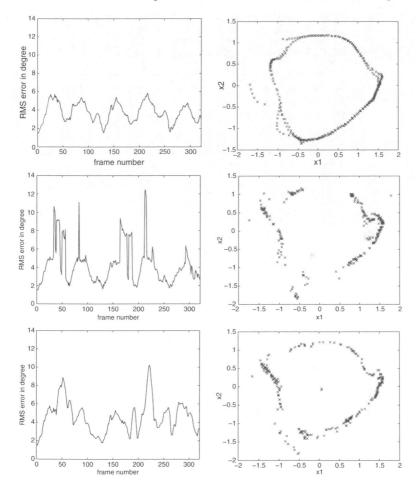

Fig. 7.13. Mean angular pose RMS errors and 2D latent space trajectories. First row: tracking using our NDS model. Seconde row: original GPLVM tracking. Third row: tracking using simple dynamics in the pose space.

perfect. However the overall result sequence depicts a plausible walking motion that agrees with the observed images.

It is also interesting to note that in a number of tracking experiments it was sufficient to carry a very small number of particles (~ 1) in the point-based tracker of Algorithm 3 and Algorithm 5. In most cases all particles clustered in a small portion of the motion subspace $\mathcal{X}$, even in ambiguous situations induced by silhouette-based features. This indicates that the presence of dynamics had an important role in disambiguating statically similar poses.

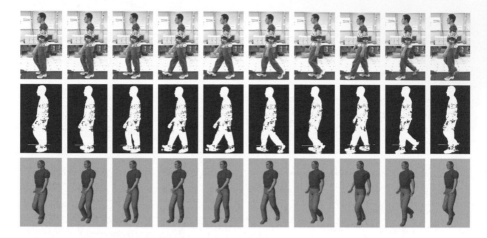

Fig. 7.14. First row: input real walking images. Second row: image silhouettes. Third row: images of the reconstructed 3D pose.

7.7.3 Complexity Comparison

Our experimental results show that Model 1 and Model 2 exhibit similar accuracies of estimating poses. The distinctive difference between the two models is the complexity in the learning and inference stages. The complexity in computing the objective function in the GPLVM is proportional to the dimension of a observation (pose) space. For the approximation of Model 1, a 44 dimensional pose is the observation for the two GP models, $P(\mathbf{Y}|\mathbf{X})$ and $P(\mathbf{Y}|\mathbf{Z})$. However, for Model 2 the pose is the observation for only one GP model $P(\mathbf{Y}|\mathbf{X})$ and the observation of the other GP model $P(\mathbf{Z}|\mathbf{X})$ (Alt Moments) has only 10 dimensions. This makes learning of Model 2 less complex than learning of Model 1. In addition, in inference only the latent variable (e.g. 3-dimension) is optimized in Model 2 while the optimization in Model 1 deals with both the latent variable and the pose (3-dimensions + 44-dimension in our experiments). As a result, Model 2 requires significantly fewer iterations of the nonlinear optimization search, leading to potentially more suitable algorithm for real-time tracking.

7.8 Conclusions

We proposed a novel method for embedding of sequences into subspaces of dynamic models. In particular, we propose a family of marginal AR (MAR) subspaces that describe all stable AR models. We show that a generative nonlinear dynamic system (NDS) can then be learned from a hybrid of Gaussian (latent) process models and MAR priors, a marginal NDS (MNDS). As a consequence, learning of NDS models and state estimation/tracking can be formulated in this new context. Several synthetic examples demonstrate the potential utility of the NDS framework and display its advantages over traditional static methods in dynamic domains. We also test the proposed approach on the problem of the 3D human figure tracking in sequences of monocular images. Our preliminary results indicate that dynamically constructed embeddings using NDS can resolve ambiguities during tracking that may plague static as well as less principled dynamic approaches.

In our future work we plan to extend the set of evaluations and gather more insight into theoretical and computational properties of MNDS with linear and nonlinear MARs. In particular, our ongoing experiments address the posterior multimodality in the embedded spaces, an issue relevant to point-based trackers.

As we noted in our experiments on 3D human figure tracking, the style problem should be resolved for ultimate automatic human tracking. Tensor decomposition has been introduced as a way to extract the motion signature capturing the distinctive pattern of movement of any particular individual [2]. It would be a promising direction to extend this linear decomposition technique to the novel nonlinear decomposition.

The computation cost for learning MNDS model is another issue to be consider. The fact that the cost is proportional to the number of data points indicates that we can utilize the sparse prior of data for more efficient leaning. We also plan to extend the NDS formalism to collections of dynamic models using the switching dynamics approaches as a way of modelling a general and diverse family of temporal processes.

Appendix: MAR Gradient

Log-likelihood of MAR model is, using Equation (7.4) and leaving out the constant term,

$$L = \frac{N}{2} \log |\mathbf{K}_{xx}| + \frac{1}{2} tr \left\{ \mathbf{K}_{xx}^{-1} \mathbf{X} \mathbf{X}' \right\} \tag{7.11}$$

with $\mathbf{K}_{xx} = \mathbf{K}_{xx}(\mathbf{X}, \mathbf{X})$ defined in Equation (7.5). Gradient of L with respect to $\mathbf{X}$ is

$$\frac{\partial L}{\partial \mathbf{X}} = \frac{\partial \mathbf{X}_\Delta}{\partial \mathbf{X}} \frac{\partial L}{\partial \mathbf{K}_{xx}} \frac{\partial \mathbf{K}_{xx}}{\partial \mathbf{X}_\Delta} + \frac{\partial L}{\partial \mathbf{X}} \bigg|_{\mathbf{X}_\Delta}. \tag{7.12}$$

$\mathbf{X}_\Delta$ can be written as a linear operator on $\mathbf{X}$,

$$\mathbf{X}_\Delta = \Delta \cdot \mathbf{X}, \quad \Delta = \begin{bmatrix} \mathbf{0}_{(T-1) \times 1} & \mathbf{I}_{(T-1) \times (T-1)} \\ \mathbf{0} & \mathbf{0}_{1 \times (T-1)} \end{bmatrix}, \tag{7.13}$$

where $\mathbf{0}$ and $\mathbf{I}$ denote zero vectors and identity matrices of sizes specified in the subscripts. It is now easily follows that

$$\frac{\partial L}{\partial \mathbf{X}} = \Delta' \left(N \mathbf{K}_{xx}^{-1} - \mathbf{K}_{xx}^{-1} \mathbf{X} \mathbf{X}' \mathbf{K}_{xx}^{-1} \right) \Delta \cdot \mathbf{X} + \mathbf{K}_{xx}^{-1} \mathbf{X}. \tag{7.14}$$

References

1. Agarwal A. and Triggs B. 3D Human Pose from Silhouettes by Relevance Vector Regression, CVPR, Vol. 2 (2004) 882–888
2. Vasilescu M.A.O. Human Motion Signatures: Analysis, Synthesis, Recognition, ICPR, Vol 3 (2002) 30456
3. Belkin M. and Niyogi P. Laplacian Eigenmaps for Dimensionality Reduction and Data Representation, Neural Computation, Vol. 15(6) (2003) 1373–1396
4. Brand M. Shadow Puppetry, CVPR, Vol. 2 (1999) 1237–1244
5. Elgammal A. and Lee C. Inferring 3D Body Pose from Silhouettes using Activity Manifold Learning, CVPR, Vol. 2 (2004) 681–688
6. Girko V. L. Circular Law, Theory of Probability and its Applications, Vol. 29 (1984) 694–706
7. Grochow K., Martin S. L., Hertzmann A., and Popovic Z. Style-based Inverse Kinematics, SIGGRAPH (2004) 522–531
8. Jolliffe I.T. Principal Component Analysis, Springer (1986)
9. Lawrence N.D. Gaussian Process Latent Variable Models for Visualisation of High Dimensional Data, NIPS (2004)
10. Lawrence N.D. Probabilistic Non-linear Principal Component Analysis with Gaussian Process Latent Variable Models, Journal of Machne Learning Research, Vol 6 (2005) 1783–1816
11. Mardia K.V., Kent J., and Bibby M. Multivariate Analysis, Academic Press (1979)
12. Ormoneit D., Sidenbladh H., Black M.J., and Hastie T. Learning and Tracking Cyclic Human, NIPS (2001) 894–900
13. Rosales R. and Sclaroff S. Specialized Mappings and the Estimation of Human Body Pose from a Single Image, Workshop on Human Motion (2000) 19–24
14. Roweis S.T. and Saul L.K. Nonlinear Dimensionality Reduction by Locally Linear Embedding, Science, Vol. 290 (2000) 2323–2326
15. Sidenbladh H., Black M.J., and Fleet D.J. Stochastic Tracking of 3D Human Figures Using 2D Image Motion, ECCV (2000) 702–718
16. Sminchisescu C. and Jepson A. Generative modelling for Continuous Non-linearly Embedded Visual Inference, ICML, ACM Press (2004)
17. Sminchisescu C., Kanaujia A., Li Z., and Metaxas D. Conditional Visual Tracking in Kernel Space, NIPS (2005)
18. Sminchisescu C., Kanaujia A., Li Z., and Metaxas D. Discriminative Density Propagation for 3D Human Motion Estimation, CVPR, Vol. 1 (2005) 390–397
19. Tenenbaum J.B., de Silva V., and Langford J.C. A Global Geometric Framework for Nonlinear Dimensionality Reduction, Science, Vol. 290 (2000) 2319–2323
20. Tian T. P., Li R., and Sclaroff S. Articulated Pose Estimation in a Learned Smooth Space of Feasible Solutions, Workshop on Learning in CVPR (2005) 50
21. Urtasun R., Fleet D.J., and Fua P. Monocular 3D Tracking of the Golf Swing, CVPR, Vol. 2 (2005) 1199
22. Urtasun R., Fleet D.J., Hertzmann A., and Fua P. Priors for People Tracking from Small Training Sets, ICCV, Vol. 1 (2005) 403–410
23. Wang J.M., Fleet D.J., and Hertzmann A. Gaussian Process Dynamical Models, NIPS (2005)

24. Wang Q., Xu G., and Ai H. Learning Object Intrinsic Structure for Robust Visual Tracking, CVPR, Vol. 2 (2003) 227–233
25. Williams C.K.I. and Barber D. Bayesian Classification with Gaussian Processes, PAMI, Vol. 20(12), (1998) 1342–1351
26. http://www.hid.ri.cmu.edu/Hid/databases.html
27. http://mocap.cs.cmu.edu/

3D Human Motion Analysis in Monocular Video: Techniques and Challenges

Cristian Sminchisescu

TTI-C, University of Chicago Press
1427 East 60th Street, Chicago, IL 60637
crismin@nagoya.uchicago.edu

Summary. Extracting meaningful 3D human motion information from video sequences is of interest for applications like intelligent human–computer interfaces, biometrics, video browsing and indexing, virtual reality or video surveillance. Analyzing videos of humans in unconstrained environments is an open and currently active research problem, facing outstanding scientific and computational challenges. The proportions of the human body vary largely across individuals, due to gender, age, weight or race. Aside from this variability, any single human body has many degrees of freedom due to articulation and the individual limbs are deformable due to moving muscle and clothing. Finally, real-world events involve multiple interacting humans occluded by each other or by other objects and the scene conditions may also vary due to camera motion or lighting changes. All these factors make appropriate models of human structure, motion and action difficult to construct and difficult to estimate from images. In this chapter we give an overview of the problem of reconstructing *3D human motion* using sequences of images acquired with a *single video camera*. We explain the difficulties involved, discuss ways to address them using generative and discriminative models and speculate on open problems and future research directions.

8.1 The Problem

The problem we address is the reconstruction of full-body 3D human motion in monocular video sequences. This can be formulated either as an *incremental* or as a *batch* problem. In incremental methods, images are available one at a time and one updates estimates of the human pose after each new image observation. This is known as filtering. Batch approaches estimate the pose at each timestep, using a sequence of images, prior and posterior to it. This is known as smoothing.

It is legitimate to ask why one should restrict attention to only one camera, as opposed to several, in order to attack an already difficult 3D inference problem? The answers are both practical and philosophical. On the practical side, often only a single image sequence is available, when processing

B. Rosenhahn et al. (eds.), Human Motion – Understanding, Modelling, Capture, and
Animation, 185–211.

and reconstructing movie footage, or when cheap devices are used as interface tools devoted to gesture or activity recognition. A more stringent practical argument is that, even when multiple cameras are available, general 3D reconstruction is complicated by occlusion from other people or scene objects. A robust human motion perception system has to necessarily deal with incomplete, ambiguous and noisy measurements. Fundamentally, these difficulties persist irrespective of how many cameras are used. From a philosophical viewpoint, reconstructing 3D structure using only one eye or a photograph is something that we, as humans, can do. We do not yet know how much is direct computation on 'objective' image information, and how much is prior knowledge in such skills, or how are these combined. But it is probably their conjunction that makes biological vision systems flexible and robust, despite being based on one eye or many. By attacking the 'general' problem instead of focusing on problem simplifications, we hope to make progress towards identifying components of such robust and efficient visual processing mechanisms.

Two general classes of strategies can be used for 3D inference:(i) *Generative (top-down) methods* optimize volumetric and appearance-based 3D human models for good alignment with image features. The objective is encoded as an observation likelihood or cost function with optima (ideally) centered at correct pose hypotheses; (ii) *Conditional (bottom-up) methods (also referred as discriminative or recognition-based)* predict human poses directly from images, typically using training sets of (pose, image) pairs. Difficulties exist in each case. Some of them, like data association are generic. Others are specific to the class of techniques used: optimizing generative models is expensive and many solutions may exist, some of which spurious, because human appearance is difficult to model accurately and because the problem is nonlinear; discriminative methods need to model complex multivalued image-to-3D (inverse) relations.

8.2 Difficulties

Extracting monocular 3D human motion poses several difficulties that we review. Some are inherent to the use of a single camera, others are generic computer vision difficulties that arise in any complex image understanding problem.

Depth 3D–2D projection ambiguities: Projecting the 3D world into images suppresses depth information. This difficulty is fundamental in computer vision. Inferring the world from *only one camera*, firmly places our research in the class of science dealing with inverse and ill-posed problems [5]. The non-uniqueness of solution when estimating human pose in monocular images is apparent in the 'forward-backward ambiguities' produced when positioning the human limbs, symmetrically, forwards or backwards, with respect to the camera 'rays of sight' (see Figure 8.1). Reflecting the limb angles in the

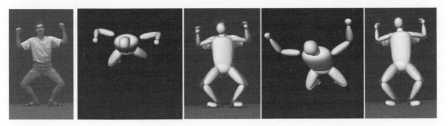

Fig. 8.1. Reflective Ambiguities (a, b, c, d, e). Original image (a). Two very different configurations of a 3D model (b and d) have image projections that align well with the contour of the imaged human subject (c and e).

frontoparallel plane leaves the image unchanged to first order. For generative models, ambiguities can lead to observation likelihood functions with multiple peaks of somewhat comparable magnitude. The distinction between a global and a local optimum becomes narrow – in this case, we are interested in all optima that are sufficiently good. For discriminative models, the ambiguities lead to multivalued image–pose relations that defeat function approximations based on neural networks or regression. The ambiguity is temporally persistent both under general smooth dynamical models [48] and under dynamics learned from typical human motions [47].

High-dimensional representation: Reconstructing 3D human motion raises the question as of what information is to be recovered and how to represent it. A priori, a model where the 3D human is discretized as densely as possible, with a set of 3D point coordinates, with independent structure and motion is as natural as any other, and could be the most realistic one. Nevertheless, in practice, this would be difficult to constrain since it has excess degrees of freedom for which the bare monocular images cannot account. Representing the human as a blob with centroid coordinates is the opposite extreme, that can be efficient and simpler to estimate at the price of not being particularly informative for 3D reasoning.[1] Consequently, a middle-ground has to be found. At present, this selection is based mostly on intuition and on facts from human structural anatomy. For 3D human tracking the preferred choice remains a kinematic representation with a skeletal structure covered with 'flesh' of more or less complex type (cones, cylinders, globally deformable surfaces). For motion estimation, the model can have, depending on the level of detail, in the order of 30–60 joint angle variables – enough to reproduce a reasonable class of human motions with accuracy. However, estimation in high-dimensional spaces is computationally expensive, and exhaustive or random search is practically infeasible. Existing algorithms rely on approximations or problem-dependent heuristics: temporal coherency, dynamical

---

[1] Apart from tractability constraints, the choice of a representation is also application dependent. For many applications, a hierarchy of models with different levels of complexity, depending on context, may be the most appropriate.

models, and symmetries (e.g., hypotheses generated using forward–backward flips of limbs, from a given configuration). From a statistical perspective, more rigorous is to follow a learned data-driven approach, i.e., a minimal representation with intrinsic dimension based on its capacity to synthesize the variability of human shapes and poses present in the tracking domain. Sections §8.4.3 and §8.5.1 discuss techniques for learning low-dimensional models and for estimating their intrinsic dimensionality.

Appearance modelling, clothing: Not operating with a anatomically accurate human body models is in most applications offset by outer clothing that deforms. This exhibits strong variability in shape and appearance, both being difficult to model.

Physical constraints: Physically inspired models based on kinematic and volumetric parameterizations can be used to reason about the physical constraints of real human bodies. For consistency, the body parts have to not penetrate eachother and the joint angles should only have limited intervals of variation (see Figure 8.2). For estimation, the presence of constraints is both good and bad news. The good news is that the admissible state space volume is smaller than initially designed, because certain regions are not reachable, and many physically unrealistic solutions may be pruned. The bad news is that handling the constraints automatically is nontrivial, especially for continuous optimization methods used in generative models.

Self-occlusion: Given the highly flexible structure of an articulated human body, self-occlusion between different body parts occurs frequently in monocular views and has to be accounted for. Self-occlusion is an observation

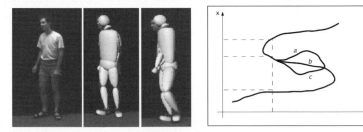

Fig. 8.2. *(Left)* Physical constraint violations when joint angle limits or body part non-penetration constraints are not enforced. *(Right)* Illustrative example of ambiguities during dynamic inference, for a model with 1D state x and observation r. The S-like distribution implies that multiple state hypotheses (shown in dashed) may exists for certain observations. The ambiguity persists for observations sequences commonly falling under each individual 'S-branch' (up, middle, bottom), see also Figure 8.6. The close loops created by the splitting-merging of trajectories a, b, and c abstract real imaging situations, as in Figure 8.1, see also [48]. Due to their loopy nature, these ambiguities cannot be resolved even when considering long observation timescales.

ambiguity (see section below). Several aspects are important. First is occlusion detection or prediction, so as to avoid the misattribution of image measurements to occluded model regions that have not generated any contribution to image appearance. The second aspect is the management of uncertainty in the position of the body parts that are not visible. Improperly handled this can produce singularities. It is appropriate to use prior-knowledge acquired during learning in order to constrain the uncertainty of unobserved body parts, based on the state of visible ones. Missing data is filled-in using learned correlations typically observed in natural human motions.

For generative models, occlusion raises the additional problem of constructing of an observation likelihood that realistically reflects the probability of different configurations under partial occlusion and viewpoint change. Independence assumptions are often used to fuse likelihoods from different measurements, but this conflicts with occlusion, which is a relatively coherent phenomenon. For realistic likelihoods, the probabilities of both occlusion and measurement have to be incorporated, but this makes the computations intractable.

General unconstrained motions: Humans move in diverse, but also highly structured ways. Certain motions have a repetitive structure like running or walking, others represent 'cognitive routines' of various levels of complexity, e.g., gestures during a discussion, or crossing the street by checking for cars to the left and to the right, or entering one's office in the morning, sitting down and checking e-mail. It is reasonable to think that if such routines could be identified in the image, they would provide strong constraints for tracking and reconstruction with image measurements serving merely to adjust and fine tune the estimate. However, human activities are not simply preprogrammed – they are parameterized by many cognitive and external unexpected variables (goals, locations of objects or obstacles) that are difficult to recover from images and several activities or motions are often combined.

Kinematic singularities: These arise when the kinematic Jacobian looses rank and the associated numerical instability can lead to tracking failure. An example is the nonlinear rotation representation used for kinematic chains, for which no singularity-free minimal representation exists.[2]

Observation ambiguities: Ambiguities arise when a subset of the model state cannot be directly inferred from image observations. They include but are by no means limited to kinematic ambiguities. Observability depends on the design of the observation model and image features used. (Prior knowledge becomes important and the solutions discussed for self-occlusion are applicable.) For instance when an imaged limb is straight and an edge-based observation likelihood is used with a symmetric body part model, rotations around the limb's own axis cannot be observed – the occluding contour changes little when the limb rotates around its own axis. Only when the elbow moves the

---

[2] Nonsingular overparameterizations exist, but they are not unique.

uncertain axial parameter values can be constrained. This may *not* be ambiguous under an intensity-based model, where the texture flow can make the rotation observable.

Data association ambiguities: Identifying which image features belong to the person and which to the background is a general vision difficulty known as data association. For our problem this is amplified by distracting clutter elements that resemble human body parts, e.g., various types of edges, ridges or pillars, trees, bookshelves, encountered in man-made and natural environments.

Lighting and motion blur: Lighting changes form another source of variability whenever image features based on edge or intensity are used. Artificial edges are created by cast shadows and inter-frame lighting variations could lead to complicated, difficult to model changes in image texture. For systems with a long shutter time, or during rapid motion, image objects appear blurred or blended with the background at motion boundaries. This has impact on the quality of both static feature extraction methods, and of frame to frame algorithms, such as the ones that compute the optical flow.

8.3 Approaches: Generative and Conditional Models

Approaches to tracking and modelling can be broadly classified as **generative** and **discriminative**. They are similar in that both require a state representation $\mathbf{x}$, here a 3D human model with kinematics (joint angles) or shape (surfaces or joint positions), and both use a set of image features as observations $\mathbf{r}$ for state inference. Often, a training set, $\mathcal{T} = \{(\mathbf{r}_i, \mathbf{x}_i) \mid i = 1 \ldots N\}$ sampled from the *joint distribution* is available. (For unsupervised problems, samples from *only* the state or *only* the observation distribution may be available to use.) The computational goal for both approaches is common: the conditional distribution, or a point estimate, for the model state, given observations.[3] Clearly, an important design choice is the state representation and the observation descriptor. The state should have representation and dimensionality well calibrated to the variability of the task, whereas the observation descriptor is subject to selectivity–invariance trade-offs: it needs to capture not only discriminative, subtle image detail, but also the strong, stable dependencies necessary for learning and generalization. Currently, these are by and large, obtained by combining a priori design and off-line unsupervised learning. But once decided upon, the representation (model state + observation descriptor) is no longer free, but known and fixed for subsequent learning

---

[3] This classification and statement of purpose is quite general. Methods may deviate from it in a way or another and shortcuts may be taken. But this should not undermine the usefulness of a framework for formal reasoning where to state the assumptions made and the models used, as well as the circumstances when these are expected to perform optimally (see Figure 8.3).

and inference stages. This holds notwithstanding of the method type, be it generative or discriminative.

Generative algorithms typically model the joint distribution using a constructive form of the observer – the observation likelihood, with maxima ideally centered at correct pose hypotheses. Inference involves complex state space search in order to locate the likelihood peaks, using either non-linear optimization or sampling. Bayes' rule is then used to compute the state conditional from the observation model and the state prior. Learning can be both supervised and unsupervised. This includes state priors [8, 13, 21, 44], low-dimensional models [47, 64] or learning the parameters of the observation model, e.g., texture, ridge or edge distributions, using problem-dependent, natural image statistics [38, 42]. Temporal inference is framed in a clear probabilistic and computational framework based on mixture filters or particle filters [12, 13, 23, 44, 56, 57, 59].

It has been argued that generative models can flexibly reconstruct complex unknown motions and can naturally handle problem constraints. It has been counter-argued that both flexibility and modelling difficulties lead to expensive, uncertainn = inference [13, 43, 48, 57], and a constructive form of the observer is both difficult to build and somewhat indirect with respect to the task, which requires conditional state estimation and not conditional observation modelling. These arguments motivate the complementary study of **discriminative algorithms** [2, 18, 34, 37, 41, 63] which model and predict the state conditional directly in order to simplify inference. Prediction however involves missing (state) data, unlike learning which is supervised. But learning is also difficult because modelling perceptual data requires adequate representations of highly multimodal distributions. The presence of multiple solutions in the image-to-pose mapping implies that, strictly, this is multivalued and cannot be functionally or globally approximated. However, several authors made initial progress using single hypothesis schemes [2, 18, 34, 41, 63]. E.g., nearest-neighbor [34, 41, 63] and regression [2, 18] have been used with good results. Others used mixture models [2, 37] to cluster the joint distribution of (observation, state) pairs and fitted function approximators (neural network or regressor) to each partition. In §8.5, we will review our BM^3E, a formal probabilistic model based on mixture of experts and conditional temporal chains [49, 51, 52].

Notation: We discuss generative and conditional models based on the graphical dependency in Figure 8.3. These have continuous temporal states $\mathbf{x}_t$, observations $\mathbf{r}_t$, observation model $p(\mathbf{r}_t|\mathbf{x}_t)$, and dynamics $p(\mathbf{x}_t|\mathbf{x}_{t-1})$, $t = 1\ldots T$ (for generative models). For conditional models, we model the conditional state distribution $p(\mathbf{x}_t|\mathbf{r}_t)$ and a previous state/current observation-based density $p(\mathbf{x}_t|\mathbf{x}_{t-1},\mathbf{r}_t)$. $\mathbf{X}_t = (\mathbf{x}_1, \mathbf{x}_2, \ldots, \mathbf{x}_t)$ is the model joint state estimated based on a time series of observations $\mathbf{R}_t = (\mathbf{r}_1, \ldots, \mathbf{r}_t)$.

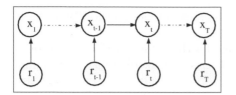 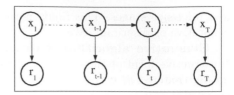

Fig. 8.3. A conditional/discriminative temporal chain model *(a, left)* reverses the direction of the arrows that link the state and the observation, compared with a generative one *(b, right)*. The state conditionals $p(\mathbf{x}_t|\mathbf{r}_t)$ or $p(\mathbf{x}_t|\mathbf{x}_{t-1}, \mathbf{r}_t)$ can be learned using training pairs and directly predicted during inference. Instead, a generative approach *(b)* will model and estimate $p(\mathbf{r}_t|\mathbf{x}_t)$ and do a more complex probabilistic inversion to compute $p(\mathbf{x}_t|\mathbf{r}_t)$ via Bayes' rule. Shaded nodes reflect variables that are not modeled but conditioned upon.

8.4 Generative Methods

Consider a nonlinear generative model $p_{\boldsymbol{\theta}}(\mathbf{x}, \mathbf{r})$ with $d = \dim(\mathbf{x})$, and parameters $\boldsymbol{\theta}$. Without loss of generality, assume a robust observation model:

$$p_{\boldsymbol{\theta}}(\mathbf{r}|\mathbf{x}) = (1 - w) \cdot \mathcal{N}(\mathbf{r}; \mathcal{G}(\mathbf{x}), \Sigma_{\boldsymbol{\theta}}) + o_{\boldsymbol{\theta}} \cdot w \qquad (8.1)$$

This corresponds to a mixture of a Gaussian having mean $\mathcal{G}(\mathbf{x})$ and covariance $\Sigma_{\boldsymbol{\theta}}$, and a uniform background of outliers $o_{\boldsymbol{\theta}}$ with proportions given by w. The outlier process is truncated at large values, so the mixture is normalizable.

In our case, the state space $\mathbf{x}$ represents human joint angles, the parameters $\boldsymbol{\theta}$ may include the Gaussian observation noise covariance, the weighting of outliers, the human body proportions, etc. $\mathcal{G}_{\boldsymbol{\theta}}(\mathbf{x})$ is a nonlinear transformation that predicts human contours, internal edges and possibly appearance (it includes nonlinear kinematics, occlusion analysis and perspective projection), according to consistent kinematic constraints. Alternatively, we also use an equivalent energy-based model – the maxima in probability or the minima in energy have similar meaning and are used interchangeably:

$$p_{\boldsymbol{\theta}}(\mathbf{x}, \mathbf{r}) = p_{\boldsymbol{\theta}}(\mathbf{r}|\mathbf{x})p(\mathbf{x}) = \frac{1}{Z_{\boldsymbol{\theta}}(\mathbf{x}, \mathbf{r})} \exp(-E_{\boldsymbol{\theta}}(\mathbf{x}, \mathbf{r})) \qquad (8.2)$$

$$E_{\boldsymbol{\theta}}(\mathbf{x}, \mathbf{r}) = -\log[(1 - w)\mathcal{N}(\mathbf{r}; \mathcal{G}(\mathbf{x}), \Sigma_{\boldsymbol{\theta}}) + o_{\boldsymbol{\theta}}w] + E_{\boldsymbol{\theta}}(\mathbf{x}) - \log Z_{\boldsymbol{\theta}}(\mathbf{x}, \mathbf{r}) \quad (8.3)$$

with prior $E_{\boldsymbol{\theta}}(\mathbf{x})$ and normalization constant $Z_{\boldsymbol{\theta}}(\mathbf{x}, \mathbf{r}) = \int_{(\mathbf{x}, \mathbf{r})} \exp(-E_{\boldsymbol{\theta}}(\mathbf{x}, \mathbf{r}))$. Notice that $Z_{\boldsymbol{\theta}}(\mathbf{x}) = \int_{\mathbf{r}} \exp(-E_{\boldsymbol{\theta}}(\mathbf{x}, \mathbf{r}))$ can be easily computed by sampling from the mixture of Gaussian and uniform outlier distribution, but computing $Z_{\boldsymbol{\theta}}(\mathbf{x}, \mathbf{r})$ and $Z_{\boldsymbol{\theta}}(\mathbf{r}) = \int_{\mathbf{x}} \exp(-E_{\boldsymbol{\theta}}(\mathbf{x}, \mathbf{r})$ is intractable because the averages are taken w.r.tlet@tokeneonedotthe unknown state distribution.[4]

---

[4] The choice of predicted and measured image features, hence the exact specification of the observation model, albeit very important, will not be further discussed.

8.4.1 Density Propagation Using Generative Models

For filtering, we compute the optimal state distribution $p(\mathbf{x}_t|\mathbf{R}_t)$, conditioned by observations $\mathbf{R}_t$ up to time t. The recursion can be derived as [20, 22, 24, 25, 46] (Figure 8.3b):

$$p(\mathbf{x}_t|\mathbf{R}_t) \;=\; \frac{1}{p(\mathbf{r}_t|\mathbf{R}_{t-1})} p(\mathbf{r}_t|\mathbf{x}_t) \int p(\mathbf{x}_t|\mathbf{x}_{t-1}) \, p(\mathbf{x}_{t-1}|\mathbf{R}_{t-1}) \mathbf{dx}_{t-1} \qquad (8.4)$$

The joint distribution factorizes as:

$$p(\mathbf{X}_T, \mathbf{R}_T) = p(\mathbf{x}_1) \prod_{t=2}^{T} p(\mathbf{x}_t|\mathbf{x}_{t-1}) \prod_{t=1}^{T} p(\mathbf{r}_t|\mathbf{x}_t) \qquad (8.5)$$

8.4.2 Optimization and Temporal Inference Algorithms

Several general-purpose sampling and optimization algorithms have been proposed in order to efficiently search the high-dimensional human pose space. In a temporal framework the methods keep a running estimate of the posterior distribution over state variable (either sample-based or mixture-based) and update it based on new observations. This works time-recursively, the starting point(s) for the current search being obtained from the results at the previous time step, perhaps according to some noisy dynamical model. To the (often limited) extent that the dynamics and the image matching cost are statistically realistic, Bayes-law propagation of a probability density for the true state is possible. For linearized unimodal dynamics and observation models under least squares/Gaussian noise, this leads to Extended Kalman Filtering. For likelihood-weighted random sampling under general multimodal dynamics and observation models, bootstrap filters [20] or CONDENSATION [23] result. In either case various model parameters must be tuned and it sometimes happens that physically implausible settings are needed for acceptable performance. In particular, to control mistracking caused by correspondence errors, selection of slightly incorrect inverse kinematics solutions, and similar model identification errors, visual trackers often require exaggerated levels of dynamical noise. The problem is that even quite minor errors can pull the state estimate a substantial distance from its true value, especially if they persist over several time steps. Recovering from such an error requires a state space jump greater than any that a realistic random dynamics is likely to provide, whereas using an exaggeratedly noisy dynamics provides an easily controllable degree of local randomization that often allows the mistracked estimate to jump back onto the right track. Boosting the dynamical noise does have the side effect of reducing the information propagated from past observations, and hence, increasing the local uncertainty associated with each mode. But this is a small penalty to pay for reliable tracking lock, and in any case the loss of accuracy is often minor in visual tracking, where weak dynamical models (i.e., short integration times: most of the state information comes from current observations and dynamical details are unimportant) are common. The critical component in most nowday trackers remains the method that searches the observation likelihood at a given timestep based on initializations from the previous one.

General search algorithms: Importance sampling [43] and annealing [13, 35] have been used to construct layered particle filters which sample with increased sensitivity to the underlying observation likelihood in order to better focus samples in probable regions. Methods based on Hybrid Monte Carlo [12, 17, 55] use the gradient of the sampling distribution in order to generate proposals that are accepted more frequently during a Markov Chain Monte Carlo simulation. Hyperdynamic Sampling [55] modifies the sampling distribution based on its local gradient and curvature in order to avoid undesirable trapping in local optima. This creates bumps in the regions of negative curvature in the core of the maxima. Samples are specifically repelled towards saddle-points, so to make inter-maxima transitions occur more frequently. Hyperdynamic Sampling is complementary and can be used in conjunction with both Hybrid-Monte Carlo and/or annealing. Non-parametric belief propagation [44, 59] progressively computes partial sample-based state estimates at each level of a temporal (or spatial, e.g., body-like structured) graphical model. It uses belief propagation and fits compact mixture approximations to the sample-estimated conditional posteriors at each level along the way.

Eigenvector Tracking and Hypersurface Sweeping [54] are saddle-point search algorithms. They can start at any given local minimum and climb uphill to locate a first-order saddle point – a stable point with only one negative curvature, hence a local maximum in one state space dimension and a local minimum in all the other dimensions. From the saddle it is easy to slide downhill to a nearby optimum using gradient descent and recursively resume the search. For high-dimensional problems many saddle points with different patterns of curvature exist, but the first-order ones are potentially the most useful. They are more likely to lead to low-cost nearby local minima because, from any given one, only one dimension is climbed uphill.

Problem specific algorithms: Covariance Scaled Sampling (CSS) [56] is a probabilistic method which represents the posterior distribution of hypotheses in state space as a mixture of long-tailed Gaussian-like distributions whose weights, centers and scale matrices ('covariances') are obtained as follows. Random samples are generated, and each is optimized (by nonlinear local optimization, respecting any joint constraints, etc.) to maximize the local posterior likelihood encoded by an image- and prior-knowledge-based cost function. The optimized likelihood value and position give the weight and center of a new component, and the inverse Hessian of the log-likelihood gives a scale matrix that is well adapted to the contours of the cost function, even for very ill-conditioned problems like monocular human tracking. However, when sampling, particles are deliberately scattered more widely than a Gaussian of this scale matrix (covariance) would predict, in order to probe more deeply for alternative minima.

Kinematic Jump Sampling (KJS) [57] is a domain-specific sampler, where each configuration of the skeletal kinematic tree has an associated *interpretation tree* – the tree of all fully or partially assigned 3D skeletal configurations that can be obtained from the given one by forwards/backwards flips. The tree contains only, and generically all, configurations that are image-consistent in the sense that their joint centers have the same image projections as the given one. (Some of these may still be inconsistent with other constraints: joint limits, body self-intersection or occlusion). The interpretation tree is constructed by traversing the kinematic tree from the root to the leaves. For each link, we construct the 3D sphere centered on the currently hypothesized position of the link's root, with radius equal to link length. This sphere is pierced by the camera ray of sight through the observed image position of the link's endpoint to give (in general) two possible 3D positions of the endpoint that are consistent with the image observation and the hypothesized parent position (see Figure 8.1). Joint angles are then recovered for each position using simple closed-form inverse kinematics. KJS can be used in conjunction with CSS in order to handle data association ambiguities. Both CSS and KJS can be used in conjunction with non-linear mixture smoothers [48] in order to optimally estimate multiple human joint angle *trajectory hypotheses* based on video sequences.

8.4.3 Learning

We review unsupervised and supervised methods for learning generative human models. These are applicable to obtain both model representations (state and observation) and parameters.

Learning Representations

Unsupervised methods have recently been used to learn state representations that are lower-dimensional, hence, better adapted for encoding the class of human motions in a particular domain, e.g., walking, running, conversations or jumps [31, 47, 64]. We discuss methods trained on sequences of high-dimensional joint angles obtained from human motion capture, but other representations, e.g., joint positions can be used. The goal is to reduce standard computations like visual tracking in the human joint angle state space – referred here as *ambient space*, to better constrained low-dimensional spaces referred as *perceptual* (or *latent*). Learning couples otherwise independent variables, so changes in *any* of the perceptual coordinates change *all* the ambient high-dimensional variables (Figure 8.4). The advantage of perceptual representations is that image measurements collected at *any* of the human body parts constrain *all* the body parts. This is useful for inference during partial visibility or self-occlusion. A disadvantage of perceptual representations is the loss of physical interpretation – joint angle limit constraints are simple to express and easy to enforce as per variable, localized inequalities in ambient space, but hard to separate in a perceptual space, where they involve (potentially complex) relations among all variables. The following aspects are important when designing latent variable models:

1. *Global perceptual coordinate system:* To make optimization efficient in a global coordinate system is necessary. This can be obtained with any of several dimensionality reduction methods including Laplacian Eigenmaps, ISOMAP, LLE, etc. [4, 14, 39, 61]. The methods represent the training set as a graph with local connections based on Euclidean distances between high-dimensional points. Local embeddings aim to preserve the local geometry of the dataset whereas ISOMAP conserves the global geometry (the geodesics on the manifold approximated as shortest paths in the graph). Learning the perceptual representation involves embedding the graph with minimal distortion. Alternatively the perceptual space can be represented with a mixture of low-dimensional local models with separate coordinate systems. In this case, one either has to manage the transition between coordinate systems by stitching their boundaries, or to align, post hoc, the local models in a global coordinate system. The procedure is more complex and the coordinates not used to estimate the alignment, or out of sample coordinates, may still not be unique. This makes global optimization based on gradient methods nontrivial.

2. *Preservation of intrinsic curvature:* The ambient space may be intrinsically curved due to the physical constraints of the human body or occlusion [15]. To preserve the structure of the ambient space when embedding, one needs to use methods that preserve the local geometry. e.g., Laplacian eigenmaps, LLE or Hessian embeddings [4, 14, 39]. ISOMAP would not be adequate, because geodesics running around a curved, inadmissible ambient region, will be mapped, at curvature loss, to straight lines in perceptual space.

3. *Intrinsic dimensionality:* It is important to select the optimal number of dimensions of a perceptual model. Too few will lead to biased, restricted models that cannot capture the variability of the problem. Too many dimensions will lead to high variance estimates during inference. A useful sample-based method to estimate the intrinsic dimensionality is based on the Hausdorff dimension, and measures the rate of growth in the number of neighbors of a point as the size of its neighborhood increases. At the well-calibrated dimensionality, the increase should be exponential in the intrinsic dimension. This is illustrated in Fig. 8.4, which shows analysis of walking data obtained using human motion capture. Figure 8.4(a) shows Hausdorff estimates for the intrinsic dimensionality: $d = \lim_{r \to 0} \frac{\log N(r)}{\log(1/r)}$, where r is the radius of a sphere centered at each point, and $N(r)$ are the number of points in that

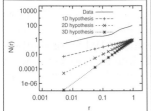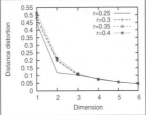

Fig. 8.4. Analysis of walking data. **(a)** Estimates of intrinsic dimensionality based on the Hausdorff dimension. **(b)** Geometric distortion vs. neighborhood size for a Laplacian embedding method. **(c)** Embedding of a walking data set of 2500 samples in 2d. Also shown, the Gaussian mixture prior (3 stdev), modelling the data density in perceptual space.

neighborhood (the plot is averaged over many nearby points). The slope of the curve in the linear domain $0.01 - 1$ corresponds roughly to a 1D hypothesis. Figure 8.4(b) plots the embedding distortion, computed as the normalized Euclidean SSE over each neighborhood in the training set. Here, 5–6 dimensions appear sufficient for a model with low-distortion.

4. *Continuous generative model:* Continuous optimization in a low-dimensional, perceptual space based on image observations requires not only a global coordinate system but also a global continuous mapping between the perceptual and observation spaces. Assuming the high-dimensional ambient model is continuous, the one obtained by reducing its dimensionality should also be. For example, a smooth mapping between the perceptual and the ambient space can be estimated using function approximation (e.g., kernel regression, neural networks) based on high-dimensional points in both spaces (training pairs are available once the embedding is computed). A perceptual continuous generative model enables the use of continuous methods for high-dimensional optimization [12, 56–58]. Working in perceptual spaces indeed targets dimensionality reduction but for many complex processes, even reduced representations would still have large dimensionality (e.g., 10D–15D) – efficient optimizers are still necessary.

5. *Consistent estimates* impose not only a prior on probable regions in perceptual space, as measured by the typical training data distribution, but also the separation of holes produced by insufficient sampling from genuine intrinsic curvature, e.g., due to physical constraints. The inherent sparsity of high-dimensional training sets makes the disambiguation difficult, but analytic expressions can be derived using a prior transfer approach. Ambient constrains can be related to perceptual ones, under a change of variables. If physical constraints are given as priors in ambient space $p_a(\mathbf{x}_a)$ and there exist a continuous perceptual-to-ambient mapping $\mathbf{x}_a = \mathbf{F}(\mathbf{x}), \forall \mathbf{x}$, with Jacobian $\mathbf{J_F}$, an equivalent prior in latent space is:

$$p(\mathbf{x}) \propto p_a(\mathbf{F}(\mathbf{x}))\sqrt{|\mathbf{J_F J_F^{\top}}|} \qquad (8.6)$$

Low-dimensional generative models based on principles (*1*)–(*5*) (or a subset of them) have been convincingly demonstrated for 3D human pose estimation [31, 47, 64].

Learning Parameters

Generative models are based on normalized probabilities parameterized by $\boldsymbol{\theta}$, that may encode the proportions of the human body, noise variances, feature weighting in the observation model, or the parameters of the dynamical model. For inference, the normalization is not important. For learning, the normalizer is essential in order to ensure that inferred model state distributions peak in the correct regions when presented with typical image data. Here, we only review learning methods for a static generative model $p_{\boldsymbol{\theta}}(\mathbf{x}, \mathbf{r})$, learning in video will instead use the joint distribution at multiple timesteps $p_{\boldsymbol{\theta}}(\mathbf{X}_T, \mathbf{R}_T)$. It is convenient to work with probabilistic quantities given as Boltzmann distributions, with uniform state priors, c.f. (8.2). Assuming a supervised training set of state-observation pairs, $\{\mathbf{x}^i, \mathbf{r}^i\}_{i=1...N}$, one can use Maximum Likelihood to optimize the model parameters using a free energy cost function:

$$\mathcal{F} = -\frac{1}{N}\sum_{n=1}^{N} \log p_{\boldsymbol{\theta}}(\mathbf{x}^n, \mathbf{r}^n) = \langle E_{\boldsymbol{\theta}}(\mathbf{x}, \mathbf{r}) \rangle_{data} + \log Z_{\boldsymbol{\theta}}(\mathbf{x}, \mathbf{r}) \qquad (8.7)$$

To minimize the free energy we need to compute its gradients:

$$\frac{d\mathcal{F}}{d\boldsymbol{\theta}} = \left\langle \frac{dE_{\boldsymbol{\theta}}(\mathbf{x}, \mathbf{r})}{d\boldsymbol{\theta}} \right\rangle_{data} - \left\langle \frac{dE_{\boldsymbol{\theta}}(\mathbf{x}, \mathbf{r})}{d\boldsymbol{\theta}} \right\rangle_{model} \qquad (8.8)$$

where the second term is equal to the negative derivative of the log-partition function w.r.tlet@tokeneonedot$\boldsymbol{\theta}$. Note that the only difference between the two terms in (8.8) is the distribution used to average the energy derivative. In the first term we use the empirical distribution, i.e., we simply average over the available data-set. In the second term however we average over the model distribution as defined by the current setting of the parameters. Computing the second average analytically is typically too complicated, and approximations are needed.[5] An unbiased estimate can be obtained by replacing the integral by a sample average, where the sample is to be drawn from the model $p_{\boldsymbol{\theta}}(\mathbf{x}, \mathbf{r})$. Any of the approximate optimization or inference methods described in §8.4.2 can be used. The goal of learning is to update the model parameters in order to make the training data likely. Normalizing using the partition function $Z_{\boldsymbol{\theta}}$ ensures discrimination: making the true solution likely automatically makes

---

[5] The problem is simpler if the prior energy $E_{\boldsymbol{\theta}}(\mathbf{x})$ is fixed and not learned and only the 'easier' partition function $Z_{\boldsymbol{\theta}}(\mathbf{x})$ needs to be computed. The problem remains hard ($Z_{\boldsymbol{\theta}}(\mathbf{r})$) for a hybrid conditional model expressed using generative energies.

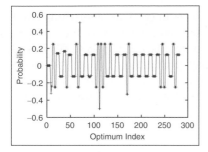 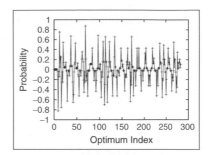

Fig. 8.5. We show the trajectory probability through each optimum of the observation model at each timestep in a video sequence before (left) and after ML learning (right). The video films a person walking towards a camera and doing a bow [48]. The time is unfolded on the x-axis and we switch sign in between successive timesteps for visualization (the values are all normally positive). Before learning, the temporal trajectory distribution collapses to fewer components in regions where the uncertainty of the model-image matching cost diminishes, but is multimodal and has high entropy. The distribution has lower entropy after learning, showing the usefulness of this procedure. The ambiguity diminishes significantly, but does not disappear. The entropy of the state posterior after learning reflects some of the limits of modelling and gives intuition about run-time speed and accuracy.

the incorrect competing solutions unlikely. ML learning iteratively reshapes the model state probability distribution to (at least!) infer the correct result on the training set. Results obtained using this learning method to estimate the parameters of a generative model (noise variances, weighting of the image features and the variance of a Gaussian dynamical model) are shown in Figure 8.5. This corresponds to the video sequence in [48], which films a person walking towards the camera and doing a bow.

8.5 Conditional and Discriminative Models

In this section we describe BM^3E, a Conditional $\underline{B}$ayesian $\underline{M}$ixture of $\underline{E}$xperts $\underline{M}$arkov $\underline{M}$odel for probabilistic estimates in discriminative visual tracking. The framework applies to temporal, uncertain inference for *continuous state-space models*, and represents the bottom-up counterpart of pervasive top-down generative models estimated with Kalman filtering or particle filtering (§8.4).[6] But instead of inverting a generative observation model at runtime, we learn to cooperatively predict complex state distributions directly from descriptors encoding image observations. These are integrated in a conditional graphical model in order to enforce temporal smoothness constraints

---

[6] Unlike most generative models, systems based on BM^3E can automatically initialize and recover from failure – an important feature for reliable 3D human pose tracking.

and allow a principled management of uncertainty. The algorithms combine sparsity, mixture modelling, and nonlinear dimensionality reduction for efficient computation in high-dimensional continuous state spaces. We introduce two key technical aspects: (1) The density propagation rules for *discriminative inference* in continuous, temporal chain models; (2) Flexible algorithms for *learning* feedforward, multimodal state distributions based on compact, conditional Bayesian mixture of experts models.

8.5.1 The BM^3E Model

Discriminative Density Propagation

We work with a conditional model having chain structure, as in Figure 8.3a. The filtered density can be derived using the conditional independence assumptions in the graphical model in Figure 8.3a [33, 51, 52]:

$$p(\mathbf{x}_t|\mathbf{R}_t) = \int p(\mathbf{x}_t|\mathbf{x}_{t-1},\mathbf{r}_t)p(\mathbf{x}_{t-1}|\mathbf{R}_{t-1})\mathbf{dx}_{t-1} \qquad (8.9)$$

The conditional joint distribution for T timesteps is:

$$p(\mathbf{X}_T|\mathbf{R}_T) = p(\mathbf{x}_1|\mathbf{r}_1) \prod_{t=2}^{T} p(\mathbf{x}_t|\mathbf{x}_{t-1},\mathbf{r}_t) \qquad (8.10)$$

In fact, (8.9) and (8.10) can be derived even more generally, based on a predictive conditional that depends on a larger window of observations up to time t [49], but the advantage of these models has to be contrasted to: (*i*) Increased amount of data required for training due to higher dimensionality. (*ii*) Increased difficulty to generalize due to sensitivity to timescale and/or alignment with a long sequence of past observations.

In practice, one can model $p(\mathbf{x}_t|\mathbf{x}_{t-1},\mathbf{r}_t)$ as a conditional Bayesian mixture of M experts (c.f let@tokeneonedot §8.5.1). The prior $p(\mathbf{x}_{t-1}|\mathbf{R}_{t-1})$ is also represented as a Gaussian mixture with M components. To compute the filtered posterior, one needs to integrate M^2 pairwise products of Gaussians analytically, and use mixture of Gaussian simplification and pruning methods to prevent the posterior from growing exponentially [46, 48].

A discriminative corrective conditional $p(\mathbf{x}_t|\mathbf{x}_{t-1},\mathbf{r}_t)$ can be more sensitive to incorrect previous state estimates than 'memoryless' distributions like $p(\mathbf{x}_t|\mathbf{r}_t)$. However we assume, as in any probabilistic approach, that the training and testing data are representative samples from the true underlying distributions in the domain. In practice, for improved robustness it is straightforward to include an importance sampler based on $p(\mathbf{x}_t|\mathbf{r}_t)$ to Equation. (8.9) – as necessary for initialization or for recovery from transient failure. Equivalently, a model based on a mixture of memoryless and dynamic distributions can be used.

Conditional Bayesian Mixture of Experts Model

This section describes the methodology for learning multimodal conditional distributions for discriminative tracking ($p(\mathbf{x}_t|\mathbf{r}_t)$ and $p(\mathbf{x}_t|\mathbf{x}_{t-1},\mathbf{r}_t)$ in §8.5.1). Many perception problems like 3D reconstruction require the computation of inverse, intrinsically multivalued mappings. The configurations corresponding to different static or dynamic estimation ambiguities are peaks in the (multi-modal) conditional state distribution (Figure 8.6). To represent them, we use several 'experts' that are simple function approximators. The experts transform their inputs[7] to output predictions, combined in a probabilistic mixture model based on Gaussians centered at their mean value. The model is consistent across experts and inputs, i.e., the mixing proportions of the experts reflect the distribution of the outputs in the training set and they sum to 1 for every input. Some inputs are predicted competitively by multiple experts and have multimodal state conditionals. Other 'unambiguous' inputs are predicted by a single expert, with the others effectively switched-off, having negligible probability (see Figure 8.6). This is the rationale behind a *conditional* Bayesian mixture of experts, and provides a powerful mechanism for contextually modelling complex multimodal distributions. Formally this is described by:

$$Q_{\boldsymbol{\nu}}(\mathbf{x}|\mathbf{r}) = p(\mathbf{x}|\mathbf{r},\mathbf{W},\boldsymbol{\Omega},\boldsymbol{\lambda}) = \sum_{i=1}^{M} g(\mathbf{r}|\boldsymbol{\lambda}_i)p(\mathbf{x}|\mathbf{r},\mathbf{W}_i,\boldsymbol{\Omega}_i^{-1}) \qquad (8.11)$$

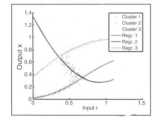 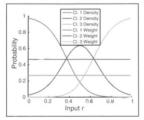 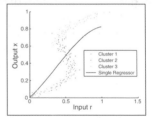

Fig. 8.6. An illustrative dataset [6] consists of about 250 values of x generated uniformly in $(0,1)$ and evaluated as $r = x + 0.3\sin(2\pi x) + \epsilon$, with ϵ drawn from a zero mean Gaussian with standard deviation 0.05. Notice that $p(x|r)$ is multimodal. *(a) Left* shows the data colored by the posterior membership probability h (assignment of points to experts) of three expert kernel regressors. *(b) Middle* shows the gates g (8.12), as a function of the input, but also the three uniform probabilities (of the joint distribution) that are computed by a clusterwise regressor [37]. *(c) Right* shows how a single kernel regressor cannot represent a multivalued dependency (it may either average the different values or commit to an arbitrary one, depending on the kernel parameters).

---

[7] The 'inputs' can be either observations $\mathbf{r}_t$, when modelling $p(\mathbf{x}_t|\mathbf{r}_t)$ or observation-state pairs $(\mathbf{x}_{t-1},\mathbf{r}_t)$ for $p(\mathbf{x}_t|\mathbf{x}_{t-1},\mathbf{r}_t)$. The 'output' is the state throughout. Notice that temporal information is used to learn $p(\mathbf{x}_t|\mathbf{x}_{t-1},\mathbf{r}_t)$.

where:

$$g(\mathbf{r}|\boldsymbol{\lambda}_i) = \frac{f(\mathbf{r}|\boldsymbol{\lambda}_i)}{\sum_{k=1}^{M} f(\mathbf{r}|\boldsymbol{\lambda}_k)} \tag{8.12}$$

$$p(\mathbf{x}|\mathbf{r}, \mathbf{W}_i, \boldsymbol{\Omega}_i) = \mathcal{N}(\mathbf{x}|\mathbf{W}_i\boldsymbol{\Phi}(\mathbf{r}), \boldsymbol{\Omega}_i^{-1}) \tag{8.13}$$

Here $\mathbf{r}$ are input or predictor variables, $\mathbf{x}$ are outputs or responses, g are *input dependent* positive gates, computed in terms of functions $f(\mathbf{r}|\boldsymbol{\lambda}_i)$, parameterized by $\boldsymbol{\lambda}_i$. f needs to produce gates g within $[0,1]$, the exponential and the softmax functions being natural choices: $f_i(\mathbf{r}|\boldsymbol{\lambda}_i) = \exp(\boldsymbol{\lambda}_i^{\top}\mathbf{r})$. Notice how g are normalized to sum to 1 for consistency, by construction, for any given input $\mathbf{r}$. We choose p as Gaussians (8.13) with covariances $\boldsymbol{\Omega}_i^{-1}$, centered at different expert predictions, here kernel ($\boldsymbol{\Phi}$) regressors with weights $\mathbf{W}_i$. Both the experts and the gates are learned using sparse Bayesian methods, which provide an automatic relevance determination mechanism [32, 62] to avoid overfitting and encourage compact models with fewer nonzero weights for efficient prediction. The parameters of the model, including experts and gates are collectively stored in $\boldsymbol{\nu} = \{(\mathbf{W}_i, \boldsymbol{\alpha}_i, \boldsymbol{\Omega}_i, \boldsymbol{\lambda}_i, \boldsymbol{\beta}_i) \mid i = 1 \ldots M\}$.

Learning the conditional mixture of experts involves two layers of optimization. As in many prediction problems, one optimizes the parameters $\boldsymbol{\nu}$ to maximize the log-likelihood of a data set, $\mathcal{T} = \{(\mathbf{r}_i, \mathbf{x}_i) \mid i = 1 \ldots N\}$, i.e., the accuracy of predicting $\mathbf{x}$ given $\mathbf{r}$, averaged over the data distribution. For learning, we use a double-loop EM algorithm. This proceeds as follows. In the *E-step* we estimate the posterior over assignments of training points to experts (there is one hidden variable h for each expert-training pair). This gives the probability that the expert i has generated the data n, and requires knowledge of both inputs and outputs. In the *M-step*, two optimization problems are solved: one for each expert and one for its gate. The first learns the expert parameters $(\mathbf{W}_i, \boldsymbol{\Omega}_i)$, based on training data $\mathcal{T}$, weighted according to the current h estimates (the covariances $\boldsymbol{\Omega}_i$ are estimated from expert prediction errors [66]). The second optimization teaches the gates g how to predict h.[8] The solutions are based on ML-II, with greedy (expert weight) subset selection. This strategy aggressively sparsifies the experts by eliminating inputs with small weights after each iteration [62, 68]. The approximation can can be interpreted as a limiting series of variational approximations (Gaussians with decreasing variances), via dual forms in weight space [68]. **Inference** (state prediction) is straightforward using (8.11). The result is a conditional mixture distribution with components and mixing probabilities that are input-dependent. In Figure 8.6 we explain the model using an illustrative toy example, and show the relation with clusterwise and (single-valued) regression.

---

[8] Prediction based on the input *only* is essential for output prediction (state inference), where membership probabilities h cannot be computed because the output is missing.

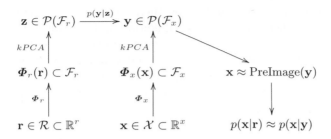

Fig. 8.7. A learned *conditional Bayesian mixture of low-dimensional kernel-induced experts* predictor to compute $p(\mathbf{x}|\mathbf{r}) \equiv p(\mathbf{x}_t|\mathbf{r}_t), \forall t$. (One can similarly learn $p(\mathbf{x}_t|\mathbf{x}_{t-1}, \mathbf{r}_t)$, with input $(\mathbf{x}, \mathbf{r})$ instead of $\mathbf{r}$ – here we illustrate only $p(\mathbf{x}|\mathbf{r})$ for clarity.) The input $\mathbf{r}$ and the output $\mathbf{x}$ are decorrelated using Kernel PCA to obtain $\mathbf{z}$ and $\mathbf{y}$ respectively. The kernels used for the input and output are $\boldsymbol{\Phi}_r$ and $\boldsymbol{\Phi}_x$, with induced feature spaces $\mathcal{F}_r$ and $\mathcal{F}_x$, respectively. Their principal subspaces obtained by kernel PCA are denoted by $\mathcal{P}(\mathcal{F}_r)$ and $\mathcal{P}(\mathcal{F}_x)$, respectively. A conditional Bayesian mixture of experts $p(\mathbf{y}|\mathbf{z})$ is learned using the low-dimensional representation $(\mathbf{z}, \mathbf{y})$. Using learned local conditionals of the form $p(\mathbf{y}_t|\mathbf{z}_t)$ or $p(\mathbf{y}_t|\mathbf{y}_{t-1}, \mathbf{z}_t)$, temporal inference can be efficiently performed in a *low-dimensional kernel induced state space* (see (8.9) where $\mathbf{y} \leftarrow \mathbf{x}$ and $\mathbf{z} \leftarrow \mathbf{r}$). For visualization and error measurement, the filtered density $p(\mathbf{y}_t|\mathbf{Z}_t)$ can be mapped back to $p(\mathbf{x}_t|\mathbf{R}_t)$ using a pre-image calculation.

Learning Conditional Bayesian Mixtures over Kernel Induced State Spaces

For many human visual tracking tasks, low-dimensional models are appropriate, because the components of the human state and of the image observation vector exhibit strong correlations, hence, low intrinsic dimensionality. In order to efficiently model conditional mappings between high-dimensional spaces with strongly correlated dimensions, we rely on kernel nonlinear dimensionality reduction and conditional mixture prediction, as introduced in §8.5.1. One can use nonlinear methods like kernel PCA [40, 67] and account for the structure of the problem, where both the inputs and the outputs are likely to be low-dimensional and their mapping multivalued (Figure 8.7). Since temporal inference is performed in the low-dimensional kernel induced state space, backtracking to high-dimensions is only necessary for visualization or error reporting.

8.6 Learning Joint Generative-Recognition Models

In the previous sections we have reviewed both generative (top-down) and conditional (bottom-up, recognition) models. Despite being a natural way to model the appearance of complex articulated structures, the success of generative models (§8.4)) has been partly shadowed because it is computational demanding to infer the distribution on their hidden states (human joint angles) and because their parameters are unknown and variable across many real scenes. In turn, conditional models are simple to understand and fast, but often need a generative model for training and could be blind-sighted by the lack of feedback for self-assessing accuracy. In summary, what appears to be necessary is a mechanism to consistently integrate top-down and bottom-up processing: the flexibility of 3D generative modelling (represent a large set of possible poses of human body parts, their correct occlusion and foreshortening relationships and their consistency with the image evidence) with the speed and simplicity of feed-forward processing. In this section we sketch one possible way to meet these requirements based on a bidirectional model with both recognition and generative sub-components – see [53] for details. *Learning* the parameters alternates self-training stages in order to maximize the probability of the observed evidence (images of humans). During one step, the recognition model is trained to invert the generative model using samples drawn from it. In the next step, the generative model is trained to have a state distribution close to the one predicted by the recognition model. At local equilibrium, which is guaranteed, the two models have consistent, registered parameterizations. During *online inference*, the estimates can be driven mostly by the fast recognition model, but may include generative (consistency) feedback.

The goal of both learning and inference is to maximize the probability of the evidence (observation) under the data generation model:

$$\log p_\theta(\mathbf{r}) = \log \int_{\mathbf{x}} p_\theta(\mathbf{x}, \mathbf{r}) = \log \int_{\mathbf{x}} Q_\nu(\mathbf{x}|\mathbf{r}) \frac{p_\theta(\mathbf{x}, \mathbf{r})}{Q_\nu(\mathbf{x}|\mathbf{r})} \tag{8.14}$$

$$\geq \int_{\mathbf{x}} Q_\nu(\mathbf{x}|\mathbf{r}) \log \frac{p_\theta(\mathbf{x}, \mathbf{r})}{Q_\nu(\mathbf{x}|\mathbf{r})} = KL(Q_\nu(\mathbf{x}|\mathbf{r})||p_\theta(\mathbf{x}, \mathbf{r})) \tag{8.15}$$

which is based on Jensen's inequality [25], and KL is the Kullback–Leibler divergence between two distributions. For learning, (8.14) will sum over the observations in the training set, omitted here for clarity. We have introduced a variational distribution Q_ν and have selected it to be exactly the recognition model. This is the same as maximizing a lower bound on the log-marginal (observation) probability of the generative model, with equality when $Q_\nu(\mathbf{x}|\mathbf{r}) = p_\theta(\mathbf{x}|\mathbf{r})$.

$$\log p_\theta(\mathbf{r}) - KL(Q_\nu(\mathbf{x}|\mathbf{r})||p_\theta(\mathbf{x}|\mathbf{r})) = KL(Q_\nu(\mathbf{x}|\mathbf{r})||p_\theta(\mathbf{x}, \mathbf{r})) \tag{8.16}$$

According to (8.14) and (8.16), optimizing a variational bound on the observed data is equivalent to minimizing the KL divergence between the state distribution inferred by the generative model $p(\mathbf{x}|\mathbf{r})$ and the one predicted by the recognition model $Q_\nu(\mathbf{x}|\mathbf{r})$. This is equivalent to minimizing the KL divergence between the recognition distribution and the joint distribution $p_\theta(\mathbf{x}, \mathbf{r})$ – the cost function we work with:

Algorithm for Bidirectional Model Learning

E-step: $\nu^{k+1} = \arg\max_\nu \mathcal{L}(\nu, \boldsymbol{\theta}^k)$

Train the *recognition* model using samples from the current *generative* model.

M-step: $\boldsymbol{\theta}^{k+1} = \arg\max_{\boldsymbol{\theta}} \mathcal{L}(\nu^{k+1}, \boldsymbol{\theta})$

Train the *generative* model to have state posterior close to the one predicted by the current *recognition* model.

Fig. 8.8. Variational Expectation-Maximization (VEM) algorithm for jointly learning a generative and a recognition model.

$$KL(Q_\nu(\mathbf{x}|\mathbf{r})\|p_{\boldsymbol{\theta}}(\mathbf{x}, \mathbf{r})) = -\int_{\mathbf{x}} Q_\nu(\mathbf{x}|\mathbf{r}) \log Q_\nu(\mathbf{x}|\mathbf{r}) \qquad (8.17)$$

$$+ \int_{\mathbf{x}} Q_\nu(\mathbf{x}|\mathbf{r}) \log p_{\boldsymbol{\theta}}(\mathbf{x}, \mathbf{r}) = \mathcal{L}(\nu, \boldsymbol{\theta}) \qquad (8.18)$$

The cost $\mathcal{L}(\nu, \boldsymbol{\theta})$ balances two conflicting goals: assign values to states that have high probability under the generative model (the second term), but at the same time be as uncommitted as possible (the first term measuring the entropy of the recognition distribution). The gradient-based learning algorithm is summarized in Figure 8.8 and is guaranteed to converge to a locally optimal solution for the parameters. The procedure is, in principle, self-supervised (one has to only provide the image of a human *without* the corresponding 3D human joint angle values), but one can initialize by training the recognition and the generative models separately using techniques described in §8.4 and §8.5.

Online inference (3D reconstruction and tracking) is straightforward using the E-step in Figure 8.8. But for efficiency one can work only with the recognition model c.f.let@tokeneonedot (8.11) and only do generative inference (full E-step) when the recognition distribution has high entropy. The model then effectively switches between a discriminative density propagation rule [51,52] and a generative propagation rule [13,24,42,47]. This offers a natural 'exploitation-exploration' or prediction-search tradeoff. An integrated 3D temporal predictor based on the model operates similarly to existing 2D object detectors. It searches the image at different locations and uses the recognition model to hypothesize 3D configurations. Feedback from the generative model helps to downgrade incorrect competing 3D hypotheses and to decide on the detection status (human or not) at the analyzed image sub-window. In Figure 8.9 we show results of this model for the *automatic* reconstruction of 3D human motion in environments with background clutter. The framework provides a uniform treatment of human detection, 3D initialization and 3D recovery from transient failure.

Fig. 8.9. Automatic human detection and 3D reconstruction using a learned generative-recognition model that combines bottom-up and top-down processing [53]. This shows some of difficulties of *automatically* detecting people and reconstructing their 3D poses in the real world. The background is cluttered, the limb constrast is often low, and there is occlusion from other objects (e.g., the chair) or people.

8.7 Training Sets and Representation

It is difficult to obtain ground truth for 3D human motion and even harder to train using many viewpoints or lighting conditions. In order to gather data one can use packages like Maya (Alias Wavefront) with realistically rendered computer graphics human surface models, animated using human motion capture [2,18,37,41,47,51,52,63]. 3D human data capture databases have emerged more recently for both motion capture [1, 38] and for human body laser-scans [3]. Alternatively, datasets based on photo-realistic multicamera human reconstruction algorithms can be used [10]. The human representation ($\mathbf{x}$) is usually based on an articulated skeleton with spherical joints, and may have 30–60 d.o.flet@tokeneonedot.

8.8 Challenges and Open Problems

One of the main challenges for the human motion sensing community today is to automatically understand people *in vivo*. We need to find where the people are, infer their poses, recognize what they do and perhaps what objects do they use or interact with. However, many of the existing human tracking systems tend to be complex to build and computationally expensive. The human structural and appearance models used are often built off-line and learned only to a limited extent. The algorithms cannot seamlessly deal with high structural variability, multiple interacting people and severe occlusion or lighting changes, and the resulting full body reconstructions are often qualitative yet not photorealistic. An entirely convincing transition between the laboratory and the real world remains to be realized.

In the long run, in order to build reliable human models and algorithms for complex, large-scale tasks, it is probable that learning will play a major role. Central themes are likely to be the choice of representation and its generalization properties, the role of bottom-up and top-down processing, and the importance of efficient search methods. Exploiting the problem structure and the scene context can be critical in order to limit inferential ambiguities. Several directions may be fruitful to investigate in order to advance existing algorithms:

- The role of representation. Methods to automatically extract complex, possibly hierarchical models (of structure, shape, appearance and dynamics) with the optimal level of complexity for various tasks, from typical, supervised and unsupervised datasets. Models that can gracefully handle partial views and multiple levels of detail.
- Cost functions adapted for learning human models with good generalization properties. Algorithms that can learn reliably from small training sets.
- Relative advantages of bottom-up (discriminative, conditional) and top-down (generative) models and ways to combine them for initialization and for recovery from tracking failure.
- Inference methods for multiple people and for scenes with complex data association. Algorithms and models able to reliably handle occlusion, clutter and lighting changes. The relative advantages of 2D and 3D models and ways to jointly use them.
- The role of context in resolving ambiguities during state inference. Methods for combining recognition and reconstruction.

Acknowledgments

This work has been supported in part by the National Science Foundation under award IIS-0535140. We thank our collaborators: Allan Jepson, Dimitris Metaxas, Bill Triggs and Atul Kanaujia, for their support with different parts of the research described in this chapter.

References

1. CMU Human Motion Capture DataBase. Available online at http://mocap.cs. cmu.edu/search.html, 2003.
2. Agarwal A. and Triggs B. Monocular human motion capture with a mixture of regressors. In *Workshop on Vision for Human Computer Interaction*, 2005.
3. Allen B., Curless B., and Popovic Z. The space of human body shapes: reconstruction and parameterization from range scans. In *SIGGRAPH*, 2003.
4. Belkin M. and Niyogi P. Laplacian Eigenmaps and Spectral Techniques for Embedding and Clustering. In *Advances in Neural Information Processing Systems*, 2002.
5. Bertero M., Poggio T., and Torre V. Ill-posed Problems in Early Vision. *Proc. of IEEE*, 1988.
6. Bishop C. and Svensen M. Bayesian mixtures of experts. In *Uncertainty in Artificial Intelligence*, 2003.

7. Blake A. and Isard M. *Active Contours*. Springer, 2000.
8. Brand M. Shadow Puppetry. In *IEEE International Conference on Computer Vision*, pp. 1237–44, 1999.
9. Bregler C. and Malik J. Tracking People with Twists and Exponential Maps. In *IEEE International Conference on Computer Vision and Pattern Recognition*, 1998.
10. Carranza J., Theobalt C., Magnor M., and Seidel H.-P. Free-viewpoint video of human actors. In *SIGGRAPH*, 2003.
11. Cham T. and Rehg J. A Multiple Hypothesis Approach to Figure Tracking. In *IEEE International Conference on Computer Vision and Pattern Recognition*, vol 2, pp. 239–245, 1999.
12. Choo K. and Fleet D. People Tracking Using Hybrid Monte Carlo Filtering. In *IEEE International Conference on Computer Vision*, 2001.
13. Deutscher J., Blake A., and Reid I. Articulated Body Motion Capture by Annealed Particle Filtering. In *IEEE International Conference on Computer Vision and Pattern Recognition*, 2000.
14. Donoho D. and Grimes C. Hessian Eigenmaps: Locally Linear Embedding Techniques for High-dimensional Data. *Proceeding of the National Acadamy of Arts and Sciences*, 2003.
15. Donoho D. and Grimes C. When Does ISOMAP Recover the Natural Parameterization of Families of Articulated Images? Technical report, Dept. of Statistics, Stanford University, 2003.
16. Drummond T. and Cipolla R. Real-time Tracking of Highly Articulated Structures in the Presence of Noisy Measurements. In *IEEE International Conference on Computer Vision*, 2001.
17. Duane S., Kennedy A.D., Pendleton B.J., and Roweth D. Hybrid Monte Carlo. *Physics Letters B*, 195(2):216–222, 1987.
18. Elgammal A. and Lee C. Inferring 3d body pose from silhouettes using activity manifold learning. In *IEEE International Conference on Computer Vision and Pattern Recognition*, 2004.
19. Gavrila D. The Visual Analysis of Human Movement: A Survey. *Computer Vision and Image Understanding*, 73(1):82–98, 1999.
20. Gordon N., Salmond D., and Smith A. Novel Approach to Non-linear/Non-Gaussian State Estimation. *IEE Proceedings F*, 1993.
21. Howe N., Leventon M., and Freeman W. Bayesian Reconstruction of 3D Human Motion from Single-Camera Video. *Advances in Neural Information Processing Systems*, 1999.
22. Isard M. and Blake A. A Smoothing Filter for CONDENSATION. In *European Conference on Computer Vision*, 1998.
23. Isard M. and Blake A. CONDENSATION – Conditional Density Propagation for Visual Tracking. *International Journal of Computer Vision*, 1998.
24. Isard M. and Blake A. Icondensation: Unifying low-level and high-level tracking in a stochastic framework. In *European Conference on Computer Vision*, 1998.
25. Jordan M. *Learning in Graphical Models*. MIT Press, 1998.
26. Kakadiaris I. and Metaxas D. Model-Based Estimation of 3D Human Motion with Occlusion Prediction Based on Active Multi-Viewpoint Selection. In *IEEE International Conference on Computer Vision and Pattern Recognition*, pp. 81–87, 1996.

27. Kehl R., Bray M., and Gool L.V. Full body tracking from multiple views using stochastic sampling. In *IEEE International Conference on Computer Vision and Pattern Recognition*, 2005.

28. Lan X. and Huttenlocher D. Beyond trees: common factor models for 2d human pose recovery. In *IEEE International Conference on Computer Vision*, 2005.

29. Lee H.J. and Chen Z.. Determination of 3D Human Body Postures from a Single View. *Computer Vision, Graphics and Image Processing*, 30:148–168, 1985.

30. Lee M. and Cohen I. Proposal maps driven mcmc for estimating human body pose in static images. In *IEEE International Conference on Computer Vision and Pattern Recognition*, 2004.

31. Li R., Yang M., Sclaroff S., and Tian T. Monocular Tracking of 3D Human Motion with a Coordianted Mixture of Factor Analyzers. In *European Conference on Computer Vision*, 2006.

32. Mackay D. Bayesian Interpolation. *Neural Computation*, 4(5):720–736, 1992.

33. McCallum A., Freitag D., and Pereira F. Maximum entropy Markov models for information extraction and segmentation. In *International Conference on Machine Learning*, 2000.

34. Mori G. and Malik J. Estimating Human Body Configurations Using Shape Context Matching. In *European Conference on Computer Vision*, 2002.

35. Neal R. Annealed Importance Sampling. *Statistics and Computing*, 11:125–139, 2001.

36. Ramanan D. and Sminchisescu C. Training Deformable Models for Localization. In *IEEE International Conference on Computer Vision and Pattern Recognition*, 2006.

37. Rosales R. and Sclaroff S. Learning Body Pose Via Specialized Maps. In *Advances in Neural Information Processing Systems*, 2002.

38. Roth S., Sigal L., and Black M. Gibbs Likelihoods for Bayesian Tracking. In *IEEE International Conference on Computer Vision and Pattern Recognition*, 2004.

39. Roweis S. and Saul L. Nonlinear Dimensionality Reduction by Locally Linear Embedding. *Science*, 2000.

40. Schölkopf B., Smola A. and Müller K. Nonlinear Component Analysis as a Kernel Eigenvalue Problem. *Neural Computation*, 10:1299–1319, 1998.

41. Shakhnarovich G., Viola P., and Darrell T. Fast Pose Estimation with Parameter Sensitive Hashing. In *IEEE International Conference on Computer Vision*, 2003.

42. Sidenbladh H. and Black M. Learning Image Statistics for Bayesian Tracking. In *IEEE International Conference on Computer Vision*, 2001.

43. Sidenbladh H., Black M., and Fleet D. Stochastic Tracking of 3D Human Figures Using 2D Image Motion. In *European Conference on Computer Vision*, 2000.

44. Sigal L., Bhatia S., Roth S., Black M., and Isard M. Tracking Loose-limbed People. In *IEEE International Conference on Computer Vision and Pattern Recognition*, 2004.

45. Sminchisescu C. Consistency and Coupling in Human Model Likelihoods. In *IEEE International Conference on Automatic Face and Gesture Recognition*, pages 27–32, Washington DC, 2002.

46. Sminchisescu C. and Jepson A. Density propagation for continuous temporal chains. Generative and discriminative models. Technical Report CSRG-401, University of Toronto, October 2004.

47. Sminchisescu C. and Jepson A. Generative modelling for Continuous Non-Linearly Embedded Visual Inference. In *International Conference on Machine Learning*, pp. 759–766, Banff, 2004.
48. Sminchisescu C. and Jepson A. Variational Mixture Smoothing for Non-Linear Dynamical Systems. In *IEEE International Conference on Computer Vision and Pattern Recognition*, vol 2, pp. 608–615, Washington DC, 2004.
49. Sminchisescu C., Kanaujia A., Li Z., and Metaxas D. Learning to reconstruct 3D human motion from Bayesian mixtures of experts. A probabilistic discriminative approach. Technical Report CSRG-502, University of Toronto, October, 2004.
50. Sminchisescu C., Kanaujia A., Li Z., and Metaxas D. Conditional models for contextual human motion recognition. In *IEEE International Conference on Computer Vision*, vol 2, pp. 1808–1815, 2005.
51. Sminchisescu C., Kanaujia A., Li Z., and Metaxas D. Discriminative Density Propagation for 3D Human Motion Estimation. In *IEEE International Conference on Computer Vision and Pattern Recognition*, vol 1, pp. 390–397, 2005.
52. Sminchisescu C., Kanaujia A. and Metaxas D. BM^3E: Discriminative Density Propagation for Visual Tracking. In *IEEE Transactions on Pattern Analysis and Machine Intelligence*, 2007.
53. Sminchisescu C., Kanaujia A., and Metaxas D. Learning Joint Top-down and Bottom-up Processes for 3D Visual Inference. In *IEEE International Conference on Computer Vision and Pattern Recognition*, 2006.
54. Sminchisescu C. and Triggs B. Building Roadmaps of Local Minima of Visual Models. In *European Conference on Computer Vision*, vol 1, pp. 566–582, Copenhagen, 2002.
55. Sminchisescu C. and Triggs B. Hyperdynamics Importance Sampling. In *European Conference on Computer Vision*, vol 1, pp. 769–783, Copenhagen, 2002.
56. Sminchisescu C. and Triggs B. Estimating Articulated Human Motion with Covariance Scaled Sampling. *International Journal of Robotics Research*, 22(6):371–393, 2003.
57. Sminchisescu C. and Triggs B. Kinematic Jump Processes for Monocular 3D Human Tracking. In *IEEE International Conference on Computer Vision and Pattern Recognition*, vol 1, pp. 69–76, Madison, 2003.
58. Sminchisescu C. and Welling M. Generalized Darting Monte-Carlo. In *9th International Conference on Artificial Intelligence and Statistics*, 2007.
59. Sudderth E., Ihler A., Freeman W., and Wilsky A. Non-parametric belief propagation. In *IEEE International Conference on Computer Vision and Pattern Recognition*, 2003.
60. Taylor C.J. Reconstruction of Articulated Objects from Point Correspondences in a Single Uncalibrated Image. In *IEEE International Conference on Computer Vision and Pattern Recognition*, pp. 677–684, 2000.
61. Tenenbaum J., Silva V., and Langford J. A Global Geometric Framewok for Nonlinear Dimensionality Reduction. *Science*, 2000.
62. Tipping M. Sparse Bayesian learning and the Relevance Vector Machine. *Journal of Machine Learning Research*, 2001.
63. Tomasi C., Petrov S., and Sastry A. 3d tracking = classification + interpolation. In *IEEE International Conference on Computer Vision*, 2003.
64. Urtasun R., Fleet D., Hertzmann A., and Fua P. Priors for people tracking in small training sets. In *IEEE International Conference on Computer Vision*, 2005.

65. Wachter S. and Nagel H. Tracking Persons in Monocular Image Sequences. *Computer Vision and Image Understanding*, 74(3):174–192, 1999.
66. Waterhouse S., Mackay D., and Robinson T. Bayesian Methods for Mixtures of Experts. In *Advances in Neural Information Processing Systems*, 1996.
67. Weston J., Chapelle O., Elisseeff A., Schölkopf B., and Vapnik V. Kernel Dependency Estimation. In *Advances in Neural Information Processing Systems*, 2002.
68. Wipf D., Palmer J., and Rao B. Perspectives on Sparse Bayesian Learning. In *Advances in Neural Information Processing Systems*, 2003.

9

Spatially and Temporally Segmenting Movement to Recognize Actions

Richard Green

Computer Science Department
University of Canterbury
Christchurch, New Zealand

Summary. This chapter presents a Continuous Movement Recognition (CMR) framework which forms a basis for segmenting continuous human motion to recognize actions as demonstrated through the tracking and recognition of hundreds of skills from gait to twisting summersaults. A novel 3D color clone-body-model is dynamically sized and texture mapped to each person for more robust tracking of both edges and textured regions. Tracking is further stabilized by estimating the joint angles for the next frame using a forward smoothing Particle filter with the search space optimized by utilizing feedback from the CMR system. A new paradigm defines an alphabet of dynemes being small units of movement, to enable recognition of diverse actions. Using multiple Hidden Markov Models, the CMR system attempts to infer the action that could have produced the observed sequence of dynemes.

9.1 Introduction

One of the biggest hurdles in human–computer interaction is the current inability for computers to recognize human activities. We introduce hierarchical Bayesian models of concurrent movement structures for temporally segmenting this complex articulated human motion. An alphabet of these motion segments are then used for recognizing activities to enable applications to extend augmented reality and novel interactions with computers.

Research into computer vision based tracking and recognizing human movement has so far been mostly limited to gait or frontal posing [52]. This chapter presents a *Continuous Movement Recognition* (CMR) framework which forms a basis for the general analysis and recognition of continuous human motion as demonstrated through tracking and recognition of hundreds of skills from gait to twisting somersault. A novel 3D color clone-body-model is dynamically sized and texture mapped to each person for more robust tracking of both edges and textured regions. Tracking is further stabilized by estimating the joint angles for the next frame using a forward smoothing Particle filter with the search space optimized by utilizing feedback from the CMR system. A new paradigm defines an alphabet of dynemes, units of full-body

213

B. Rosenhahn et al. (eds.), Human Motion – Understanding, Modelling, Capture, and Animation, 213–241.

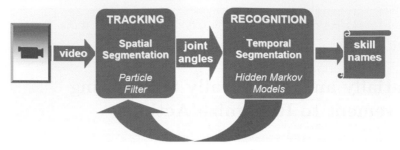

Fig. 9.1. Overview of the continuous movement recognition framework.

movement skills, to enable recognition of diverse skills. Using multiple Hidden Markov Models, the CMR system attempts to infer the human movement skill that could have produced the observed sequence of dynemes. The novel clone-body-model and dyneme paradigm presented in this chapter enable the CMR system to track and recognize hundreds of full-body movement skills thus laying the basis for effective human-computer interactions associated with full-body movement activity recognition.

Human movement is commercially tracked[1] by requiring subjects to wear joint markers/identifiers, an approach witch has the disadvantage of significant set up time. Such a marker based approach to tracking has barely changed since it was developed in the 1970s. Such commercial marker based tracking systems produce motion capture files (*mocap files*) of joint angles. An important aspect of computer vision is the analysis of motion and other chapters in this book use mocap files for both analysis ("MoCap for Interaction Environments" by Daniel Grest in Chapter 14) and validation ("Recognition and Synthesis of Actions using a Dynamic Bayes Network" by Volker Krger in Chapter 3). Chapter 20 "Automatic Classification and Retrieval of Motion Capture Data" (by Meinard Mueller) also investigates using mocap files to recognize skills based on 3D marker coordinates without using any image processing.

Using approaches free of the markers used to generate mocap files, computer vision research into tracking and recognizing full-body human motion has so far been mainly limited to gait or frontal posing [52]. Various approaches for tracking the whole body have been proposed in the image processing literature using a variety of 2D and 3D shape models and image models as listed in Table 9.1.

The approaches described above determine body/part orientation by tracking only the edges or same-color regions. To improve tracking accuracy and robustness by also tracking the textured colors within regions, this chapter describes a clone-body-model.

---

[1] Commercially   available   trackers   are   listed   at   www.hitl.washington.edu/scivw/tracker-faq.html.

Table 9.1. Comparison of different human body models.

	Stick model	2D contour	3D volume	Feature based
Karaulova et al., 2000	✓			
Guo et al., 1994	✓			
Wren et al., 2000	✓			
Iwai et al., 1999	✓			
Luo et al., 1992	✓			
Yaniz et al., 1998	✓			
Schrotter G. et al., 2005	✓			
Remondino F. 2003	✓			
Theobalt C. et al., 2002	✓			
Bregler C., 2004	✓			
Silaghi et al., 1998	✓			
Leung and Yang, 1995		✓		
Chang and Huang, 1996		✓		
Niyogi and Adelson, 1994		✓		
Black and Yaccob, 1996		✓		
Kameda et al., 1993		✓		
Kameda et al., 1995		✓		
Hu et al., 2000		✓		
Rohr, 1994			✓	
Wachter and Nagel, 1999			✓	
Rehg and Kanade, 1995			✓	
Kakadiaris, 1996			✓	
Goddard, 1994			✓	
Bregler and Malik, 1998			✓	
Mark J, 2004			✓	
Munkelt et al., 1998			✓	
Delamarre, 1999			✓	
Urtasun R. and Fua P. 2004			✓	
Huang Y. and Huang T. 2002			✓	
Bhatia S. et al., 2004			✓	
Luck et al., 2001			✓	
Polana and Nelson, 1994				✓
Yang R. et al., 2001				✓
Segen and Pingali, 1996				✓
Jang and Choi, 2000				✓
Rosales and Sclaroff, 1999				✓
Krinidis M. et al., 2005				✓
Li D. et al., 2005				✓
Nguyen et al., 2001				✓

This model is dynamically sized and texture mapped to each person, enabling of both edge and region tracking to occur. No previous approaches use such a method as can be seen in Table 9.1. The prediction of joint angles for the next frame is cast as an estimation problem, which is solved using a Particle filter with forward smoothing. This approach optimizes the huge search space related to calculating so many particles for these 32 degrees of freedom (DOF) used in our model (see Table 9.3) by utilizing feedback from the recognition process.

Human–computer interactions will become increasingly effective as computers more accurately recognize and understand full-body movement in terms of everyday activities. Stokoe began recognizing human movement in the 1970s by constructing sign language gestures (signs) from hand location, shape and movement and assumed that these three components occur concurrently with no sequential contrast (independent variation of these components within a single sign). Ten years later Liddel and Johnson used sequential contrast and introduced the movement-hold model. In the early 1990s Yamato et al began using Hidden Markov Models (HMMs) to recognize tennis strokes. Recognition accuracy rose as high as 99.2% in Starner and Pentland's work in 1996. Constituent components of movement have been named cheremes [69], phonemes [74] and movemes [8].

Most movement recognition research (Table 9.2) has been limited to frontal posing of a constrained range of partial-body motion. By contrast, this chapter describes a computer vision based framework that recognizes continuous full-body motion of hundreds of different movement skills (Figure 9.2). The full-body movement skills in this study are constructed from an alphabet of 35 dynemes – the smallest contrastive dynamic units of human movement.

Table 9.2. Human movement recognition research.

	Template matching	State based
Cui and Weng, 1997	✓	
Polana and Nelson, 1994	✓	
Boyd and Little, 1997	✓	
Bobick and Davis, 1996	✓	
Davis and Bobick, 1997	✓	
Silaghi et al., 1998	✓	
Collins R., 2002	✓	
Zhong H., 2004	✓	
Starner and Pentland, 1995		✓
Yamato et al., 1992		✓
Brand et al., 1997		✓
Bregler, 1997		✓
Campbell and Bobick, 1995		✓
Gao J. and Shi J., 2004		✓
Chen D. et al., 2004		✓
Bauckhage C. et al., 2004		✓
Kumar S. 2005		✓
Green, 2004		✓

Fig. 9.2. CMR system tracking and recognizing a sequence of movement skills.

Using a novel framework of multiple HMMs the recognition process attempts to infer the human movement skill that could have produced the observed sequence of dynemes. This dyneme approach has been inspired by the paradigm of the phoneme as used by the continuous speech recognition research community where pronunciation of the English language is constructed from approximately 50 phonemes which are the smallest contrastive phonetic units of human speech.

9.2 Tracking

Various approaches for tracking the whole body have been proposed in the image processing literature. They can be distinguished by the representation of the body as a stick figure, 2D contour or volumetric model and by their dimensionality being 2D or 3D. Volumetric 3D models have the advantage of being more generally valid with self occlusions more easily resolved [5]. They also allow 3D joint angles to be able to be more directly estimated by mapping 3D body models onto a given 2D image. Most volumetric approaches model body parts using generalized cylinders [57] or super-quadratics [54]. Some extract features [78], others fit the projected model directly to the image [57] and in the Chapter 15 by Lars Mndermann, he uses an individual, laser scanned model.

9.2.1 Clone-Body-Model

Cylindrical, quadratic and ellipsoidal [30] body models of previous studies do not contour accurately to the body, thus decreasing tracking stability. To overcome this problem, in this research 3D clone-body-model regions are sized and texture mapped from each body part by extracting features during the initialization phase. This clone-body-model has a number of advantages over previous body models:

- It allows for a larger variation of somatotype (from ectomorph to endomorph), gender (cylindrical trunks do not allow for breasts or pregnancy) and age (from baby to adult).
- Exact sizing of clone-body-parts enables greater accuracy in tracking edges, rather than the nearest best fit of a cylinder.
- Texture mapping of clone-body-parts increases region tracking and orientation accuracy over the many other models which assume a uniform color for each body part.
- Region patterns, such as the ear, elbow and knee patterns, assist in accurately fixing orientation of clone-body-parts.

The *clone-body-model* proposed in this chapter is a 3D model of the body consisting of a set of clone-body-parts, connected by joints, similar to the representations proposed by Badler [3]. Clone-body-parts include the head, clavicle, trunk, upper arms, forearms, hands, thighs, calves and feet. Degrees of freedom are modeled for gross full body motion (Table 9.3). Degrees of freedom supporting finer resolution movements are not yet modeled, including the radioulnar (forearm rotation), interphalangeal (toe), metacarpophalangeal (finger) and carpometacarpal (thumb) joint motions.

Each *clone-body-part* is a 3D model of a part of the body consisting of a rigid spine with pixels radiating out (Figure 9.3) with up to three DOF for each joint linking the clone-body-parts. Each pixel represents a point on the surface $b()$ of a clone-body-part. Associated with each pixel is: cylindrical coordinate (where d is the distance and θ is the angle), r is the radius or thickness of the clone-body-part at that point; h, s, i is the color as in hue, saturation and intensity; a_{hsi} is the accuracy of the color; a_r is the accuracy

Table 9.3. Degrees of freedom associated with each joint.

Joint	DOF
Neck (atlantoaxial)	3
Shoulder	3*
Clavicle	1*
Vertebrae	3
Hip	3*
Elbow	1*
Wrist	2*
Knee	1*
Ankle	2*
double for left and right	**32 total**

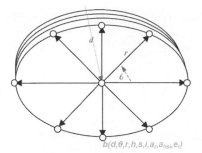

Fig. 9.3. Clone-body-model consisting of clone-body-parts which have a cylindrical coordinate system of surface points.

Fig. 9.4. Clone-body-model example rotating through 360 degrees.

of the radius; and e_r is the elasticity of radius inherent in the body part at that point. Although each point on a clone-body-part is defined by cylindrical coordinates, the radius varies in a cross section to exactly follow the contour of the body as shown in Figure 9.4.

Automated initialization assumes only one person is walking upright in front of a static background initially with gait being a known movement model. Anthropometric data [30] is used as a Gaussian prior for initializing the clone-body-part proportions with left-right symmetry of the body used as a stabilizing guide from 50th percentile proportions. Such constraints on the relative size of clone-body-parts and on limits and neutral positions of joints help to stabilize initializations. Initially a low accuracy is set for each clone-body-part with the accuracy increasing as structure from motion resolves the relative proportions. For example, a low color and high radius accuracy is initially set for pixels near the edge of a clone-body-part, high color and low radius accuracy for other near side pixels and a low color and low radius accuracy is set for far side pixels. The ongoing temporal resolution following self-occlusions enables increasing radius and color accuracy. Breathing, muscle flexion and other normal variations of body part radius are accounted for by the radius elasticity parameter.

9.2.2 Kinematic Model

The *kinematic model* tracking the position and orientation of a person relative to the camera entails projecting or aligning 3D clone-body-model parts with a

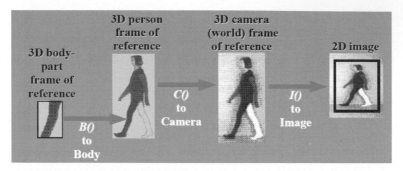

Fig. 9.5. Three homogeneous transformation functions $B()$, $C()$, $I()$ project a point from a clone-body-part onto a pixel in the 2D image.

2D image of the person using a *kinematic chain* of three chained homogeneous transformation matrices as illustrated in Figure 9.5:

$$p(x, b) = I_i(x, C_i(x, B_i(x, b))) \qquad (9.1)$$

where x is a parameter vector calculated for optimum alignment of the projected body model with the image of the person in the ith video frame, B is the Body frame-of-reference transformation, C is the Camera frame-of-reference transformation, I is the Image frame-of-reference transformation, b is a body-part surface point, p is a pixel in 2D frame of video [32]. The three transformations, $B(x)$, $C(x)$ and $I(x)$ which are chained together into a single transformation in equation 9.1, are illustrated separately in Figure 9.5. $B()$ maps each surface point of each 3D clone-body-model part onto the full 3D clone-body frame-of-reference to support joint angles. $C()$ then maps this full clone-body frame-of-reference into the 3D space within which the person is moving to support the six degrees of freedom of the body as a whole. Finally, $I()$ maps the camera's perspective of the person in a 3D space onto the actual image of a person to support alignment of the clone-body-model with that person's pose.

Joint angles are used to track the location and orientation of each body part, with the range of joint angles being constrained by limiting the DOF associated with each joint. A simple motion model of constant angular velocity for joint angles is used in the kinematical model. Each DOF is constrained by anatomical joint-angle limits, body-part inter-penetration avoidance and joint-angle equilibrium positions modeled with Gaussian stabilizers around their equilibria. To stabilize tracking, the joint angles are predicted for the next frame. The calculation of joint angles, for the next frame, is cast as an estimation problem which is solved using a Particle filter (Condensation algorithm).

9.2.3 Particle Filter

The *Particle Filter* is used in computer vision to address the problem of tracking objects moving in front of cluttered backgrounds [33], [34] where it is difficult to segment an object from the background in each video frame. The filter attempts to predict where an object will be in the next frame to improve tracking stability. The Particle filter's output at a given time-step (video frame) estimates a number of possible different postions for an object, rather than just a single estimate of one position (and covariance) as in a Kalman filter. This allows the Particle filter to maintain multiple hypotheses and thus be robust to distracting cluttered background by tracking a number of possible positions with a probability assigned to each position. That is, in each frame of video, z, the Particle Filter is tracking 32 joint angles, where these joint angles represent the state, x, of the object being tracked. Also, for each joint angle, there is a set, s, of alternate values.

With a *motion vector* of about 32 joint angles (32 DOF) to be determined for each frame of video, there is the potential for exponential complexity when evaluating such a high dimensional search space. MacCormick [49] proposed Partitioned Sampling and Sullivan [68] proposed Layered Sampling to reduce the search space by partitioning it for more efficient particle filtering. Although Annealed Particle Filtering [15] is an even more general and robust solution, it struggles with efficiency which Deutscher [16] improves with Partitioned Annealed Particle Filtering.

The Particle Filter is a simpler algorithm than the more popular Kalman Filter. Moreover despite its use of random sampling, which is often thought to be computationally inefficient, the Particle filter can run in real-time. This is because tracking over time maintains relatively tight distributions for shape at successive time steps and particularly so given the availability of accurate learned models of shape and motion from the human-movement-recognition (CMR) system. In this chapter, the Particle filter attempts to estimate the joint angles in the next video frame (State Density) using the joint angles from the previous frame (Prior Density), using the kinematic/body models (Process Density) and using image data from the next video frame (Observation Density).

So here the particle filter has:

Three probability distributions in the problem specification:

1. *Prior density* $p(x_{t-1}|z_{t-1})$ for the state x_{t-1}, where x_{t-1}=joint angles in previous frame, z_{t-1}
2. *Process density* $p(x_t|x_{t-1})$ for kinematic and clone-body-models, where x_{t-1}=previous frame, x_t=next frame
3. *Observation density* $p(z_{t-1}|x_{t-1})$ for image z_{t-1} in previous frame

One probability distribution in the solution specification:

- *State Density* $p(x_t|Z_t)$: where x_t is the joint angles in next frame Z_t

1. **Prior density**: Sample s_t from the prior density $p(x_{t-1}|z_{t-1})$ where x_{t-1}=joint angles in previous frame, z_{t-1}. The sample set consists of possible alternate values for joint angles. When tracking through background clutter or occlusion, a joint angle may have N alternate possible values (samples) s with respective weights w, where prior density,

$$p(x_{t-1}|z_{t-1}) \approx S_{t-1} = (s^{(n)}, w_{t-1})$$

where $n = 1..N$ is a sample set, S_{t-1} is the sample set for the previous frame, $w^{(n)}$ is the $n^{(th)}$ weight of the $n^{(th)}$ sample $s^{(n)}$

For the next frame, a new sample, s_t is selected from a sample in the previous frame, s_{t-1}.

2. **Process density**: Predict s_t from the process density $p(x_t|x_{t-1} = s_t)$. Joint angles are predicted for the next frame using the kinematic model, body model and error minimization. A joint angle, s in the next frame is predicted by sampling from the *process density*, $p(x_t|x_{t-1} = s)$ which encompasses the kinematic model, clone-body-model and cost function minimization. In this prediction step both edge and region information are used. The edge alignment error is calculated by directly matching the image gradients with the expected body model edge gradients (sum of squares of edge differences in Equation 9.2). The region alignment error is calculated by directly matching the values of pixels in the image with those on the clone-body-model's 3D color texture map (Equation 9.3). The prediction step involves minimizing the cost functions (measurement likelihood density). That is, the prediction step involves minimizing errors aligning the edges of the body model with the image and aligning the texture map of the body model with the image:

edge alignment error E_e using edge information (sum of squares of difference between edges in image and model):

$$E_e(S_t) = \frac{1}{2 n_e v_e} \sum_{x,y} (|\nabla i_t(x,y)| - m_t(x,y,S_t))^2 + 0.5(S-S_t)^T C_t^{-1}(S-S_t) \rightarrow minS_t \tag{9.2}$$

region alignment error E_r using region information (sum of squares of difference between model texture map and image pixels):

$$E_r(S_t) = \frac{1}{2 n_r v_r} \sum_{j=1}^{n_r} (\nabla i_t[p_j(S_t)] - i_{t-1}[p_j(S_{t-1})])^2 + E_e(S_t) \rightarrow minS_t \tag{9.3}$$

where it represents an image (of size x pixels by y pixels) at time t, m_t is the model gradients at time t, n_e is the number of edge values summed, v_e is the edge variance, n_r is the number of region values summed, v_r is the region variance, p_j is the image pixel coordinate of the jth surface point on a clone-body-part. Use of both cost functions at the same time requires optimization in a Pareto sense to avoid only one function being optimal in the end.

3. **Observation density**: Measure and weigh the new position in terms of the observation density, $p(z_t|x_t)$. Weights $w_t = p(z_t|x_t = s_t)$ are estimated and then weights

$$\sum_{n}^{n} w^{(n)} = 1$$

are normalized. The new position in terms of the observation density, $p(z_t|x_t)$ is then measured and weighed with *forward smoothing*:

- Smooth weights w_t over $1..t$, for n trajectories
- Replace each sample set with its n trajectories (s_t, w_t) for $1..t$
- Re-weight all $w^{(n)}$ over $1..t$

Trajectories of different estimates for the same joint angle tend to merge within 10 frames:

- $O(N_t)$ storage prunes down to $O(N)$

In this research, feedback from the CMR system utilizes the large training set of skills to achieve an even larger reduction of the search space. In practice, human movement is found to be highly efficient for most activities, with minimal DOFs rotating at any one time. The equilibrium positions and physical limits of each DOF further stabilize and minimize the dimensional space. With so few DOFs to track at any one time, a minimal number of particles are required, significantly raising the efficiency of the tracking process. Such highly constrained movement results in a sparse domain of motion projected by each motion vector. Because the temporal variation of related joints and other parameters also contains information that helps the recognition process infer dynemes, the system computes and appends the temporal derivatives and second derivatives of these features to form the final motion vector. Hence, the motion vector includes joint angles (32 DOF), body location and orientation (6 DOF), centre of mass (3 DOF), principle axis (2 DOF) all with first and second derivatives.

9.3 Recognition

To simplify the design, it is assumed that the CMR system contains a limited set of possible human movement skills. This approach restricts the search for possible skill sequences to those skills listed in the skill model, which lists the candidate skills and provides dynemes – an alphabet of granules of human motion – for the composition of each skill. The current skill model contains hundreds of skills where the length of the skill sequence being performed is unknown. If M represents the number of human movement skills in the skill model, the CMR system could hypothesize M^N possible skill sequences for a skill sequence of length N. However these skill sequences are not equally likely to occur due to the biomechanical constraints of human motion. For example, the skill sequence stand jump lie is much more likely than stand lie jump (as it is difficult to jump from a lying down position). Given an observed sequence of Ψ motion vectors $y = (y_1...y_\Psi)$, the recognition process attempts

224   R. Green

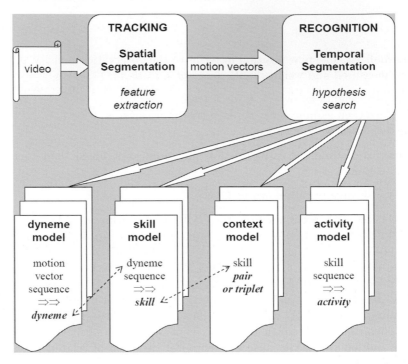

Fig. 9.6. The dyneme, skill, context and activity models construct a hypothesis for interpreting a video sequence.

to find the skill sequence $S_1^N = (S_1..S_\Psi)$ that maximizes this skill sequence's probability:

$$S_1^N = \arg\max_{S_1^N} p(s_1^N | y_1^\Psi) \equiv \arg\max_{S_1^N} p(y_1^\Psi | s_1^N) p(s_1^N) \qquad (9.4)$$

This approach applies Bayes' law and ignores the denominator term to maximize the product of two terms: the probability of the motion vectors given the skill sequence and the probability of the skill sequence itself. The CMR framework described by this Equation 9.4 is illustrated below in Figure 9.6 where, using motion vectors from the tracking process, the recognition process uses the dyneme, skill, context and activity models to construct a hypothesis for interpreting a video sequence.

In the tracking process, motion vectors are extracted from the video stream. In the recognition process, the search hypothesizes a probable movement skill sequence using four models:

1. *Dyneme model* models the relationship between the motion vectors and the dynemes.
2. *Skill model* defines the possible movement skills that the search can hypothesize, representing each movement skill as a linear sequence of dynemes.

3. *Context model* models the semantic structure of movement by modelling the probability of sequences of skills simplified to only triplets or pairs of skills as discussed in Section 3.3 below.
4. *Activity model* defines the possible human movement activities that the search can hypothesize, representing each activity as a linear sequence of skills (not limited to only triplets or pairs as in the context model).

Three principle components comprise the basic hypothesis search: a dyneme model, a skill model and a context model.

9.3.1 Dyneme Model

As the phoneme is a phonetic unit of human speech, so the *dyneme* is a dynamic unit of human motion. The word *dyneme* is derived from the Greek *dynamikos* "powerful", from *dynamis* "power", from *dynasthai* "to be able" and in this context refers to motion. This is similar to the phoneme being derived from *phono* meaning sound and with *eme* inferring the smallest contrastive unit. Thus *dyn-eme* is the smallest contrastive unit of movement. The movement skills in this study are constructed from an alphabet of 35 dynemes which HMMs used to recognize the skills. This approach has been inspired by the paradigm of the phoneme as used by the continuous speech recognition research community where pronunciation of the English language is constructed from approximately 50 phonemes

The dyneme can also be understood as a type of movement notation. An example of a similar movement notation system is that used in dance. Many dance notation systems have been designed over the centuries. Since 1928, there has been an average of one new notation system every 4 years [32]. Currently, there are two prominent dance notation systems in use: Labanotation and Benesh.

Although manual movement notation systems have been developed for dance, computer vision requires an automated approach where each human movement skill has clearly defined temporal boundaries. Just as it is necessary to isolate each letter in cursive handwriting recognition, so it is necessary in the computer vision analysis of full-body human movement to define when a dyneme begins and ends. This research defined an alphabet of dynemes by deconstructing (mostly manually) hundreds of movement skills into their correlated lowest common denominator of basic movement patterns.

Although there are potentially an infinite number of movements the human body could accomplish, there are a finite number ways to achieve motion in any direction. For simplicity, consider only xy motion occurring in a frontoparallel plane:

- x translation caused by:
  - Min–max of hip flexion/extension – e.g., gait, crawl
  - Min–max of hip abduction/adduction or lateral flexion of spine – e.g. cartwheel

- – Min–max of shoulder flexion – e.g., walk on hands, drag-crawl
- – Rotation about the transverse (forward roll) or antero-posterior (cartwheel) or longitudinal (rolling) axes
- – Min–max foot rotation – e.g., isolated feet based translation
- – Min–max waist angle – e.g., inch-worm
- • y translation caused by:
  - – Min–max Center of Mass (COM) – e.g., jump up, crouch down
- • no x or y translation:
  - – Motion of only one joint angle – e.g., head turn
  - – Twist – rotation about the longitudinal axis – e.g., pirouette

The number of dynemes depends on the spatial-temporal resolution threshold. A dyneme typically encapsulates diverse fine granules of motion. A gait step dyneme for example, encompasses diverse arm motions (shoulder and elbow angular displacements, velocities and accelerations) where some arm movements have a higher probability of occurring during the step dyneme than others.

A *hidden Markov model* offers a natural choice for modelling human movement's stochastic aspects. HMMs function as probabilistic finite state machines: The model consists of a set of states and its topology specifies the allowed transitions between them. At every time frame, a HMM makes a probabilistic transition from one state to another and emits a motion vector with each transition.

Figure 9.7 shows a HMM for a dyneme where the *state transition probabilities* $p1 = p(y|1)$, $p2 = p(y|2)$, $p3 = p(y|3)$ govern the possible transitions between states of 1-p1, 1-p2, 1-p3 respectively. A set of state transition probabilities – $p1$, $p2$ and $p3$ – governs the possible transitions between states. They specify the probability of going from one state at time t to another state at time $t + 1$. The motion vectors emitted while making a particular transition represent the characteristics for the human movement at that point, which vary corresponding to different executions of the dyneme. A *probability distribution* or *probability density function* models this variation. The functions – $p(y|1)$, $p(y|2)$ and $p(y|3)$ – can be different for different transitions. These distributions are modeled as parametric distributions – a mixture of multidimensional Gaussians. The HMM shown in Figure 9.7 consists of three states.

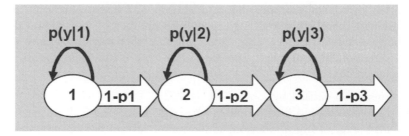

Fig. 9.7. Hidden Markov Model for a dyneme.

The dyneme's execution starts from the first state and makes a sequence of transitions to eventually arrive at the third state. The duration of the dyneme equals the number of video frames required to complete the transition sequence. The three transition probabilities implicitly specify a probability distribution that governs this duration. If any of these transitions exhibit high self-loop probabilities, the model spends more time in the same state, consequently taking longer to go from the first to the third state. The probability density functions associated with the three transitions govern the sequence of output motion vectors. A fundamental operation is the computation of the likelihood that a HMM produces a given sequence of motion vectors. For example, assume that the system extracted T motion vectors from human movement corresponding to the execution of a single dyneme and that the system seeks to infer which dyneme from a set of 35 was performed. The procedure for inferring the dyneme assumes that the ith dyneme was executed and finds the likelihood that the HMM for this dyneme produced the observed motion vectors. If the sequence of HMM states is known, the probability of a sequence of motion vectors can be easily computed. In this case, the system computes the likelihood of the tth motion vector, y_t, using the probability density function for the HMM state being active at time t. The likelihood of the complete set of T motion vectors is the product of all these individual likelihoods. However, because the actual sequence of transitions is not known, the likelihood computation process sums all possible state sequences. Given that all HMM dependencies are local, efficient formulas can be derived for performing these calculations recursively [36].

With various dynemes overlapping, a hierarchy of dynemes is required to clearly define the boundary of each granule of motion and so define a high level movement skill as the construction of a set of dynemes. For example, a somersault with a full-twist rotates 360° about the transverse axis in the somersault and 360° about the longitudinal axis in the full-twist. This twisting-somersault is then an overlap of two different rotational dynemes. Whole body rotation is more significant than a wrist flexion when recognizing a skill involving full body movement. To this end, dynemes have different weights in the Skill Model (HMM) to support the following descending hierarchy of five dyneme categories:

- Full body rotation
- COM motion (including flight)
- Static pose
- Weight transfer
- Hierarchy of DOFs

Each category of motion is delineated by a pause, min, max, or full, half, quarter rotations. For example, a COM category of dyneme is illustrated in Figure 9.8a where each running step is delimited by COM minima. A full 360° rotation of the principle axis during a cartwheel in Figure 9.8b illustrates a rotation dyneme category.

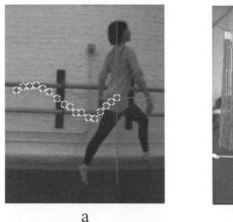 

a   b

Fig. 9.8. COM parameters during running and principle-axis parameters through a cartwheel.

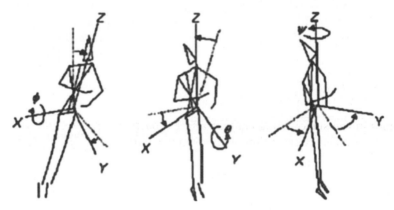

Fig. 9.9. Rotating about the x-axis/Transverse axis (somersault), Y axis/ Antero-posterior axis (cartwheel), and the z-axis/Longitudinal axis (turn/twist).

The 35 dynemes are:

- Full body rotation: 3 dynemes
  - 3 rotation DOF dynemes (rotations about the transverse, antero-posterior and longitudinal axes) as shown in Figure 9.9.
- COM motion: 4 dynemes
  - 3 translational DOF dynemes and 1 flight dyneme (special case of vertical)
- Static pose: 1 dyneme
  - 1 stationary dyneme
- Weight transfer: 2 dynemes
  - 2 dynemes for left weight transfer (left step) and right weight transfer (right step)

- Hierarchy of DOFs: 25 dynemes (included left and right)
  - 6 hip dynemes
  - 2 knee dynemes
  - 6 shoulder dynemes
  - 2 elbow dynemes
  - 3 neck (atlantoaxial) dynemes
  - 2 vertebrae dynemes
  - 4 ankle dynemes

This simplified list of dynemes could actually be thought of as a total of 66 dynemes if direction is taken into account with an extra three rotation dynemes (anti/clock wise), an extra three translation dynemes (up/down/left/right/backward/forward) and an extra 25 joint dynemes (flexion/extension). Notice also that although 32 DOF were required for accurate tracking, not all DOF were needed to recognize skills in the training set. (Omitted are 4 wrist DOF, 2 clavicle DOF and 1 vertebrae DOF.)

A dyneme model computes the probability of motion vector sequences under the assumption that a particular skill produced the vectors. Given the inherently stochastic nature of human movement, individuals do not usually perform a skill in exactly the same way twice. The variation in a dyneme's execution manifests itself in three ways: duration, amplitude and phase variations. Further, dynemes in the surrounding context can also cause variations in a particular dyneme's duration, amplitude and phase relationships, a phenomenon referred to in this chapter as *coexecution*. Hence, in some cases the dynemes in the surrounding context affect a particular dyneme's motion vector sequence. This coexecution phenomenon is particularly prevalent in poorly executed movement skills. The system models coexecution by assuming that the density of the observations depends on both the specific dyneme and the surrounding dynemes. However, modelling every dyneme in every possible context generates a prohibitively large number of densities to be modeled. For example, if the dyneme alphabet consists of 35 dynemes, and the system models every dyneme in the context of its immediately surrounding neighbors, it would need to model 42,875 densities. Consequently, the approach taken here clusters the surrounding dynemes into a few equivalence classes of categories, thus reducing the densities that require modelling [36].

9.3.2 Skill Model

The typical *skill model* shown in Table 9.4 lists each skill's possible executions, constructed from dynemes. An individual movement skill can have multiple forms of execution which complicates recognition.

The system chooses the skill model on a task-dependent basis, trading off skill-model size with skill coverage. Although a search through many videos can easily find dyneme sequences representing commonly used skills in various sources, unusual skills in highly specific situations may require manual

Table 9.4. Typical minimal dyneme skill model (with the skill walk having two alternative executions).

Movement skill	Dyneme
Walk	step (right), step (left)
Walk	step (left), step (right)
Handstand from stand	step, rotate-fwd (180^0)
Jump	knee-extension, COM-flight
Backward somersault	knee-extension, COM-flight, rotate-bwd (360^0)

specification of the dyneme sequence. In fact the initial definition of skills in terms of dynemes involved extensive manual specification in this research.

9.3.3 Context Model

The search for the most likely skill sequence in Equation 9.4 requires the computation of two terms, $p(y_1^T|s_1^N)$ and $p(s_1^N)$. The second of these computations is the *context model* which assigns a probability to a sequence of skills, s_1^N. The simplest way to determine such a probability would be to compute the relative frequencies of different skill sequences. However, the number of different sequences grows exponentially with the length of the skill sequence, making this approach infeasible.

A typical approximation assumes that the probability of the current skill depends on the previous one or two skills only, so that the computation can approximate the probability of the skill sequence as:

$$p(s_1^N) \approx p(s_1)p(s_2|s_1) \prod_{i=3}^{i=N} p(s_i|s_{i-1}, s_{i-2}) \tag{9.5}$$

where $p(s_i|s_{i-1}, s_{i-2})$ can be estimated by counting the relative frequencies of skill triplets:

$$p(s_i|s_{i-1}, s_{i-2}) \approx \mu(s_i, s_{i-1}, s_{i-2})/\mu(s_{i-1}, s_{i-2}) \tag{9.6}$$

Here, refers to the associated event's relative frequency. This context model was trained using hundreds of skills to estimate $p(s_i|s_{i-1}, s_{i-2})$. Even then, many skill pairs and triplets do not occur in the training videos, so the computation must smooth the probability estimates to avoid zeros in the probability assignment [18].

9.3.4 Training

Before using a HMM to compute the likelihood values of motion vector sequences, the HMMs must be trained to estimate the model's parameters. This process assumes the availability of a large amount of training data, which consists of the executed skill sequences and corresponding motion vectors extracted from the video stream. The *maximum likelihood (ML) estimation process* training paradigm is used for this task. Given a skill sequence and corresponding motion vector sequence, the ML estimation process tries to choose the HMM parameters that maximize the training motion vectors' likelihood computed using the HMM for the correct skill sequence. If y_1^T represents a sequence of T motion vectors, and s_1^N represents the corresponding correct skill sequence, then the ML estimate of the parameters of the HMM, $\widehat{\theta}$ is

$$\widehat{\theta}_{ML} = \arg\max_{\theta} \log[p_\theta(y_1^T|s_1^N)] \qquad (9.7)$$

The system begins the training process by constructing a HMM for the correct skill sequence. First, it constructs the HMMs for each skill by concatenating the HMMs for the dynemes that compose that skill. Then it concatenates the skill HMMs to form the HMM for the complete skill sequence where the transitional probabilities between connecting states for these HMMs are set to one and those between non-connecting states are set to zero. For example, the HMM for the sequence, *"skip"* would be the concatenation of the two dynemes, *"step, hop"*.

The training process assumes that the system can generate the motion vectors y_1^T by traversing the HMM from its initial state to its final state in T time frames. However, because the system cannot trace the actual state sequence, the ML estimation process assumes that this state sequence is hidden and averages all possible state sequence values since the number of states in the analyzed motion vector sequence is unknown and should be determined by searching the optimal values. By using x_t to denote the hidden state at time t, the system can express the maximization of Equation 9.7 in terms of the parameters of the HMM's hidden states, $\widehat{\theta}$ as follows:

$$\widehat{\theta}_{ML} = \arg\max_{\widehat{\theta}} \sum_{t=1}^{T} \sum_{S_t} p_\theta(x_t|y_1^T) \log[p_{\widehat{\theta}}(y_t|x_t)] \qquad (9.8)$$

The system uses an iterative process to solve Equation 9.8, where each iteration involves an expectation step and a maximization step. The first step (expectation step) involves the computation of $p_\theta(x_t|y_1^T)$, which is the posterior probability, or count of a state, conditioned on all the motion vectors. The system then uses the current HMM parameter estimates and the Forward-Backward algorithm [36] to perform this computation. The second step (maximization step) involves choosing the parameter $\widehat{\theta}$ to maximize Equation 9.5. Using a Gaussian probability density functions, the computation derives closed-form expressions for this step.

9.3.5 Hypothesis Search

The hypothesis search seeks the skill sequence with the highest likelihood given the model's input features and parameters [36]. Because the number of skill sequences increases exponentially with the skill sequence's length, the search might seem at first to be an intractable problem for anything other than short skill sequences from a small lexicon of skills. However, because the model has only local probabilistic dependencies the system can incrementally search through the hypothesis in a left-to-right fashion and discard most candidates with no loss in optimality [36].

Although the number of states in the context model can theoretically grow as the square of the number of skills in the skill model, many skill triplets never actually occur in the training data. The smoothing operation backs off to skill pair and single skill estimators, substantially reducing size. To speed up the recursive process, the system conducts a *beam search*, which makes additional approximations such as retaining only hypotheses that fall within threshold of the maximum score in any time frame.

Given a time-series, the *Viterbi*[1] algorithm computes the most probable hidden state sequence; the *forward–backward algorithm* computes the data likelihood and expected sufficient statistics of hidden events such as state transitions and occupancies. These statistics are used in *Baum-Welch parameter re-estimation* to maximize the likelihood of the model given the data. The expectation-maximization (EM) algorithm for HMMs consists of forward-backward analysis and Baum-Welch re-estimation iterated to convergence at a local likelihood maximum.

Brand [6] replaced the Baum-Welch formula with parameter estimators that minimize entropy to avoid the local optima. However, with hundreds of movement skill samples it is felt that the research in this chapter avoided this pitfall with a sufficiently large sample size. Viterbi alignment is applied to the training data followed by Baum-Welch re-estimation. Rather than the rule based grammar model common in speech processing, a context model is trained from the movement skill data set. The *Hidden Markov Model Tool Kit*[1] (HTK) version 3.2 (current stable release) is used to support these dyneme, skill and context models.

The HTK is a portable toolkit for building and manipulating hidden Markov models. HTK is primarily used for speech recognition research although it has been used for numerous other applications including research into speech synthesis, character recognition, gesture recognition and DNA sequencing. HTK is in use at hundreds of sites worldwide. HTK consists of a set of library modules and tools available in C source form. The tools provide sophisticated facilities for speech analysis, HMM training, testing and results

---

[1] An excellent discussion of HMMs and application of Viterbi alignment and Baum-Welch re-estimation can be found in the extensive HTK documentation of HTK-Book: http://htk.eng.cam.ac.uk/docs/docs.shtml

analysis. The software supports HMMs using both continuous density mixture Gaussians and discrete distributions and can be used to build complex HMM systems.

9.4 Performance

Hundreds of skills were tracked and classified using a 2 GHz dual core Pentium with 2GB RAM processing 24 bit color within the Microsoft DirectX 9 environment under Windows XP. The video sequences were captured with a JVC DVL-9800 digital video camera at 30 fps, 720 by 480 pixel resolution. Each person moved in front of a stationary camera with a static background and static lighting conditions. Only one person was in frame at any one time. Tracking began when the whole body was visible which enabled initialization of the clone-body-model.

The skill error rate quantifies CMR system performance by expressing, as a percentage, the ratio of the number of skill errors to the number of skills in the reference training set. Depending on the task, CMR system skill error rates can vary by an order of magnitude. The CMR system results are based on a set of a total of 840 movement patterns, from walking to twisting somersaults. From this, an independent test set of 200 skills were selected leaving 640 in the training set. Training and testing skills were performed by the same subjects. These were successfully tracked, recognized and evaluated with their respective biomechanical components quantified where a skill error rate of 4.5% was achieved.

Recognition was processed using the (Microsoft owned) Cambridge University Engineering Department HMM Tool Kit (HTK) with 96.8% recognition accuracy on the training set alone and a more meaningful 95.5% recognition accuracy for the independent test set where H=194, D=7, S=9, I=3, N=200 (H=correct, D=Deletion, S=Substitution, I=Insertion, N=test set, Accuracy=(H-I)/N). 3.5% of the skills were ignored (deletion errors) and 4.5% were incorrectly recognized as other skills (substitution errors). There was only about 1.5% insertion errors - that is incorrectly inserting/recognizing a skill between other skills.

As mentioned previously, the HTK performed Viterbi alignment on the training data followed by Baum-Welch re-estimation with a context model for the movement skills. Although the recognition itself was faster than real-time at about 120 fps, the tracking of 32 DOF with particle filtering was computationally expensive using up to 16 seconds per frame.

Figure 9.10 illustrates the CMR system recognizing the sequence of skills *stretch* and *step*, *cartwheel*, *step* and *step* from continuous movement. In each picture, *four tiles* display CMR processing steps:

1. Principle axis through the body
2. Body frame of reference (normalized to the vertical)

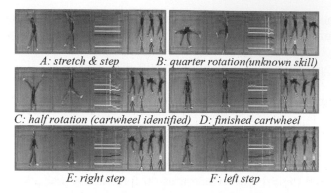

A: stretch & step   *B: quarter rotation(unknown skill)*

C: half rotation (cartwheel identified)   *D: finished cartwheel*

E: right step   *F: left step*

Fig. 9.10. CMR system recognizing stretching into a cartwheel followed by gait steps.

a   b   c

Fig. 9.11. CMR system tracking through motion blur of right calf and foot segments during a flic-flac (back-handspring).

3. Motion vectors are graphed (subset displayed)
4. Recognizing *step*, *stretch* and *cartwheel* indicated by stick figures with respective snapshots of the skills

As each skill is recognized a snapshot of the corresponding pose is displayed in the fourth tile. Below each snapshot is a stick figure representing an internal identification of the recognized skill. Notice that the cartwheel is not recognized after the first quarter rotation. Only after the second quarter rotation is the skill identified as probably a cartwheel.

Motion blurring lasted about 10 frames on average with the effect of perturbing joint angles within the blur envelope as shown in Figure 9.11 where, a: motion blur of right calf and foot segments, b: alternative particles (knee angles) for the right calf location, and c: expected value of the distrubution $\sum_n w_t s_t$. Given a reasonably accurate angular velocity, it was possible to sufficiently de-blur the image. There was minimal motion blur arising from rotation about the longitudinal axis during a double twisting somersault due to a low surface velocity tangential to this axis from minimal radius with limbs held close to a straight body shape. This can be seen in Figure 9.12 where the arms exhibit no blurring from twisting rotation, contrasted with motion blurred legs due to a higher tangential velocity of the somersault rotation.

Fig. 9.12. Minimal motion blur arising from rotation about the longitudinal axis during a double twisting somersault.

An elongated trunk with disproportionate short legs is the body-model consequence of the presence of a skirt – the clone-body-model failed to initialize for tracking due to the variance of body-part proportions exceeding an acceptable threshold. The CMR system also failed for loose clothing. Even with smoothing, joint angles surrounded by baggy clothes permutated through unexpected angles within an envelope sufficiently large as to invalidate the tracking and recognition.

9.5 Summary

As described in this chapter, recognition of human movement skills was successfully processed using the Cambridge University HMM Tool Kit. Probable movement skill sequences were hypothesized using the recognition process framework of four integrated models - dyneme, skill, context and activity models. The 95.5% recognition accuracy ($H = 194, D = 7, S = 9, I = 3, N = 200$) validated this framework and the dyneme paradigm. However, the 4.5% error rate attained in this research is not yet evaluating a natural world environment nor is this a real-time system with up to 16 seconds to process each frame. The CMR system did achieve 95.5% recognition accuracy for the independent test set of 200 skills which encompassed a much larger diversity of full-body movement than any previous study. Although this 95.5% recognition rate was not as high as the 99.2% accuracy Starner and Pentland [66] achieved recognizing 40 signs, a larger test sample of 200 skills were evaluated in this chapter. With a larger training set, lower error rates are expected. Generalization to a user independent system encompassing partial body movement domains such as sign language should be attainable. To progress towards this goal, the following improvements seem most important:

- Expand the dyneme model to improve discrimination of more subtle movements in partial-body domains. This could be achieved by either expanding the dyneme alphabet or having domain dependent dyneme alphabets layered hierarchically below the full-body movement dynemes.

- Expand the clone-body-model to include a complete hand-model for enabling even more subtle movement domains such as finger signing and to better stabilize the hand position during tracking.
- Use a multi-camera or multi-modal vision system such as infra-red and visual spectrum combinations to better disambiguate the body parts in 3D and track the body in 3D.
- More accurately calibrate all movement skills with multiple subjects performing all skills on an accurate commercial tracking system recording multiple camera angles to improve on depth of field ambiguities. Such calibration would also remedy the qualitative nature of tracking results from computer vision research in general.
- Enhance tracking granularity using cameras with higher resolution, frame rate and lux sensitivity.

So far movement domains with exclusively partial-body motion such as sign language have been ignored. Incorporating partial-body movement domains into the full-body skill recognition system is an interesting challenge. Can the dyneme model simply be extended to incorporate a larger alphabet of dynemes or is there a need for sub-domain dyneme models for maximum discrimination within each domain? The answers to such questions may be the key to developing a general purpose unconstrained skill recognition system.

References

1. Abdelkader, M., R. Chellappa, Q. Zheng, and A. Chan: Integrated Motion Detection and Tracking for Visual Surveillance, In *Proc. Fourth IEEE International Conference on Computer Vision Systems*, pp. 28–36, 2006
2. Aggarwal A., S. Biswas, S. Singh, S. Sural, and A. Majumdar: Object Tracking Using Background Subtraction and Motion Estimation in MPEG Videos, In *Proc. Asian Conference on Computer Vision*, pp. 121–130, 2006.
3. Badler, N., C. Phillips and B. Webber: *Simulating Humans*. Oxford University Press, New York, pp. 23–65, 1993.
4. Bauckhage C., M. Hanheide, S. Wrede and G. Sagerer: A Cognitive Vision System for Action Recognition in Office Environment, In *Proc. IEEE Computer Society Conference on Computer Vision and Pattern Recognition*, pp. 827–833, 2004.
5. Bhatia S., L. Sigal, M. Isard, and M. Black: 3D Human Limb Detection using Space Carving and Multi-view Eigen Models, In *Proc. Second IEEE International Conference on Computer Vision Systems* 2004.
6. Brand, M., and V. Kettnaker: Discovery and segmentation of activities in video, *IEEE Transactions on Pattern Analysis and Machine Intelligence*, 22(8), 2000.
7. Bregler C.: Twist Based Acquisition and Tracking of Animal and Human Kinematics, *International Journal of Computer Vision*, 56(3):179–194, 2004.
8. Bregler, C.:Learning and Recognizing Human Dynamics in Video Sequences, In *Proc. IEEE Conference on Computer Vision and Pattern Recognition*, CVPR, 1997.

9. Bregler, C. and J. Malik: Tracking people with twists and exponential maps, In *Proc. IEEE Conference on Computer Vision and Pattern Recognition*, CVPR, pp. 8–15, 1998.
10. Campos T.: 3D Hand and Object Tracking for Intention Recognition. DPhil Transfer Report, Robotics Research Group, Department of Engineering Science, University of Oxford, 2003.
11. Cham, T., and J. Rehg: A Multiple Hypothesis Approach to Figure Tracking, In *Proc. IEEE Conference on Computer Vision and Pattern Recognition*, CVPR, pp. 239–245, 1999.
12. Chen D., J. Yang, H.: Towards Automatic Analysis of Social Interaction Patterns in a Nursing Home Environment from Video, In *Proc. ACM Multimedia Information Retrieval*, pp. 283–290, 2004.
13. Daugman, J.: How Iris Recognition Works, In *Proc. IEEE Conference on ICIP*, 2002.
14. Demirdjian D., T. Ko, and T. Darrell: Untethered Gesture Acquisition and Recognition for Virtual World Manipulation, In *Proc. International Conference on Virtual Reality*, 2005.
15. Deutscher, J., A. Blake, I. Reid; Articulated Body Motion Capture by Annealed Particle Filtering, In *Proc. IEEE Conference on Computer Vision and Pattern Recognition*, CVPR, 2: 1144–1149, 2000.
16. Deutscher, J., A. Davison, and I. Reid: Automatic Partitioning of High Dimensional Search Spaces Associated with Articulated Body Motion Capture, In *Proc. IEEE Conference on Computer Vision and Pattern Recognition*, CVPR, 2: 669–676, 2001.
17. Drummond, T., and R. Cipolla: Real-time Tracking of Highly Articulated Structures in the Presence of Noisy Measurements, In *Proc. IEEE International Conference on Computer Vision*, ICCV, 2: 315–320, 2001.
18. Elias H., O. Carlos, and S. Jesus: Detected motion classification with a double-background and a neighborhood-based difference, *Pattern Recognition Letters*, 24(12): 2079–2092, 2003.
19. Fang G., W. Gao and D. Zhao: Large Vocabulary Sign Language Recognition Based on Hierarchical Decision Trees, In *Proc. International Conference on Multimodal Interfaces*, pp. 301–312, 2003
20. Ferryman J., A. Adams, S. Velastin, T. Ellis, P. Emagnino, and N. Tyler: REASON: Robust Method for Monitoring and Understanding People in Public Spaces. Technological Report, Computational Vision Group, University of Reading, 2004.
21. Gao H.: Tracking Small and Fast Objects in Noisy Images. Masters Thesis. Computer Science Department, University of Canterbury, 2005.
22. Gao J. and J. Shi: Multiple Frame Motion Inference Using Belief Propagation, In *Proc. IEEE International Conference on Automatic Face and Gesture Recognition*, 2004.
23. Gavrila, D. and L. Davis: 3-D model-based tracking of humans in action: a multi-view approach, In *Proc. IEEE Conference on Computer Vision and Pattern Recognition*, CVPR, 73–80, 1996.
24. Goncalves, L., E. Di Bernardo, E. Ursella and P. Perona: Monocular Tracking of the Human Arm in 3D, In *Proc. IEEE International Conference on Computer Vision*, ICCV, 764–770, 1995.

25. Green R. and L. Guan: Quantifying and Recognising Human Movement Patterns from Monocular Video Images - Part I: A New Framework for Modelling Human Motion, *IEEE Transactions on Circuits and Systems for Video Technology*, 14(2): 179–190, 2004.

26. Green R. and L. Guan: Quantifying and Recognising Human Movement Patterns from Monocular Video Images - Part II: Application to Biometrics. *IEEE Transactions on Circuits and Systems for Video Technology*, 14(2): 191–198, 2004.

27. Grobel, K. and M. Assam: Isolated Sign Language Recognition Using Hidden Markov Models, In *Proc. IEEE International Conference on Systems, Man and Cybernetics*, pp. 162–167, Orlando, 1997.

28. Grossmann E., A. Kale and C. Jaynes: Towards Interactive Generation of "Ground-truth" in Background Subtraction from Partially Labelled Examples, In *Proc. IEEE Workshop on VS PETS*, 2005.

29. Grossmann E., A. Kale, C. Jaynes and S. Cheung: Offline Generation of High Quality Background Subtraction Data, In *Proc. British Machine Vision Conference*, 2005.

30. Herbison-Evans, D., R. Green and A. Butt: Computer Animation with NUDES in Dance and Physical Education, *Australian Computer Science Communications*, 4(1): 324–331, 1982.

31. Hogg, D.: Model-based vision: A program to see a walking person, *Image and Vision Computing*, 1(1): 5–20, 1983.

32. Hutchinson-Guest, A.: *Choreo-Graphics; A Comparison of Dance Notation Systems from the Fifteenth Century to the Present*, Gordon and Breach, New York, 1989.

33. Isard, M. and A. Blake: Visual Tracking by Stochastic Propagation of Conditional Density, In *Proc. Fourth European Conference on Computer Vision*, pp. 343–356, Cambridge, 1996.

34. Isard, M. and A. Blake: A Mixed-state Condensation Tracker with Automatic Model Switching, In *Proc. Sixth International Conference on Computer Vision*, pp. 107–112, 1998.

35. Jaynes C., A. Kale, N. Sanders, and E. Grossman: The Terrascope Dataset: A Scripted Multi-Camera Indoor Video Surveillance Dataset with Ground-truth, In *Proc. IEEE Workshop on VS PETS*, 2005.

36. Jelinek, F.: *Statistical Methods for Speech Recognition*, MIT Press, Cambridge, 1999.

37. Jeong K. and C. Jaynes: Moving Shadow Detection Using a Combined Geometric and Color Classification Approach, In *Proc. IEEE Motion, Breckenridge*, 2005.

38. Ju, S., M. Black and Y. Yacoob: Cardboard People: A Parameterized Model of Articulated Motion, In *Proc. IEEE International Conference on Automatic Face and Gesture Recognition*, pp. 38–44, 1996.

39. Kadous, M.: Machine recognition of Auslan signs using PowerGloves: Towards large-lexicon recognition of sign language, In *Proc. Workshop on the Integration of Gesture in Language and Speech*, pp. 165–74, Applied Science and Engineering Laboratories, Newark, 1996.

40. Kakadiaris, I. and D. Metaxas: Model-based estimation of 3D human motion with occlusion based on active multi-viewpoint selection, *IEEE Conference on Computer Vision and Pattern Recognition*, pp. 81–87, 1996.

41. Krinidis M., N. Nikolaidis and I. Pitas: Feature-Based Tracking Using 3D Physics-Based Deformable Surface. Department of Informatics, Aristotle University of Thessaloniki, 2005.
42. Kumar S.: Models for Learning Spatial Interactions in Natural Images for Context-Based Classification. Phd Thesis, The Robotics Institute School of Computer Science Carnegie Mellon University, 2005.
43. Leventon, M. and W. Freeman: Bayesian estimation of 3-d human motion from an image sequence, Technical Report 98–06, Mitsubishi Electric Research Lab, Cambridge, 1998.
44. Li D., D. Winfield and D. Parkhurst: Starburst: A hybrid algorithm for video-based eye tracking combining feature-based and model-based approaches. Technical Report of Human Computer Interaction Program, Iowa State University, 2005.
45. Liang, R. and M. Ouhyoung: A Real-time Continuous Gesture Recognition System for Sign Language, In *Proc. Third International Conference on Automatic Face and Gesture Recognition*, pp. 558–565, Nara, 1998.
46. Liddell, S. and R. Johnson: American Sign Language: the phonological base, *Sign Language Studies*, 64: 195–277, 1989.
47. Liebowitz, D. and S. Carlsson: Uncalibrated Motion Capture Exploiting Articulated Structure Constraints, In *Proc. IEEE International Conference on Computer Vision*, ICCV, 2001.
48. Lukowicz, P., J. Ward, H. Junker, M. Stager, G. Troster, A. Atrash, and T. Starner: Recognising Workshop Activity Using Body Worn Microphones and Accelerometers, In *Proc. Second International Conference on Pervasive Computing*, pp. 18–22, 2004.
49. MacCormick, J. and M. Isard: Partitioned Sampling, Articulated Objects and Interface-quality Hand Tracking, In *Proc. European Conference on Computer Vision*, 2: 3–19, 2000.
50. Makris D.: Learning an Activity Based Semantic Scene Model. PhD Thesis, School of Engineering and Mathematical Science, City University, 2004.
51. Mark J. Body Tracking from Single-Camera Video. Technical Report of Mitsubishi Electric Research Laboratories, 2004.
52. Moeslund, T. and E. Granum: A survey of computer vision-based human motion capture, *Computer Vision and Image Understanding*, 18: 231–268, 2001.
53. Nam Y. and K. Wohn: Recognition of space-time hand-gestures using hidden Markov model, *ACM Symposium on Virtual Reality Software and Technology*, 1996.
54. Pentland, A. and B. Horowitz: Recovery of nonrigid motion and structure, *IEEE Transactions on PAMI*, 13:730–742, 1991.
55. Pheasant, S. *Bodyspace. Anthropometry, Ergonomics and the Design of Work*, Taylor & Francis, 1996.
56. Plnkers, R. and P. Fua: Articulated Soft Objects for Video-based Body Modelling, In *Proc. IEEE International Conference on Computer Vision*, ICCV, pp. 394–401, 2001.
57. Rehg, J. and T. Kanade: Model-based Tracking of Self-occluding Articulated Objects, In *Proc. Fifth International Conference on Computer Vision*, pp. 612–617, 1995.
58. Remondino F. and A. Roditakis: Human Figure Reconstruction and Modelling from Single Image or Monocular Video Sequence, In *Proc. Fourth International Conference on 3D Digital Image and Modelling*, 2003.

59. Ren, J., J. Orwell, G. Jones, and M. Xu: A General Framework for 3D Soccer Ball Estimations and Tracking, *IEEE Transactions on Image Processing*, 24–27, 2004.
60. Rittscher, J., A. Blake and S. Roberts: Towards the automatic analysis of complex human body motions, *Image and Vision Computing*, 20(12): 905–916, 2002.
61. Rohr, K. Towards model-based recognition of human movements in image sequences, *CVGIP - Image Understanding*, 59(1):94–115, 1994.
62. Rosales, R. and S. Sclaroff: Inferring Body Pose Without Tracking Body Parts, In *Proc. IEEE Conference on Computer Vision and Pattern Recognition*, CVPR, 2000.
63. Schlenzig, J., E. Hunter, and R. Jain: Recursive Identification of Gesture Inputers Using Hidden Markov Models, In *Proc. Applications of Computer Vision*, 187–194, 1994.
64. Schrotter G., A. Gruen, E. Casanova, and P. Fua: Markerless Model Based Surface Measurement and Motion Tracking, In *Proc. Seventh conference on Optical 3D Measurement Techniques*, Zurich, 2005.
65. Sigal, L., S. Bhatia S., Roth, M. Black, and M. Isard: Tracking Loose-limbed People, In *Proc. IEEE Conference on Computer Vision and Pattern Recognition*, 2004.
66. Starner, T. and A. Pentland: Real-time American Sign Language recognition from video using Hidden Markov Models, Technical Report 375, MIT Media Laboratory, 1996.
67. Stokoe, W.: *Sign Language Structure: An Outline of the Visual Communication System of the American Deaf, Studies in Linguistics*: Chapter 8. Linstok Press, Silver Spring, MD, 1960. Revised 1978.
68. Sullivan, J., A. Blake, M. Isard, and J. MacCormick: Object Localization by Bayessian Correlation, In *Proc. International Conference on Computer Vision*, 2: 1068–1075, 1999.
69. Tamura, S., and S. Kawasaki: Recognition of sign language motion images, *Pattern Recognition*, 31: 343–353, 1988.
70. Taylor, C. Reconstruction of Articulated Objects from Point Correspondences in a Single Articulated Image, In *Proc. IEEE Conference on Computer Vision and Pattern Recognition*, CVPR, pp. 586–591, 2000.
71. Urtasun R. and P. Fua: (2004) 3D Human Body Tracking using Deterministic Temporal Motion Models, Technical Report of Computer Vision Laboratory, EPFL, Lausanne, 2004.
72. Vogler, C. and D. Metaxas: Adapting hidden Markov Models for ASL Recognition by Using Three-dimensional Computer Vision Methods, In *Proc. IEEE International Conference on Systems, Man and Cybernetics*, pp. 156–161, Orlando, 1997.
73. Vogler, C. and D. Metaxas: ASL Recognition Based on a Coupling Between HMMs and 3D Motion Analysis, In *Proc. IEEE International Conference on Computer Vision*, pp. 363–369, Mumbai, 1998.
74. Vogler, C. and D. Metaxas: Toward scalability in ASL recognition: breaking down signs into phonemes, *Gesture Workshop 99*, Gif-sur-Yvette, 1999.
75. Wachter, S. and H. Nagel, Tracking of persons in monocular image sequences, *Computer Vision and Image Understanding*, 74(3):174–192, 1999.
76. Waldron, M. and S. Kim, Isolated ASL sign recognition system for deaf persons, *IEEE Transactions on Rehabilitation Engineering*, 3(3):261–71, 1995.

77. Wang, J., G. Lorette, and P. Bouthemy, Analysis of Human Motion: A Modelbased Approach, In *Proc. Scandinavian Conference on Image Analysis*, 2:1142–1149, 1991.
78. Wren, C., A. Azarbayejani, T. Darrell and A. Pentland, "Pfinder: Real-time tracking of the human body", *IEEE Transactions on PAMI*, 19(7):780–785, 1997.
79. Yamato, J., J. Ohya, and K. Ishii, Recognizing Human Action in Time-sequential Images Using Hidden Markov Models, In *Proc. IEEE International Conference on Computer Vision*, pp. 379–385, 1992.
80. Zhong H., J. Shi, and M. Visontai: Detecting Unusual Activity in Video, In *Proc.IEEE Computer Society Conference on Computer Vision and Pattern Recognition*, 2004.

10

Topologically Constrained Isometric Embedding

Guy Rosman, Alexander M. Bronstein, and Michael M. Bronstein, and Ron Kimmel

Technion – Israel Institute of Technology, Department of Computer Science, Haifa 32000, Israel

Summary. Presented is an algorithm for nonlinear dimensionality reduction that uses both local(short) and global(long) distances in order to learn the intrinsic geometry of manifolds with complicated topology. Since our algorithm matches nonlocal structures, it is robust even to strong noise. We show experimental results on both synthetic and real data demonstrating the advantages of our approach over state-of-the-art manifold learning methods.

10.1 Introduction

Analysis of high-dimensional data is encountered in numerous pattern recognition applications. It appears that often a small number of dimensions is needed to explain the high-dimensional data. Dimensionality reduction methods such as *principal components analysis* (PCA) [15] and *multidimensional scaling* (MDS) [5] are often used to obtain a low-dimensional representation of the data. This is a commonly used pre-processing stage in pattern recognition.

While methods like PCA assume the existence of a linear map between the data points and the parametrization space, such a map often does not exist. Applying linear dimensionality reduction methods to nonlinear data may result in a distorted representation. *Nonlinear dimensionality reduction* (NLDR) methods attempt to describe a given high-dimensional data set of points as a low-dimensional manifold, by a nonlinear map preserving certain properties of the data. This kind of analysis has applications in numerous fields, such as color perception, pathology tissue analysis [11], enhancement of MRI images [13], face recognition [7], and biochemistry [24], to mention a few. In the field of motion understanding, several methods use NLDR techniques [22, 29], including works presented in this book, see Chapter 2.

As the input data, we assume to be given N points in the M-dimensional Euclidean space, $\{\mathbf{z}_i\}_{i=1}^N \subset \mathbb{R}^M$. The points constitute vertices of a proximity graph with the set of edges E; the points $\mathbf{z}_i, \mathbf{z}_j$ are neighbors if $(i, j) \in E$. The data points are samples of an m-dimensional manifold $\mathcal{M} \subset \mathbb{R}^M$, where

B. Rosenhahn et al. (eds.), *Human Motion – Understanding, Modelling, Capture, and Animation*, 243–262.

$M \gg m$. The manifold is assumed to have a parametrization, represented by the smooth bijective map $\varphi : \mathcal{C} \subset \mathbb{R}^m \to \mathcal{M}$.

The goal is to find a set of points $\{\mathbf{x}_i\}_{i=1}^N \subset \mathbb{R}^m$ representing the parametrization. We will use the $N \times m$ matrix $\mathbf{X}$ representing the coordinates of the points in the parametrization space. Many NLDR techniques attempt finding an m-dimensional representation for the data, while preserving some *local* invariant. Another class of algorithms preserves *global* invariants, like the *geodesic distances* d_{ij}, approximated as shortest paths on the proximity graph.

The *locally linear embedding* (LLE) algorithm [30] expresses each point as a linear combination of its neighbors,

$$\mathbf{z}_i = \sum_{j:(i,j)\in E} w_{ij}\mathbf{z}_j,$$

and minimizes the least-squares deviation from this local relation. This gives rise to a minimal eigenvalue problem

$$\mathbf{X}^* = \underset{\mathbf{X}^T\mathbf{X}=\mathbf{I}}{\operatorname{argmin}} \operatorname{trace}(\mathbf{X}^T\mathbf{M}\mathbf{X}),$$

where $\mathbf{M}$ is an $N \times N$ sparse matrix with elements

$$m_{ij} = \begin{cases} 1 - w_{ij} - w_{ji} + \sum_k w_{ki}w_{kj} & i = j, \\ -w_{ij} - w_{ji} + \sum_k w_{ki}w_{kj} & i \neq j. \end{cases}$$

A similar approach is the *Laplacian eigenmaps* algorithm proposed by Belkin and Niyogi in [1]. It solves the following minimum eigenvalue problem,

$$\mathbf{X}^* = \underset{\substack{\mathbf{X}^T\mathbf{D}\mathbf{X}=\mathbf{I} \\ \mathbf{X}^T\mathbf{D}\mathbf{1}=\mathbf{0}}}{\operatorname{argmin}} \operatorname{trace}(\mathbf{X}^T\mathbf{L}\mathbf{X}),$$

where $\mathbf{1}$ is an all-ones vector, and $\mathbf{L}$ is the Laplacian of the graph, an $N \times N$ sparse matrix with elements

$$l_{ij} = \begin{cases} w_{ij} & (i,j) \in E \text{ and } i = j, \\ -\sum_{k\neq i} w_{ki} & i = j, \\ 0 & \text{else}, \end{cases}$$

and $\mathbf{D}$ is a diagonal matrix with elements $d_{ii} = -l_{ii}$. The weights w_{ij} can be selected, for example, as $w_{ij} = \exp\{-\frac{1}{\sigma^2}\|\mathbf{x}_i - \mathbf{x}_j\|_2^2\}$. The Laplacian $\mathbf{L}$ is a discrete approximation of the Laplace–Beltrami operator and its eigenfunctions are locally flat. This geometric interpretation is valid if the sampling of the manifold is uniform.

The *diffusion maps* framework [11] tries to generalize the work of Belkin and Niyogi and other local methods. Writing the quadratic expression in $\mathbf{X}$ in a slightly different way,

$$\mathbf{X}^* = \underset{\mathbf{X}}{\operatorname{argmin}} \sum_{k=1}^{m} \sum_{i=1}^{N} a_i(\mathbf{x}^k),$$

where $\mathbf{x}^k$ is the kth column of $\mathbf{X}$, one can think of $a_i(\mathbf{x})$ as of an operator, applied coordinate-wise, and measuring the local variation of the result in the neighborhood of the point $\mathbf{x}_i$, independently for each dimension. In general, $a_i(\mathbf{x})$ can be defined to be an arbitrary quadratic positive semidefinite form on the columns of $\mathbf{X}$. For example, choosing $a_i(\mathbf{x}^k)$ to measure the Frobenius norm of the Hessian of the kth coordinate of the result at the ith point, one obtains the *Hessian locally linear embedding* (HLLE) algorithm [20]. Another choice is the Laplace–Beltrami operator normalized with respect to the sampling density. This choice gives rise to an algorithm acts locally on the graph and provides a global coordinate system that mimics the diffusion distances in the graph [27].

Semidefinite embedding [37] maximizes the variance in the data set while keeping the local distances unchanged. The problem is formulated and solved as a semidefinite programming (SDP) [2] problem, under constraints reflecting invariance to translation and local isometry to Euclidean space,

$$\mathbf{X}^* = \underset{\mathbf{K} \succeq 0}{\arg \max} \operatorname{trace}(\mathbf{K}) \quad \text{s.t.} \quad \begin{cases} \sum_j k_{ij} = 0, \\ (i,j) \in E, \\ k_{ii} - k_{ij} - k_{ji} + k_{jj} = \delta_{ij}, \end{cases}$$

where $k_{ij} = \langle \mathbf{x}_i, \mathbf{x}_j \rangle$ are elements of the *Gram matrix* $\mathbf{K}$, and δ_{ij} are the geodesic distances. The second constraint preserves the local isometry property. Yet, the computational cost of solving an SDP problem is $O(N^6)$ [2], which is prohibitive even in medium-scale problems. Attempts to overcome it by using *landmarks* [36] still incurs high computational complexity.

Tensor voting [28] is another local method, using a local voting mechanism to estimate, in a robust way, the local data dimensionality. It can determine the local dimensionality even for objects that have a spatially varying dimensionality. Although this method does not recover a global parametrization of the manifold, it may be used as a preprocessing stage for other algorithms.

Unlike local approaches, the *Isomap* algorithm [31,34] considers both local and *global* invariants – the lengths of geodesics between the points on the manifold. Short geodesics are assumed to be equal to Euclidean distances, and longer ones are approximated as shortest path lengths on the proximity graph, using standard graph search methods like the Dijkstra's algorithm [12, 14]. Isomap then uses *multidimensional scaling* (MDS) [5] attempting to find an m-dimensional Euclidean representation of the data, such that the Euclidean distances between points are as close as possible to the corresponding geodesic ones, for example, using the least squares criterion (referred to as *stress*),

$$\mathbf{X}^* = \underset{\mathbf{X} \in \mathbb{R}^{N \times m}}{\operatorname{argmin}} \sum_{i<j} w_{ij} \left(d_{ij}(\mathbf{X}) - \delta_{ij}\right)^2,$$

where $d_{ij}(\mathbf{X}) = \|\mathbf{x}_i - \mathbf{x}_j\|_2$ is the Euclidean distance between points $\mathbf{x}_i$ and $\mathbf{x}_j$ in $\mathbb{R}^m$.

The main advantage of Isomap is that it uses global geometric invariants, which are less sensitive to noise compared to local ones. Yet, its underlying assumption is that $\mathcal{M}$ is *isometric* to $\mathcal{C} \subset \mathbb{R}^m$ with the *induced metric* $d_\mathcal{C}$, that is, $\delta(\mathbf{z}_i, \mathbf{z}_j) = d_{\mathbb{R}^m}(\mathbf{x}_i, \mathbf{x}_j)$ for all $i, j = 1, ..., N$. If $\mathcal{C}$ is convex, the restricted metric $d_{\mathbb{R}^m}|_\mathcal{C}$ coincides with the induced metric $d_\mathcal{C}$ and Isomap succeeds recovering the parametrization of $\mathcal{M}$. Otherwise, $\mathcal{C}$ has no longer Euclidean geometry and MDS cannot be used. The assumption of convexity of $\mathcal{C}$ appears to be too restrictive, as many data manifolds have complicated topology which violates this assumption. Donoho and Grimes [19] showed examples of data in which $\mathcal{C}$ is nonconvex, and pointed out that Isomap fails in such cases.

Here, we suggest a solution based on removing pairs of points inconsistent with the convexity assumption. Our approach, hereinafter referred to as the *topologically constrained isometric embedding* (TCIE), allows handling data manifolds of arbitrary topology. In Section 10.2, we introduce a new algorithm for that goal, and prove that it rejects inconsistent geodesics. Section 10.3 discusses the numerical implementation of the algorithm and suggests ways to speed up its convergence. In Section 10.4 we demonstrate our approach on real and synthetic data.

10.2 Topologically Constrained Isometric Embedding

As we mentioned, the Isomap algorithm assumes that the parametrization $\mathcal{C}$ of $\mathcal{M}$ is a convex subset of $\mathbb{R}^m$, and relies on the isometry assumption to find the map from $\mathcal{M}$ to the metric space $(\mathcal{C}, d_\mathcal{C})$ by means of MDS (the stress in the solution will be zero). MDS can be used because $d_\mathcal{C} = d_{\mathbb{R}^m}|_\mathcal{C}$ due to the convexity assumption. In the case when $\mathcal{C}$ is nonconvex, this is not necessarily true, as there may exist pairs of points for which $d_\mathcal{C} \neq d_{\mathbb{R}^m}|_\mathcal{C}$. We call such pairs *inconsistent*. An example of such a pair is shown in Figure 10.1. We denote the set of all consistent pairs by

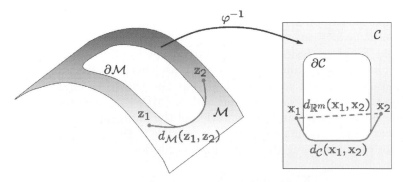

Fig. 10.1. Example of inconsistent $\mathbf{x}_1$ and $\mathbf{x}_2$, for which the line connecting them in $\mathbb{R}^m$ (dashed line) is shorter than the geodesic $g_\mathcal{C}(\mathbf{x}_1, \mathbf{x}_2)$ (dotted curve).

$$P = \{(i,j) : d_{\mathcal{C}}(\mathbf{x}_i, \mathbf{x}_j) = d_{\mathbb{R}^m}|_{\mathcal{C}}(\mathbf{x}_i, \mathbf{x}_j)\} \subseteq I \times I.$$

In the TCIE algorithm, we find a subset $\bar{P} \subseteq P$ of pairs of points that will be consistently used in the MDS problem. The algorithm goes as follows:

---

1 Compute the $N \times N$ matrix of *geodesic distances* $\boldsymbol{\Delta} = (\delta_{ij})$.
2 Detect the boundary points $\partial\mathcal{M}$ of the data manifold.
3 Detect a subset of *consistent distances* according to the following criterion,

$$\bar{P}_1 = \{(i,j) : c(\mathbf{z}_i, \mathbf{z}_j) \cap \partial\mathcal{M} = \emptyset\}, \tag{1.1}$$

or

$$\bar{P}_2 = \{(i,j) : \delta(\mathbf{z}_i, \mathbf{z}_j) \leq \delta(\mathbf{z}_j, \partial\mathcal{M}) + \delta(\mathbf{z}_i, \partial\mathcal{M})\}, \tag{1.2}$$

where $\delta(\mathbf{z}, \partial\mathcal{M}) = \inf_{\mathbf{z}' \in \partial\mathcal{M}} \delta(\mathbf{z}, \mathbf{z}')$ denotes the distance from $\mathbf{z}$ to the boundary.
4 Solve the MDS problem for consistent pairs only,

$$\mathbf{X}^* = \operatorname*{argmin}_{\mathbf{X} \in \mathbb{R}^{N \times m}} \sum_{(i,j) \in \bar{P}} (d_{ij}(\mathbf{X}) - \delta_{ij})^2.$$

We now describe the main stages of the algorithm:

Detection of Boundary Points

Step 2 in our algorithm is the detection of boundary points. There exist many algorithms that detect boundaries in point clouds, see, e.g. [6, 21]. For two-dimensional manifolds, we use the following heuristic that worked well even for relatively sparsely sampled manifolds. Essentially, our method assumes the point i and its two opposite neighbors, j, k, to be a part of a curve along the boundary. It then tries to find points that are placed outside of this boundary on both sides of it, violating the conjecture. This idea can be extended for $m > 2$ as well, in the following way:

The second method assumes an isotropic distribution of points in the neighborhood reconstructed by MDS and seems to work well in practice for three-dimensional manifolds. Directions in this method are selected according to neighboring points, avoiding the need to artificially determine the normal direction.

Besides the proposed heuristics, other boundary detection algorithms can be used [18, 28]. In practice, if the intrinsic dimension of the manifold is large, many samples are required for boundary detection.

---

1 **for** $i = 1, ..., N$ **do**
2   Find the set $\mathcal{N}(i)$ of the K nearest neighbors of the point i.
3   Apply MDS to the $K \times K$ matrix $\boldsymbol{\Delta}_K = (\delta_{kl \in \mathcal{N}(i)})$ and obtain a set of local coordinates $\mathbf{x}'_1, ..., \mathbf{x}'_K \in \mathbb{R}^m$.
4   **for** $j, k \in \mathcal{N}(i)$ *such that* $\frac{\langle \mathbf{x}'_j - \mathbf{x}'_i, \mathbf{x}'_k - \mathbf{x}'_i \rangle}{\|\mathbf{x}'_j - \mathbf{x}'_i\| \cdot \|\mathbf{x}'_k - \mathbf{x}'_i\|} \approx -1$ **do**
5   Mark the pair (j, k) as `valid`.
6   **if** $|\mathbf{x}' : \frac{\langle \mathbf{x}' - \mathbf{x}'_i, \mathbf{v}_l \rangle}{\|\mathbf{x}' - \mathbf{x}'_i\|} \approx 1| \geq \tau_a |\mathcal{N}(i)|$ *for all* $l = 1, ..., m - 1$ **then**
7   Label the pair (j, k) as `satisfied`. (here $\mathbf{v}_l$ denotes the lth vector of an orthonormal basis of the subspace of $\mathbb{R}^m$ orthogonal to $\mathbf{x}'_j - \mathbf{x}'_k$).
8   **end**
9   **end**
10   **if** *the ratio of* `satisfied` *to* `valid` *pairs is smaller than threshold* τ_b **then**
11   Label point i as `boundary`.
12   **end**
13 **end**

---

1 **for** $i = 1, ..., N$ **do**
2   Find the set $\mathcal{N}(i)$ of the K nearest neighbors of the point i.
3   Apply MDS to the $K \times K$ matrix $\boldsymbol{\Delta}_K = (\delta_{kl \in \mathcal{N}(i)})$ and obtain a set of local coordinates $\mathbf{x}'_1, ..., \mathbf{x}'_K \in \mathbb{R}^m$.
4   **for** $j = 1, ..., K$ **do**
5   **if** $\frac{|\{\mathbf{x} \in \mathbb{R}^m : \langle \mathbf{x}'_i - \mathbf{x}'_j, \mathbf{x} - \mathbf{x}'_i \rangle > 0\}|}{|\{\mathbf{x} \in \mathbb{R}^m : \langle \mathbf{x}'_i - \mathbf{x}'_j, \mathbf{x} - \mathbf{x}'_i \rangle \leq 0\}|} \leq \tau_a$ **then**
6   mark j as `candidate`.
7   **end**
8   **end**
9 **end**
10 **if** *the number of* `candidate` *points is larger than* τ_b **then**
11   Label point i as `boundary`.
12 **end**

Detection of Inconsistent Geodesics

The first consistency criterion requires to check whether geodesics touch the boundary. Once we have detected the boundary points, we use a modification of the Dijkstra algorithm [12, 14], in the following manner: while extending the shortest path from source i to point j, towards point j's neighbors: for each neighbor k of point j, check if the currently known shortest path from i to point k is longer than the route through j. If so, update the path length to the shorter path, as is done in the Dijkstra algorithm, but in addition, mark the newly updated path as an inconsistent if either (i) the path from i to j is a path marked to be removed, or (ii) j is a boundary point, and the path from

i to k through j travels through more than one point. The second condition protects paths with a boundary end point from being removed. This way we eliminate only geodesics that do not end at the boundary but rather touch it and continue their journey. Removal is done after the path length calculation by assigning zero weight to the measured distance. Similar modifications can be made to the Bellman–Ford and Floyd algorithms, or other algorithms [25]. We note that for the second criterion, detection of inconsistent geodesic is trivially done by looking at the geodesic distances.

SMACOF Algorithm

A way to include only consistent point pairs is to use *weighted stress*,

$$\mathbf{X}^* = \underset{\mathbf{X}}{\operatorname{argmin}} \sum_{i=0, i<j} w_{ij}(d_{ij}(\mathbf{X}) - \delta_{ij})^2,$$

where $w_{ij} = 1$ if $(i, j) \in \bar{P}$, $\bar{P}$ can be set according to either the first criterion or the second critterion, and $w_{ij} = 0$ otherwise. This allows us, by choosing the right weights, to minimize the error only for consistent geodesics.

The geodesics that were not marked as inconsistent during the Dijkstra algorithm have their weight set to one. We also allow a positive weight for short geodesics, in order to keep the connectivity of the manifold as a graph, even at boundary points. All other geodesics have their weight set to zero. We then use the SMACOF algorithm [5] to minimize the weighted stress.

We note that the correctness of these conditions depends on the assumption that our manifold is isometric to a subregion of an Euclidean space, similarly to the underlying assumption of Isomap.

10.2.1 Theoretical Analysis

In this section we discuss a continuous case, in which the manifold is sampled with non-zero density. We assume the same assumptions on sampling density and uniformity made by Bernstein et al. [3], who proved the convergence of the graph distances approximation, used by the Isomap algorithm, to the geodesic distances on the manifold. Also note that the requirement of a positive density function prevents problems that may occur in geodesics approximated by a graph when the surface is sampled in a specific regular pattern. In our case, there is also the question of whether or not we remove too many geodesics. The answer is related to the topology of the manifold.

In the continuous case, our algorithm approximates an isometry between $\mathcal{M}$ with the geodesic metric δ and $\mathcal{C} \subset \mathbb{R}^m$ with the induced metric $d_{\mathcal{C}}$. Our criteria always select consistent distances, as shown in the following propositions:

Proposition 1 $\bar{P}_1 = \{(i,j) : c(\mathbf{z}_i, \mathbf{z}_j) \cap \partial\mathcal{M} = \emptyset\} \subseteq P.$

Proof. Let $(i,j) \in \bar{P}_1$. To prove the proposition, it is sufficient to show that the pair of points (i,j) is consistent, i.e., $(i,j) \in P$. Let $c_{\mathcal{M}}(\mathbf{z}_1, \mathbf{z}_2)$ be the geodesic connecting $\mathbf{z}_i$ and $\mathbf{z}_j$ in $\mathcal{M}$, and let $c_{\mathcal{C}}(\mathbf{x}_1, \mathbf{x}_2)$ be its image under φ^{-1} in $\mathcal{C}$. Since $c(\mathbf{z}_i, \mathbf{z}_j) \cap \partial\mathcal{M} = \emptyset$ and because of the isometry, $c_{\mathcal{C}}(\mathbf{x}_i, \mathbf{x}_j) \subset \mathrm{int}(\mathcal{C})$.

Assume that (i,j) is inconsistent. This implies that $d_{\mathcal{C}}(\mathbf{x}_i, \mathbf{x}_j) > d_{\mathbb{R}^m}(\mathbf{x}_i, \mathbf{x}_j)$, i.e., that the geodesic $c_{\mathcal{C}}(\mathbf{x}_i, \mathbf{x}_j)$ is not a straight line. Therefore, there exists a point $x \in c_{\mathcal{C}}(\mathbf{x}_i, \mathbf{x}_j)$, in whose proximity $c_{\mathcal{C}}(\mathbf{x}_i, \mathbf{x}_j)$ is not a straight line. Since $c_{\mathcal{C}}(\mathbf{x}_i, \mathbf{x}_j) \subset \mathrm{int}(\mathcal{C})$, there exists a ball $B_\epsilon(\mathbf{x})$ with the Euclidean metric $d_{\mathbb{R}^m}$ around $\mathbf{x}$ of radius $\epsilon > 0$. Let us take two points on the segment of the geodesic within the ball, $\mathbf{x}', \mathbf{x}'' \in c_{\mathcal{C}}(\mathbf{x}_i, \mathbf{x}_j) \cap B_\epsilon(\mathbf{x})$. The geodesic $c_{\mathcal{C}}(\mathbf{x}', \mathbf{x}'')$ coincides with the segment of $c_{\mathcal{C}}(\mathbf{x}_i, \mathbf{x}_j)$ between $\mathbf{x}', \mathbf{x}''$. Yet, this segment is not a straight line, therefore we can shorten the geodesic by replacing this segment with $c_{\mathbb{R}^m}(\mathbf{x}', \mathbf{x}'')$, in contradiction to the fact that $c_{\mathcal{C}}(\mathbf{x}_i, \mathbf{x}_j)$ is a geodesic. Therefore, $(i,j) \in P$. □

Therefore, for every geodesic in $\mathcal{M}$ which was not detected as touching the boundary, the image under φ^{-1} is a line, which is approximated correctly by the MDS procedure. In the more general case, where $(\mathcal{M}, d_{\mathcal{M}})$ is not isometric to a subregion of Euclidean space, the second criterion we have presented ensures that if the manifold is isometric to a subregion $\mathcal{C}$ of a space $\mathcal{C}'$ with Riemannian metric, we only use geodesics for which the induced and the restricted metric identify.

Assume we have a pair of points for which the induced and the restricted metric on $\mathcal{C}$ are not the same. Therefore, the geodesic in $\mathcal{C}'$ must cross the boundary of $\mathcal{C}$, resulting in the inequality

$$d_{\mathcal{C}'}(\mathbf{x}_1, \mathbf{x}_2) > d_{\mathcal{C}}(\mathbf{x}_1, \partial\mathcal{C}) + d_{\mathcal{C}}(\mathbf{x}_2, \partial\mathcal{C}).$$

Using the second criterion, replacing the right-hand side we have

$$d_{\mathcal{C}'}(\mathbf{x}_1, \mathbf{x}_2) > d_{\mathcal{C}}(\mathbf{x}_1, \mathbf{x}_2).$$

Resulting in a contradiction to the definition of induced metric.

Note that for a parametrization manifold $\mathcal{C}'$ with an arbitrary Riemannian metric, the MDS procedure would not be able to give us the correct mapping. This would require the use of another procedure, as is done in [8]. The second criterion may be of use in cases where the metric on $\mathcal{C}'$ is close to Euclidean, and yet we only want to use geodesics which stay in $\mathcal{C}$.

10.2.2 Complexity Analysis

The algorithm involves several computationally demanding procedures,

- Boundary detection – $O(N^2)$.
- Distance computation – $O(N^2 \log N)$.

- Stress minimization – $O(N^2)$ per iteration. While the number of SMACOF iterations is not invariant to the number of samples, in practice it rises slowly, depending on the topology of the manifold and the noise level.

10.3 Implementation Considerations

For determining the shortest paths we used the Dijkstra algorithm implementation supplied by Tenenbaum, et al. [34], for the Isomap code. We added the detection of geodesics touching boundary points. The rest of the algorithm was implemented in MATLAB. In practice, the Dijkstra algorithm takes less than 10% of the total running time for 2000 points. Solving the MDS optimization problem consumes most of the time.

10.3.1 Numerical Properties and Convergence

The optimization problem solved is a nonconvex one, and as such is liable to local convergence if convex optimization methods are employed [35]. In our experiments, we have seen that removing more distances from the stress function caused the problem to be more sensitive to local minima. An example of one such local minimum, encountered when flattening the Swiss-role with a hole example, is shown in Figure 10.2. It appears as a fold over, or "flip". In general, the number of remaining weights depends on the surface topology, as well as the number of sampled points in the surface.[1]

We reduce the risk of convergence to a local minimum by starting from a classical scaling (as mentioned by Trosset et al. [23]) or unweighted least-squares scaling solution. This allows the algorithm to avoid some of the local minima. Although the solutions found by classical scaling and least square scaling may differ, under the assumption of correct distance approximation, the solutions are similar.

Using the unweighted LS-MDS problem to avoid local minima, and then gradually changing the problem solved into the weighted one is in the flavor of graduated non-convexity [4], although the problem remains nonconvex.

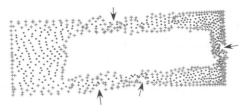

Fig. 10.2. Example of a local minimizer of the weighted MDS problem. Ripples along the boundaries of the rectangle are marked with arrows.

---

[1] Typically, in our experiments **W** contained between 6% to 18% nonzero weights.

10.3.2 Convergence Acceleration by Vector Extrapolation Methods

To speed up the convergence of the SMACOF iterations, we employ *vector extrapolation*. These methods use a sequence of solutions at subsequent iterations of the optimization algorithm and extrapolate the limit solution of the sequence. While these algorithms were derived assuming a linear iterative scheme, in practice, they work well also for nonlinear schemes, such as some processes in computational fluid dynamics [32]. For further details, we refer to [10, 16, 33].

The main idea of vector extrapolation is, given a sequence of solutions $\mathbf{X}^{(k)}$ from iterations $k = 0, 1, ...,$ to approximate the limit $\lim_{k\to\infty} \mathbf{X}^{(k)}$, which should coincide with the optimal solution $\mathbf{X}^*$. The extrapolation $\hat{\mathbf{X}}^{(k)}$ is constructed as an affine combination of last $K + 1$ iterates, $\mathbf{X}^{(k)}, ..., \mathbf{X}^{(k+K)}$

$$\hat{\mathbf{X}}^{(k)} = \sum_{j=0}^{K} \gamma_j \mathbf{X}^{(k+j)}; \qquad \sum_{j=0}^{K} \gamma_j = 1.$$

The coefficients γ_j can be determined in various ways. In the *reduced rank extrapolation* (RRE) method, γ_j are obtained by the solution of the minimization problem,

$$\min_{\gamma_0,...,\gamma_K} \left\| \sum_{j=0}^{K} \gamma_j \Delta\mathbf{X}^{(k+j)} \right\|, \quad \text{s.t.} \sum_{j=0}^{K} \gamma_j = 1,$$

where $\Delta\mathbf{X}^{(k)} = \mathbf{X}^{(k+1)} - \mathbf{X}^{(k)}$. In the *minimal polynomial extrapolation* (MPE) method,

$$\gamma_j = \frac{c_j}{\sum_{i=0}^{K} c_i}, \quad j = 0, 1, ..., K,$$

where c_i arise from the solution of the minimization problem,

$$\min_{c_0,..,c_{K-1}} \left\| \sum_{j=0}^{K} c_j \Delta\mathbf{X}^{(k+j)} \right\|, \quad c_K = 1,$$

which in turn can be formulated as a linear system [33].

10.3.3 Convergence Acceleration by Multiscale Optimization

Another way to accelerate the solution of the MDS problem is using *multiresolution* (MR) methods [9]. The main idea is subsequently approximating the solution by solving the MDS problem at different resolution levels. At each level, we work with a *grid* consisting of points with indices $\Omega_L \subset \Omega_{L-1} \subset ... \subset$

$\Omega_0 = \{1, ..., N\}$, such that $|\Omega_l| = N_l$. At the lth level, the data is represented as an $N_l \times N_l$ matrix $\mathbf{\Delta}_l$, obtained by extracting the rows and columns of $\mathbf{\Delta}_0 = \mathbf{\Delta}$, corresponding to the indices Ω_l. The solution $\mathbf{X}_l^*$ of the MDS problem on the lth level is transferred to the next level $l - 1$ using an *interpolation operator* P_l^{l-1}, which can be represented as an $N_{l-1} \times N_l$ matrix.

---

1 Construct the hierarchy of grids $\Omega_0, ..., \Omega_L$ and interpolation operators $P_1^0, ..., P_L^{L-1}$.

2 Start with some initial $\mathbf{X}_L^{(0)}$ at the coarsest grid, and $l = L$.

3 While $l \geq 0$ **do**

4   Solve the lth level MDS problem

$$\mathbf{X}_l^* = \underset{\mathbf{X}_l \in \mathbb{R}^{N_l \times m}}{\operatorname{argmin}} \sum_{i,j \in \Omega_l} w_{ij}(d_{ij}(\mathbf{X}_l) - \delta_{ij})^2$$

using SMACOF iterations initialized with $X_l^{(0)}$.

5   Interpolate the solution to the next resolution level, $\mathbf{X}_{l-1}^{(0)} = P_l^{l-1}(\mathbf{X}_l^*)$

6      $l \longleftarrow l - 1$

7 End

We use a modification of the *farthest point sampling* (FPS) [17] strategy to construct the grids, in which we add more points from the boundaries, to allow correct interpolation of the fine grid using the coarse grid elements. We use linear interpolation with weights determined using a least squares fitting problem with regularization made to ensure all available nearest neighbors are used.

The multiresolution scheme can be combined with vector extrapolation by employing MPE or RRE methods at each resolution level. In our experiments we used the RRE method, although in practice, for the SMACOF algorithm, both the MPE and the RRE algorithms gave comparable results, giving us a threefold speedup. A comparison of the convergence with and without vector extrapolation and multiresolution methods is shown in Figure 10.3. The stress values shown are taken from the problem shown in Figure 10.6. Major spikes in the stress function of the appropriate method's graph indicate a change in the resolution level with inaccurate interpolation.

10.4 Results

We tested our algorithm on the Swiss roll surface with a large rectangular hole, sampled at 1200 points. Flattening was performed for points sampled on the manifold with additive independently identically distributed Gaussian noise in each coordinate of each point. The various instances of the surface with noise are shown in Figure 10.4. We compare the proposed algorithm to Isomap,

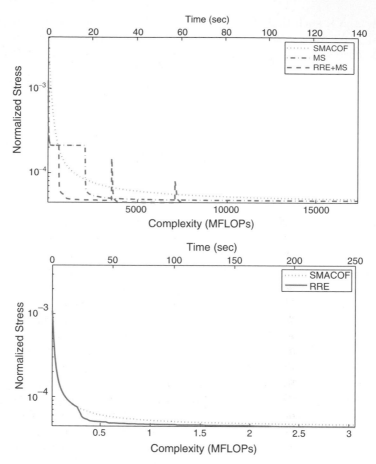

Fig. 10.3. Convergence (in terms of stress value) of basic SMACOF (dotted), SMA-COF with multiresolution acceleration (solid), SMACOF with RRE (dash-dotted) and SMACOF with both RRE and multiscale (dashed), in terms of CPU time and MFLOPS. CPU time is approximated. Convergence was stopped at the same relative change of stress value.

Original Swiss-roll $\sigma = 0.015$ $\sigma = 0.05$

Fig. 10.4. Left to right: Swiss roll surface without noise, and contaminated by Gaussian noise with $\sigma = 0.015$ and $\sigma = 0.05$, and the spiral surface. Detected boundary points are shown in red.

Fig. 10.5. The planar surface cut as a spiral. Detected boundary points are shown in red.

LLE, Laplacian eigenmaps, diffusion maps and Hessian LLE, in Figures 10.6–10.8[2]. Our algorithm finds a representation of the manifold, with relatively small distortion. Adding i.i.d. Gaussian noise to the coordinates of the sampled points, our method remains accurate compared to other popular algorithms that exhibit large distortions. This can be seen, for example, for 1200 points, with $\sigma = 0.05$, in Figure 10.8, where for comparison, $\mathrm{diam}(\mathcal{M}) \approx 6$. The algorithm was allowed to converge until the relative change in the weighted stress was below some threshold. Tests with higher noise levels were also performed and the performance boosting due to the RRE method is quite consistent. Using multiscale further reduces the computational cost of the solution, by a factor of 2, for the problem shown in the example. We note that the speedup depends on both the manifold topology and the problem size, among other factors, but such a reduction in computational effort is typical for all the problems we have tested.

We also tested our algorithm on a planar patch cut in the form of a spiral. Ideally, a correct solution can be achieved by linear methods such as PCA on the embedding space coordinates, and the large number of geodesics removed (only 6% of the geodesic distances remained) makes a worst case scenario

---

[2] We used the same number of neighboring points (12) in all algorithms, except for the diffusion maps algorithm, where we used value of the diffusion distance constant according to

$$\sigma^2 = \frac{2}{N} \sum_{i=0}^{N} \min_{j} \|x_i - x_j\|^2.$$

This is the same rule used by Lafon ([26], p. 33) up to a constant. The larger diffusion distance gave us more robust results. More specifically, the values of ϵ we used in the noiseless example and with $\sigma = 0.015$, $\sigma = 0.05$ were 3.897×10^{-3}, 3.998×10^{-3} and 8.821×10^{-3} respectively.

As for α, the parameter used by Lafon et al. [11] to specify various types of diffusion maps, we show here the results for $\alpha = 1$, though other values of α were also tested.

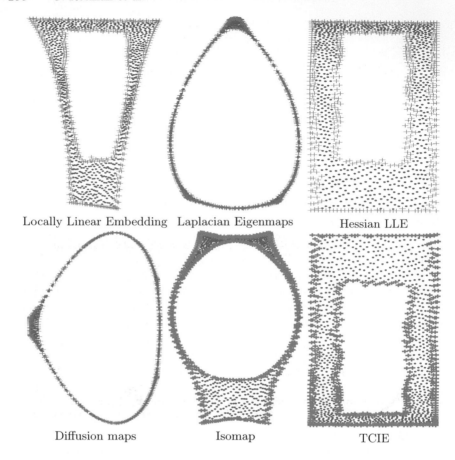

Fig. 10.6. Embedding of the swiss roll (without noise), as produced by LLE, Laplacian eigenmaps, Hessian LLE, diffusion maps, Isomap, and our algorithm. Detected boundary points are shown as red pluses.

for our algorithm. In practice, as Figure 10.9 shows, the proposed algorithm introduces just a minor distortion whereas other algorithms fail to extract the structure of the manifold.

Although in practical cases the data manifold is not necessarily isometric to a subregion of a low-dimensional Euclidean space, our algorithm appears to be able to produce meaningful results in image analysis applications. Figure 10.10 demonstrates the recovery of gaze direction of a person from a sequence of gray-scale images. Assuming that facial pose and expressions do not change significantly, images of the area of the eyes form a manifold approximately parameterized by the direction of the gaze. Similar to previous image manifold experiments [34], we use Euclidean distances between the row-stacked images as the distance measure. In order to reduce the effect of head movement, simple block matching was used.

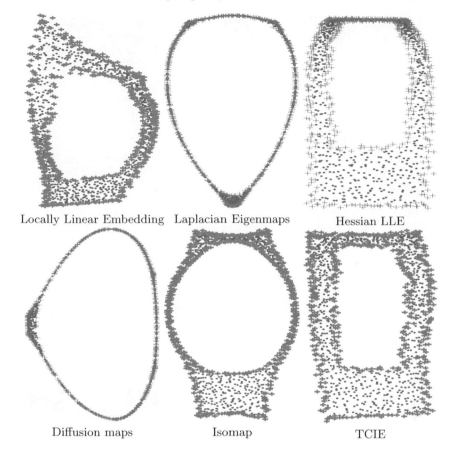

Locally Linear Embedding   Laplacian Eigenmaps   Hessian LLE

Diffusion maps   Isomap   TCIE

Fig. 10.7. Embedding of the swiss roll contaminated by Gaussian noise with $\sigma =$ 0.015, as produced by LLE, Laplacian eigenmaps, Hessian LLE, diffusion maps, Isomap, and our algorithm. Detected boundary points are shown as red pluses.

10.5 Conclusions

We presented a new global method for nonlinear dimensionality reduction. We showed that using a careful selection of geodesics we can, robustly, flatten non-convex manifolds. Since the proposed method uses global information it is less sensitive to noise than local ones, as shown in the examples. In addition, in order to allow the algorithm to run at the same time scale as local methods, we show how vector extrapolation methods such as MPE and RRE can be used to accelerate the solution of nonlinear dimensionality reduction problems.

In future work we consider extending our results to non-Euclidean spaces. We would like to improve its computational efficiency using multigrid [9] methods. We plan to introduce a version of the algorithm which is more robust to changes in the sampling density of the manifold. It is similar in spirit to con-

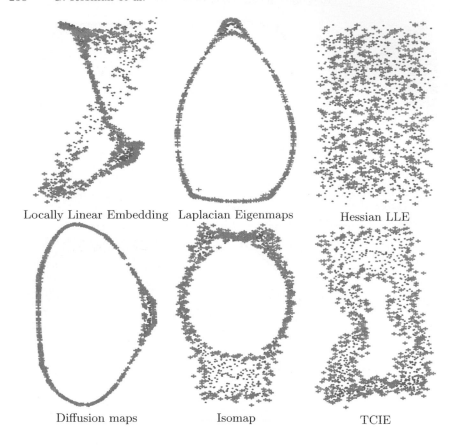

258   G. Rosman et al.

Locally Linear Embedding   Laplacian Eigenmaps   Hessian LLE

Diffusion maps   Isomap   TCIE

Fig. 10.8. Embedding of a 2D manifold contaminated by Gaussian noise with $\sigma = 0.05$, as produced by LLE, Laplacian eigenmaps, Hessian LLE, diffusion maps, Isomap, and our algorithm. Detected boundary points are shown as red pluses.

cepts introduced in the Laplacian eigenmaps algorithm [11]. Furthermore, the boundary detection algorithms that play a key role in the elimination process will be further explored. Finally, we note that the main limitation of the proposed algorithm is its memory complexity, and we are currently searching for ways to reduce this limitation.

Acknowledgments

We would like to thank Professor Avram Sidi for his advice regarding vector extrapolation algorithms. We would also like to thank the respective owners of the code of various nonlinear dimensionality reduction methods demonstrated in this publication.

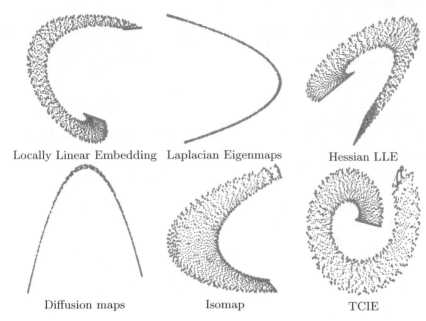

Locally Linear Embedding   Laplacian Eigenmaps   Hessian LLE

Diffusion maps   Isomap   TCIE

Fig. 10.9. Embedding of a 2D manifold in the shape of a flat spiral, as produced by LLE, Laplacian eigenmaps, Hessian LLE, diffusion maps, Isomap, and our algorithm. Detected boundary points are shown as red pluses.

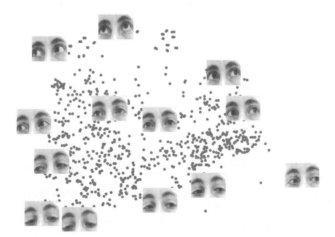

Fig. 10.10. The intrinsic coordinates of the image manifold of the eyes area with different gaze directions, as mapped by our algorithm.

References

1. Belkin M. and Niyogi P. Laplacian eigenmaps and spectral techniques for embedding and clustering. In T. G. Dietterich, S. Becker, and Z. Ghahramani, editors, *Advances in Neural Information Processing Systems*, volume 14, pp. 585–591, MIT Press, Cambridge, MA, 2002.
2. Ben-Tal A. and Nemirovski A. S. *Lectures on Modern Convex Optimization: Analysis, Algorithms, and Engineering Applications*. Society for Industrial and Applied Mathematics, Philadelphia, PA, 2001.
3. Bernstein M., de Silva V., Langford J. C., and Tenenbaum J. B. Graph approximations to geodesics on embedded manifolds. Technical report, Stanford University, January 2001.
4. Blake A. and Zisserman A. *Visual Reconstruction*. The MIT Press, London, 1987.
5. Borg I. and Groenen P. *Modern Multidimensional Scaling: Theory and Applications*. Springer, New York, 1997. xviii+471 pp.
6. Boult T. E. and Kender J. R. Visual surface reconstruction using sparse depth data. In *Computer Vision and Pattern Recognition*, volume 86, pp. 68–76, 1986.
7. Bronstein A. M., Bronstein M. M., and Kimmel R. Three-dimensional face recognition. *International Journal of Computer Vision (IJCV)*, 64(1):5–30, August 2005.
8. Bronstein A. M., Bronstein M. M., and Kimmel R. Generalized multidimensional scaling: a framework for isometry-invariant partial surface matching. *Proceedings of the National Academy of Sciences*, 103(5):1168–1172, January 2006.
9. Bronstein M. M., Bronstein A. M., and Kimmel R. Multigrid multidimensional scaling. *Numerical Linear Algebra with Applications (NLAA)*, 13(2-3):149–171, March–April 2006.
10. Cabay S. and Jackson L. Polynomial extrapolation method for finding limits and antilimits of vector sequences. *SIAM Journal on Numerical Analysis*, 13(5):734–752, October 1976.
11. Coifman R. R., Lafon S., Lee A. B., Maggioni M., Nadler B., Warner F., and Zucker S. W. Geometric diffusions as a tool for harmonic analysis and structure definition of data. *Proceedings of the National Academy of Sciences*, 102(21):7426–7431, May 2005.
12. Cormen T. H., Leiserson C. E., and Rivest R. L. *Introduction to Algorithms*. MIT Press and McGraw-Hill, 1990.
13. Diaz R. M. and Arencibia A. Q., editors. *Coloring of DT-MRI FIber Traces using Laplacian Eigenmaps*, Las Palmas de Gran Canaria, Spain, Springer, February 24–28 2003.
14. Dijkstra E. W. A note on two problems in connection with graphs. *Numerische Mathematik*, 1:269–271, 1959.
15. Duda R. O., Hart P. E., and Stork D. G. *Pattern Classification and Scene Analysis*. Wiley-Interscience, 2nd edn, 2000.
16. Eddy R. P. Extrapolationg to the limit of a vector sequence. In P. C. Wang, editor, *Information Linkage Between Applied Mathematics and Industry*, Academic Press, pp. 387–396. 1979.
17. Eldar Y., Lindenbaum M., Porat M., and Zeevi Y. The farthest point strategy for progressive image sampling. *IEEE Transactions on Image Processing*, 6(9):1305–1315, September 1997.

18. Freedman D. Efficient simplicial reconstructions of manifolds from their samples. *IEEE Transactions on Pattern Analysis and Machine Intelligence*, 24(10): 1349–1357, October 2002.
19. Grimes C. and Donoho D. L. When does isomap recover the natural parameterization of families of articulates images? Technical Report 2002-27, Department of Statistics, Stanford University, Stanford, CA 94305-4065, 2002.
20. Grimes C. and Donoho D. L. Hessian eigenmaps: Locally linear embedding techniques for high-dimensional data. *Proceedings of the National Academy of Sciences*, 100(10):5591–5596, May 2003.
21. Guy G. and Medioni G. Inference of surfaces, 3D curves and junctions from sparse, noisy, 3D data. *IEEE Transactions on Pattern Analysis and Machine Intelligence*, 19(11):1265–1277, November 1997.
22. Hu N., Huang W., and Ranganath S. Robust attentive behavior detection by non-linear head pose embedding and estimation. In A. Leonardis, H. Bischof and A. Pinz, editors, *Proceedings of the 9th European Conference Computer Vision (ECCV)*, volume 3953, pp. 356–367, Graz, Austria, Springer. May 2006.
23. Kearsley A., Tapia R., and Trosset M. W. The solution of the metric stress and sstress problems in multidimensional scaling using newton's method. *Computational Statistics*, 13(3):369–396, 1998.
24. Keller Y., Lafon S., and Krauthammer M. Protein cluster analysis via directed diffusion. In *The fifth Georgia Tech International Conference on Bioinformatics*, November 2005.
25. Kimmel R. and Sethian J. A. Computing geodesic paths on manifolds. *Proceedings of the National Academy of Sciences*, 95(15):8431–8435, July 1998.
26. Lafon S. *Diffusion Maps and Geometric Harmonics*. Ph.D. dissertation, Graduate School of Yale University, May 2004.
27. Lafon S. and Lee A. B. Diffusion maps and coarse-graining: A unified framework for dimensionality reduction, graph partitioning and data set parameterization. *IEEE transactions on Pattern Analysis and Machine Intelligence*, 2006. To Appear.
28. Mordohai P. and Medioni G. Unsupervised dimensionality estimation and manifold learning in high-dimensional spaces by tensor voting. In *Proceedings of International Joint Conference on Artificial Intelligence*, pp. 798–803, 2005.
29. Pless R. Using Isomap to explore video sequences. In *Proceedings of the 9th International Conference on Computer Vision*, pp. 1433–1440, Nice, France, October 2003.
30. Roweis S. T. and Saul L. K. Nonlinear dimensionality reduction by locally linear embedding. *Science*, 290:2323–2326, 2000.
31. Schwartz E. L., Shaw A., and Wolfson E. A numerical solution to the generalized mapmaker's problem: Flattening nonconvex polyhedral surfaces. *IEEE Transactions on Pattern Analysis and Machine Intelligence*, 11:1005–1008, November 1989.
32. Sidi A. Efficient implementation of minimal polynomial and reduced rank extrapolation methods. *Journal of Computational and Applied Mathematics*, 36 (3):305–337, 1991.
33. Smith D. A., Ford W. F., and Sidi A. Extrapolation methods for vector sequences. *SIAM Review*, 29(2):199–233, June 1987.
34. Tenenbaum J. B., de Silva V., and Langford J. C. A global geometric framework for nonlinear dimensionality reduction. *Science*, 290(5500):2319–2323, December 2000.

35. Trosset M. and Mathar R. On the existence of nonglobal minimizers of the stress criterion for metric multidimensional scaling. *American Statistical Association: Proceedings Statistical Computing Section*, pp. 158–162, November 1997.
36. Weinberger K. Q., Packer B. D., and Saul L. K. Nonlinear dimensionality reduction by semidefinite programming and kernel matrix factorization. In *Proceedings of the 10th International Workshop on Artificial Intelligence and Statistics*, Barbados, January 2005.
37. Weinberger K. Q. and Saul L. K. Unsupervised learning of image manifolds by semidefinite programming. In *Proceedings of the IEEE Conference on Computer Vision and Pattern Recognition*, vol. 2, pp. 988–995, Washington DC, IEEE Computer Society, 2004.

Part III

2D–3D Tracking

Part II

3D-3D Treatment

11

Contours, Optic Flow, and Prior Knowledge: Cues for Capturing 3D Human Motion in Videos

Thomas Brox[1], Bodo Rosenhahn[2], and Daniel Cremers[1]

[1] CVPR Group, University of Bonn
  Römerstr. 164, 53117 Bonn, Germany
[2] MPI for Computer Science
  Stuhlsatzenhausweg 85, 66123 Saarbrücken, Germany

Summary. Human 3D motion tracking from video is an emerging research field with many applications demanding highly detailed results. This chapter surveys a high quality generative method, which employs the person's silhouette extracted from one or multiple camera views for fitting an a priori given 3D body surface model. A coupling between pose estimation and contour extraction allows for reliable tracking in cluttered scenes without the need of a static background. The optic flow computed between two successive frames is used for pose prediction. It improves the quality of tracking in case of fast motion and/or low frame rates. In order to cope with unreliable or insufficient data, the framework is further extended by the use of prior knowledge on static joint angle configurations.

11.1 Introduction

Tracking of humans in videos is a popular research field with numerous applications ranging from automated surveillance to sports movement analysis. Depending on applications and the quality of video data, there are different approaches with different objectives. In many people tracking methods, for instance, only the position of a person in the image or a region of interest is sought. Extracting more detailed information is often either not necessary or very difficult due to image resolution.

In contrast to such model-free tracking methods, the present chapter is concerned with the detailed fitting of a given 3D model to video data. The model consists of the body surface and a skeleton that contains predefined joints [8, 26]. Given the video data from one or more calibrated cameras, one is interested in estimating the person's 3D pose and the joint angles. This way, the tracking becomes an extended 2D–3D pose estimation problem, where additionally to the person's rigid body motion one is interested in some restricted kind of deformation, namely the motion of limbs.

B. Rosenhahn et al. (eds.), Human Motion – Understanding, Modelling, Capture, and
Animation, 265–293.

Applications of this kind of tracking are sports movement and clinical analysis, as well as the recording of motion patterns for animations in computer graphics. The state-of-the-art for capturing human motion is currently defined by large industrial motion capture systems with often more than 20 cameras. These systems make use of markers attached to the person's body in order to allow for a fast and reliable image processing. Often the reliability of the results is further improved by manually controlling the matching of markers. Such systems are described in Chapter 16.

While results of marker-based motion capturing systems are very trustworthy, markers need to be attached, which is sometimes not convenient. Moreover the manual supervision of marker matching can be very laborious. For these reasons, one is interested in marker-less motion capturing, using the appearance of the person as a natural marker.

While markers in marker-based systems have been designed for being easy to identify, finding correct correspondences of points in marker-free systems is not as simple. A sensible selection of the right feature to be tracked is important. A typical way to establish point correspondences is to concentrate on distinctive patches in the image and to track these patches, for instance, with the KLT tracker [56] or a tracker based on the so-called SIFT descriptor [37]. However, patch-based tracking typically only works reliably if the appearance of the person contains sufficiently textured areas.

An alternative feature, particularly for tracking people with non-textured clothing, is the silhouette of the person. Early approaches have been based on edge detectors and have tried to fit the 3D model to dominant edges [26]. Since image edges are not solely due to the person's silhouette, the most relevant problem of such approaches is their tendency to get stuck in local optima. Sophisticated optimization techniques have been suggested in order to attenuate this problem [62]. Nowadays, silhouette based tracking usually relies on background subtraction. Assuming both a static camera and a static background, the difference between the current image and the background image efficiently yields the foreground region. Apart from the restrictive assumptions, this approach works very well and is frequently employed for human tracking [1, 25, 60].

In [53] a contour-based method to 3D pose tracking has been suggested that does not impose such strict assumptions on the scene. Instead, it demands dissimilarity of the foreground and background region, which is a typical assumption in image segmentation. In order to deal with realistic scenarios where persons may also wear nonuniform cloths and the background is cluttered, the dissimilarity is defined in a texture feature space, and instead of homogeneous regions, the model expects only *locally* homogeneous regions. The main difference to other pose tracking methods, however, is the coupling between feature extraction and estimation of the pose parameters. In a joint optimization one seeks the pose parameters that lead to the best fit of the contour in the image. Vice versa, one seeks a segmentation that fits the image data *and* resembles the projected surface model. Due to this coupling, the contour extraction is

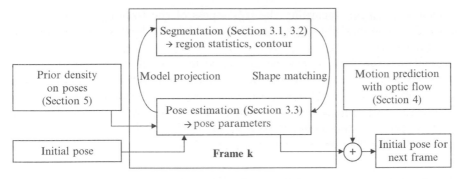

Fig. 11.1. System overview: the core is the coupled contour extraction and pose estimation. The motion between frames is predicted by optic flow in order to ensure a close initialization in the next frame. Pose configurations are constrained by a prior density estimated from training samples.

much more reliable than in a conventional two-step approach, where the contour is computed independently from the pose estimation task. We will survey the method in Section 11.3.

Although this way of integrating contours into pose estimation is more robust than the edge-based approach, it is still a local optimization method that can get stuck in local optima in case of fast motion. To alleviate these effects, it is common practice in tracking to predict the pose of the tracked object in the coming frame. A prediction is usually computed by simply extrapolating the motion between the last two frames to the next frame. In a more subtle way, learning based approaches incorporate auto-regressive models or nonlinear subspace methods based on training sequences to accomplish this task [1, 17, 25, 57, 65].

Another possibility to predict the pose parameters in the next frame is by the optic flow. Optic flow based tracking is similar to patch-based tracking, though instead of patches one tries to match single points under certain smoothness assumptions. With a reliable optic flow estimation method, one can predict rather large displacements [10]. In combination with the contour-based method, one obtains a system that can handle fast motions and is free from error accumulation, which is a severe problem for optic flow or patch-based tracking. A similar concept has been presented earlier in [21] and [38] in combination with edges instead of contours and different optic flow estimation techniques. The pose prediction by means of optic flow and how the flow can be efficiently computed is explained in Section 11.4.

Since 3D human tracking is generally an ill-posed problem with many solutions explaining the same data, methods suffer enormously from unreliable data. Therefore, in recent years, it has become more and more popular to exploit prior assumptions about typical human poses and motion patterns [11, 59, 65]. In Section 11.5 it will be described how the tracking model can be constrained to prefer solutions that are close to familiar poses. The impact of such a constraint regarding the robustness of the technique to disturbed image data is remarkable. See also Chapters 2 and 8 for learning techniques in human tracking.

Although the tracking system described in this chapter comprises many advanced methods, there is still much room for extensions or alternative approaches. In Section 11.6 we discuss issues such as running times, auto-initialization, dynamical pose priors, and cloth tracking including cursors to other chapters in this book or to seminal works in the literature. A brief summary of the chapter is given in Section 11.7.

11.2 Human Motion Representation with Twists and Kinematic Chains

A human body can be modeled quite well by means of a kinematic chain. A kinematic chain is a set of (usually rigid) bodies interconnected by joints. For example, an arm consists of an upper and lower arm segment and a hand, with the shoulders, elbow and wrist as interconnecting joints. For a proper representation of joints and transformations along kinematic chains in the human tracking method, we use the exponential representation of rigid body motions [42], as suggested in [7, 8].

Every 3D rigid motion can be represented in exponential form

$$\boldsymbol{M} = \exp(\theta\hat{\xi}) = \exp\begin{pmatrix} \hat{\omega} & \boldsymbol{v} \\ 0_{3\times 1} & 0 \end{pmatrix} \tag{11.1}$$

where $\theta\hat{\xi}$ is the matrix representation of a twist $\xi \in se(3) = \{(\boldsymbol{v}, \hat{\omega}) | \boldsymbol{v} \in \mathbb{R}^3, \hat{\omega} \in so(3)\}$, with $so(3) = \{\boldsymbol{A} \in \mathbb{R}^{3\times 3} | \boldsymbol{A} = -\boldsymbol{A}^T\}$. The Lie algebra $so(3)$ is the tangential space of the 3D rotations at the origin. Its elements are (scaled) rotation axes, which can either be represented as a 3D vector

$$\theta\omega = \theta \begin{pmatrix} \omega_1 \\ \omega_2 \\ \omega_3 \end{pmatrix}, \text{ with } \|\omega\|_2 = 1 \tag{11.2}$$

or as a skew symmetric matrix

$$\theta\hat{\omega} = \theta \begin{pmatrix} 0 & -\omega_3 & \omega_2 \\ \omega_3 & 0 & -\omega_1 \\ -\omega_2 & \omega_1 & 0 \end{pmatrix}. \tag{11.3}$$

In fact, $\boldsymbol{M}$ is an element of the Lie group $SE(3)$, known as the group of direct affine isometries. A main result of Lie theory is that to each Lie group there exists a Lie algebra which can be found in its tangential space by derivation and evaluation at its origin. Elements of the Lie algebra therefore correspond to infinitesimal group transformations. See [42] for more details. The corresponding Lie algebra to $SE(3)$ is denoted as $se(3)$.

A twist contains six parameters and can be scaled to $\theta\xi$ with a unit vector ω. The parameter $\theta \in \mathbb{R}$ corresponds to the motion velocity (i.e., the rotation velocity and pitch). The one-parameter subgroup $\Phi_{\hat{\xi}}(\theta) = \exp(\theta\hat{\xi})$ generated by this twist corresponds to a screw motion around an axis in space. The six twist components can either be represented as a 6D vector

$$\theta\xi = \theta(\omega_1, \omega_2, \omega_3, v_1, v_2, v_3)^T$$
$$\text{with } \|\omega\|_2 = \|(\omega_1, \omega_2, \omega_3)^T\|_2 = 1, \tag{11.4}$$

or as a 4×4 matrix

$$\theta\hat{\xi} = \theta \begin{pmatrix} 0 & -\omega_3 & \omega_2 & v_1 \\ \omega_3 & 0 & -\omega_1 & v_2 \\ -\omega_2 & \omega_1 & 0 & v_3 \\ 0 & 0 & 0 & 0 \end{pmatrix}. \tag{11.5}$$

To reconstruct a group action $M \in SE(3)$ from a given twist, the exponential function $\exp(\theta\hat{\xi}) = \sum_{k=0}^{\infty} \frac{(\theta\hat{\xi})^k}{k!} = M \in SE(3)$ must be computed. This can be done efficiently by using the Rodriguez formula [42].

In this framework, joints are expressed as special screws with no pitch. They have the form $\theta_j\hat{\xi}_j$ with known $\hat{\xi}_j$ (the location of the rotation axes as part of the model representation) and unknown joint angle θ_j. A point on the jth joint can be represented as consecutive evaluation of exponential functions of all involved joints,

$$X_i' = \exp(\theta\hat{\xi}_{RBM})(\exp(\theta_1\hat{\xi}_1)\ldots\exp(\theta_j\hat{\xi}_j)X_i) \tag{11.6}$$

The human body motion is then defined by a parameter vector $\xi :=$ (ξ_{RBM}, Θ) that consists of the 6 parameters for the global twist ξ_{RBM} (3D rotation and translation) and the joint angles $\Theta := (\theta_1, \ldots, \theta_N)$.

11.3 Contour-based Pose Estimation

In this section, we survey the coupled extraction of the contour and the estimation of the pose parameters by means of this contour. For better understanding we start in Section 11.3.1 with the simple segmentation case that is not yet related to the pose parameters. In the end, the idea is to find pose parameters in such a way that the projected surface leads to a region that is homogeneous according to a certain statistical model. The statical region model will be explained and motivated in Section 11.3.2. In Section 11.3.3 we then bend the bow to pose estimation by introducing the human model as a 3D shape prior into the segmentation functional. This leads to a matching of 2D shapes. From the point correspondences of this matching, one can derive 2D–3D correspondences and finally estimate the pose parameters from these.

11.3.1 Contour Extraction with Level Sets

Level set representation of contours: The contour extraction is based on variational image segmentation with level sets [23, 44], in particular region-based active contours [15, 20, 46, 64]. Level set formulations of the image segmentation problem have several advantages. One is the convenient embedding of a 1D curve into a 2D, image-like structure. This allows for a convenient and sound interaction between constraints that are imposed on the contour itself and constraints that act on the regions separated by the contour. Moreover, the level set representation yields the inherent capability to model topological changes. This can be an important issue, for instance, when the person is partially occluded and the region is hence split into two parts, or if the pose of legs or arms leads to topological changes of the background.

In the prominent case of a segmentation into foreground and background, a level set function $\Phi \in \Omega \mapsto \mathbb{R}$ splits the image domain Ω into two regions Ω_1 and Ω_2, with $\Phi(x) > 0$ if $x \in \Omega_1$ and $\Phi(x) < 0$ if $x \in \Omega_2$. The zero-level line thus marks the boundary between both regions, i.e., it represents the person's silhouette that is sought to be extracted.

Optimality criteria and corresponding energy functional: As optimality criteria for the contour we want the data within one region to be similar and the length of the contour to be as small as possible. Later in Section 11.3.3 we will add similarity to the projected surface model as a further criterion. The model assumptions can be expressed by the following energy functional [15, 66]:

$$E(\Phi) = - \int_{\Omega} \Big(H(\Phi) \log p_1 + (1 - H(\Phi)) \log p_2 \Big) \, dx + \nu \int_{\Omega} |\nabla H(\Phi)| \, dx \,(11.7)$$

where $\nu > 0$ is parameter that weights the similarity against the length constraint, and $H(s)$ is a regularized Heaviside function with $\lim_{s \to -\infty} H(s) = 0$, $\lim_{s \to \infty} H(s) = 1$, and $H(0) = 0.5$. It indicates to which region a pixel belongs. Chan and Vese suggested two alternative functions in [15]. The particular choice of H is not decisive. We use the error function, which has the convenient property that its derivative is the Gaussian function.

Minimizing the first two terms in (11.7) maximizes the likelihood given the probability densities p_1 and p_2 of values in Ω_1 and Ω_2, respectively. The third term penalizes the length of the contour, what can be interpreted as a log-prior on the contour preferring smooth contours. Therefore, minimizing (11.7) maximizes the total a-posteriori probability of all pixel assignments.

Minimization by gradient descent: For energy minimization one can apply a gradient descent. The Euler–Lagrange equation (11.7) leads to the following update equation[1]:

---

[1] As the probability densities in general also depend on the contour there may appear additional terms depending on the statistical model. For global Gaussian densities, however, the terms are zero, and for other models they have very little influence on the result, so they are usually neglected.

$$\partial_t \Phi = H'(\Phi) \left(\log \frac{p_1}{p_2} + \nu \nabla^\top \left(\frac{\nabla \Phi}{|\nabla \Phi|} \right) \right) \tag{11.8}$$

where $H'(s)$ is the derivative of $H(s)$ with respect to its argument. Applying this evolution equation to some initialization Φ^0, and given the probability densities p_i, which are defined in the next section, the contour converges to the next local minimum for the numerical evolution parameter $t \to \infty$.

11.3.2 Statistical Region Models

An important factor for the contour extraction process is how the probability densities $p_i : \mathbb{R} \to [0,1]$ are modeled. This model determines what is considered similar or dissimilar. There is on one hand the choice of the feature space, e.g. gray value, RGB, texture, etc., and on the other hand the parametrization of the probability density function.

Texture features: Since uniformly colored cloths without texture are in general not realistic, we adopt the texture feature space proposed in [12]. It comprises $M = 5$ feature channels I_j for gray scale images, and $M = 7$ channels if color is available. The color channels are considered in the CIELAB color space. Additionally to gray value and color, the texture features in [12] encode the texture magnitude, orientation, and scale, i.e., they provide basically the same information as the frequently used responses of Gabor filters [24]. However, the representation is less redundant, so 4 feature channels substitute 12–64 Gabor responses. Alternatively, Gabor features can be used at the cost of larger computation times. In case of people wearing uniform cloths and the background also being more or less homogeneous, one can also work merely with the gray value or color in order to increase computation speed.

Channel independence: The probability densities of the M feature channels are assumed to be independent, thus the total probability density can be composed of the densities of the separate channels:

$$p_i = \prod_{j=1}^{M} p_{ij}(I_j) \qquad i = 1, 2. \tag{11.9}$$

Though assuming channel independence is merely an approximation, it keeps the density model tractable. This is important, as the densities have to be estimated from a limited amount of image data.

Density models of increasing complexity: There are various possibilities how to model channel densities. In [15] a simple piecewise constant region model is suggested, which corresponds to a Gaussian density with fixed standard deviation. In order to admit different variations in the regions, it is advisable to use at least a full Gaussian density [66], a generalized Laplacian [28], or a Parzen estimate [32, 54]. While more complex density models can represent more general distributions, they also imply the estimation of more parameters which generally leads to a more complex objective function.

Local densities: Nevertheless, for the task of human tracking, we advocate the use of a more complex region model, in particular a Gaussian density that is estimated using only values in a local neighborhood of a point instead of values from the whole region. Consequently, the probability density is no longer fixed for one region but varies with the position. Local densities have been proposed in [31,53]. Segmentation with such densities has been shown to be closely related to the piecewise smooth Mumford–Shah model [13,41]. Formally, the density is modeled as

$$p_{ij}(s,x) = \frac{1}{\sqrt{2\pi}\sigma_{ij}(x)} \exp\left(\frac{(s-\mu_{ij}(x))^2}{2\sigma_{ij}(x)^2}\right). \tag{11.10}$$

The parameters $\mu_{ij}(x)$ and $\sigma_{ij}(x)$ are computed in a local Gaussian neighborhood K_ρ around x by:

$$\mu_{ij}(x) = \frac{\int_{\Omega_i} K_\rho(\zeta-x)I_j(\zeta)\,d\zeta}{\int_{\Omega_i} K_\rho(\zeta-x)\,d\zeta} \qquad \sigma_{ij}(x) = \frac{\int_{\Omega_i} K_\rho(\zeta-x)(I_j(\zeta)-\mu_{ij}(x))^2\,d\zeta}{\int_{\Omega_i} K_\rho(\zeta-x)\,d\zeta} \tag{11.11}$$

where ρ denotes the standard deviation of the Gaussian window. In order to have enough data to obtain reliable estimates for the parameters $\mu_{ij}(x)$ and $\sigma_{ij}(x)$, we choose $\rho = 12$.

Taking advantage of local dissimilarity of foreground and background: The idea behind the local density model is the following: in realistic scenarios, the foreground and background regions are rarely globally dissimilar. For instance, the head may have a different color than the shirt or the trousers. If the same colors also appear in the background, it is impossible to accurately distinguish foreground and background by means of a standard global region distribution. Locally, however, foreground and background can be easily distinguished. Although the local density model is too complex to detect the desired contour in an image without a good contour initialization and further restrictions on the contour's shape, we are in a tracking scenario, i.e., the result from the previous frame always provides a rather good initialization. Moreover, in the next section a shape constraint is imposed on the contour that keeps it close to the projection of the surface model. Also note, that we still have a statistical region based model, which yields considerably less local optima than previous edge-based techniques. The results in Figure 11.2 and Figure 11.4 show that local region statistics provide more accurate contours and thus allow for a more reliable estimate of the 3D pose.

Optimization with EM: Estimating both the probability densities p_{ij} and the region contour works according to the *expectation-maximization principle*

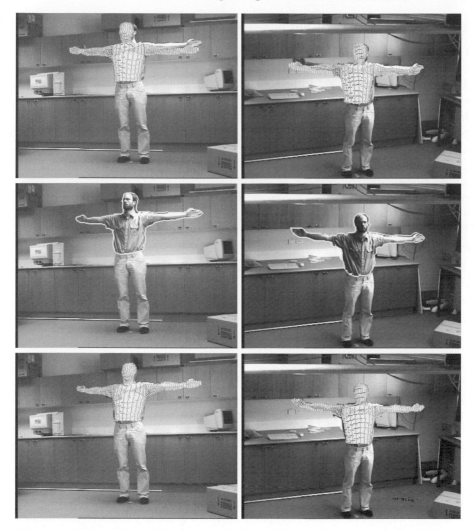

Fig. 11.2. Human pose estimation with the coupled contour-based approach given two camera views. **First row:** Initialization of the pose. The projection to the images is used as contour initialization. **Second row:** Estimated contour after 5 iterations. **Third row:** Estimated pose after 5 iterations.

[22,39]. Having the level set function initialized with some partitioning Φ^0, the probability densities in these regions can be approximated. With the probability densities, on the other hand, one can compute an update on the contour according to (11.8), leading to a further update of the probability densities, and so on. In order to attenuate the dependency on the initialization, one can apply a continuation method in a coarse-to-fine manner [6].

11.3.3 Coupled Estimation of Contour and Pose Parameters

Bayesian inference: So far, only the person's silhouette in the image has been estimated, yet actually we are interested in the person's pose parameters. They can be estimated from the contour in the image, but also vice versa the surface model with given pose parameters can help to determine this contour. In a Bayesian setting this joint estimation problem can be written as the maximization of

$$p(\Phi, \xi | I) = \frac{p(I|\Phi, \xi)p(\Phi|\xi)p(\xi)}{p(I)} \tag{11.12}$$

where Φ indicates the contour given as a level set function, ξ the set of pose parameters, and I the given image(s). This formula imposes a shape prior on Φ given the pose parameters, and it imposes a prior on the pose parameters. For the moment we will use a uniform prior for $p(\xi)$, effectively ignoring this factor, but we will come back to this prior later in Section 11.5.

Joint energy minimization problem: Assuming that the appearance in the image is completely determined by the contour with no further (hidden) dependence on ξ, we can set $p(I|\Phi, \xi) \equiv p(I|\Phi)$. Minimizing the negative logarithm of (11.12) then leads to the following energy minimization problem:

$$E(\Phi, \theta\xi) = -\log p(\Phi, \xi | I)$$

$$= -\int_\Omega \left(H(\Phi) \log p_1 + (1 - H(\Phi)) \log p_2 \right) dx + \nu \int_\Omega |\nabla H(\Phi)| \, dx$$

$$+ \lambda \underbrace{\int_\Omega (\Phi - \Phi_0(\xi))^2 \, dx}_{\text{Shape}} + \text{const.} \tag{11.13}$$

One recognizes the energy from (11.7) with an additional term that imposes the shape constraint on the contour and relates at the same time the contour to the sought pose parameters. The parameter $\lambda \geq 0$ introduced here determines the variability of the estimated contour Φ from the projected surface model Φ_0. Φ_0 is again a level set function and it is obtained by projecting the surface to the image plane (by means of the known projection matrice) and by applying a signed distance transform to the resulting shape. The signed distance transform assigns each point x of Φ_0 the Euclidean distance of x to the closest contour point. Points inside the projected region get positive sign, points outside this region, get negative sign.

Alternating optimization: In order to minimize (11.13) for both the contour and the pose parameters, an alternating scheme is proposed. First, the pose parameters are kept fixed and the energy is minimized with respect to the contour. Afterwards, the contour is retained and one optimizes the energy for the pose parameters. In the tracking scenario, with the initial pose being already close to the desired solution, only few (2–5) iterations are sufficient for convergence.

Optimization with respect to the contour: Since the shape term is modeled in the image domain, minimization of (11.13) with respect to Φ is straightforward and leads to the gradient descent equation

$$\partial_t \Phi = H'(\Phi) \left(\log \frac{p_1}{p_2} + \nu \nabla^\top \left(\frac{\nabla \Phi}{|\nabla \Phi|} \right) \right) + 2\lambda \left(\Phi_0(\xi) - \Phi \right). \quad (11.14)$$

One can observe that the shape term pushes Φ towards the projected surface model, while on the other hand, Φ is still influenced by the image data ensuring homogeneous regions according to the statistical region model.

Optimization with respect to the pose parameters: Optimization with respect to the pose parameters needs more care, since the interaction of the model with the contour in the image involves a projection. At the same time, the variation of the projected shape with a certain 3D transformation is quite complex. Principally, the 3D transformation can be estimated from a set of 2D–3D point correspondences in a least squares setting, as will be explained later in this section. Since we know how 2D points in Φ_0 correspond to 3D points on the surface (Φ_0 was constructed by projecting these 3D points), 2D–3D point correspondences can be established by matching points of the two 2D shapes Φ and Φ_0.

Shape matching: For minimizing the shape term in (11.13) with respect to the pose parameters, we look for a transformation in 2D that can account for the projections of all permitted transformations in 3D. Therefore, we choose a nonparametric transformation, in particular a smooth displacement field $\mathbf{w}(\mathbf{x}) := (u(\mathbf{x}), v(\mathbf{x}))$ and formulate the shape term as

$$E(u, v) = \int_\Omega (\Phi(\mathbf{x}) - \Phi_0(\mathbf{x} + \mathbf{w}))^2 + \alpha(|\nabla u|^2 + |\nabla v|^2) \, d\mathbf{x}. \quad (11.15)$$

where $\alpha \geq 0$ is a regularization parameter that steers the influence of the regularization relative to the matching criterion. The considered transformation is very general and, hence, can handle the projected transformations in 3D. The regularization ensures a smooth displacement field, which corresponds to penalizing shape deformations. Furthermore, it makes the originally ill-posed matching problem well-posed.

Optic flow estimation problem: A closer look at (11.15) reveals strong connections to optic flow estimation. In fact, the energy is a nonlinear version of the Horn–Schunck functional in [29]. Consequently, the matching problem can be solved using a numerical scheme known from optic flow estimation. We will investigate this scheme more closely in Section 11.4.

Alternative matching via ICP: Alternatively, one can match the two shapes by an iterated closest point (ICP) algorithm [4]. As Φ and Φ_0 are both Euclidean distance images, this is closely related to minimization of (11.15) for $\alpha \to 0$. In [51] it has been shown empirically that the combination of point correspondences from both methods is beneficial for pose estimation.

Inverse projection and Plücker lines: After matching the 2D contours, the remaining task is to estimate from the nonparametric 2D transformation a 3D transformation parameterized by the sought vector $\xi = (\xi_{RBM}, \Theta)$. For this purpose, the 2D points are changed into 3D entities. For the points in Φ this means that their projection rays need to be constructed. A projection ray contains all 3D points that, when projected to the image plane, yield a zero distance to the contour point there. Hence, for minimizing the distance in the image plane, one can as well minimize the distance between the model points and the rays reconstructed from the corresponding points.

There exist different ways to represent projection rays. As we have to minimize distances between correspondences, it is advantageous to use an implicit representation for a 3D line. It allows instantaneously to determine the distance between a point and a line.

An implicit representation of projection rays is by means of so-called *Plücker lines* [55, 63]. A Plücker line $L = (n, m)$ is given as a unit vector n and a moment m with $m = x \times n$ for a given point x on the line. The incidence of a point x on a line $L = (n, m)$ can then be expressed as

$$x \times n - m = 0. \tag{11.16}$$

Parameter estimation by nonlinear least squares: This equation provides an error vector and we seek the transformation $\xi = (\xi_{RBM}, \Theta)$ that minimizes the norm of this vector over all correspondences. For $j = \mathcal{J}(x_i)$ being the joint index of a model point x_i, the error to be minimized can be expressed as

$$\sum_i \| \Pi\left(\exp(\hat{\xi}_{RBM}) \exp(\theta_1 \hat{\xi}_1) \ldots \exp(\theta_{\mathcal{J}(x_i)} \hat{\xi}_{\mathcal{J}(x_i)}) x_i \right) \times n_i - m_i \|_2^2, \tag{11.17}$$

where Π is the projection of the homogeneous 4D vector to a 3D vector by neglecting the homogeneous component (which is 1), and the symbol $\times$ denotes the cross product.

Linearization: The minimization problem in (11.17) is a least squares problem. Unfortunately, however, the equations are non-quadratic due to the exponential form of the transformation matrices. For this reason, the transformation matrix is linearized and the pose estimation procedure is iterated, i.e., the nonlinear problem is decomposed into a sequence of linear problems. This is achieved by

$$\exp(\theta \hat{\xi}) = \sum_{k=0}^{\infty} \frac{(\theta \hat{\xi})^k}{k!} \approx I + \theta \hat{\xi} \tag{11.18}$$

with I as identity matrix.

This results in

$$((I + \theta \hat{\xi} + \theta_1 \hat{\xi}_1 \ldots + \theta_{\mathcal{J}(x_i)} \hat{\xi}_{\mathcal{J}(x_i)}) X_i)_{3 \times 1} \times n_i - m_i = 0 \tag{11.19}$$

with the unknown pose parameters ξ acting as linear components. This equation can be reordered into the form $\boldsymbol{A}(\theta\xi_{RBM}, \theta_1 \ldots \theta_N)^T = \boldsymbol{b}$. Collecting a set of such equations (each is of rank two) leads to an overdetermined linear system of equations, which can be solved using, for example, the Householder algorithm. The Rodriguez formula can be applied to reconstruct the group action from the estimated parameter vector ξ. The 3D points can be transformed and the process is iterated until it converges.

Multiple camera views: The method can easily be extended to make use of multiple camera views if all cameras are calibrated to the same world coordinate system. The point correspondences, obtained by projecting the surface model to all images and extracting contours there, can be combined in a joint system of equations. The solution of this system is the least squares fit of the model to the contours in all images. Due to the coupling of contour and pose estimation, also the contour extraction can benefit from the multiview setting. This is demonstrated in the comparison depicted in Figures 11.2 and 11.3.

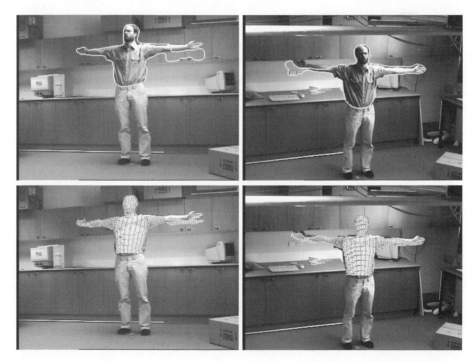

Fig. 11.3. Result with a two-step approach, i.e., extraction of the contours from the images followed by contour-based pose estimation. The same initialization as in Figure 11.2 was used. **Top row:** Estimated contour. **Bottom row:** Estimated pose. As pose and contour are not coupled, the contour extraction cannot benefit from the two camera views. Moreover, as the contour is not bound to the surface model, it can run away.

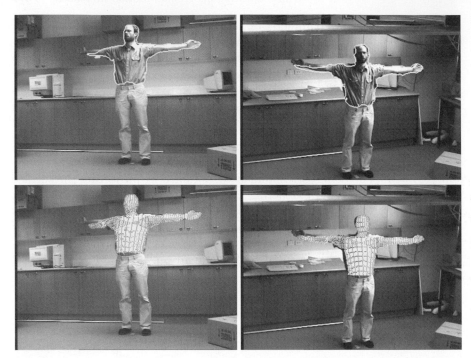

Fig. 11.4. Result with a global Parzen estimator instead of the suggested local region statistics. The same initialization as in Figure 11.2 was used. **Top row:** Estimated contour. **Bottom row:** Estimated pose. Local differences between foreground and background are not modeled. With the global model, the right arm of the person better fits to the background.

11.4 Optic Flow for Motion Prediction

The contour-based tracking explained in the previous section demands a pose initialization that is close enough to obtain reasonable estimates of the region statistics. For high frame rates and reasonably slow motion, the result from the previous frame is a sufficiently good initialization. For very fast motion or small frame rates, however, it may happen that limbs have moved too far and the method is not able to recapture them starting with the result from the previous frame. This problem is illustrated in the first row of Figure 11.5.

A remedy is to improve the initialization by predicting the pose parameters in the successive frame. The most simple approach is to compute the velocity from the results in the last two frames and to assume that the velocity stays constant. However, it is obvious that this assumption is not satisfied at all times and can lead to predictions that are even much worse than the initialization with the latest result. Auto-regressive models are much more reliable. They predict the new state from previous ones by means of a parametric model estimated from a set of training data.

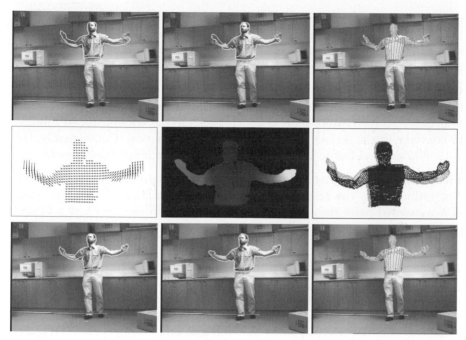

Fig. 11.5. Motion prediction by optic flow and its relevance. **First row:** Initialization with the pose from the previous frame (left). Due to fast motion, the initialization is far from the correct contour. Consequently, contour extraction (center) and tracking (right) fail. **Second row:** Optic flow field as arrow (left) and color plot (center) and prediction computed from this flow field (right). The brighter mesh shows the old pose, the dark mesh the predicted one. **Third row:** Like first row, but now the initialization is from the pose predicted by the optic flow.

Pose estimation from optic flow: In this chapter, we focus on an image-driven prediction by means of optic flow. We assume that the pose has been correctly estimated in frame t, and we are now interested in a prediction of the pose in frame $t+1$ given the images in t and $t+1$. For this prediction, we need to compute the optic flow, which provides 2D–2D correspondences between points in the images. As the 2D–3D correspondences in frame t are known, we obtain a set of 2D–3D point correspondences between the new frame $t+1$ and the model. From these, the pose of the model in frame $t+1$ can be computed by solving a sequence of linear systems, as described by Equation (11.19) in the previous section.

Accumulation of errors: The inherent assumption of knowing the correct pose of the model in frame t is in fact not exactly satisfied. In practice, there will be inaccuracies in the estimated pose. This results in the accumulation of errors when using only model-free schemes based on the optic flow or feature tracking. However, the contour-based pose estimation from the previous section, which directly derives correspondences between the image and the model, does not suffer from this problem. It is able to correct errors from the previous frame or from the estimated optic flow. For this reason, error accumulation is not an issue in the system described here. A result obtained with the optic flow model detailed below is shown in Figure 11.5.

Optic flow model: The remaining open question is how to compute the optic flow. The main goal here is to provide a prediction that brings the initialization closer to the correct pose in order to allow the contour-based method to converge to the correct solution in case of fast motion. Consequently, the optic flow method has to be able to deal with rather large displacements.

First assumption: gray value constancy: The basic assumption for optic flow estimation is the gray value constancy assumption, i.e., the gray value of a translated point does not change between the frames. With $\mathbf{w} := (u, v)$ denoting the optic flow, this can be expressed by

$$I(\mathbf{x} + \mathbf{w}, t + 1) - I(\mathbf{x}, t) = 0. \tag{11.20}$$

This equation is also called the *optic flow constraint*. Due to nonlinearity in $\mathbf{w}$, it is usually linearized by a Taylor expansion to yield

$$I_x u + I_y v + I_t = 0, \tag{11.21}$$

where subscripts denote partial derivatives. The linearization may be applied if displacements are small. For larger displacements, however, the linearization is not a good approximation anymore. Therefore, it has been suggested to minimize the original constraint in (11.20) [43] and to postpone all linearizations to the numerical scheme [2,9], which comes down to so-called warping schemes [3,5,40]. These schemes can deal with rather large displacements and, therefore, are appropriate for the problem at hand.

Second assumption: smooth flow field: The gray value constancy assumption alone is not sufficient for a unique solution. Additional constraints have to be introduced. Here we stick to the constraint of a smooth flow field, as suggested in [29]. It leads to the following energy minimization problem

$$E(u, v) = \int_\Omega (I(\mathbf{x}, t) - I(\mathbf{x}+\mathbf{w}, t+1))^2 + \alpha(|\nabla u|^2 + |\nabla v|^2) \, \mathbf{dx} \rightarrow \min \tag{11.22}$$

that can be solved with variational methods. Note that exactly the same problem appeared in Section 11.3.3 for matching two contours via (11.15). Thus we can use almost the same scheme for computing the optic flow between images and for shape matching.

Noise and brightness changes: When matching two images, one has to expect noise and violations of the gray value constancy assumption. These effects have to be taken into account in the optic flow model. In order to deal with noise, one can apply a robust function $\Psi(s^2) = \sqrt{s^2 + 0.001^2}$ to the first term in (11.22) [5, 40]. This has the effect that outliers in the data have less influence on the estimation result.

Robustness to brightness changes can be obtained by assuming constancy of the gradient [9]:

$$\nabla I(\mathbf{x} + \mathbf{w}, t + 1) - \nabla I(\mathbf{x}, t) = 0. \tag{11.23}$$

With both assumptions together, one ends up with the following energy:

$$
\begin{aligned}
E(u, v) = & \int_{\Omega_1} \Psi\big((I(\mathbf{x}, t) - I(\mathbf{x} + \mathbf{w}, t + 1))^2\big) \, \mathbf{dx} \\
& + \gamma \int_{\Omega_1} \Psi\big((\nabla I(\mathbf{x}, t) - \nabla I(\mathbf{x} + \mathbf{w}, t + 1))^2\big) \, \mathbf{dx} \\
& + \alpha \int_{\Omega_1} (|\nabla u|^2 + |\nabla v|^2) \, \mathbf{dx}.
\end{aligned}
\tag{11.24}
$$

Note that the domain is restricted to the foreground region Ω_1, since we are only interested in correspondences within this region anyway. This masking of the background region has the advantage that it considers the most dominant motion discontinuities, which would otherwise violate the smoothness assumption of the optic flow model. Moreover, it allows for cropping the images to reduce the computational load.

Euler–Lagrange equations: According to the calculus of variations, a minimizer of (11.24) must fulfill the Euler–Lagrange equations

$$
\begin{aligned}
\Psi'(I_z^2) I_x I_z + \gamma \Psi'(I_{xz}^2 + I_{yz}^2)(I_{xx} I_{xz} + I_{xy} I_{yz}) - \alpha \, \Delta u = 0 \\
\Psi'(I_z^2) I_y I_z + \gamma \Psi'(I_{xz}^2 + I_{yz}^2)(I_{yy} I_{yz} + I_{xy} I_{xz}) - \alpha \, \Delta v = 0
\end{aligned}
\tag{11.25}
$$

with reflecting boundary conditions, $\Delta := \partial_{xx} + \partial_{yy}$, and the following abbreviations:

$$
\begin{aligned}
I_x &:= \partial_x I(\mathbf{x} + \mathbf{w}, t + 1), \\
I_y &:= \partial_y I(\mathbf{x} + \mathbf{w}, t + 1), \\
I_z &:= I(\mathbf{x} + \mathbf{w}, t + 1) - I(\mathbf{x}, t), \\
I_{xx} &:= \partial_{xx} I(\mathbf{x} + \mathbf{w}, t + 1), \\
I_{xy} &:= \partial_{xy} I(\mathbf{x} + \mathbf{w}, t + 1), \\
I_{yy} &:= \partial_{yy} I(\mathbf{x} + \mathbf{w}, t + 1), \\
I_{xz} &:= \partial_x I(\mathbf{x} + \mathbf{w}, t + 1) - \partial_x I(\mathbf{x}, t), \\
I_{yz} &:= \partial_y I(\mathbf{x} + \mathbf{w}, t + 1) - \partial_y I(\mathbf{x}, t).
\end{aligned}
\tag{11.26}
$$

Numerical scheme: The nonlinear system of equations in (11.25) can be solved with the numerical scheme proposed in [9]. It consists of two nested fixed point iterations for removing the nonlinearities in the equations. The outer iteration is in $\mathbf{w}^k$. It is combined with a downsampling strategy in order to better approximate the global optimum of the energy. Starting with the initialization $\mathbf{w} = 0$, a new estimate is computed as $\mathbf{w}^{k+1} = \mathbf{w}^k + (du^k, dv^k)^\top$.

In each iteration one has to solve for the increment (du^k, dv^k). Ignoring here the term for the gradient constancy, which can be derived in the same way, the system to be solved in each iteration is

$$\Psi'(I_z^k)\left(I_x^k du^k + I_y^k dv^k + I_z^k\right) I_x^k - \alpha\Delta(u^k + du^k) = 0$$
$$\Psi'(I_z^k)\left(I_x^k du^k + I_y^k dv^k + I_z^k\right) I_y^k - \alpha\Delta(v^k + dv^k) = 0$$

(11.27)

If Ψ' is constant, this is the case for the shape matching problem in (11.15), (11.27) is already a linear system of equations and can be solved directly with an efficient iterative solver like SOR. If Ψ' depends on (du, dv), however, we have to implement a second fixed point iteration, now in $(du^{k,l}, dv^{k,l})$ to remove the remaining nonlinearity. Each inner iteration computes a new estimate of Ψ' from the most recent $(du^{k,l}, dv^{k,l})$. As Ψ' is kept fixed in each such iteration, the resulting system is linear in $(du^{k,l}, dv^{k,l})$ and can be solved with SOR. With a faster multigrid solver, it is even feasible to compute the optic flow in real time [14]. However, in the scenario here, where the contour-based part is far from real-time performance, the difference to an SOR solver is probably not worth the effort.

11.5 Prior Knowledge of Joint Angle Configurations

The method surveyed in Sections 11.3 and 11.4 incorporates, apart from the input images, also prior knowledge explicitly given by the 3D shape model and the position of the joints. It has been demonstrated that this prior knowledge plays an important role when seeking the contours. This is in accordance with findings in previous works on segmentation methods incorporating 2D shape priors [18, 19, 36]. In particular when the object of interest is partially occluded, the use of shape priors improves the results significantly.

While the method in Sections 11.3 and 11.4 includes a prior on the contour (for given pose parameters), it does not incorporate a prior on the pose parameters yet. Knowing the effects of prior shape knowledge, one expects similarly large improvements when using knowledge about familiar poses. It is intuitively clear that many poses are a-priori impossible or very unlikely, and that a successful technique for human tracking should exclude such solutions. Indeed, recent works on human pose estimation focus a lot on this issue [11, 57, 59, 65]. Their results confirm the relevance of pose priors for reliable tracking.

Integrating the prior via the Bayesian formula: The Bayesian formalism in (11.12) provides the basis for integrating such prior knowledge into the tracking technique. For convenience we repeat the formula:

$$p(\Phi, \xi|I) = \frac{p(I|\Phi, \xi)p(\Phi|\xi)p(\xi)}{p(I)} \rightarrow \max.$$

(11.28)

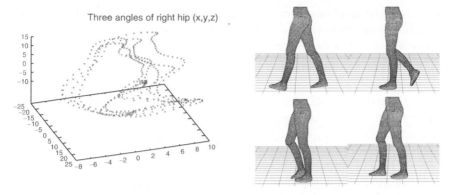

Fig. 11.6. Left: Visualization of the training data obtained from two walking sequences. Only a 3D projection (the three joint angles of the right hip) is shown. **Right:** Some training samples applied to a leg model.

While the prior $p(\xi)$ has been ignored so far, the goal of this section is to learn a probability density from training samples and to employ this density in order to constrain the pose parameters.

As the prior should be independent from the global translation and rotation of the body in the training sequences, a uniform prior is applied to the global twist parameters ξ_{RBM}. Only the probability density for the joint angle vector $p(\Theta)$ is learned and integrated into the tracking framework.

Nonparametric density estimation: Figure 11.6 visualizes training data for the legs of a person from two walking sequences obtained by a marker-based tracking system with a total of 480 samples. Only a projection to three dimensions (the three joint angles of the right hip) is shown.

There are many possibilities to model probability densities from such training samples. The most common way is a parametric representation by means of a Gaussian density, which is fully described by the mean and covariance matrix of the training samples. Such representations, however, tend to over-simplify the sample data. Although Figure 11.6 shows only a projection of the full configuration space, it is already obvious from this figure that pose configurations in a walking motion cannot be described accurately by a Gaussian density.

In order to cope with the non-Gaussian nature of the configuration space, [11] have advocated a nonparametric density estimate by means of the Parzen–Rosenblatt estimator [47,50]. It approximates the probability density by a sum of kernel functions centered at the training samples. A common kernel is the Gaussian function, which leads to:

$$p(\Theta) = \frac{1}{\sqrt{2\pi}\sigma N} \sum_{i=1}^{N} \exp\left(-\frac{(\Theta_i - \Theta)^2}{2\sigma^2}\right) \qquad (11.29)$$

where N is the number of training samples. Note that (11.29) does not involve a projection but acts on the conjoint configuration space of all angles. This means, also the interdependency between joint angles is taken into account.

Choice of the kernel width: The Parzen estimator involves the kernel width σ as a tuning parameter. Small kernel sizes lead to an accurate representation of the training data. On the other hand, unseen test samples close to the training samples may be assigned a too small probability. Large kernel sizes are more conservative, leading to a smoother approximation of the density, which in the extreme case comes down to a uniform distribution. Numerous works on how to optimally choose the kernel size are available in the statistics literature [58]. In our work, we fix σ as the maximum nearest neighbor distance between all training samples, i.e., the next sample is always within one standard deviation. This choice is motivated from the fact that our samples stem from a smooth sequence of poses.

Energy minimization: Taking the prior density into account leads to an additional term in the energy (11.13) that constrains the pose parameters to familiar configurations:

$$E_{\text{Prior}} = -\log(p(\xi)). \tag{11.30}$$

The gradient descent of (11.30) in Θ reads

$$\partial_t \Theta = -\frac{\partial E_{\text{Prior}}}{\partial \Theta} = \frac{\sum_{i=1}^{N} w_i(\Theta_i - \Theta)}{\sigma^2 \sum_{i=1}^{N} w_i} \tag{11.31}$$

$$w_i := \exp\left(-\frac{|\Theta_i - \Theta|^2}{2\sigma^2}\right). \tag{11.32}$$

Obviously, this equation draws the pose to the next local maximum of the probability density. It can be directly integrated into the linear system (11.19) from Section 11.3.3. For each joint j, an additional equation $\theta_j^{k+1} = \theta_j^k + \tau \partial_t \theta_j^k$ is appended to the linear system. In order to achieve an equal weighting of the image against the prior, the new equations are weighted by the number of point correspondences obtained from the contours. The step size parameter $\tau = 0.125\sigma^2$ yielded empirically stable results.

Regularization: The prior obviously provides a regularization of the equation system. Assume a foot is not visible in any camera view. Without prior knowledge, this would automatically lead to a singular system of equations, since there are no correspondences that generate any constraint equation with respect to the joint angles at the foot. Due to the interdependency of the joint angles, the prior equation draws the joint angles of the invisible foot to the most probable solution given the angles of all visible body parts.

Robustness to partial occlusions: Apart from providing unique solutions, the prior also increases the robustness of the tracking in case of unreliable data, as demonstrated in Figure 11.7. Instead of nonsensically fitting the bad

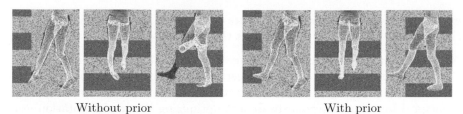

Without prior   With prior

Fig. 11.7. Relevance of the learned configurations for the tracking stability. Occlusions locally disturb the image-driven pose estimation. This can finally cause a global tracking failure. The prior couples the body parts and seeks the most familiar configuration given all the image data.

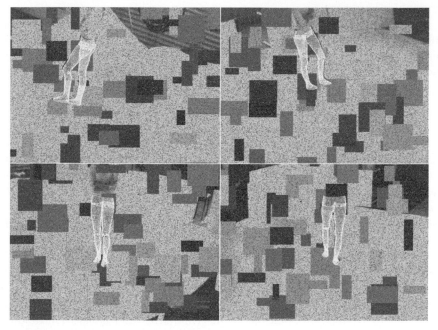

Fig. 11.8. Pose estimates in a sample frame disturbed by 50 varying rectangles with random position, size, and gray value and 25% uncorrelated pixel noise.

data, the method seeks a familiar solution that fits the data best. Another example is shown in Figure 11.8 where, additionally to 25% uniform noise, 50 rectangles of random position, size, and gray value were placed in each image.

11.6 Discussion

The human tracking system described in the preceding sections is based only on few assumptions on the scene and works quite reliably, as shown for rigid bodies in [10, 53] and humans in [52] as well as in this chapter. Further experiments with the same technique are contained in the next chapter. Nevertheless, there are still lots of challenges that shall be discussed in this section.

Running time: One of these challenges is a reduction of the running time. Currently, with a 2 GHz laptop, the method needs around 50s /per frame for 384×280 stereo images. Even though one could at least obtain a speedup of factor 4 by using faster hardware and optimizing the implementation, the method is not adequate for real-time processing. The main computational load is caused by the iterative contour and pose estimation and the approximation of region statistics involved therein. More considerable speedups may be achieved by using the parallelism in these operations via an implementation on graphics hardware.

However, most applications of 3D human motion tracking do not demand real-time performance but high accuracy. Sports movement analysis and modelling of motion patterns for computer graphics are run in batch mode anyway. Thus, improving the running time would mainly reduce hardware costs and improve user interaction.

Auto-initialization: Trying to automatically initialize the pose in the first frame is another interesting challenge. So far, a quite accurate initialization of the pose is needed. For this kind of detection task, the proposed framework seems less appropriate, as it is difficult to detect silhouettes in cluttered images. For object detection, patch based methods have already proven their strength. Thus they can probably solve this task more efficiently. Auto-initialization has, for instance, been demonstrated in [45] for rigid bodies. Works in the scope of human tracking can be found in [1, 25, 49, 60, 61]. Some of these approaches even use silhouettes for the initialization. However, in these cases the contour must be easy to extract from the image data. This is feasible, for instance, with background subtraction if the background is static. The advantage of such discriminative tracking is the possibility to reinitialize after the person has been lost due to total occlusion or the person moving out of all camera views. Combinations of discriminative and generative models, as suggested in [61], are discussed in Chapter 8.

Clothed people: In nearly all setups, the subjects have to wear a body suit to ensure an accurate matching between the silhouettes and the surface models of the legs. Unfortunately, body suits may be uncomfortable to wear in contrast to loose clothing (shirts, shorts, skirts, etc.). The subjects also move slightly different in body suits compared to being in clothes since all body parts (even unfavored ones) are clearly visible. The incorporation of cloth models would ease the subjects and also simplify the analysis of outdoor scenes and arbitrary sporting activities. A first approach in this direction is presented in Chapter 12.

Prior knowledge on motion dynamics: In Section 11.5, a prior on the joint angle vector has been imposed. This has lead to a significant improvement in the tracking reliability given disturbed or partially occluded input images. However, the prior is on the static pose parameters only. It does not take prior information about motion patterns, i.e. the dynamics, into account. Such dynamical priors can be modeled by regression methods such as linear regression or Gaussian processes [48]. In the ideal case, the model yields a probability density, which allows the sound integration in a Bayesian framework [17]. Recently, nonlinear dimensionality reduction methods have become very popular in the context of motion dynamics.

Subspace learning: The idea of dimensionality reduction methods is to learn a mapping between the original, high-dimensional space of pose parameters and a low-dimensional manifold in this space. Solutions are expected to lie only on this manifold, i.e., the search space has been considerably reduced. The motivation for this procedure is the expected inherent low-dimensional structure in a human motion pattern. For instance, the pattern of walking is basically a closed one-dimensional loop of poses when modeled on an adequate, however complex, manifold. Linear projection methods like PCA can be supposed to only insufficiently capture all the limb movements in motion patterns. Nonlinear methods like *Gaussian process latent variable models* (GPLVM), ISOMAP, or others have been shown to be more adequate [25, 27, 35, 65]. See also Chapter 2 and Chapter 10 for more detailed insights.

While dimensionality reduction can successfully model a single motion pattern like walking, running, jumping, etc., it is doubtful that the same concept still works if the model shall contain multiple such patterns. Even though each single pattern may be one- or two-dimensional, the combination of patterns is not. Hence, one has to employ a mixture model with all the practical problems concerning the choice of mixture components and optimization. In case of multiple motion patterns, it may thus be beneficial to define models in the original high-dimensional space, as done, e.g., in the last chapter for static pose priors. This way, one knows for sure that all different patterns can be distinguished. Dealing with the arising high dimensionality when dynamics are included, however, remains a challenging open problem.

11.7 Summary

This chapter has presented a generative Bayesian model for human motion tracking. It includes the joint estimation of the human silhouette and the body pose parameters. The estimation is constrained by a static pose prior based on nonparametric Parzen densities. Furthermore, the pose in new frames is predicted by means of optic flow computed in the foreground region. The approach demands a predefined surface model, the positions of the joints, an initialization of the pose in the first frame, and a calibration of all cameras to the same world coordinate system. In return one obtains reliable estimates of all pose parameters without error accumulation. There is no assumption of a static background involved. Instead, the foreground and background regions are supposed to be locally different. Due to the pose prior, the method can cope with partial occlusions of the person. We also discussed further extensions, in particular the use of image patches for initial pose detection and the integration of dynamical priors.

Acknowledgments

We acknowledge funding of our research by the project CR250/1 of the German Research Foundation (DFG) and by the Max-Planck Center for visual computing and communication.

Appendix: Semiautomatic Acquisition of a Body Model

As most model-based human tracking methods, also the approach in this chapter is based on a model that consists of multiple rigid parts interconnected by joints. Basically, this body model has to be designed manually. Thus, often one can find quite simplistic stick figures based on ellipsoidal limbs in the literature. In this subsection, we briefly describe a method that allows to construct a more accurate surface model by means of four key views of a person as shown in Figure 11.9.

Body separation. After segmentation we separate the arms from the torso of the model. Since we only generate the upper torso, the user can define a bottom line of the torso by clicking on the image. Then we detect the arm pits and the neck joint from the *front view* of the input image. The arm pits are

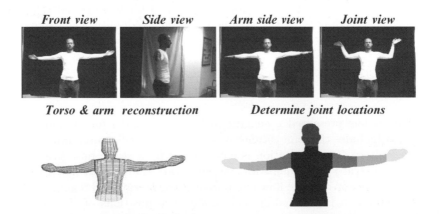

Fig. 11.9. Steps for semiautomatically deriving a body model from four input images.

simply given by the two lowermost corners of the silhouette which are not at the bottom line and exceed a preset angle threshold. The position of the neck joint can be found when moving along the boundary of the silhouette from an upper shoulder point to the head. The narrowest x-slice of the silhouette gives the neck joint.

Joint localization. After this rough segmentation of the human torso we detect the positions of the arm joints. We use a special reference frame (*joint view* in Figure 11.9) that allows to extract arm segments. To gain the length of the hands, upper arms, etc. we first apply a skeletonization procedure. Skeletonization [33] is a process of reducing object pixels in a binary image to a skeletal remnant that largely preserves the extent and connectivity of the original region while eliminating most of the original object pixels. Then we use the method presented in [16] to detect corners of the skeleton to identify joint positions of the arms.

Since the center of the elbow joint is not at the center of the arm but beneath, the joint localizations need to be refined. For this reason, we shift the joint position aiming at correspondence with the human anatomy. The resulting joint locations are shown in the middle right image of Figure 11.9.

Surface mesh reconstruction. For surface mesh reconstruction we assume calibrated cameras in nearly orthogonal views. Then a shape-from-silhouettes approach [34] is applied. We detect control points for each slice and interpolate them by a B-spline curve using the DeBoor algorithm. We start with one slice of the first image and use its edge points as the first two reference points. They are then multiplied with the fundamental matrix of the first to the second camera, and the resulting epipolar lines are intersected with the second silhouette resulting in two more reference points. The reference points are intersected leading to four control points in 3D space.

For arm generation we use a similar scheme for building a model: We use two other reference frames (input images 2 and 3 in Figure 11.9). Then the arms are aligned horizontally and we use the fingertip as starting point on both arms. These silhouettes are sliced vertically to obtain the width and height of each arm part. The arm patches are then connected to the mid plane of the torso.

References

1. Agarwal A. and Triggs B. Recovering 3D human pose from monocular images. *IEEE Transactions on Pattern Analysis and Machine Intelligence*, 28(1):44–58, Jan. 2006.
2. Alvarez L., Weickert J., and Sánchez J. Reliable estimation of dense optical flow fields with large displacements. *International Journal of Computer Vision*, 39(1):41–56, Aug. 2000.
3. Anandan P. A computational framework and an algorithm for the measurement of visual motion. *International Journal of Computer Vision*, 2:283–310, 1989.

4. Besl P. and McKay N. A method for registration of 3D shapes. *IEEE Transactions on Pattern Analysis and Machine Intelligence*, 12:239–256, 1992.
5. Black M.J. and Anandan P. The robust estimation of multiple motions: parametric and piecewise smooth flow fields. *Computer Vision and Image Understanding*, 63(1):75–104, Jan. 1996.
6. Blake A. and Zisserman A. *Visual Reconstruction.* MIT Press, Cambridge, MA, 1987.
7. Bregler C. and Malik J. Tracking people with twists and exponential maps. In *Proc. IEEE Computer Society Conference on Computer Vision and Pattern Recognition*, pp. 8–15, Santa Barbara, California, 1998.
8. Bregler C., Malik J. and Pullen K. Twist based acquisition and tracking of animal and human kinematics. *International Journal of Computer Vision*, 56(3):179–194, 2004.
9. Brox T., Bruhn A., Papenberg N., and Weickert J. High accuracy optical flow estimation based on a theory for warping. In T.Pajdla and J.Matas, editors, *Proc.8th European Conference on Computer Vision*, volume 3024 of *LNCS*, pp. 25–36. Springer, May 2004.
10. Brox T., Rosenhahn B., Cremers D., and Seidel H.-P. High accuracy optical flow serves 3-D pose tracking: exploiting contour and flow based constraints. In A.Leonardis, H.Bischofand A.Prinz, editors, *Proc.European Conference on Computer Vision*, volume 3952 of *LNCS*, pp. 98–111, Graz, Austria, Springer, May 2006.
11. Brox T., Rosenhahn B., Kersting U., and Cremers D. Nonparametric density estimation for human pose tracking. In K.F. et al., editor, *Pattern Recognition*, volume 4174 of *LNCS*, pp. 546–555, Berlin, Germany, Sept. 2006. Springer.
12. Brox T. and Weickert J. A TV flow based local scale estimate and its application to texture discrimination. *Journal of Visual Communication and Image Representation*, 17(5):1053–1073, Oct. 2006.
13. Brox T. and Cremers D. On the statistical interpretation of the piecewise smooth Mumford-Shah functional. In *Scale Space and Variational Methods in Computer Vision*, volume 4485 of *LNCS*, pp. 203–213 Springer, 2007.
14. Bruhn A. and Weickert J. Towards ultimate motion estimation: Combining highest accuracy with real-time performance. In *Proc.10th International Conference on Computer Vision*, pp. 749–755. IEEE Computer Society Press, Beijing, China, Oct. 2005.
15. Chan T. and Vese L. Active contours without edges. *IEEE Transactions on Image Processing*, 10(2):266–277, Feb. 2001.
16. Chetverikov D. A simple and efficient algorithm for detection of high curvature points. In N.Petkov and M.Westenberg, editors, *Computer Analysis of Images and Patterns*, volume 2756 of *LNCS*, pp. 746–753, Groningen, Springer, 2003.
17. Cremers D. Dynamical statistical shape priors for level set based tracking. *IEEE Transactions on Pattern Analysis and Machine Intelligence*, 28(8):1262–1273, Aug. 2006.
18. Cremers D., Kohlberger T. and Schnörr C. Shape statistics in kernel space for variational image segmentation. *Pattern Recognition*, 36(9):1929–1943, Sept. 2003.
19. Cremers D., Osher S., and Soatto S. Kernel density estimation and intrinsic alignment for shape priors in level set segmentation. *International Journal of Computer Vision*, 69(3):335–351, 2006.

20. Cremers D., Rousson M., and Deriche R. A review of statistical approaches to level set segmentation: integrating color, texture, motion and shape. *International Journal of Computer Vision*, 72(2):195–215, 2007.
21. DeCarlo D. and Metaxas D. Optical flow constraints on deformable models with applications to face tracking. *International Journal of Computer Vision*, 38(2):99–127, July 2000.
22. Dempster A., Laird N., and Rubin D. Maximum likelihood from incomplete data via the EM algorithm. *Journal of the Royal Statistical Society series B*, 39:1–38, 1977.
23. Dervieux A. and Thomasset F. A finite element method for the simulation of Rayleigh–Taylor instability. In R.Rautman, editor, *Approximation Methods for Navier–Stokes Problems*, volume 771 of *Lecture Notes in Mathematics*, pp. 145–158. Berlin, Springer, 1979.
24. Dunn D., Higgins W.E. and Wakeley J. Texture segmentation using 2-D Gabor elementary functions. *IEEE Transactions on Pattern Analysis and Machine Intelligence*, 16(2):130–149, Feb. 1994.
25. Elgammal A. and Lee C. Inferring 3D body pose from silhouettes using activity manifold learning. In *Proc.International Conference on Computer Vision and Pattern Recognition*, pp. 681–688, Washington DC, 2004.
26. Gavrila D. and Davis L. 3D model based tracking of humans in action: a multiview approach. In *ARPA Image Understanding Workshop*, pp. 73–80, Palm Springs, 1996.
27. Grochow K., Martin S.L., Hertzmann A., and Popović Z. Style-based inverse kinematics. In *ACM Transactions on Graphics (Proc.SIGGRAPH)*, volume23, pp. 522–531, 2004.
28. Heiler M. and Schnörr C. Natural image statistics for natural image segmentation. *International Journal of Computer Vision*, 63(1):5–19, 2005.
29. Horn B. and Schunck B. Determining optical flow. *Artificial Intelligence*, 17:185–203, 1981.
30. Horprasert T., Harwood D., and Davis L. A statistical approach for real-time robust background subtraction and shadow detection. In *International Conference on Computer Vision, FRAME-RATE Workshop*, Kerkyra, Greece, 1999. Available at www.vast.uccs.edu/~tboult/FRAME.
31. Kadir T. and Brady M. Unsupervised non-parametric region segmentation using level sets. In *Proc.Ninth IEEE International Conference on Computer Vision*, volume 2, pp. 1267–1274, 2003.
32. Kim J., Fisher J., Yezzi A., Cetin M., and Willsky A. A nonparametric statistical method for image segmentation using information theory and curve evolution. *IEEE Transactions on Image Processing*, 14(10):1486–1502, 2005.
33. Klette R. and Rosenfeld A. *Digital Geometry–Geometric Methods for Digital Picture Analysis*. Morgan Kaufmann, San Francisco, 2004.
34. Klette R., Schlüns K., and Koschan A. *Computer Vision. Three-Dimensional Data from Images*. Singapore, Springer, 1998.
35. Lawrence N.D. Gaussian process latent variable models for visualisation of high dimensional data. In *Neural Information Processing Systems 16*.
36. Leventon M.E., Grimson W.E.L., and Faugeras O. Statistical shape influence in geodesic active contours. In *Proc.2000 IEEE Computer Society Conference on Computer Vision and Pattern Recognition (CVPR)*, volume 1, pp. 316–323, Hilton Head, SC, June 2000.

37. Lowe D. Distinctive image features from scale-invariant keypoints. *International Journal of Computer Vision*, 60(2):91–110, 2004.
38. Marchand E., Bouthemy P., and Chaumette F. A 2D-3D model-based approach to real-time visual tracking. *Image and Vision Computing*, 19(13):941–955, Nov. 2001.
39. McLachlan G. and Krishnan T. *The EM Algorithm and Extensions*. Wiley series in probability and statistics. Wiley, 1997.
40. Mémin E. and Pérez P. Dense estimation and object-based segmentation of the optical flow with robust techniques. *IEEE Transactions on Image Processing*, 7(5):703–719, May 1998.
41. Mumford D. and Shah J. Optimal approximations by piecewise smooth functions and associated variational problems. *Communications on Pure and Applied Mathematics*, 42:577–685, 1989.
42. Murray R., Li Z., and Sastry S. *Mathematical Introduction to Robotic Manipulation*. CRC Press, Baton Rouge, 1994.
43. Nagel H.-H. and Enkelmann W. An investigation of smoothness constraints for the estimation of displacement vector fields from image sequences. *IEEE Transactions on Pattern Analysis and Machine Intelligence*, 8:565–593, 1986.
44. Osher S. and Sethian J.A. Fronts propagating with curvature-dependent speed: Algorithms based on Hamilton–Jacobi formulations. *Journal of Computational Physics*, 79:12–49, 1988.
45. Özuysal M., Lepetit V., Fleuret F., and Fua P. Feature harvesting for tracking-by-detection. In *Proc.European Conference on Computer Vision*, volume 3953 of *LNCS*, pp. 592–605. Graz, Austria, Springer, 2006.
46. Paragios N. and Deriche R. Geodesic active regions: A new paradigm to deal with frame partition problems in computer vision. *Journal of Visual Communication and Image Representation*, 13(1/2):249–268, 2002.
47. Parzen E. On the estimation of a probability density function and the mode. *Annals of Mathematical Statistics*, 33:1065–1076, 1962.
48. Rasmussen C.E. and Williams C.K.I. *Gaussian Processes for Machine Learning*. MIT Press, Cambridge, MA, 2006.
49. Rosales R. and Sclaroff S. Learning body pose via specialized maps. In *Proc. Neural Information Processing Systems*, Dec. 2001.
50. Rosenblatt F. Remarks on some nonparametric estimates of a density function. *Annals of Mathematical Statistics*, 27:832–837, 1956.
51. Rosenhahn B., Brox T., Cremers D., and Seidel H.-P. A comparison of shape matching methods for contour based pose estimation. In R.Reulke, U.Eckhardt, B.Flach, U.Knauer and K.Polthier, editors, *Proc.International Workshop on Combinatorial Image Analysis*, volume 4040 of *LNCS*, pp. 263–276, Berlin, Germany, Springer, June 2006.
52. Rosenhahn B., Brox T., Kersting U., Smith A., Gurney J., and Klette R. A system for marker-less motion capture. *Künstliche Intelligenz*, (1):45–51, 2006.
53. Rosenhahn B., Brox T., and Weickert J.. Three-dimensional shape knowledge for joint image segmentation and pose tracking. *International Journal of Computer Vision*, 73(3):243–262, July 2007.
54. Rousson M., Brox T., and Deriche R. Active unsupervised texture segmentation on a diffusion based feature space. In *Proc.International Conference on Computer Vision and Pattern Recognition*, pp. 699–704, Madison, WI, June 2003.

55. Shevlin F. Analysis of orientation problems using Plücker lines. In *International Conference on Pattern Recognition (ICPR)*, volume 1, pp. 685–689, Brisbane, 1998.

56. Shi J. and Tomasi C. Good features to track. In *Proc.International Conference on Computer Vision and Pattern Recognition*, pp. 593–600, 2004.

57. Sidenbladh H., Black M., and Sigal L. Implicit probabilistic models of human motion for synthesis and tracking. In A. Heyden, G. Sparr, M. Nielsen and P. Johansen, editors, *Proc.European Conference on Computer Vision*, volume 2353 of *LNCS*, pp. 784–800. Springer, 2002.

58. Silverman B.W. *Density Estimation for Statistics and Data Analysis*. Chapman & Hall, New York, 1986.

59. Sminchisescu C. and Jepson A. Generative modelling for continuous non-linearly embedded visual inference. In *Proc.International Conference on Machine Learning*, 2004.

60. Sminchisescu C., Kanaujia A., Li Z., and Metaxas D. Discriminative density propagation for 3D human motion estimation. In *Proc.International Conference on Computer Vision and Pattern Recognition*, pp. 390–397, 2005.

61. Sminchisescu C., Kanaujia A., and Metaxas D. Learning joint top-down and bottom-up processes for 3D visual inference. In *Proc.International Conference on Computer Vision and Pattern Recognition*, pp. 1743–1752, 2006.

62. Sminchisescu C. and Triggs B. Estimating articulated human motion with covariance scaled sampling. *International Journal of Robotics Research*, 22(6):371–391, 2003.

63. Sommer G., editor. *Geometric Computing with Clifford Algebra: Theoretical Foundations and Applications in Computer Vision and Robotics*. Berlin, Springer, 2001.

64. Tsai A., Yezzi A., and Willsky A. Curve evolution implementation of the Mumford-Shah functional for image segmentation, denoising, interpolationand magnification. *IEEE Transactions on Image Processing*, 10(8):1169–1186, 2001.

65. Urtasun R., Fleet D.J., and Fua P. 3D people tracking with Gaussian process dynamical models. In *Proc.International Conference on Computer Vision and Pattern Recognition*, pp. 238–245. IEEE Computer Society Press, 2006.

66. Zhu S.-C. and Yuille A. Region competition: unifying snakes, region growing, and Bayes/MDL for multiband image segmentation. *IEEE Transactions on Pattern Analysis and Machine Intelligence*, 18(9):884–900, Sept. 1996.

12

Tracking Clothed People

Bodo Rosenhahn[1], Uwe G. Kersting[2], Katie Powell[2], T. Brox[3],
and Hans-Peter Seidel[1]

[1] Max Planck Institute for Informatics, Stuhlsatzhausenweg 85
   D-66123 Saarbrücken, Germany
[2] Department of Sport and Exercise Science
   The University of Auckland, New Zealand
[3] CVPR Group, University of Bonn, Germany

Summary. This chapter presents an approach for motion capturing (MoCap) of dressed people. A cloth draping method is embedded in a silhouette-based MoCap system and an error functional is formalized to minimize image errors with respect to silhouettes, pose and kinematic chain parameters, the cloth draping components and external forces. Furthermore, Parzen-Rosenblatt densities on static pose configurations are used to stabilize tracking in highly noisy image sequences. We report on various experiments with two types of clothes, namely a skirt and a pair of shorts. Finally we compare the angles of the MoCap system with results from a commercially available marker-based tracking system. The experiments show, that we are less than one dregree above the error range of marker-based tracking systems, though body parts are occluded with cloth.

12.1 Introduction

Classical motion capture (MoCap) comprises techniques for recording the movements of real objects such as humans or animals [35]. In biomechanical settings, it is aimed at analyzing captured data to quantify the movement of body segments, e.g., for clinical studies, diagnostics of orthopaedic patients or to help athletes to understand and improve their performances. It has also grown increasingly important as a source of motion data for computer animation. Surveys on existing methods for MoCap can be found in [11, 21, 22]. Well-known and commercially available marker-based tracking systems exist, e.g., those provided by Motion Analysis, Vicon or Simi [20]. The use of markers comes along with intrinsic problems, e.g., incorrect identification of markers, tracking failures, the need for special laboratory environments and lighting conditions and the fact that people may not feel comfortable with markers attached to the body. This can lead to unnatural motion patterns. As well, marker-based systems are designed to track the motion of the markers themselves, and thus it must be assumed that the recorded motion of the markers

B. Rosenhahn et al. (eds.), Human Motion – Understanding, Modelling, Capture, and Animation, 295–317.

is identical to the motion of the underlying human segments. Since human segments are not truly rigid, this assumption may cause problems, especially in highly dynamic movements typically seen in sporting activities. For these reasons, marker-less tracking is an important field of research that requires knowledge in biomechanics, computer vision and computer graphics.

Typically, researchers working in the area of computer vision prefer simplified human body models for MoCap, e.g., stick, ellipsoidal, cylindric or skeleton models [3, 4, 10, 13, 19]. In computer graphics advanced object modelling and texture mapping techniques for human motions are well-known [5, 6, 17, 36], but image processing or pose estimation techniques (if available) are often simplified.

In [9] a shape-from-silhouettes approach is applied to track human beings and incorporates surface point clouds with skeleton models. One of the subjects even wears a pair of shorts, but the cloth is not explicitly modeled and simply treated as rigid component. Furthermore, the authors just perform a quantitative error analysis on synthetic data, whereas in the present study a second (commercial) marker-based tracking system is used for comparison.

A recent work of us [28] combines silhouette-based pose estimation with more realistic human models: These are represented by free-form surface patches and local morphing along the surface patches is applied to gain a realistic human model within silhouette-based MoCap. Also a comparison with a marker-based system is performed indicating a stable system. In this setup, the subjects have to wear a body suit to ensure an accurate matching between the silhouettes and the surface models of the legs. Unfortunately, body suits may be uncomfortable to wear in contrast to loose clothing (shirts, shorts, skirts, etc.). The subjects also move slightly different in body suits compared to being in clothes since all body parts (even unfavorable ones) are clearly visible. The incorporation of cloth models would also simplify the analysis of outdoor scenes and arbitrary sporting activities. It is for these reasons that we are interested in a MoCap system which also incorporates cloth models. A first version of our approach has been presented in [29].

Cloth draping [12, 14, 18, 34] is a well-known research topic in computer graphics. Virtual clothing can be moved and rendered so that it blends seamlessly with motion and appearance in movie scenes. The motion of fabrics is determined by bending, stretching and shearing parameters, as well as external forces, aerodynamic effects and collisions. For this reason the estimation of cloth simulation parameters is essential and can be done by video [2,7,24,25] or range data [16] analysis. Existing approaches can be roughly divided into geometrically or physically based ones. Physical approaches model cloth behavior by using potential and kinetic energies. The cloth itself is often represented as a particle grid in a spring-mass scheme or by using finite elements [18]. Geometric approaches [34] model cloths by using other mechanics theories which are often determined empirically. These methods can be very fast computationally but are often criticized as being not very appealing visually.

The chapter is built upon the foundations and basic tracking system described in Chapter 11. This comprises techniques for image segmentation with

level sets, pose estimation of kinematic chains, shape registration based on
ICP or optic flow, motion prediction by optic flow, and a prior on joint angle
configurations. The focus of this work is now the embedding of a clothing
model within the MoCap system. To make the chapter self-contained, we
repeat foundations in the next section. In Section 12.3, we continue with introducing a kinematically motivated cloth draping model and a deformation
model, which allow deformation of the particle mesh of a cloth with respect to
oncoming external forces. The proposed draping method belongs to the class
of geometric approaches [34] for cloth draping. The advantages for choosing
this class are twofold: Firstly, we need a model which supports time efficiency,
since cloth draping is needed in one of the innermost loops for minimization of
the used error functional. Secondly, it should be easy to implement and based
on the same parametric representation as the used free-form surface patches.
This allows a direct integration into the MoCap system. In Section 12.4 we
will explain how to minimize the cloth draping and external forces within an
error functional for silhouette-based MoCap. This allows us to determine joint
positions of the legs even if they are partially occluded (e.g., by skirts). We
present MoCap results of a subject wearing a skirt and a pair of shorts and
perform a quantitative error analysis. Section 12.5 concludes with a summary.

12.1.1 Contributions

In this chapter we inform about the following main contributions:

1. A so-called kinematic cloth draping method is proposed. It belongs to the
   class of geometric cloth draping methods and is well suited to be embedded
   in a MoCap-system due to the use of a joint model.
2. The cloth draping is extended by including a deformation model which
   allows to adapt the cloth draping to external forces, the scene dynamics
   or speed of movement.
3. The main contribution is to incorporate the cloth draping algorithm in
   a silhouette-based MoCap system. This allows for determining the joint
   configurations even when parts of the person are covered with fabrics (see
   Figure 12.1).
4. Finally we perform a quantitative error analysis. This is realized by comparing the MoCap-results with a (commercially available) marker-based
   tracking system. The analysis shows that we get stable results and can
   compete with the error range of marker-based tracking systems.

12.2 Foundations: Silhouette-based MoCap

This work is based on a marker-less MoCap system [28, 30] (Figure 12.2). In
this system, the human being is represented in terms of free-form surface
patches, joint indices are added to each surface node and the joint positions

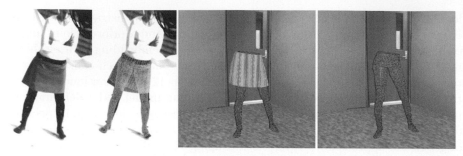

Fig. 12.1. (A) Input: A multi-view image sequence (4 cameras, one cropped image is shown). (B) The algorithm determines the cloth parameters and joint configuration of the underlined leg model. (C) Cloth and leg configuration in a virtual environment. (D) Plain leg configuration.

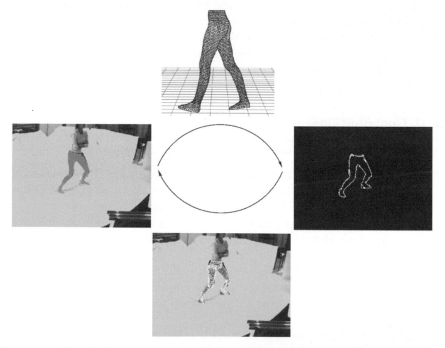

Fig. 12.2. The MoCap system in [28]: **Top:** The model of the person is assumed. **Left:** The object contours are extracted in the input images. **Right:** These are used for correspondence pose estimation. The pose result is applied as shape prior for the segmentation process and the process is iterated.

are assumed. This allows to generate arbitrary body configurations, steered through joint angles. The corresponding counterparts in the images are 2D silhouettes: These are used to reconstruct 3D ray bundles and a spatial distance constraint is minimized to determine the position and orientation of the surface mesh and the joint angles. In this section we will give a brief summary of the MoCap system. These foundations are needed later to explain concisely, where and how the cloth draping approach is incorporated. A more detailed survey can be found in Chapter 11.

12.2.1 Silhouette Extraction

In order to estimate the pose from silhouettes, these silhouettes have to be extracted first, which comes down to a classical segmentation problem. In the system described here, the segmentation is based on a level set representation of contours.

A level set function $\Phi \in \Omega \mapsto \mathbb{R}$ splits the image domain Ω into the foreground region Ω_1 and background region Ω_2 with $\Phi(x) > 0$ if $x \in \Omega_1$ and $\Phi(x) < 0$ if $x \in \Omega_2$. The zero-level line thus marks the boundary between both regions. In order to make the representation unique, the level set functions are supposed to be signed distance functions.

Both regions are analyzed with respect to their feature distribution. The feature space may contain, e.g., gray value, color, or texture features. The key idea is to evolve the contour such that the two regions maximize the a-posteriori probability. Usually, one assumes a priori a smooth contour, but more sophisticated shape priors can be incorporated as well, as shown in Section 12.2.4. Maximization of the posterior can be reformulated as minimization of the following energy functional:

$$E(\Phi, p_1, p_2) = -\int_{\Omega} \left(H(\Phi(x)) \log p_1 + (1 - H(\Phi(x))) \log p_2 + \nu |\nabla H(\Phi(x))| \right) dx,$$

$$(12.1)$$

where $\nu > 0$ is a weighting parameter and $H(s)$ is a regularized version of the Heaviside function, e.g., the error function. The probability densities p_i describe the region model. We use a local Gaussian distribution. These densities are estimated according to the *expectation-maximization principle*. Having the level set function initialized with some contour, the probability densities within the two regions can be estimated. Contour and probability densities are then updated in an iterative manner. An illustration can be seen in Figure 12.3. The left picture depicts the initialization of the contour. The right one shows the estimated (stationary) contour after 50 iterations. As can be seen, the legs and the skirt are well extracted, but there are some problems in the area of the feet region caused by shadows. Incorporation of a shape prior greatly reduces such effects.

12.2.2 Pose Estimation

Assuming an extracted image contour and the silhouette of the projected surface mesh, the closest point correspondences between both contours are used

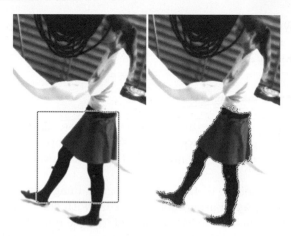

Fig. 12.3. Silhouette extraction based on level set functions. Left: Initial segmentation. Right: Segmentation result.

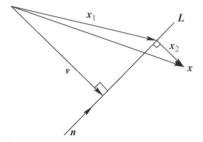

Fig. 12.4. Comparison of a 3D point x with a 3D line L.

to define a set of corresponding 3D lines and 3D points. Then a 3D point-line-based pose estimation algorithm for kinematic chains is applied to minimize the spatial distance between both contours: For point-based pose estimation each line is modeled as a 3D Plücker line $L_i = (n_i, m_i)$, with a (unit) direction n_i and moment m_i [23]. There exist different ways to represent projection rays. As we have to minimize distances between correspondences, it is advantageous to use an implicit representation for a 3-D line. It allows instantaneously to determine the distance between a point and a line. A Plücker line $L = (n, m)$ is given as a unit vector n and a moment m with $m = x \times n$ for a given point x on the line. An advantage of this representation is its uniqueness (apart from possible sign changes). Moreover, the incidence of a point x on a line $L = (n, m)$ can be expressed as

$$x \in L \Leftrightarrow x \times n - m = 0. \tag{12.2}$$

This equation provides us with an error vector. Let $L = (n, m)$, with $m = v \times n$ as shown in Figure 12.4, and $x = x_1 + x_2$, with $x \notin L$ and $x_2 \perp n$.

Since $x_1 \times n = m$, $x_2 \perp n$, and $\|n\| = 1$, we have

$$\|x \times n - m\| = \|x_1 \times n + x_2 \times n - m\| = \|x_2 \times n\| = \|x_2\| \qquad (12.3)$$

where $\|\cdot\|$ denotes the Euclidean norm. This means that $x \times n - m$ in (12.2) results in the (rotated) perpendicular error vector to line L.

The 3D rigid motion is expressed as exponential form

$$M = \exp(\theta\hat{\xi}) = \exp\begin{pmatrix} \theta\hat{\omega} & \theta v \\ 0_{3\times 1} & 0 \end{pmatrix} \qquad (12.4)$$

where $\theta\hat{\xi}$ is the matrix representation of a twist $\xi \in se(3) = \{(v,\hat{\omega})|v \in \mathbb{R}^3, \hat{\omega} \in so(3)\}$, with $so(3) = \{A \in \mathbb{R}^{3\times 3}|A = -A^T\}$. The Lie algebra $so(3)$ is the tangential space of the 3D rotations. Its elements are (scaled) rotation axes, which can either be represented as a 3D vector or skew symmetric matrix,

$$\theta\omega = \theta\begin{pmatrix} \omega_1 \\ \omega_2 \\ \omega_3 \end{pmatrix}, \text{ with } \|\omega\|_2 = 1 \quad \text{ or } \quad \theta\hat{\omega} = \theta\begin{pmatrix} 0 & -\omega_3 & \omega_2 \\ \omega_3 & 0 & -\omega_1 \\ -\omega_2 & \omega_1 & 0 \end{pmatrix}. \quad (12.5)$$

A twist ξ contains six parameters and can be scaled to $\theta\xi$ for a unit vector ω. The parameter $\theta \in \mathbb{R}$ corresponds to the motion velocity (i.e., the rotation velocity and pitch). For varying θ, the motion can be identified as screw motion around an axis in space. The six twist components can either be represented as a 6D vector or as a 4×4 matrix,

$$\theta\xi = \theta(\omega_1, \omega_2, \omega_3, v_1, v_2, v_3)^T, \|\omega\|_2 = 1, \quad \theta\hat{\xi} = \theta\begin{pmatrix} 0 & -\omega_3 & \omega_2 & v_1 \\ \omega_3 & 0 & -\omega_1 & v_2 \\ -\omega_2 & \omega_1 & 0 & v_3 \\ 0 & 0 & 0 & 0 \end{pmatrix}. \quad (12.6)$$

To reconstruct a group action $M \in SE(3)$ from a given twist, the exponential function $\exp(\theta\hat{\xi}) = \sum_{k=0}^{\infty} \frac{(\theta\hat{\xi})^k}{k!} = M \in SE(3)$ must be computed. This can be done efficiently by using the Rodriguez formula [23].

For pose estimation the reconstructed Plücker lines are combined with the screw representation for rigid motions:

Incidence of the transformed 3D point X_i with the 3D ray $L_i = (n_i, m_i)$ can be expressed as

$$(\exp(\theta\hat{\xi})X_i)_{3\times 1} \times n_i - m_i = 0. \qquad (12.7)$$

Since $\exp(\theta\hat{\xi})X_i$ is a 4D vector, the homogeneous component (which is 1) is neglected to evaluate the cross product with n_i. Then the equation is linearized and iterated, see [28].

Joints are expressed as special screws with no pitch of the form $\theta_j\hat{\xi}_j$ with known $\hat{\xi}_j$ (the location of the rotation axes is part of the model) and unknown joint angle θ_j. The constraint equation of an ith point on a jth joint has the form

$$(\exp(\theta\hat{\xi})\exp(\theta_1\hat{\xi_1})\ldots\exp(\theta_j\hat{\xi_j})X_i)_{3\times 1} \times n_i - m_i = 0 \qquad (12.8)$$

which is linearized in the same way as the rigid body motion itself. It leads to three linear equations with the six unknown pose parameters and j unknown joint angles.

12.2.3 Shape Registration

The goal of shape registration can be formulated as follows: Given a certain distance measure; the task is to determine one transformations that leads to the minimum distance between shapes. A very popular shape matching method working on such representations is the iterated closest point (ICP) algorithm [1]. Given two finite sets P and Q of points. The (original) ICP algorithm calculates a rigid transformation T and attempts to ensure $TP \subseteq Q$.

1. **Nearest point search**: for each point $p \in P$ find the closest point $q \in Q$.
2. **Compute registration**: determine the transformation T that minimizes the sum of squared distances between pairs of closest points (p, q).
3. **Transform**: apply the transformation T to all points in set P.
4. **Iterate**: repeat step 1 to 3 until the algorithm converges.

This algorithm converges to the next local minimum of the sum of squared distances between closest points. A good initial estimate is required to ensure convergence to the sought solution. Unwanted solutions may be found if the sought transformation is too large, e.g., many shapes have a convergence radius in the area of 20° [8], or if the point sets do not provide sufficient information for a unique solution.

The original ICP algorithm has been modified in order to improve the rate of convergence and to register partially overlapping sets of points. Zhang [37] uses a modified cost function based on robust statistics to limit the influence of outliers. Other approaches aim at the avoidance of local minima during registration subsuming the use of Fourier descriptors [31], color information [15], or curvature features [33].

The advantages of ICP algorithms are obvious: they are easy to implement and will provide good results, if the sought transformation is not too large [8]. For our tracking system we compute correspondences between points on image silhouettes to the surface mesh with the ICP algorithm presented in [31].

In Chapter 11 and [27] it is further explained how an alternative matching procedure, by using the optic flow can be used to improve the convergence rate and convergence radius. In this work we make use of both matchers to register a surface model to an image silhouette.

12.2.4 Combined Pose Estimation and Segmentation

Since segmentation and pose estimation can both benefit from each other, it is sensible to couple both problems in a joint optimization problem. To this end, the energy functional for image segmentation in (12.1) is extended by an additional term that integrates the surface model:

$$E(\Phi, \theta\xi) = -\int_\Omega \left(H(\Phi) \log p_1 + (1 - H(\Phi)) \log p_2 \right) dx + \nu \int_\Omega |\nabla H(\Phi)| \, dx$$

$$+ \lambda \underbrace{\int_\Omega (\Phi - \Phi_0(\theta\xi))^2 \, dx}_{\text{Shape}} . \tag{12.9}$$

The quadratic error measure in the shape term has been proposed in the context of 2D shape priors, e.g., in [32]. The prior $\Phi_0 \in \Omega \to \mathbb{R}$ is assumed to be represented by the signed distance function. This means in our case, $\Phi_0(x)$ yields the distance of x to the silhouette of the projected object surface.

Given the contour Φ, the pose estimation method from Section 12.2.2 minimizes the shape term in (12.9). Minimizing (12.9) with respect to the contour Φ, on the other hand, leads to the gradient descent equation

$$\partial_t \Phi = H'(\Phi) \left(\log \frac{p_1}{p_2} + \nu \nabla^\top \left(\frac{\nabla \Phi}{|\nabla \Phi|} \right) \right) + 2\lambda \left(\Phi_0(\theta\xi) - \Phi \right). \tag{12.10}$$

The total energy is minimized by iterating both minimization procedures. Both iteration steps minimize the distance between Φ and Φ_0. While the pose estimation method draws Φ_0 towards Φ, thereby respecting the constraint of a rigid motion, (12.10) in return draws the curve Φ towards Φ_0, thereby respecting the data in the image.

12.2.5 Quantitative Error Analysis

A lack of many studies (e.g. [19]) is that only a visual feedback about the pose result is given, by overlaying the pose result with the image data.

To enable a quantitative error analysis, we use a commercial marker-based tracking system for a comparison. Here, we use the Motion Analysis software [20], with an 8-Falcon-camera system. For data capture we use the Eva 3.2.1 software and the Motion Analysis Solver Interface 2.0 for inverse kinematics computing. In this system a human has to wear a body suit and retroflective markers are attached to it. Around each camera is a strobe light led ring and a red-filter is in front of each lens. This gives very strong image signals of the markers in each camera. These are treated as point markers which are reconstructed in the eight-camera system. Figure 12.5 shows a screen shot of the Motion Analysis system. The system is calibrated by using a wand-calibration method. Due to the filter in front of the images we had to use a

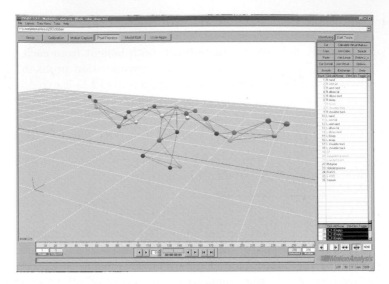

Fig. 12.5. Screen shot of the used EVA solver from motion analysis.

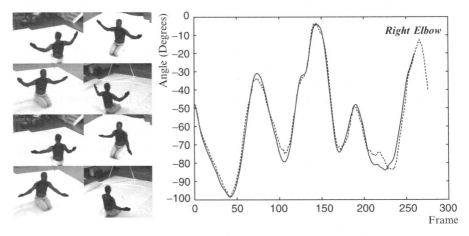

Fig. 12.6. Tracked arms: The angle diagrams show the elbow values of the motion analysis system (dotted) and the silhouette system (solid).

second camera setup which provides *real* image data. This camera system is calibrated by using a calibration cube. After calibration, both camera systems are calibrated with respect to each other. Then we generate a stick-model from the point markers including joint centers and orientations. This results in a complete calibrated setup we use for a system comparison.

Figure 12.6 shows the first test sequence, where the subject is just moving the arms forwards and backwards. The diagram on the right side shows the estimated angles of the right elbow. The marker results are given as dotted

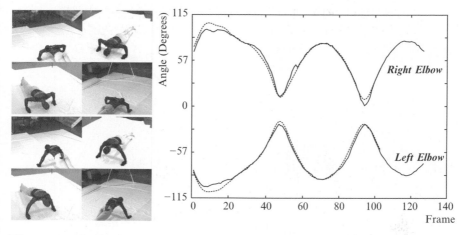

Fig. 12.7. Tracked Push-ups: The angle diagrams show the elbow values of the motion analysis system (dotted) and the silhouette system (solid).

lines and the silhouette results in solid lines. The overall error between both angles diagrams is 2.3 degrees, including the tracking failure between frames 200 till 250.

Figure 12.7 shows the second test sequence, where the subject is performing a series of push-ups. Here the elbow angles are much more characteristic and also well comparable. The overall error is 1.7 degrees. Both sequences contain partial occlusions in certain frames. But this can be handled from the algorithm.

In [26] eight biomechanical measurement systems are compared (including the Motion Analysis system). A rotation experiment is performed, which shows that the RMS[1] errors are typically within three degrees. Our error measures fit in this range quite well.

12.3 Kinematic Cloth Draping

To integrate a clothing model in the MoCap system, we decided to use a geometric approach. The main reason is that cloth draping is needed in one of the innermost loops for pose estimation and segmentation. Therefore it must be very fast. In our case we need around 400 iterations for each frame to converge to a solution. A cloth draping algorithm in the area of seconds would require hours to calculate the pose of one frame and weeks for a whole sequence.

We decided to model the skirt as a string-system with underlined kinematic chains: The main principle is visualized on the left in Figure 12.8 for a piece of cloth falling on a plane. The piece of cloth is represented as a particle grid,

---

[1] Root mean square.

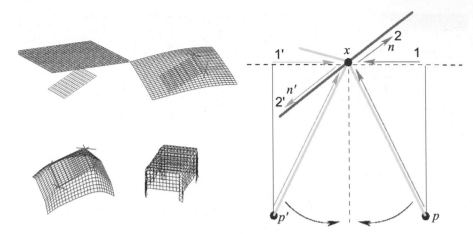

Fig. 12.8. The cloth draping principle. Joints are used to deform the cloth while draping on the surface mesh.

a set of points with known topology. While lowering the cloth, the distance of each cloth point to the ground plane is determined. If the distance between one point on the cloth to the surface is below a threshold, the point is set as a fixed-point, see the top right image on the left of Figure 12.8. Now the remaining points are not allowed to *fall* downwards anymore. Instead, for each point, the nearest fixed-point is determined and a joint (perpendicular to the particle point) is used to rotate the free point along the joint axis through the fixed point. The used joint axes are marked as blue lines in Figure 12.8. The image on the right in Figure 12.8 shows the geometric principle to determine the twist for rotation around a fixed point: The blue line represents a mesh of the rigid body, x is the fixed point and the (right) pink line segment connects x to a particle p of the cloth. The direction between both points is projected onto the y-plane of the fixed point (1). The direction is then rotated around 90 degrees (2), leading to the rotation axis n. The point pairs $(n, x \times n)$ are the components of the twist, see Equation (12.6). While lowering the cloth, free particles not touching a second rigid point, will swing below the fixed point (e.g. p'). This leads to an opposite rotation (indicated with (1'), (2') and n') and the particle swings back again, resulting in a natural swinging draping pattern. The draping velocity is steered through a rotation velocity θ, which is set to 2 degrees during iteration. Since all points either become fixed points, or result in a stationary configuration while swinging backwards and forwards, we constantly use 50 iterations to drape the cloth. The remaining images on the left in Figure 12.8 show the ongoing draping and the final result. Figure 12.9 shows the cloth draping steps of a skirt model.

Figure 12.10 shows example images of a skirt and a pair of shorts falling on the leg model. The skirt is modeled as a 2-parametric mesh model. Due to the use of general rotations, the internal distances in the particle mesh cannot

Fig. 12.9. Draping of the skirt model.

Fig. 12.10. Cloth draping of a skirt and shorts in a simulation environment.

Fig. 12.11. Reconstraining the skirts' length.

change with respect to one of these dimensions, since a rotation maintains the distance between the involved points. However, this is not the case for the second sampling dimension. For this reason, the skirt needs to be reconstrained after draping. This is visualized in Figure 12.11: If a stretching parameter is exceeded, the particles are reconstrained to minimal distance to each other. This is only done for the non-fixed points (i.e., for those which are not touching the skin). It results in a better appearance especially for certain leg configurations.

Figure 12.11 shows that even the creases are maintained. In this case, shorts are simpler since they are modeled as cylinders, transformed together with the legs and then draped.

To improve the dynamic behavior of clothing during movements, we also add external forces to the cloth draping. We continue with the cloth-draping in the following way: dependent on the direction of a force we determine a joint on the nearest fixed point for each free point on the surface mesh with the joint direction being perpendicular to the force direction. Now we rotate

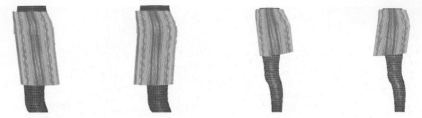

Fig. 12.12. External forces, e.g., virtual wind on the shorts (left) and the skirt (right). Visualized is frontal and backward wind.

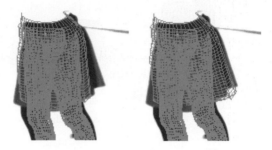

Fig. 12.13. Left: Overlaid pose result without external forces. Right: Overlaid pose result including external forces.

the free point around this axis dependent on the force amount (expressed as an angle) or until the cloth is touching the underlying surface. Figure 12.12 shows examples of the shorts and skirt with frontal or backward virtual wind acting as external force. The external forces and their directions are later part of the minimization function during pose tracking. Figure 12.13 visualizes the effect of the used deformation model. Since the motion dynamics of the cloth are determined dynamically, we need no information about the cloth type or weight since they are implicitly determined from the minimized cloth dynamics in the image data; we only need the measurements of the cloth.

12.4 Combined Cloth Draping and MoCap

The assumptions are as follows: We assume the representation of a subject's lower torso (i.e., for the hip and legs) in terms of free-form surface patches. We also assume known joint positions along the legs. Furthermore we assume the wearing of a skirt or shorts with known measures. The person is walking or stepping in a four-camera setup. These cameras are triggered and calibrated with respect to one world coordinate system. The task is to determine the pose of the model and the joint configuration. For this we minimize the image error between the projected surface meshes to the extracted image silhouettes. The

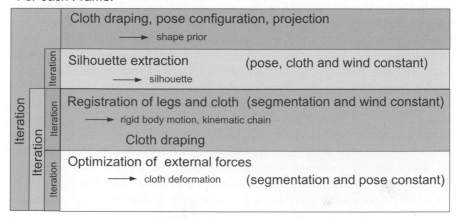

Fig. 12.14. The basic algorithm for combined cloth draping and motion capturing.

unknowns are the pose, kinematic chain and the cloth parameters (external forces, cloth thickness, etc.). The task can be represented as an error functional as follows:

$$E(\varPhi, p_1, p_2, \theta\xi, \theta_1, \ldots, \theta_n, c, w) =$$

$$\underbrace{-\int_\Omega \big(H(\varPhi)\log p_1 + (1 - H(\varPhi))\log p_2 + \nu|\nabla H(\varPhi)|\big)\, dx}_{\text{segmentation}}$$

$$\underbrace{+ \lambda \int_\Omega (\varPhi - \varPhi_0(\; \underbrace{\theta\xi, \theta_1, \ldots, \theta_n}_{\text{pose and kinematic chain,}} \;,\; \underbrace{c, w}_{\text{external forces}} \;))dx}_{\text{shape error}}$$

Due to the large number of parameters and unknowns we decided for an iterative minimization scheme, see Figure 12.14: Firstly, the pose, kinematic chain and external forces are kept constant, while the error functional for the segmentation (based on $\varPhi, p_1, p_2$) is minimized (Section 12.2.1). Then the segmentation and external forces are kept constant while the pose and kinematic chain are determined to fit the surface mesh and the cloth to the silhouettes (Section 12.2.2). Finally, different directions of external forces are sampled to refine the pose result (Section 12.3). Since all parameters influence each other, the process is iterated until a steady state is reached. In our experiments, we always converged to a local minimum.

12.5 Experiments

For the experiments we used a four-camera set up and grabbed image sequences of the lower torso with different motion patterns: The subject was asked to wear the skirt and the shorts while performing walking, leg crossing and turning, knee bending and walking with knees pulled up. We decided on these different patterns, since they are not only of importance for medical studies (e.g., walking), but they are also challenging for the cloth simulator, since the cloth is partially stretched (knee pulling sequence) or hanging down loosely (knee bending). The turning and leg crossing sequence is interesting due to the higher occlusions. Figure 12.15 shows some pose examples for the subject wearing the skirt (top) and shorts (bottom). The pose is visualized by overlaying the projected surface mesh onto the images. Just one of the four camera views is shown. Each sequence consists of 150–240 frames. Figure 12.16 visualizes the stability of our approach: While grabbing the images, a couple of frames were stored completely wrong. These sporadic outliers can be compensated from our algorithm, and a few frames later (see the image on the right) the pose is correct. Figure 12.17 shows leg configurations in a virtual environment. The position of the body and the joints reveal a natural configuration.

Finally, the question about the stability arises. To answer this question, we attached markers to the subject and tracked the sequences simultaneously with the commercially available Motion Analysis system [20]. The markers are attached to the visible parts of the leg and are not disturbed by the cloth, see Figure 12.18. We then compare joint angles for different sequences with the results of the marker-based system, similar to Section 12.2.5. The

Fig. 12.15. Example sequences for tracking clothed people. **Top row**: walking, leg crossing, knee bending and knee pulling with a skirt. **Bottom row**: walking, leg crossing, knee bending and knee pulling with shorts. The pose is determined from four views (just one of the views is shown, images are cropped).

Fig. 12.16. Error during grabbing the images.

Fig. 12.17. Example leg configurations of the sequences. The examples are taken from the subject wearing the shorts (blue) and the skirt (red) (leg crossing, walking, knee bending, knee pulling).

Fig. 12.18. The set-up for quantitative error analysis: Left/middle, the subject with attached markers. The cloth does not interfere with tracking the markers. Right: A strobe light camera of the used Motion Analysis system.

overall errors for both types of cloth varies between 1.5 and 4.5 degrees, which indicates a stable result, see [26]. Table 12.1 summarizes the deviations.

The diagrams in Figure 12.19 shows the overlay of the knee angles for two skirt and two shorts sequences. The two systems can be identified by the smooth curves from the Motion Analysis system and unsmoothed curves (our system).

Table 12.1. Deviations of the left and right knee for different motion sequences.

	Skirt			Shorts	
Sequence	Left knee	Right knee	Sequence	Left knee	Right knee
Dancing	3.42	2.95	Dancing	4.0	4.0
Knee-up	3.22	3.43	Knee-up	3.14	4.42
Knee bending	3.33	3.49	Knee bending	2.19	3.54
Walking	2.72	3.1	Walking	1.52	3.38

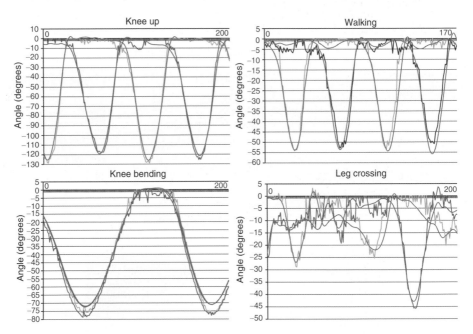

Fig. 12.19. Left: Knee angles from sequences wearing the shorts. **Right**: Knee angles from sequences wearing the skirt. **Top left**: Angles of the knee up sequence. **Bottom left**: Angles of the knee bending sequence. **Top right**: Angles of the walking sequence. **Bottom right**: Angles of the leg crossing sequence.

12.5.1 Prior Knowledge on Angle Configurations

Chapter 11 also introduced the use of nonparametric density estimates to build additional constraint equations to enforce the algorithm to converge to familiar configurations. It further can be seen as the embedding of soft-constraints to penalize configurations which are uncommon (e.g., to move the arm through the body) and they regularize the equations which results in guaranteed nonsingular system of equations. These advantages can also be seen from the experiments in Figures 12.20, 12.21 and 12.22:

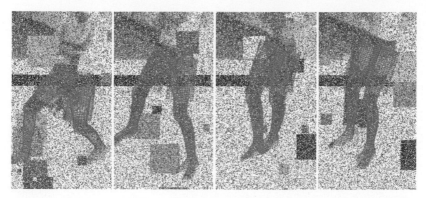

Fig. 12.20. Scissors sequence (cropped images, one view is shown): The images have been distorted with 60% uncorrelated noise, rectangles of random size and gray values and a black stripe across the images.

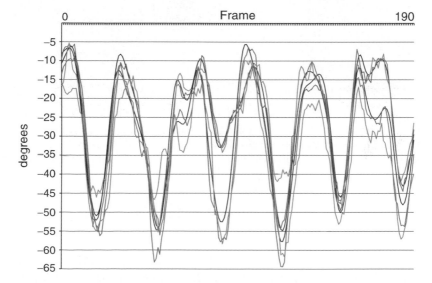

Fig. 12.21. Quantitative error analysis of the scissors sequence (the knee angles are shown). Black: the results of the Motion Analysis system. Blue: the outcome from our markerless system (deviation: 3.45 and 2.18 degrees). Red: The results from our system for the highly disturbed image data (deviation: 6.49 and 3.0 degrees).

Figure 12.20 shows results of a scissors sequence. The images have been distorted with 60% uncorrelated noise, rectangles of random size and gray values and a black stripe across the images. Due to prior knowledge of scissor jumps, the algorithm is able to track the sequence successfully. The diagram in Figure 12.21 quantifies the result by overlaying the knee angles for the marker results (black) with the undisturbed result (blue) and highly noised (red) image data. The deviation for the plain data is 3.45 and 2.18 degrees,

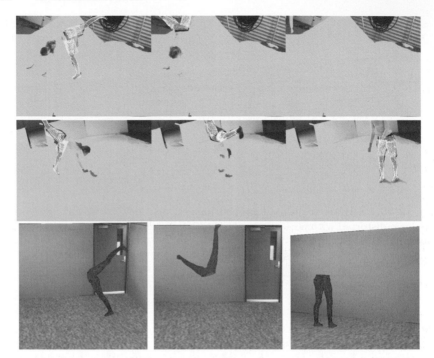

Fig. 12.22. Cartwheel sequence in a Lab environment: The legs are not visible in some key frames, but the prior knowledge allows to give the legs a natural (most likely) configuration. The top images shows two views and three example frames of the sequence. The bottom images show the leg configuration in a virtual environment.

respectively and for the highly disturbed sequence we get 6.49 and 3.0 degrees. Due to the high amount of noise added to the image data, we consider the outcome as a good result.

Figure 12.22 shows results from a cartwheel sequence. During tracking the sequence the legs are not visible in some key frames, but the prior knowledge allows to give the legs a natural (most likely) configuration. The top images shows two views and three example frames of the sequence. The bottom images show the leg configuration in a virtual environment.

12.6 Summary

The contribution presents an approach for motion capture of clothed people. To achieve this we extend a silhouette-based motion capture system, which relies on image silhouettes and free-form surface patches of the body with a cloth draping procedure. We employ a geometric approach based on kinematic chains. We call this cloth draping procedure kinematic cloth draping.

This model is very well suited to be embedded in a motion capture system since it allows us to minimize the cloth draping parameters (and external forces) within the same error functional such as the segmentation and pose estimation algorithm. Due to the number of unknowns for the segmentation, pose estimation, joints and cloth parameters, we decided on an iterative solution. The experiments with a skirt and shorts show that the formulated problem can be solved. We are able to determine joint configurations and pose parameters of the kinematic chains, though they are considerably covered with clothes. Indeed, we use the cloth draping appearance in images to recover the joint configuration and simultaneously determine dynamics of the cloth. Furthermore, Parzen-Rosenblatt densities of static joint configurations are used to generate constraint equations which enables a tracking of persons in highly noisy or corrupted image sequences.

To quantify the results, we performed an error analysis by comparing our method with a commercially available marker-based tracking system. The experiments show that we are close to the error range of marker-based tracking systems [26].

Applications are straightforward: The motion capture results can be used to animate avatars in computer animations, and the angle diagrams can be used for the analysis of sports movements or clinical studies. The possibility of wearing loose clothes is much more comfortable for many people and enables a more natural motion behavior. The presented extension also allows us to analyze outdoor activities, e.g., soccer or other team sports.

For future works we plan to extend the cloth draping model with more advanced ones [18] and we will compare different draping approaches and parameter optimization schemes in the motion capturing setup.

Acknowledgments

This work has been supported by the Max-Planck Center for Visual Computing and Communication.

References

1. Besl P. and McKay N. A method for registration of 3D shapes. *IEEE Transactions on Pattern Analysis and Machine Intelligence*, 12:239–256, 1992.
2. Bhat K.S., Twigg C.D., Hodgins J.K., Khosla P.K., Popovic Z. and Seitz S.M. Estimating cloth simulation parameters from video. In D. Breen and M. Lin, editors, *Proc. ACM SIGGRAPH/Eurographics Symposium on Computer Animation*, pages 37–51, 2003.
3. Bregler C. and Malik J. Tracking people with twists and exponential maps. In *Proc. Computer Vision and Pattern Recognition*, pages 8–15. Santa Barbara, California, 1998.

4. Bregler C., Malik J. and Pullen K. Twist based acquisition and tracking of animal and human kinetics. *International Journal of Computer Vision*, 56(3):179–194, 2004.

5. Carranza J., Theobalt C., Magnor M.A. and Seidel H.-P. Free-viewpoint video of human actors. In *Proc. SIGGRAPH 2003*, pages 569–577, 2003.

6. Chadwick J.E., Haumann D.R. and Parent R.E. Layered construction for deformable animated characters. *Computer Graphics*, 23(3):243–252, 1989.

7. Chafri H., Gagalowicz A. and Brun R. Determination of fabric viscosity parameters using iterative minimization. In A. Gagalowicz and W. Philips, editors, *Proc. Computer Analysis of Images and Patterns*, volume 3691 of *Lecture Notes in Computer Science*, pages 789–798. Springer, Berlin, 2005.

8. Chetverikov D., Stepanov D. and Krsek P. Robust Euclidean alignment of 3D point sets: The trimmed iterative closest point algorithm. *Image and Vision Computing*, 23(3):299–309, 2005.

9. Cheung K.M., Baker S. and Kanade T. Shape-from-silhouette across time: Part ii: Applications to human modelling and markerless motion tracking. *International Journal of Computer Vision*, 63(3):225–245, 2005.

10. Fua P., Plänkers R. and Thalmann D. Tracking and modelling people in video sequences. *Computer Vision and Image Understanding*, 81(3):285–302, March 2001.

11. Gavrilla D.M. The visual analysis of human movement: A survey. *Computer Vision and Image Understanding*, 73(1):82–92, 1999.

12. Haddon J., Forsyth D. and Parks D. The appearance of clothing. http://http.cs.berkeley.edu/haddon/clothingshade.ps, June 2005.

13. Herda L., Urtasun R. and Fua P. Implicit surface joint limits to constrain video-based motion capture. In T. Pajdla and J. Matas, editors, *Proc. 8th European Conference on Computer Vision*, volume 3022 of *Lecture Notes in Computer Science*, pages 405–418. Springer Prague, May 2004.

14. House D.H., DeVaul R.W. and Breen D.E. Towards simulating cloth dynamics using interacting particles. *Clothing Science and Technology*, 8(3):75–94, 1996.

15. Johnson A.E. and Kang S.B. Registration and integration of textured 3-D data. In *Proc.International Conference on Recent Advances in 3-D Digital Imaging and modelling*, pages 234–241. IEEE Computer Society, May 1997.

16. Jojic N. and Huang T.S. Estimating cloth draping parameters from range data. In *Proc. Int. Workshop Synthetic-Natural Hybrid Coding and 3-D Imaging*, pages 73–76. Greece, 1997.

17. Magnenat-Thalmann N., Seo H. and Cordier F. Automatic modelling of virtual humans and body clothing. *Computer Science and Technology*, 19(5):575–584, 2004.

18. Magnenat-Thalmann N. and Volino P. From early draping to haute cotoure models: 20 years of research. *Visual Computing*, 21:506–519, 2005.

19. Mikic I., Trivedi M., Hunter E. and Cosman P. Human body model acquisition and tracking using voxel data. *International Journal of Computer Vision*, 53(3):199–223, 2003.

20. MoCap-System. Motion analysis: A marker-based tracking system. www.motionanalysis.com, June 2005.

21. Moeslund T.B. and Granum E. A survey of computer vision based human motion capture. *Computer Vision and Image Understanding*, 81(3):231–268, 2001.

22. Moeslund T.B., Granum E. and Krüger V. A survey of of advances in vision-based human motion capture and analysis. *Computer Vision and Image Understanding*, 104(2):90–126, 2006.
23. Murray R.M., Li Z. and Sastry S.S. *Mathematical Introduction to Robotic Manipulation*. CRC Press, Baton Rouge, 1994.
24. Povot X. *Deformation Constraints in a Mass-Spring Model to Describe Rigid Cloth Behavior* Graphics Interface '95, Canadian Human-Computer Communications Society, W. A. Davis and P. Prusinkiewicz (editors), pages 147–154, 1995.
25. Pritchard D. and Heidrich W. Cloth motion capture. *Eurographics*, 22(3):37–51, 2003.
26. Richards J. The measurement of human motion: A comparison of commercially available systems. *Human Movement Science*, 18:589–602, 1999.
27. Rosenhahn B., Brox T., Cremers D. and Seidel H.-P. A comparison of shape matching methods for contour based pose estimation. In R. Reulke, U. Eckhardt, B.Flach and U.Knauer, editors, *Proc. 11th Int. Workshop Combinatorial Image Analysis*, volume 4040 of *Lecture Notes in Computer Science*, pages 263–276. Springer, Berlin, 2006.
28. Rosenhahn B., Brox T., Kersting U., Smith A., Gurney J. and Klette R. A system for marker-less motion capture. *Künstliche Intelligenz*, (1):45–51, 2006.
29. Rosenhahn B., Kersting U., Powell K. and Seidel H.-P. Cloth x-ray: Mocap of people wearing textiles. In *Accepted: Pattern Recognition, 28th DAGM-symposium*, Lecture Notes in Computer Science, Springer, Berlin, Germany, September 2006.
30. Rosenhahn B., Kersting U., Smith A., Gurney J., Brox T. and Klette R. A system for marker-less human motion estimation. In W. Kropatsch, R. Sablatnig and A. Hanbury, editors, *Pattern Recognition, 27th DAGM-symposium*, volume 3663 of *Lecture Notes in Computer Science*, pages 230–237, Springer, Vienna, Austria, September 2005.
31. Rosenhahn B. and Sommer G. Pose estimation of free-form objects. In T. Pajdla and J. Matas, editors, *Computer Vision - Proc.8th European Conference on Computer Vision*, volume 3021 of *Lecture Notes in Computer Science*, pages 414–427. Springer, May 2004.
32. Rousson M. and Paragios N. Shape priors for level set representations. In A.Heyden, G.Sparr, M.Nielsen and P.Johansen, editors, *Computer Vision – ECCV 2002*, volume 2351 of *Lecture Notes in Computer Science*, pages 78–92. Springer, Berlin, 2002.
33. Rusinkiewicz S. and Levoy M. Efficient variants of the ICP algorithm. In *Proc.3rdIntl. Conf. on 3-D Digital Imaging and modelling*, pages 224–231, 2001.
34. Weil J. The synthesis of cloth objects. *Computer Graphics (Proc. SigGraph)*, 20(4):49–54, 1986.
35. Wikipedia. Motion capture. http://en.wikipedia.org/wiki/Motion_capture, September 2005.
36. You L. and Zhang J.J. Fast generation of 3d deformable moving surfaces. *IEEE Trans. Systems, Man and Cybernetics, Part B: Cybernetics*, 33(4):616–615, 2003.
37. Zhang Z. Iterative points matching for registration of free form curves and surfaces. *International Journal of Computer Vision*, 13(2):119–152, 1994.

13

An Introduction to Interacting Simulated Annealing

Juergen Gall, Bodo Rosenhahn, and Hans-Peter Seidel

Max-Planck Institute for Computer Science
Stuhlsatzenhausweg 85, 66123 Saarbrücken, Germany

Summary. Human motion capturing can be regarded as an optimization problem where one searches for the pose that minimizes a previously defined error function based on some image features. Most approaches for solving this problem use iterative methods like gradient descent approaches. They work quite well as long as they do not get distracted by local optima. We introduce a novel approach for global optimization that is suitable for the tasks as they occur during human motion capturing. We call the method interacting simulated annealing since it is based on an interacting particle system that converges to the global optimum similar to simulated annealing. We provide a detailed mathematical discussion that includes convergence results and annealing properties. Moreover, we give two examples that demonstrate possible applications of the algorithm, namely a global optimization problem and a multi-view human motion capturing task including segmentation, prediction, and prior knowledge. A quantative error analysis also indicates the performance and the robustness of the interacting simulated annealing algorithm.

13.1 Introduction

13.1.1 Motivation

Optimization problems arise in many applications of computer vision. In pose estimation, e.g. [28], and human motion capturing, e.g. [31], functions are minimized at various processing steps. For example, the marker-less motion capture system [26] minimizes in a first step an energy function for the segmentation. In a second step, correspondences between the segmented image and a 3D model are established. The optimal pose is then estimated by minimizing the error given by the correspondences. These optimization problems also occur, for instance, in model fitting [17, 31]. The problems are mostly solved by iterative methods as gradient descent approaches. The methods work very well as long as the starting point is near the global optimum, however, they get easily stuck in a local optimum. In order to deal with it, several random selected starting points are used and the best solution is selected in

B. Rosenhahn et al. (eds.), Human Motion – Understanding, Modelling, Capture, and Animation, 319–345.

the hope that at least one of them is near enough to the global optimum, cf. [26]. Although it improves the results in many cases, it does not ensure that the global optimum is found.

In this chapter, we introduce a global optimization method based on an interacting particle system that overcomes the dilemma of local optima and that is suitable for the optimization problems as they arise in human motion capturing. In contrast to many other optimization algorithms, a distribution instead of a single value is approximated by a particle representation similar to particle filters [10]. This property is beneficial, particularly, for tracking where the right parameters are not always exact at the global optimum depending on the image features that are used.

13.1.2 Related Work

A popular global optimization method inspired by statistical mechanics is known as simulated annealing [14,18]. Similar to our approach, a function $V \geq 0$ interpreted as energy is minimized by means of an unnormalized *Boltzmann–Gibbs measure* that is defined in terms of V and an inverse temperature $\beta > 0$ by

$$g(dx) = \exp\left(-\beta V(x)\right) \lambda(dx), \tag{13.1}$$

where λ is the Lebesgue measure. This measure has the property that the probability mass concentrates at the global minimum of V as $\beta \to \infty$.

The key idea behind simulated annealing is taking a random walk through the search space while β is successively increased. The probability of accepting a new value in the space is given by the Boltzmann–Gibbs distribution. While values with less energy than the current value are accepted with probability one, the probability that values with higher energy are accepted decreases as β increases. Other related approaches are fast simulated annealing [30] using a Cauchy–Lorentz distribution and generalized simulated annealing [32] based on Tsallis statistics.

Interacting particle systems [19] approximate a distribution of interest by a finite number of weighted random variables $X^{(i)}$ called particles. Provided that the weights $\Pi^{(i)}$ are normalized such that $\sum \Pi^{(i)} = 1$, the set of weighted particles determines a random probability measures by

$$\sum_{i=1}^{n} \Pi^{(i)} \delta_{X^{(i)}}. \tag{13.2}$$

Depending on the weighting function and the distribution of the particles, the measure converges to a distribution η as n tends to infinity. When the particles are identically independently distributed according to η and uniformly weighted, i.e. $\Pi^{(i)} = 1/n$, the convergence follows directly from the law of large numbers [3].

Interacting particle systems are mostly known in computer vision as particle filter [10] where they are applied for solving nonlinear, non-Gaussian

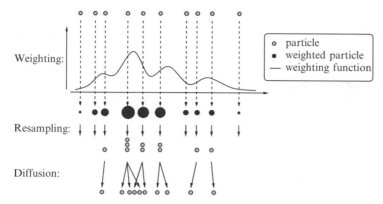

Fig. 13.1. Operation of an interacting particle system. After weighting the particles (*black circles*), the particles are resampled and diffused (*gray circles*).

filtering problems. However, these systems also apply for trapping analysis, evolutionary algorithms, statistics [19], and optimization as we demonstrate in this chapter. They usually consist of two steps as illustrated in Figure 13.1. During a selection step, the particles are weighted according to a weighting function and then resampled with respect to their weights, where particles with a great weight generate more offspring than particles with lower weight. In a second step, the particles mutate or are diffused.

13.1.3 Interaction and Annealing

Simulated annealing approaches are designed for global optimization, i.e., for searching the global optimum in the entire search space. Since they are not capable of focusing the search on some regions of interest in dependency on the previous visited values, they are not suitable for tasks in human motion capturing. Our approach, in contrast, is based on an interacting particle system that uses Boltzmann–Gibbs measures (13.1) similar to simulated annealing. This combination ensures not only the annealing property as we will show, but also exploits the distribution of the particles in the space as measure for the uncertainty in an estimate. The latter allows an automatic adaption of the search on regions of interest during the optimization process. The principle of the annealing effect is illustrated in Figure 13.2.

A first attempt to fuse interaction and annealing strategies for human motion capturing has become known as annealed particle filter [9]. Even though the heuristic is not based on a mathematical background, it already indicates the potential of such combination. Indeed, the annealed particle filter can be regarded as a special case of interacting simulated annealing where the particles are predicted for each frame by a stochastic process, see Section 13.3.1.

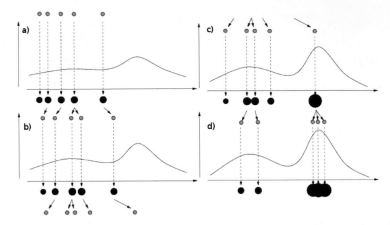

Fig. 13.2. Illustration of the annealing effect with three runs. Due to annealing, the particles migrate towards the global maximum without getting stuck in the local maximum.

13.1.4 Outline

The interacting annealing algorithm is introduced in Section 13.3.1 and its asymptotic behavior is discussed in Section 13.3.2. The given convergence results are based on Feynman–Kac models [19] which are outlined in Section 13.2. Since a general treatment including proofs is out of the scope of this introduction, we refer the interested reader to [11] or [19]. While our approach is evaluated for a standard global optimization problem in Section 13.4.1, Section 13.4.2 demonstrates the performance of interacting simulated annealing in a complete marker-less human motion capture system that includes segmentation, pose prediction, and prior knowledge.

13.1.5 Notations

We always regard E as a subspace of R^d, and let $\mathcal{B}(E)$ denote its Borel σ-algebra. $B(E)$ denotes the set of bounded measurable functions, δ_x is the Dirac measure concentrated in $x \in E$, $\| \cdot \|_2$ is the Euclidean norm, and $\| \cdot \|_\infty$ is the well-known supremum norm. Let $f \in B(E)$, μ be a measure on E, and let K be a Markov kernel on E.[1] We write

$$\langle \mu, f \rangle = \int_E f(x) \, \mu(dx), \quad \langle \mu, K \rangle(B) = \int_E K(x, B) \, \mu(dx) \quad \text{for } B \in \mathcal{B}(E).$$

---

[1] A Markov kernel is a function $K : E \times \mathcal{B}(E) \rightarrow [0, \infty]$ such that $K(\cdot, B)$ is $\mathcal{B}(E)$-measurable $\forall B$ and $K(x, \cdot)$ is a probability measure $\forall x$. An example of a Markov kernel is given in Equation (13.12). For more details on probability theory and Markov kernels, we refer to [3].

Furthermore, $U[0,1]$ denotes the uniform distribution on the interval $[0,1]$ and

$$\operatorname{osc}(\varphi) := \sup_{x,y \in E} \{|\varphi(x) - \varphi(y)|\}. \tag{13.3}$$

is an upper bound for the oscillations of f.

13.2 Feynman–Kac Model

Let $(X_t)_{t \in \mathbb{N}_0}$ be an E-valued Markov process with family of transition kernels $(K_t)_{t \in \mathbb{N}_0}$ and initial distribution η_0. We denote by P_{η_0} the distribution of the Markov process, i.e., for $t \in \mathbb{N}_0$,

$$P_{\eta_0}\left(d(x_0, x_1, \ldots, x_t)\right) = K_{t-1}(x_{t-1}, dx_t) \ldots K_0(x_0, dx_1)\, \eta_0(dx_0),$$

and by $E_{\eta_0}[\cdot]$ the expectation with respect to P_{η_0}. The sequence of distributions $(\eta_t)_{t \in \mathbb{N}_0}$ on E defined for any $\varphi \in B(E)$ and $t \in \mathbb{N}_0$ as

$$\langle \eta_t, \varphi \rangle := \frac{\langle \gamma_t, \varphi \rangle}{\langle \gamma_t, 1 \rangle}, \qquad \langle \gamma_t, \varphi \rangle := E_{\eta_0}\left[\varphi(X_t) \exp\left(-\sum_{s=0}^{t-1} \beta_s V(X_s)\right)\right],$$

is called the *Feynman–Kac model* associated with the pair $(\exp(-\beta_t V), K_t)$. The Feynman–Kac model as defined above satisfies the recursion relation

$$\eta_{t+1} = \langle \Psi_t(\eta_t), K_t \rangle, \tag{13.4}$$

where the *Boltzmann–Gibbs transformation* Ψ_t is defined by

$$\Psi_t(\eta_t)(dy_t) = \frac{E_{\eta_0}\left[\exp\left(-\sum_{s=0}^{t-1} \beta_s V(X_s)\right)\right]}{E_{\eta_0}\left[\exp\left(-\sum_{s=0}^{t} \beta_s V(X_s)\right)\right]} \exp\left(-\beta_t V_t(y_t)\right) \eta_t(dy_t).$$

The particle approximation of the flow (13.4) depends on a chosen family of Markov transition kernels $(K_{t,\eta_t})_{t \in \mathbb{N}_0}$ satisfying the compatibility condition

$$\langle \Psi_t(\eta_t), K_t \rangle := \langle \eta_t, K_{t,\eta_t} \rangle.$$

A family $(K_{t,\eta_t})_{t \in \mathbb{N}_0}$ of kernels is not uniquely determined by these conditions. As in [19, Chapter 2.5.3], we choose

$$K_{t,\eta_t} = S_{t,\eta_t} K_t, \tag{13.5}$$

where

$$\begin{aligned}
S_{t,\eta_t}(x_t, dy_t) = {}& \epsilon_t \exp\left(-\beta_t V_t(x_t)\right) \delta_{x_t}(dy_t) \\
& + (1 - \epsilon_t \exp\left(-\beta_t V_t(x_t)\right)) \Psi_t(\eta_t)(dy_t), \quad (13.6)
\end{aligned}$$

with $\epsilon_t \geq 0$ and $\epsilon_t \|\exp(-\beta_t V)\|_\infty \leq 1$. The parameters ϵ_t may depend on the current distribution η_t.

13.3 Interacting Simulated Annealing

Similar to simulated annealing, one can define an annealing scheme $0 \leq \beta_0 \leq \beta_1 \leq \ldots \leq \beta_t$ in order to search for the global minimum of an energy function V. Under some conditions that will be stated in Section 13.3.2, the flow of the Feynman–Kac distribution becomes concentrated in the region of global minima of V as t goes to infinity. Since it is not possible to sample from the distribution directly, the flow is approximated by a particle set as it is done by a particle filter. We call the algorithm for the flow approximation *interacting simulated annealing (ISA)*.

13.3.1 Algorithm

The particle approximation for the Feynman–Kac model is completely described by the Equation (13.5). The particle system is initialized by n identically, independently distributed random variables $X_0^{(i)}$ with common law η_0 determining the random probability measure $\eta_0^n := \sum_{i=1}^n \delta_{X_0^{(i)}}/n$. Since K_{t,η_t} can be regarded as the composition of a pair of selection and mutation Markov kernels, we split the transitions into the following two steps

$$\eta_t^n \xrightarrow{\ Selection\ } \check{\eta}_t^n \xrightarrow{\ Mutation\ } \eta_{t+1}^n,$$

where

$$\eta_t^n := \frac{1}{n} \sum_{i=1}^n \delta_{X_t^{(i)}}, \qquad \check{\eta}_t^n := \frac{1}{n} \sum_{i=1}^n \delta_{\check{X}_t^{(i)}}.$$

During the selection step each particle $X_t^{(i)}$ evolves according to the Markov transition kernel $S_{t,\eta_t^n}(X_t^{(i)}, \cdot)$. That means $X_t^{(i)}$ is accepted with probability

$$\epsilon_t \exp(-\beta_t V(X_t^{(i)})), \tag{13.7}$$

and we set $\check{X}_t^{(i)} = X_t^{(i)}$. Otherwise, $\check{X}_t^{(i)}$ is randomly selected with distribution

$$\sum_{i=1}^n \frac{\exp(-\beta_t V(X_t^{(i)}))}{\sum_{j=1}^n \exp(-\beta_t V(X_t^{(j)}))} \delta_{X_t^{(i)}}.$$

The mutation step consists in letting each selected particle $\check{X}_t^{(i)}$ evolve according to the Markov transition kernel $K_t(\check{X}_t^{(i)}, \cdot)$.

There are several ways to choose the parameter ϵ_t of the *selection kernel* (13.6) that defines the resampling procedure of the algorithm, cf. [19]. If

$$\epsilon_t := 0 \qquad \forall t, \tag{13.8}$$

the selection can be done by multinomial resampling. Provided that[2]

---

[2] The inequality satisfies the condition $\epsilon_t \|\exp(-\beta_t V)\|_\infty \leq 1$ for Equation (13.6).

Algorithm 6 Interacting Simulated Annealing Algorithm

Requires: parameters $(\epsilon_t)_{t\in\mathbb{N}_0}$, number of particles n, initial distribution η_0, energy function V, annealing scheme $(\beta_t)_{t\in\mathbb{N}_0}$ and transitions $(K_t)_{t\in\mathbb{N}_0}$

1. *Initialization*
   - Sample $x_0^{(i)}$ from η_0 for all i
2. *Selection*
   - Set $\pi^{(i)} \leftarrow \exp(-\beta_t V(x_t^{(i)}))$ for all i
   - For i from 1 to n:
     Sample κ from $U[0,1]$
     If $\kappa \leq \epsilon_t \pi^{(i)}$ then
     $\star$ Set $\check{x}_t^{(i)} \leftarrow x_t^{(i)}$
     Else
     $\star$ Set $\check{x}_t^{(i)} \leftarrow x_t^{(j)}$ with probability $\dfrac{\pi^{(j)}}{\sum_{k=1}^n \pi^{(k)}}$
3. *Mutation*
   - Sample $x_{t+1}^{(i)}$ from $K_t(\check{x}_t^{(i)}, \cdot)$ for all i and go to step 2

$$n \geq \sup_t \left(\exp(\beta_t \, \mathrm{osc}(V))\right),$$

another selection kernel is given by

$$\epsilon_t(\eta_t) := \frac{1}{n \, \langle \eta_t, \exp(-\beta_t V)\rangle}. \tag{13.9}$$

In this case the expression $\epsilon_t \pi^{(i)}$ in Algorithm 6 is replaced by $\pi^{(i)} / \sum_{k=1}^n \pi^{(k)}$. A third kernel is determined by

$$\epsilon_t(\eta_t) := \frac{1}{\inf\{y \in \mathbb{R} \, : \, \eta_t(\{x \in E \, : \, \exp(-\beta_t V(x)) > y\}) = 0\}}, \tag{13.10}$$

yielding the expression $\pi^{(i)} / \max_{1\leq k\leq n} \pi^{(k)}$ instead of $\epsilon_t \pi^{(i)}$.

Pierre del Moral showed in [19, Chapter 9.4] that for any $t \in \mathbb{N}_0$ and $\varphi \in B(E)$ the sequence of random variables

$$\sqrt{n}(\langle \eta_t^n, \varphi \rangle - \langle \eta_t, \varphi \rangle)$$

converges in law to a Gaussian random variable W when the selection kernel (13.6) is used to approximate the flow (13.4). Moreover, it turns out that when (13.9) is chosen, the variance of W is strictly smaller than in the case with $\epsilon_t = 0$.

We remark that the annealed particle filter [9] relies on interacting simulated annealing with $\epsilon_t = 0$. The operation of the method is illustrated by

$$\eta_t^n \xrightarrow{\ Prediction\ } \hat{\eta}_{t+1}^n \xrightarrow{\quad ISA\quad } \eta_{t+1}^n.$$

The *ISA* is initialized by the predicted particles $\hat{X}_{t+1}^{(i)}$ and performs M times the selection and mutation steps. Afterwards the particles $X_{t+1}^{(i)}$ are obtained by an additional selection. This shows that the annealed particle filter uses a simulated annealing principle to locate the global minimum of a function V at each time step.

13.3.2 Convergence

This section discusses the asymptotic behavior of the interacting simulated annealing algorithm. For this purpose, we introduce some definitions in accordance with [19] and [15].

Definition 1. *A kernel K on E is called mixing if there exists a constant $0 < \varepsilon < 1$ such that*

$$K(x_1, \cdot) \geq \varepsilon\, K(x_2, \cdot) \quad \forall x_1,\, x_2 \in E. \qquad (13.11)$$

The condition can typically only be established when $E \subset \mathbb{R}^d$ is a bounded subset, which is the case in many applications like human motion capturing. For example the (bounded) Gaussian distribution on E

$$K(x, B) := \frac{1}{Z} \int_B \exp\left(-\frac{1}{2}\, (x - y)^T\, \Sigma^{-1}\, (x - y) \right) dy, \qquad (13.12)$$

where $Z := \int_E \exp(-\frac{1}{2}\, (x - y)^T\, \Sigma^{-1}\, (x - y))\, dy$, is mixing if and only if E is bounded. Moreover, a Gaussian with a high variance satisfies the mixing condition with a larger ε than a Gaussian with lower variance.

Definition 2. *The Dobrushin contraction coefficient of a kernel K on E is defined by*

$$\beta(K) := \sup_{x_1, x_2 \in E}\; \sup_{B \in \mathcal{B}(E)}\; |K(x_1, B) - K(x_2, B)|. \qquad (13.13)$$

Furthermore, $\beta(K) \in [0, 1]$ and $\beta(K_1 K_2) \leq \beta(K_1)\,\beta(K_2)$.

When the kernel M is a composition of several mixing Markov kernels, i.e., $M := K_s K_{s+1} \ldots K_t$, and each kernel K_k satisfies the mixing condition for some ε_k, the Dobrushin contraction coefficient can be estimated by $\beta(M) \leq \prod_{k=s}^{t} (1 - \varepsilon_k)$.

The asymptotic behavior of the interacting simulated annealing algorithm is affected by the convergence of the flow of the Feynman–Kac distribution (13.4) to the region of global minima of V as t tends to infinity and by the convergence of the particle approximation to the Feynman–Kac distribution at each time step t as the number of particles n tends to infinity.

Convergence of the Flow

We suppose that $K_t = K$ is a Markov kernel satisfying the mixing condition (13.11) for an $\varepsilon \in (0, 1)$ and $\mathrm{osc}(V) < \infty$. A time mesh is defined by

$$t(n) := n(1 + \lfloor c(\varepsilon) \rfloor) \quad c(\varepsilon) := (1 - \ln(\varepsilon/2))/\varepsilon^2 \quad \text{for } n \in \mathbb{N}_0. \tag{13.14}$$

Let $0 \leq \beta_0 \leq \beta_1 \ldots$ be an annealing scheme such that $\beta_t = \beta_{t(n+1)}$ is constant in the interval $(t(n), t(n + 1)]$. Furthermore, we denote by $\breve{\eta}_t$ the Feynman–Kac distribution after the selection step, i.e. $\breve{\eta}_t = \Psi_t(\eta_t)$. According to [19, Proposition 6.3.2], we have

Theorem 1. *Let $b \in (0, 1)$ and $\beta_{t(n+1)} = (n + 1)^b$. Then for each $\delta > 0$*

$$\lim_{n \to \infty} \breve{\eta}_{t(n)} \left(V \geq V_\star + \delta \right) = 0,$$

where $V_\star = \sup\{v \geq 0; \ V \geq v \ a.e.\}$.

The rate of convergence is $d/n^{(1-b)}$ where d is increasing with respect to b and $c(\varepsilon)$ but does not depend on n as given in [19, Theorem 6.3.1]. This theorem establishes that the flow of the Feynman–Kac distribution $\breve{\eta}_t$ becomes concentrated in the region of global minima as $t \to +\infty$.

Convergence of the Particle Approximation

Del Moral established the following convergence theorem [19, Theorem 7.4.4].

Theorem 2. *For any $\varphi \in B(E)$,*

$$E_{\eta_0} \left[|\langle \eta^n_{t+1}, \varphi \rangle - \langle \eta_{t+1}, \varphi \rangle| \right] \leq \frac{2 \, \mathrm{osc}(\varphi)}{\sqrt{n}} \left(1 + \sum_{s=0}^{t} r_s \beta(M_s) \right),$$

where

$$r_s := \exp \left(\mathrm{osc}(V) \sum_{r=s}^{t} \beta_r \right),$$

$$M_s := K_s K_{s+1} \ldots K_t,$$

for $0 \leq s \leq t$.

Assuming that the kernels K_s satisfy the mixing condition with ε_s, we get a rough estimate for the number of particles

$$n \geq \frac{4 \, \mathrm{osc}(\varphi)^2}{\delta^2} \left(1 + \sum_{s=0}^{t} \left\{ \exp \left(\mathrm{osc}(V) \sum_{r=s}^{t} \beta_r \right) \prod_{k=s}^{t} (1 - \varepsilon_k) \right\} \right)^2 \tag{13.15}$$

needed to achieve a mean error less than a given $\delta > 0$.

Optimal Transition Kernel

The mixing condition is not only essential for the convergence result of the flow as stated in Theorem 1 but also influences the time mesh by the parameter ε. In view of Equation (13.14), kernels with ε close to 1 are preferable, e.g., Gaussian kernels on a bounded set with a very high variance. The right-hand side of (13.15) can also be minimized if Markov kernels K_s are chosen such that the mixing condition is satisfied for a ε_s close to 1, as shown in Figure 13.3. However, we have to consider two facts. First, the inequality in Theorem 2 provides an upper bound of the accumulated error of the particle approximation up to time $t + 1$. It is clear that the accumulation of the error is reduced when the particles are highly diffused, but it also means that the information carried by the particles from the previous time steps is mostly lost by the mutation. Secondly, we cannot sample from the measure $\check{\eta}_t$ directly, instead we approximate it by n particles. Now the following problem arises. The mass of the measure concentrates on a small region of E on one hand and, on the other hand, the particles are spread over E if ε is large. As a result we get a degenerated system where the weights of most of the particles are zero and thus the global minima are estimated inaccurately, particularly for small n. If we choose a kernel with small ε in contrast, the convergence rate of the flow is very slow. Since neither of them is suitable in practice, we suggest a *dynamic variance scheme* instead of a fixed kernel K.

It can implemented by Gaussian kernels K_t with covariance matrices Σ_t proportional to the sample covariance after resampling. That is, for a constant $c > 0$,

$$\Sigma_t := \frac{c}{n-1} \sum_{i=1}^{n} (x_t^{(i)} - \mu_t)_\rho \, (x_t^{(i)} - \mu_t)_\rho^T, \qquad \mu_t := \frac{1}{n} \sum_{i=1}^{n} x_t^{(i)}, \qquad (13.16)$$

where $((x)_\rho)_k = \max(x_k, \rho)$ for a $\rho > 0$. The value ρ ensures that the variance does not become zero. The elements off the diagonal are usually set to zero, in order to reduce computation time.

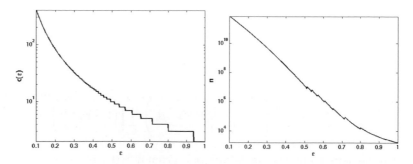

Fig. 13.3. Impact of the mixing condition satisfied for $\varepsilon_s = \varepsilon$. **Left:** Parameter $c(\varepsilon)$ of the time mesh (13.14). **Right:** Rough estimate for the number of particles needed to achieve a mean error less than $\delta = 0.1$.

Optimal Parameters

The computation cost of the interacting simulated annealing algorithm with n particles and T annealing runs is $O(n_T)$, where

$$n_T := n \cdot T. \qquad (13.17)$$

While more particles give a better particle approximation of the Feynman–Kac distribution, the flow becomes more concentrated in the region of global minima as the number of annealing runs increases. Therefore, finding the optimal values is a trade-off between the convergence of the flow and the convergence of the particle approximation provided that n_T is fixed.

Another important parameter of the algorithm is the annealing scheme. The scheme given in Theorem 1 ensures convergence for any energy function V – even for the worst one in the sense of optimization – as long as $\mathrm{osc}(V) < \infty$ but is too slow for most applications, as it is the case for simulated annealing. In our experiments the schemes

$$
\begin{aligned}
\beta_t &= \ln(t + b) \quad \text{for some } b > 1 \quad (logarithmic), & (13.18) \\
\beta_t &= (t + 1)^b \quad \text{for some } b \in (0, 1) \quad (polynomial) & (13.19)
\end{aligned}
$$

performed well. Note that in contrast to the time mesh (13.14) the schemes are not anymore constant on a time interval.

Even though a complete evaluation of the various parameters is out of the scope of this introduction, the examples given in the following section demonstrate settings that perform well, in particular for human motion capturing.

13.4 Examples

13.4.1 Global Optimization

The Ackley function [1, 2]

$$f(x) = -20 \exp\left(-0.2 \sqrt{\frac{1}{d} \sum_{i=1}^{d} x_i^2}\right) - \exp\left(\frac{1}{d} \sum_{i=1}^{d} \cos(2\pi x_i)\right) + 20 + e$$

is a widely used multimodal test function for global optimization algorithms. As one can see from Figure 13.4, the function has a global minimum at $(0, 0)$ that is surrounded by several local minima. The problem consists of finding the global minimum in a bounded subspace $E \subset \mathbb{R}^d$ with an error less than a given $\delta > 0$ where the initial distribution is the uniform distribution on E.

In our experiments, the maximal number of time steps were limited by 999, and we set $E = [-4, 4] \times [-4, 4]$ and $\delta = 10^{-3}$. The interacting simulated annealing algorithm was stopped when the Euclidean distance between the

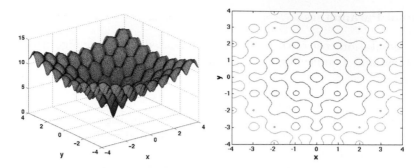

Fig. 13.4. Ackley function. Unique global minimum at $(0,0)$ with several local minima around it.

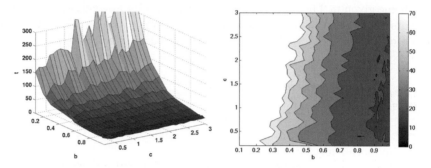

Fig. 13.5. Average time steps needed to find global minimum with error less than 10^{-3} with respect to the parameters b and c.

global minimum and its estimate was less than δ or when the limit of time steps was exceeded. All simulations were repeated 50 times and the average number of time steps needed by ISA was used for evaluating the performance of the algorithm. Depending on the chosen selection kernel (13.8), (13.9), and (13.10), we write ISA_{S1}, ISA_{S2}, and ISA_{S3}, respectively.

Using a polynomial annealing scheme (13.19), we evaluated the average time steps needed by the ISA_{S1} with 50 particles to find the global minimum of the Ackley function. The results with respect to the parameter of the annealing scheme, $b \in [0.1, 0.999]$, and the parameter of the dynamic variance scheme, $c \in [0.1, 3]$, are given in Figure 13.5. The algorithm performed best with a fast increasing annealing scheme, i.e., $b > 0.9$, and with c in the range 0.5–1.0. The plots in Figure 13.5 also reveal that the annealing scheme has greater impact on the performance than the factor c. When the annealing scheme increases slowly, i.e., $b < 0.2$, the global minimum was actually not located within the given limit for all 50 simulations.

The best results with parameters b and c for ISA_{S1}, ISA_{S2}, and ISA_{S3} are listed in Table 13.1. The optimal parameters for the three selection kernels are quite similar and the differences of the average time steps are marginal.

Table 13.1. Parameters b and c with lowest average time t for different selection kernels.

	Ackley			Ackley with noise		
	ISA_{S1}	ISA_{S2}	ISA_{S3}	ISA_{S1}	ISA_{S2}	ISA_{S3}
b	0.993	0.987	0.984	0.25	0.35	0.27
c	0.8	0.7	0.7	0.7	0.7	0.9
t	14.34	15.14	14.58	7.36	7.54	7.5

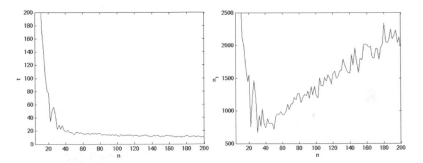

Fig. 13.6. Left: Average time steps needed to find global minimum with respect to number of particles. **Right:** Computation cost.

Table 13.2. Number of particles with lowest average computation cost for different selection kernels.

	Ackley			Ackley with noise		
	ISA_{S1}	ISA_{S2}	ISA_{S3}	ISA_{S1}	ISA_{S2}	ISA_{S3}
n	30	30	28	50	50	26
t	22.4	20.3	21.54	7.36	7.54	12.54
n_t	672	609	603.12	368	377	326.04

In a second experiment, we fixed the parameters b and c, where we used the values from Table 13.1, and varied the number of particles in the range 4–200 with step size 2. The results for ISA_{S1} are shown in Figure 13.6. While the average of time steps declines rapidly for $n \leq 20$, it is hardly reduced for $n \geq 40$. Hence, n_t and thus the computation cost are lowest in the range 20–40. This shows that a minimum number of particles are required to achieve a success rate of 100%, i.e., the limit was not exceeded for all simulations. In this example, the success rate was 100% for $n \geq 10$. Furthermore, it indicates that the average of time steps is significantly higher for n less than the optimal number of particles. The results for ISA_{S1}, ISA_{S2}, and ISA_{S3} are quite similar. The best results are listed in Table 13.2.

The ability of dealing with noisy energy functions is one of the strength of ISA as we will demonstrate. This property is very usefull for applications

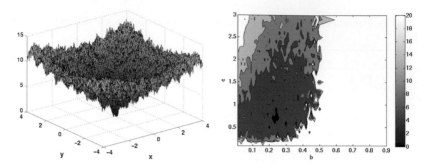

Fig. 13.7. **Left:** Ackley function distorted by Gaussian noise with standard deviation 0.5. **Right:** Average time steps needed to find global minimum with error less than 10^{-2} with respect to the parameters b and c.

where the measurement of the energy of a particle is distorted by noise. On the left-hand side of Figure 13.7, the Ackley function is distorted by Gaussian noise with standard deviation 0.5, i.e.,

$$f_W(x) := \max\{0, f(x) + W\}, \qquad W \sim N(0, 0.5^2).$$

As one can see, the noise deforms the shape of the function and changes the region of global minima. In our experiments, the ISA was stopped when the true global minimum at $(0, 0)$ was found with an accuracy of $\delta = 0.01$.

For evaluating the parameters b and c, we set $n = 50$. As shown on the right hand side of Figure 13.7, the best results were obtained by annealing schemes with $b \in [0.22, 0.26]$ and $c \in [0.6, 0.9]$. In contrast to the undistorted Ackley function, annealing schemes that increase slowly performed better than the fast one. Indeed, the success rate dropped below 100% for $b \geq 0.5$. The reason is obvious from the left-hand side of Figure 13.7. Due to the noise, the particles are more easily distracted and a fast annealing scheme diminishes the possibility of escaping from the local minima. The optimal parameters for the dynamic variance scheme are hardly affected by the noise.

The best parameters for ISA_{S1}, ISA_{S2}, and ISA_{S3} are listed in the Tables 13.1 and 13.2. Except for ISA_{S3}, the optimal number of particles is higher than it is the case for the simulations without noise. The minimal number of particles to achieve a success rate of 100% also increased, e.g., 28 for ISA_{S1}. We remark that ISA_{S3} required the least number of particles for a complete success rate, namely 4 for the undistorted energy function and 22 in the noisy case.

We finish this section by illustrating two examples of energy function where the dynamic variance schemes might not be suitable. On the left-hand side of Figure 13.8, an energy function with shape similar to the Ackley function is drawn. The dynamic variance schemes perform well for this type of function with an unique global minimum with several local minima around it. Due to the scheme, the search focuses on the region near the global minimum after

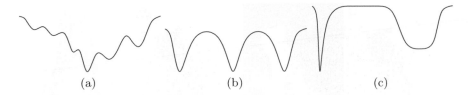

(a)   (b)   (c)

Fig. 13.8. Different cases of energy functions. **(a)** Optimal for dynamic variance schemes. An unique global minimum with several local minima around it. **(b)** Several global minima that are widely separated. This yields a high variance even in the case that the particles are near to the global minima. **(c)** The global minimum is a small peak far away from a broad basin. When all particles fall into the basin, the dynamic variance schemes focus the search on the basin.

some time steps. The second function, see Figure 8(b), has several, widely separated global minima yielding a high variance of the particles even in the case that the particles are near to the global minima. Moreover, when the region of global minima is regarded as a sum of Dirac measures, the mean is not essentially a global minimum. In the last example shown on the right hand side of Figure 13.8, the global minimum is a small peak far away from a broad basin with a local minimum. When all particles fall into the basin, the dynamic variance schemes focus the search on the region near the local minimum and it takes a long time to discover the global minimum.

In most optimization problems arising in the field of computer vision, however, the first case occurs where the dynamic variance schemes perform well. One application is human motion capturing which we will discuss in the next section.

13.4.2 Human Motion Capture

In our second experiment, we apply the interacting simulated annealing algorithm to model-based 3D tracking of the lower part of a human body, see Figure 13.9(a). This means that the 3D *rigid body motion (RBM)* and the joint angles, also called the *pose*, are estimated by exploiting the known 3D model of the tracked object. The mesh model illustrated in Figure 13.9(d) has 18 *degrees of freedom (DoF)*, namely 6 for the rigid body motion and 12 for the joint angles of the hip, knees, and feet. Although a marker-less motion capture system is discussed, markers are also sticked to the target object in order to provide a quantitative comparison with a commercial marker-based system.

Using the extracted silhouette as shown in Figure 13.9(b), one can define an energy function V which describes the difference between the silhouette and an estimated pose. The pose that fits the silhouette best takes the global minimum of the energy function, which is searched by the ISA. The estimated pose projected onto the image plane is displayed in Figure 13.9(c).

Fig. 13.9. From left to right: *(a)* Original image. *(b)* Silhouette. *(c)* Estimated pose. *(d)* 3D model.

Pose Representation

There are several ways to represent the pose of an object, e.g., Euler angles, quaternions [16], twists [20], or the axis-angle representation. The *ISA* requires from the representation that primarily the mean but also the variance can be at least well approximated. For this purpose, we have chosen the axis-angle representation of the absolute rigid body motion M given by the 6D vector $(\theta\omega, t)$ with

$$\omega = (\omega_1, \omega_2, \omega_3), \quad \|\omega\|_2 = 1 \quad \text{and} \quad t = (t_1, t_2, t_3).$$

Using the exponential, M is expressed by

$$M = \begin{pmatrix} \exp(\theta\hat{\omega}) & t \\ 0 & 1 \end{pmatrix}, \qquad \hat{\omega} = \begin{pmatrix} 0 & -\omega_3 & \omega_2 \\ \omega_3 & 0 & -\omega_1 \\ -\omega_2 & \omega_1 & 0 \end{pmatrix}. \tag{13.20}$$

While t is the absolute position in the world coordinate system, the rotation vector $\theta\omega$ describes a rotation by an angle $\theta \in \mathbb{R}$ about the rotation axis ω. The function $\exp(\theta\hat{\omega})$ can be efficiently computed by the Rodriguez formula [20].

Given a rigid body motion defined by a rotation matrix $R \in SO(3)$ and a translation vector $t \in \mathbb{R}^3$, the rotation vector is constructed according to [20] as follows: When R is the identity matrix, θ is set to 0. For the other case, θ and the rotation axis ω are given by

$$\theta = \cos^{-1}\left(\frac{trace(R) - 1}{2}\right), \qquad \omega = \frac{1}{2\sin(\theta)} \begin{pmatrix} r_{32} - r_{23} \\ r_{13} - r_{31} \\ r_{21} - r_{12} \end{pmatrix}. \tag{13.21}$$

We write $\log(R)$ for the inverse mapping of the exponential.

The mean of a set of rotations r_i in the axis-angle representation can be computed by using the exponential and the logarithm as described in [22,23]. The idea is to find a geodesic on the Riemannian manifold determined by the set of 3D rotations. When the geodesic starting from the mean rotation in the manifold is mapped by the logarithm onto the tangent space at the mean, it is a straight line starting at the origin. The tangent space is called *exponential chart*.

Hence, using the notations

$$r_2 \star r_1 = \log\left(\exp(r_2) \cdot \exp(r_1)\right), \qquad r_1^{-1} = \log\left(\exp(r_1)^T\right)$$

for the rotation vectors r_1 and r_2, the mean rotation $\bar{r}$ satisfies

$$\sum_i \left(\bar{r}^{-1} \star r_i\right) = 0. \tag{13.22}$$

Weighting each rotation with π_i, yields the least squares problem:

$$\frac{1}{2}\sum_i \pi_i \left\|\bar{r}^{-1} \star r_i\right\|_2^2 \rightarrow min. \tag{13.23}$$

The weighted mean can thus be estimated by

$$\hat{r}_{t+1} = \hat{r}_t \star \left(\frac{\sum_i \pi_i \left(\hat{r}_t^{-1} \star r_i\right)}{\sum_i \pi_i}\right). \tag{13.24}$$

The gradient descent method takes about 5 iterations until it converges.

The variance and the normal density on a Riemannian manifold can also be approximated, cf. [24]. Since, however, the variance is only used for diffusing the particles, a very accurate approximation is not needed. Hence, the variance of a set of rotations r_i is calculated in the Euclidean space.[3]

The twist representation used in [7,26] and in Chapters 11 and 12 is quite similar. Instead of a separation between the translation t and the rotation r, it describes a screw motion where the motion velocity θ also affects the translation. A twist $\hat{\xi} \in se(3)$ is represented by

$$\theta\hat{\xi} = \theta \begin{pmatrix} \hat{\omega} & v \\ 0 & 0 \end{pmatrix}, \tag{13.25}$$

where $\exp(\theta\hat{\xi})$ is a rigid body motion.

The logarithm of a rigid body motion $M \in SE(3)$ is the following transformation:

$$\theta\omega = \log(R), \qquad v = A^{-1}t, \tag{13.26}$$

where

$$A = (I - \exp(\theta\hat{\omega}))\hat{\omega} + \omega\omega^T\theta \tag{13.27}$$

is obtained from the Rodriguez formula. This follows from the fact, that the two matrices which comprise A have mutually orthogonal null spaces when $\theta \neq 0$. Hence, $Av = 0 \Leftrightarrow v = 0$.

We remark that the two representations are identical for the joints where only a rotation around a known axis is performed. Furthermore, a linearization is not needed for the *ISA* in contrast to the pose estimation as described in Chapters 11 and 12.

---

[3] $se(3)$ is the Lie algebra that corresponds to the Lie group $SE(3)$.

Pose Prediction

The *ISA* can be combined with a pose prediction in two ways. When the dynamics are modelled by a Markov process for example, the particles of the current frame can be stored and predicted for the next frame according to the process as done in [12]. In this case, the *ISA* is already initialized by the predicted particles. But when the prediction is time consuming or when the history of previous poses is needed, only the estimate is predicted. The *ISA* is then initialized by diffusing the particles around the predicted estimate. The reinitialization of the particles is necessary for example when the prediction is based on local descriptors [13] or optical flow as discussed in Chapter 11 and [5].

In our example, the pose is predicted by an autoregression that takes the global rigid body motions P_i of the last N frames into account [13]. For this purpose, we use a set of twists $\xi_i = \log(P_i P_{i-1}^{-1})$ representing the relative motions. By expressing the local rigid body motion as a screw action, the spatial velocity can be represented by the twist of the screw, see [20] for details.

In order to generate a suited rigid body motion from the motion history, a screw motion needs to be represented with respect to another coordinate system. Let $\hat{\xi} \in se(3)$ be a twist given in a coordinate frame A. Then for any $G \in SE(3)$, which transforms a coordinate frame A to B, is $G\hat{\xi}G^{-1}$ a twist with the twist coordinates given in the coordinate frame B, see [20] for details. The mapping $\hat{\xi} \longmapsto G\hat{\xi}G^{-1}$ is called the *adjoint transformation* associated with G.

Let $\xi_1 = \log(P_2 P_1^{-1})$ be the twist representing the relative motion from P_1 to P_2. This transformation can be expressed as local transformation in the current coordinate system M_1 by the adjoint transformation associated with $G = M_1 P_1^{-1}$. The new twist is then given by $\hat{\xi}_1' = G\hat{\xi}_1 G^{-1}$. The advantage of the twist representation is now, that the twists can be scaled by a factor

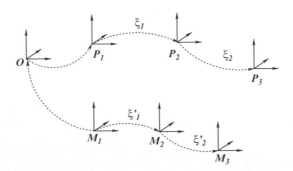

Fig. 13.10. Transformation of rigid body motions from prior data P_i in a current world coordinate system M_1. A proper scaling of the twists results in a proper damping.

$0 \leq \lambda_i \leq 1$ to damp the local rigid body motion, i.e. $\hat{\xi}_i' = G\lambda_i\hat{\xi}_iG^{-1}$. For given λ_i such that $\sum_i \lambda_i = 1$, the predicted pose is obtained by the rigid body transformation

$$\exp(\hat{\xi}_N')\exp(\hat{\xi}_{N-1}')\dots\exp(\hat{\xi}_1'). \tag{13.28}$$

Energy Function

The energy function V of a particle x, which is used for our example, depends on the extracted silhouette and on some learned prior knowledge as in [12], but it is defined in a different way.

Silhouette: First of all, the silhouette is extracted from an image by a level set-based segmentation as in [8, 27]. We state the energy functional E for convenience only and refer the reader to chapter 11 where the segmentation is described in detail. Let Ω^i be the image domain of view i and let $\Phi_0^i(\hat{x})$ be the contour of the predicted pose in Ω^i. In order to obtain the silhouettes for all r views, the energy functional $E(\hat{x}, \Phi^1, \dots, \Phi^r) = \sum_{i=1}^r E(\hat{x}, \Phi^i)$ is minimzed, where

$$E(\hat{x}, \Phi^i) = -\int H(\Phi^i)\ln p_1^i + (1 - H(\Phi^i))\ln p_2^i \, dx$$
$$+ \nu \int_{\Omega^i} |\nabla H(\Phi^i)| \, dx + \lambda \int_{\Omega^i} \left(\Phi^i - \Phi_0^i(\hat{x})\right)^2 \, dx. \tag{13.29}$$

In our experiments, we weighted the smoothness term with $\nu = 4$ and the shape prior with $\lambda = 0.04$.

After the segmentation, 3D–2D correspondences between the 3D model (X_i) and a 2D image (x_i) are established by the projected vertices of the 3D mesh that are part of the model contour and their closest points of the extracted contour that are determined by a combination of an iterated closest point algorithm [4] and an optic flow-based shape registration [25]. More details about the shape matching are given in Chapter 12. We write each correspondence as pair (X_i, x_i) of homogeneous coordinates.

Each image point x_i defines a projection ray that can be represented as Plücker line [20] determined by a unique vector n_i and a moment m_i such that $x \times n_i - m_i = 0$ for all x on the 3D line. Furthermore,

$$\|x \times n_i - m_i\|_2 \tag{13.30}$$

is the norm of the perpendicular error vector between the line and a point $x \in \mathbb{R}^3$. As we already mentioned, a joint j is represented by the rotation angle θ_j. Hence, we write $M(\omega, t)$ for the rigid body motion and $M(\theta_j)$ for the joints. Furthermore, we have to consider the kinematic chain of articulated objects. Let X_i be a point on the limb k_i whose position is influenced by s_i joints in a certain order. The inverse order of these joints is then given by the mapping ι_{k_i}, e.g., a point on the left shank is influenced by the left knee joint $\iota_{k_i}(4)$ and by the three joints of the left hip $\iota_{k_i}(3)$, $\iota_{k_i}(2)$, and $\iota_{k_i}(1)$.

Hence, the pose estimation consists of finding a pose x such that the error

$$err_S(x,i) := \left\| \left(M(\omega,t) M(\theta_{\iota_{k_i}(1)}) \dots M(\theta_{\iota_{k_i}(s_i)}) X_i \right)_{3\times 1} \times n_i - m_i \right\|_2 \quad (13.31)$$

is minimal for all pairs, where $(\cdot)_{3\times 1}$ denotes the transformation from homogeneous coordinates back to nonhomogeneous coordinates.

Prior knowledge: Using prior knowledge about the probability of a certain pose can stabilize the pose estimation as shown in [12] and [6]. The prior ensures that particles representing a familiar pose are favored in problematic situations, e.g., when the observed object is partially occluded. As discussed in Chapter 11, the probability of the various poses is learned from N training samples, where the density is estimated by a Parzen–Rosenblatt estimator [21, 29] with a Gaussian kernel

$$p_{pose}(x) = \frac{1}{(2\pi\sigma^2)^{d/2} N} \sum_{i=1}^{N} \exp\left(-\frac{\|x_i - x\|_2^2}{2\sigma^2} \right). \quad (13.32)$$

In our experiments, we chose the window size σ as the maximum second nearest neighbor distance between all training samples as in [12].

Incorporating the learned probability of the poses in the energy function has additional advantages. First, it already incorporates correlations between the parameters of a pose – and thus of a particle – yielding an energy function that is closer to the model and the observed object. Moreover, it can be regarded as a soft constraint that includes anatomical constraints, e.g., by the limited freedom of joints movement, and that prevents the estimates from self-intersections since unrealistic and impossible poses cannot be contained in the training data.

Altogether, the energy function V of a particle x is defined by

$$V(x) := \frac{1}{l} \sum_{i=1}^{l} err_S(x,i)^2 - \eta \, \ln(p_{pose}(x)), \quad (13.33)$$

where l is the number of correspondences. In our experiments, we set $\eta = 8$.

Results

In our experiments, we tracked the lower part of a human body using four calibrated and synchronized cameras. The walking sequence was simultaneously captured by a commercial marker-based system[4] allowing a quantitative error analysis. The training data used for learning p_{pose} consisted of 480 samples that were obtained from walking sequences. The data was captured by the commercial system before recording the test sequence that was not contained in the training data.

---

[4] Motion Analysis system with 8 Falcon cameras.

The *ISA* performed well for the sequence consisting of 200 frames using a polynomial annealing scheme with $b = 0.7$, a dynamic variance scheme with $c = 0.3$, and the selection kernel (13.9). Results are given in Figure 13.11 where the diagram shows a comparison of the estimated knee-joint angles with the marker based system.

The convergence of the particles towards the pose with the lowest energy is illustrated for one frame in Figure 13.12. Moreover, it shows that variance of the particles decreases with an increasing number of annealing steps. This can also be seen from Figure 13.13 where the standard deviations for four parameters, which are scaled by c, are plotted. While the variances of the hip joint and of the knee joint decline rapidly, the variance of the ankle increases

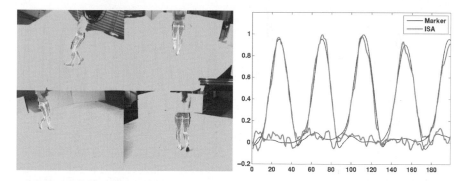

Fig. 13.11. **Left:** Results for a walking sequence captured by four cameras (200 frames). **Right:** The joint angles of the right and left knee in comparison with a marker based system.

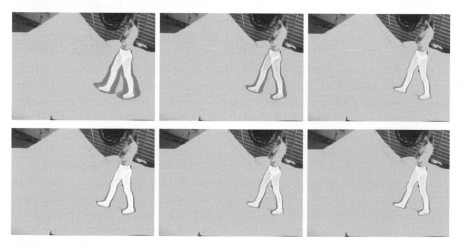

Fig. 13.12. Weighted particles at $t = 0, 1, 2, 4, 8$, and 14 of *ISA*. Particles with a higher weight are brighter, particles with a lower weight are darker. The particles converge to the pose with the lowest energy as t increases.

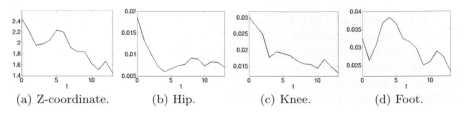

(a) Z-coordinate.   (b) Hip.   (c) Knee.   (d) Foot.

Fig. 13.13. Variance of the particles during ISA. The scaled standard deviations for the z-coordinate of the position and for three joint angles are given. The variances decrease with an increasing number of annealing steps.

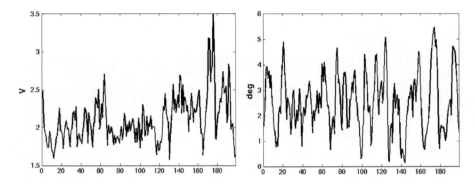

Fig. 13.14. **Left:** Energy of estimate for walking sequence (200 frames). **Right:** Error of estimate (left and right knee).

for the first steps before it decreases. This behavior results from the kinematic chain of the legs. Since the ankle is the last joint in the chain, the energy for a correct ankle is only low when also the previous joints of the chain are well estimated.

On the right hand side of Figure 13.14, the energy of the estimate during tracking is plotted. We also plotted the root-mean-square error of the estimated knee-angles for comparison where we used the results from the marker based system as ground truth with an accuracy of 3 degrees. For $n = 250$ and $T = 15$, we achieved an overall root-mean-square error of 2.74 degrees. The error was still below 3 degrees with 375 particles and $T = 7$, i.e., $n_T = 2625$. With this setting, the ISA took 7–8 s for approximately 3900 correspondences that were established in the 4 images of one frame. The whole system including segmentation, took 61 s for one frame. For comparison, the iterative method as used in Chapter 12 took 59 s with an error of 2.4 degrees. However, we have to remark that for this sequence the iterative method performed very well. This becomes clear from the fact that no additional random starting points were needed. Nevertheless, it demonstrates that the ISA can keep up even in situations that are perfect for iterative methods.

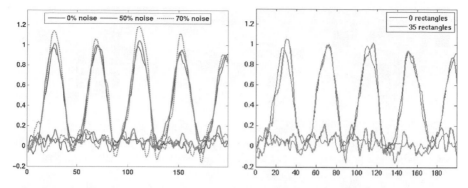

Fig. 13.15. Left: Random pixel noise. **Right:** Occlusions by random rectangles.

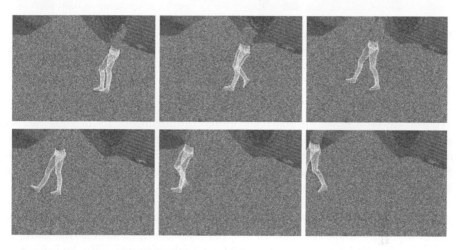

Fig. 13.16. Estimates for a sequence distorted by 70% random pixel noise. One view of frames 35, 65, 95, 125, 155, and 185 is shown.

Figures 13.16 and 13.17 show the robustness in the presence of noise and occlusions. For the first sequence, each frame was independently distorted by 70% pixel noise, i.e., each pixel value was replaced with probability 0.7 by a value uniformly sampled from the interval $[0, 255]$. The second sequence was distorted by occluding rectangles of random size, position, and gray value, where the edge lengths were in the range from 1 to 40. The knee angles are plotted in Figure 13.15. The root mean-square errors were 2.97 degrees, 4.51 degrees, and 5.21 degrees for 50% noise, 70% noise, and 35 occluding rectangles, respectively.

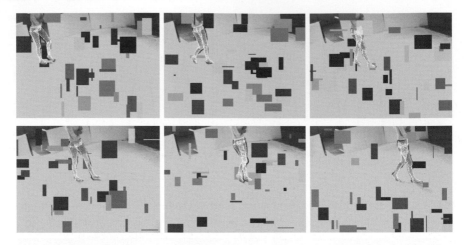

Fig. 13.17. Estimates for a sequence with occlusions by 35 rectangles with random size, color, and position. One view of frames 35, 65, 95, 125, 155, and 185 is shown.

13.5 Discussion

We introduced a novel approach for global optimization, termed interacting simulated annealing (ISA), that converges to the global optimum. It is based on an interacting particle system where the particles are weighted according to Boltzmann-Gibbs measures determined by an energy function and an increasing annealing scheme.

The variance of the particles provides a good measure of the confidence in the estimate. If the particles are all near the global optimum, the variance is low and only a low diffusion of the particles is required. The estimate, in contrast, is unreliable for particles with an high variance. This knowledge is integrated via dynamic variance schemes that focus the search on regions of interest depending on the confidence in the current estimate. The performance and the potential of ISA was demonstrated by means of two applications.

The first example showed that our approach can deal with local optima and solves the optimization problem well even for noisy measurements. However, we also provided some limitations of the dynamic variance schemes where standard global optimization methods might perform better. Since a comparison with other global optimization algorithm is out of the scope of this introduction, this will be done in future.

The application to multi-view human motion capturing, demonstrated the embedding of ISA into a complex system. The tracking system included silhouette extraction by a level-set method, a pose prediction by an autoregression, and prior knowledge learned from training data. Providing an error analysis, we demonstrated the accuracy and the robustness of the system in the presence of noise and occlusions. Even though we considered only a relative simple walking sequence for demonstration, it already indicates the potential of ISA for human motion capturing. Indeed, a comparison with an iterative approach revealed that on the one hand global optimization methods cannot perform better than local optimization methods when local optima are not problematic as it is the case for the walking sequence, but on the other hand it also showed that the ISA can keep up with the iterative method. We expect therefore that the ISA performs better for faster movements, more complex motion patterns, and human models with higher degrees of freedom. In addition, the introduced implementation of the tracking system with ISA has one essential drawback for the performance. While the pose estimation is performed by a global optimization method, the segmentation is still susceptible to local minima since the energy function (13.29) is minimized by a local optimization approach.

As part of future work, we will integrate ISA into the segmentation process to overcome the local optima problem in the whole system. Furthermore, an evaluation and a comparison with an iterative method needs to be done with sequences of different kinds of human motions and also when the segmentation is independent of the pose estimation, e.g., as it is the case for background subtraction. Another improvement might be achieved by considering correlations between the parameters of the particles for the dynamic variance schemes, where an optimal trade-off between additional computation cost and increased accuracy needs to be found.

Acknowledgments

Our research is funded by the Max-Planck Center for Visual Computing and Communication. We thank Uwe Kersting for providing the walking sequence.

References

1. Ackley D. *A Connectionist Machine for Genetic Hillclimbing.* Kluwer, Boston, 1987.
2. Bäck T. and Schwefel H.-P. An overview of evolutionary algorithms for parameter optimization. *Evolutionary Computation*, 1:1–24, 1993.
3. Bauer H. *Probability Theory.* de Gruyter, Baton Rouge, 1996.
4. Besl P. and McKay N. A Method for registration of 3-D Shapes. *IEEE Transactions on Pattern Analysis and Machine Intelligence*, 14(2):239–256, 1992.
5. Brox T., Rosenhahn B., Cremers D., and Seidel H.-P. High accuracy optical flow serves 3-D pose tracking: exploiting contour and flow based constraints. In A. Leonardis, H. Bischof, and A. Pinz, editors, *European Conference on Computer Vision (ECCV)*, LNCS 3952, pp. 98–111. Springer, 2006.
6. Brox T., Rosenhahn B., Kersting U., and Cremers D., Nonparametric density estimation for human pose tracking. In *Pattern Recognition (DAGM). LNCS* 4174, pp. 546–555. Springer, 2006.

7. Bregler C., Malik J., and Pullen K. Twist based acquisition and tracking of animal and human kinematics. *International Journal of Computer Vision*, 56:179–194, 2004.

8. Brox T., Rosenhahn B., and Weickert J. Three-dimensional shape knowledge for joint image segmentation and pose estimation. In W. Kropatsch, R. Sablatnig, A. Hanbury, editors, *Pattern Recognition (DAGM), LNCS* 3663, pp. 109–116. Springer, 2005.

9. Deutscher J. and Reid I. Articulated body motion capture by stochastic search. *International Journal of Computer Vision*, 61(2):185–205, 2005.

10. Doucet A., deFreitas N., and Gordon N., editors. *Sequential Monte Carlo Methods in Practice*. Statistics for Engineering and Information Science. Springer, New York, 2001.

11. Gall J., Potthoff J., Schnoerr C., Rosenhahn B., and Seidel H.-P. Interacting and annealing particle systems – mathematics and recipes. *Journal of Mathematical Imaging and Vision*, 2007, To appear.

12. Gall J., Rosenhahn B., Brox T., and Seidel H.-P. Learning for multi-view 3D tracking in the context of particle filters. In *International Symposium on Visual Computing (ISVC), LNCS* 4292, pp. 59–69. Springer, 2006.

13. Gall J., Rosenhahn B., and Seidel H.-P. Robust pose estimation with 3D textured models. In *IEEE Pacific-Rim Symposium on Image and Video Technology (PSIVT), LNCS* 4319, pp. 84–95. Springer, 2006.

14. Geman S. and Geman D. Stochastic relaxation, Gibbs distributions and the Bayesian restoration of images. *IEEE Transactions on Pattern Analysis and Machine Intelligence*, 6(6):721–741, 1984.

15. Gidas B. *Topics in Contemporary Probability and Its Applications*, Chapter 7: Metropolis-type Monte Carlo simulation algorithms and simulated annealing, pp. 159–232. Probability and Stochastics Series. CRC Press, Boca Raton, 1995.

16. Goldstein H. *Classical Mechanics*. Addison-Wesley, Reading, MA, second edition, 1980.

17. Grest D., Herzog D., and Koch R. Human Model Fitting from Monocular Posture Images. In G. Greiner, J. Hornegger, H. Niemann, and M. Stamminger, editors, *Vision, modelling and Visualization*. Akademische Verlagsgesellschaft Aka, 2005.

18. Kirkpatrick S., Gelatt C. Jr. and Vecchi M. Optimization by simulated annealing. *Science*, 220(4598):671–680, 1983.

19. DelMoral P. *Feynman–Kac Formulae. Genealogical and Interacting Particle Systems with Applications*. Probability and its Applications. Springer, New York, 2004.

20. Murray R.M., Li Z., and Sastry S.S. *Mathematical Introduction to Robotic Manipulation*. CRC Press, Baton Rouge, 1994.

21. Parzen E. On estimation of a probability density function and mode. *Annals of Mathematical Statistics*, 33:1065–1076, 1962.

22. Pennec X. and Ayache N. Uniform distribution, distance and expectation problems for geometric features processing. *Journal of Mathematical Imaging and Vision*, 9(1):49–67, 1998.

23. Pennec X. Computing the mean of geometric features: Application to the mean rotation. Rapport de Recherche RR–3371, INRIA, Sophia Antipolis, France, March 1998.

24. Pennec X. Intrinsic statistics on Riemannian manifolds: basic tools for geometric measurements. *Journal of Mathematical Imaging and Vision*, 25(1):127–154, 2006.
25. Rosenhahn B., Brox T., Cremers D., and Seidel H.-P. A comparison of shape matching methods for contour based pose estimation. In R. Reulke, U. Eckhardt, B. Flach, U. Knauer, and K. Polthier, editors, *11th International Workshop on Combinatorial Image Analysis (IWCIA)*, *LNCS* 4040, pp. 263–276. Springer, 2006.
26. Rosenhahn B., Brox T., Kersting U., Smith A., Gurney J., and Klette R. A system for marker-less human motion estimation. *Künstliche Intelligenz*, 1:45–51, 2006.
27. Rosenhahn B., Brox T., and Weickert J. Three-dimensional shape knowledge for joint image segmentation and pose tracking. In *International Journal of Computer Vision*, 2006, To appear.
28. Rosenhahn B., Perwass C., and Sommer G. Pose Estimation of Free-form Contours. *International Journal of Computer Vision*, 62(3):267–289, 2005.
29. Rosenblatt F. Remarks on some nonparametric estimates of a density function. *Annals of Mathematical Statistics*, 27(3):832–837, 1956.
30. Szu H. and Hartley R. Fast simulated annealing. *Physic Letter A*, 122:157–162, 1987.
31. Theobalt C., Magnor M., Schueler P., and Seidel H.-P. Combining 2D feature tracking and volume reconstruction for online video-based human motion capture. In *10th Pacific Conference on Computer Graphics and Applications*, pp. 96–103. IEEE Computer Society, 2002.
32. Tsallis C. and Stariolo D.A. Generalized simulated annealing. *Physica A*, 233:395–406, 1996.

14

Motion Capture for Interaction Environments

Daniel Grest and Reinhard Koch

Christian-Albrechts-University Kiel, Germany
Multimedia Information Processing

Summary. The accuracy of marker-less motion capture systems is comparable to marker-based systems, however the segmentation step makes strong restrictions to the capture environment, e.g., homogeneous clothing or background, constant lighting, etc. In interaction environments the background is non-static, cluttered and lighting changes often and rapidly. Stereo algorithms can provide data that is robust with respect to lighting and background and are available in real-time. Because speed is an issue, different optimization methods are compared, namely Gauss-Newton(Levenberg-Marquardt), Gradient Descent, and Stochastic Meta Descent. Experiments on human movement show the advantages and disadvantages of each method.

14.1 Introduction

Though video-based marker-tracking motion capture systems are state of the art in industrial applications, they have significant disadvantages in comparison with marker-less systems. In fact there is a growing need of marker-less systems as discussed in 15.

Accurate marker-less motion capture systems rely on images that allow segmentation of the person in the foreground. While the accuracy of such approaches is comparable to marker-based systems [9, 21], the segmentation step makes strong restrictions to the capture environment, e.g., homogenous clothing or background, constant lighting, camera setups that cover a complete circular view on the person etc. If motion of a human is to be captured in an interaction environment, there are different conditions to be dealt with.

A general problem in interaction environments is, that the interaction area should be well lighted for better camera images with less noise, while the display screens should not receive any additional light. The compromise between both is usually a rather dimly illuminated environment, as shown in Figure (14.1), where the displayed scene is clearly visible in spite of the light from the ceiling (more details in [11]). Additionally the background is cluttered

B. Rosenhahn et al. (eds.), *Human Motion – Understanding, Modelling, Capture, and Animation*, 347–376.

Fig. 14.1. The interaction area.

and non-static, which makes a segmentation of the human's silhouette difficult (like in Chapters 11 and 12).

Possible applications of full body motion capture in such an environment are detection and reconstruction of specific gestures, like pointing gestures for manipulating the virtual scene or online avatar animation for games or video conferences.

Because the environment requires real-time motion capture to make interaction possible, previous approaches to pose estimation of articulated objects are analyzed with respect to their efficiency. Most approaches rely on numerical derivatives, though the motion of points in an articulated object can be derived analytically, which increases accuracy and reduces computation time.

The framework for motion capture presented here, relies on pose estimation of articulated objects. We will at first give details about the estimation algorithm, which is not limited to applications within an interaction environment. A *Nonlinear Least Squares* approach is taken, that does not only cover previous approaches, but enhances them. Estimation from 2D–3D correspondences as in the work of Rosenhahn (12, [21]) is increased in accuracy by minimizing errors in the image plane. Numbers of parameters and computation time is reduced in comparison to other minimization methods, like Stochastic Meta Descent [3] as discussed in detail in Section 14.4.

Results on human motion estimation from depth data calculated from stereo images is presented in the end, which shows, that even complex full body motion involving 24 degrees of freedom can be captured from single view stereo sequences.

14.2 Body Models

A body model for human motion capture has different important aspects, which have to be distinguished. These are in detail:

- **The geometry or skin.** The most important aspect of the model is the skin surface. This skin model does not include any aspects of motion or animation. It just describes the appearance of the human for one specific pose. For example each body part is a cylinder or the whole body consists of one triangle mesh as obtained from laser scanners.

- **The skeleton** which defines the motion capabilities. The skeleton gives the degrees of freedom of the motion model by defining a hierarchy of joints and specific bone lengths.
- **Joint positions.** This defines the relation of skeleton and surface skin model. It can be thought of "where in the skin" the skeleton is positioned. In addition to the joint positions, it has to be specified for each element of the skin model to which joint it belongs. For "Linear Blend Skinned" models [16], each skin surface element even belongs to multiple joints with different weights.

These three parts define how a point on the surface moves with respect to joint parameters, e.g., joint angles. The definition of these parts, which gives the possibility to evaluate skin point positions for given joint values, is also referred to as "skinning" in modelling tools like MayaTM [15] or Blender [6]. The calculation of skin point positions with defined model parts is known in robotics as *forward kinematics*, because it allows to calculate the pose of a specific object part, for example the end effector, if joint angles are given.

The inverse kinematics is necessary for parameter estimation from observed data with known correspondences as in this work, or for example estimating pose from 3D marker data as provided by professional marker systems, e.g., from Vicon [24] or MetaMotion [17].

Most systems simplify the task by making the assumption, that all three aspects are known for the observed human and the problem of motion capture is reduced to estimating joint angles per frame.

The main drawback of estimating only joint angles per frame, is the limited accuracy of motion, pose and surface reconstruction.

One example, that tries to estimate skin geometry and joint angles simultaneously for each video frame is the work of Plaenkers & Fua [20]. Their goal was to very accurately reconstruct the appearance of the observed subject in each frame, e.g., including muscle bulges or skin deformations.

In motion capture systems there exists a wide variety of models used for estimation, their complexity and accuracy mainly depends on aspects like type of input data, achieved degree of accuracy, application environment etc. Examples for template skin models are from simple to complex: stick Figures (just a skeleton without skin), cylindrical [8] body parts, ellipsoidal models [4, 20], arbitrary rigid body parts [7, 21], linear blend skinned models [3]. The complexity of the skeleton varies accordingly, with up to over 30 DOF for hand models [3].

The template models used in the experiments here consist of rigid fixed body parts of arbitrary shape (see Figure 14.2), because they can be represented by a scene graph, stored in VRML format and can be effectively rendered on graphics hardware using openGL. The movement capabilities of the models are consistent with the MPEG4 [1] specification with up to 180 DOF, while the maximum number of estimated DOF is 28.

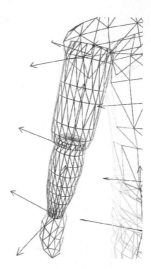

Fig. 14.2. One arm model with MPEG4 conformal joints as used in this work.

14.3 Pose Estimation

The pose of an object is defined here as its position and orientation with respect to some coordinate system. The estimation is performed by calculation of relative movement.

Capturing human motion by pose estimation of an articulated object is done in many approaches and is motivated from inverse kinematic problems in robotics, e.g., [18]. Solving the estimation problem by optimization of an objective function is also very common [13, 19, 20]. Silhouette information is usually part of this function, that tries to minimize the difference between the model silhouette and the silhouette of the real person either by background segmentation [19, 21] or image gradient [13]. Other approaches fit a 3D human body model to 3D point clouds calculated from depth maps. Stereo algorithms can provide these depth images even in real-time [27].

The pose estimation methods described in this work cover also full body inverse kinematics as known from professional modelling Tools like Maya [15] and include methods to estimate human motion from marker-based data, which are 3D marker positions over time and associated 3D points on the human. While the motion estimation from marker data is included in many modelling tools, the algorithm behind is not known.

At first pose estimation of rigid bodies is explained and extended in the latter to articulated objects. Optimization functions and their Jacobians are given for different type of input data. Beginning with 3D-point–3D-point correspondences and extended later to 2D–3D correspondences.

14.3.1 Rigid Bodies

A rigid body motion (RBM) in $\mathbb{R}^3$ is a transformation of an object, that keeps the distances between all points in the object constant. There are different formulations for rigid body motions, e.g., twists [4], which use a exponential term e^{ψ} or rotors [21], which may be seen as an extension of quaternions. Most common are transformation matrices $T \in \mathbb{R}^{4 \times 4}$ in homogenous coordinates. Given here is another description, which allows a straightforward application of the Gauss-Newton method for *Nonlinear Least Squares*.

Rotation around Arbitrary Axis

The movement of a point around an arbitrary axis in space is a circular movement on a plane in 3D space as shown in Figure 14.3. Consider the normal vector $\boldsymbol{\omega}$, which describes the direction of the axis, and the point $\mathbf{q}$ on the axis, which has the shortest distance to the origin, i.e., $\mathbf{q}$ lies on the axis and $\mathbf{q}^T \boldsymbol{\omega} = 0$, refer to Figure 14.3.

The rotation of a point $\mathbf{p}$ around that axis may then be written as

$$R_{\boldsymbol{\omega},\mathbf{q}}(\theta, \mathbf{p}) = \mathbf{p} + \sin\theta(\boldsymbol{\omega} \times (\mathbf{p} - \mathbf{q})) + (1 - \cos\theta)(\mathbf{q} - \mathbf{p_{proj}})$$
$$= \mathbf{p} + \sin\theta(\boldsymbol{\omega} \times (\mathbf{p} - \mathbf{q})) + (\mathbf{q} - \mathbf{p_{proj}}) - \cos\theta(\mathbf{q} - \mathbf{p_{proj}})$$
$$(14.1)$$

where $\mathbf{p_{proj}} = \mathbf{x} - (\mathbf{x}^T\boldsymbol{\omega})\boldsymbol{\omega}$ is the projection of $\mathbf{p}$ onto the plane through the origin with normal $\boldsymbol{\omega}$. Note that $\mathbf{q}$ is also on that plane. This expression is very useful as the derivative $\frac{\partial R_{\boldsymbol{\omega},\mathbf{q}}(\theta)}{\partial \theta}$ is easy to calculate:

$$\frac{\partial R_{\boldsymbol{\omega},\mathbf{q}}(\theta)}{\partial \theta} = \cos\theta(\boldsymbol{\omega} \times (\mathbf{p} - \mathbf{q})) + \sin\theta(\mathbf{q} - \mathbf{p_{proj}}) \qquad (14.2)$$

with the special derivative at zero:

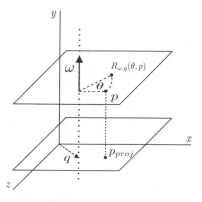

Fig. 14.3. Rotation around an arbitrary axis in space.

$$\left.\frac{\partial R_{\omega,\mathbf{q}}(\theta)}{\partial \theta}\right|_{\theta=0} = \boldsymbol{\omega} \times (\mathbf{p} - \mathbf{q}) = \boldsymbol{\omega} \times \mathbf{p} - \boldsymbol{\omega} \times \mathbf{q}_{\omega} \qquad (14.3)$$

with $\mathbf{q}_{\omega}$ denoting an arbitrary point on the axis. The cross product $\boldsymbol{\omega} \times \mathbf{q}_{\omega}$ is also known as *momentum*. For practical application the momentum is more useful, because an arbitrary point on the axis is sufficient to have a valid description. The derivative at zero gives the velocity or the tangent of the point moving on the circle.

14.3.2 Rigid Movement by Nonlinear Least Squares

In this work estimation of pose relies on correspondences. At first, estimation from 3D-point–3D-point correspondences $(\mathbf{p_i}, \tilde{\mathbf{p}}_i)$ is given. Pose estimation is understood here as calculating the rigid movement, which transforms all points $\mathbf{p_i}$ as close as possible to their corresponding points $\tilde{\mathbf{p}}_i$. For a given set of correspondences the *Nonlinear Least Squares method* can be applied to estimate that movement. A general *Nonlinear Least Squares* problem [5] reads:

$$\hat{\boldsymbol{\theta}} = \arg\min_{\theta} \sum_{i=1}^{m} (r_i(\boldsymbol{\theta}))^2$$

In case of pose estimation from corresponding 3D data points $(\mathbf{p_i}, \tilde{\mathbf{p}}_i)$, as shown in Figure 14.4, the *residual* functions r_i are of the form:

$$\mathbf{r_i}(\boldsymbol{\theta}) = (\mathbf{m}(\boldsymbol{\theta}, \mathbf{p_i}) - \tilde{\mathbf{p}}_i) \qquad (14.4)$$

with $\mathbf{r_i}(\boldsymbol{\theta}) = (r_{ix}, r_{iy}, r_{iz})^T$ and $\boldsymbol{\theta} = (\theta_x, \theta_y, \theta_z, \theta_\alpha, \theta_\beta, \theta_\gamma)^T$ there are $3m$ real-valued *residual* functions for minimization. The rigid movement function of the point $\mathbf{p_i}$ is

$$\mathbf{m}(\boldsymbol{\theta}, \mathbf{p_i}) = (\theta_x, \theta_y, \theta_z)^T + \left(R_{\omega_x}(\theta_\alpha) \circ R_{\omega_y}(\theta_\beta) \circ R_{\omega_z}(\theta_\gamma) \right)(\mathbf{p_i}) \qquad (14.5)$$

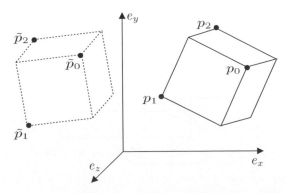

Fig. 14.4. Rigid movement of an object and corresponding points.

where $\omega_x, \omega_y, \omega_z$ denote, that the center of rotation is not necessarily the origin of the world coordinate system. The operator "∘" denotes function concatenation, such that the point is first rotated around ω_z, then around ω_y etc.

To find the minimizer $\hat{\boldsymbol{\theta}}$ Newton's method can be applied, which requires the first and second derivatives of the residual functions.

14.3.3 Jacobian for the Rigid Movement

The Jacobian matrix with its partial derivatives are given in the following for a point set with one correspondence $(\mathbf{p}, \tilde{\mathbf{p}})$. For point sets with m points the Jacobian is simply extended by additional rows for the additional correspondences.

The derivative for the residual functions of rigid movement is:

$$\frac{\partial \mathbf{r}(\boldsymbol{\theta})}{\partial \theta_j} = \frac{\partial \mathbf{m}(\boldsymbol{\theta}, \mathbf{p})}{\partial \theta_j} = \frac{\partial (\theta_x, \theta_y, \theta_z)^T + \left(R_{\omega_x}(\theta_\alpha) \circ R_{\omega_y}(\theta_\beta) \circ R_{\omega_z}(\theta_\gamma)\right)(\mathbf{p})}{\partial \theta_j}$$

$$(14.6)$$

Therefore the Jacobian for one point–point correspondence is

$$J = \begin{bmatrix} 1 & 0 & 0 & \frac{\partial m_x}{\partial \theta_\alpha} & \frac{\partial m_x}{\partial \theta_\beta} & \frac{\partial m_x}{\partial \theta_\gamma} \\ 0 & 1 & 0 & \frac{\partial m_y}{\partial \theta_\alpha} & \frac{\partial m_y}{\partial \theta_\beta} & \frac{\partial m_y}{\partial \theta_\gamma} \\ 0 & 0 & 1 & \frac{\partial m_z}{\partial \theta_\alpha} & \frac{\partial m_z}{\partial \theta_\beta} & \frac{\partial m_z}{\partial \theta_\gamma} \end{bmatrix} \tag{14.7}$$

and for the derivatives at zero:

$$\left.\frac{\partial \mathbf{m}(\boldsymbol{\theta}, \mathbf{p})}{\partial \theta_\alpha}\right|_{\boldsymbol{\theta}=0} = \boldsymbol{\omega_x} \times (\mathbf{p} - \mathbf{q_x})$$

$$\left.\frac{\partial \mathbf{m}(\boldsymbol{\theta}, \mathbf{p})}{\partial \theta_\beta}\right|_{\boldsymbol{\theta}=0} = \boldsymbol{\omega_y} \times (\mathbf{p} - \mathbf{q_y}) \tag{14.8}$$

$$\left.\frac{\partial \mathbf{m}(\boldsymbol{\theta}, \mathbf{p})}{\partial \theta_\gamma}\right|_{\boldsymbol{\theta}=0} = \boldsymbol{\omega_z} \times (\mathbf{p} - \mathbf{q_z})$$

Here $\boldsymbol{\omega_x}, \mathbf{q}_x$ denote rotation around an arbitrary point in space and $\boldsymbol{\omega_x}, \boldsymbol{\omega_y}, \boldsymbol{\omega_z}$ are assumed orthogonal. If $\boldsymbol{\omega_x}, \boldsymbol{\omega_y}, \boldsymbol{\omega_z}$ are the world coordinate axes then the three equations above are equal to the linearized rotation matrix as commonly used.

The minimizer is then found by iteratively solving

$$\boldsymbol{\theta_{t+1}} = \boldsymbol{\theta_t} - \left(J^T J\right)^{-1} J^T \mathbf{r}(\boldsymbol{\theta_t}) \tag{14.9}$$

14.3.4 Relative Movement and Derivative at Zero

The Jacobian above is only valid, if the derivative is taken at zero. This can be achieved by estimating the relative movement in each iteration step, such that $\boldsymbol{\theta_t} = 0$ and $\boldsymbol{\theta_{t+1}} = \Delta\boldsymbol{\theta_t}$.

Therefore recalculation of the world coordinates of the point $\mathbf{p}$ at each iteration is required. The corresponding observed point $\tilde{\mathbf{p}}$ stays unchanged. If the rotation center is $\mathbf{c}$, which is the intersection of $\boldsymbol{\omega}_{\mathbf{x}}, \boldsymbol{\omega}_{\mathbf{y}}, \boldsymbol{\omega}_{\mathbf{z}}$, the final movement is obtained by

$$\hat{\boldsymbol{\theta}}_{t+1} = \hat{\boldsymbol{\theta}}_t + \Delta\boldsymbol{\theta}_t \tag{14.10}$$

and

$$\mathbf{p}_{t+1} = \left(R_x(\hat{\theta}_{\alpha,t+1}) \circ R_y(\hat{\theta}_{\beta,t+1}) \circ R_z(\hat{\theta}_{\gamma,t+1}) \right)(\mathbf{p}_{t_0} - \mathbf{c}) + \mathbf{c} \tag{14.11}$$

where $\mathbf{p}_{t_0}$ is the initial point coordinate $\mathbf{p}$. In the last equation it may be more efficient to use rotation matrices for evaluating the new points than Equation (14.1).

14.3.5 Geometric Interpretation

Newton's Method approximates the objective function by a quadratic function and iterates until convergence. Shown here is is the development of the iterative approach for a rotation with angle θ around a fixed axis $(\boldsymbol{\omega}, \mathbf{q})$ and a known 3D-point–3D-point correspondence $(\tilde{\mathbf{p}}, \mathbf{p})$.

As we consider a nonlinear problem of estimating the rotation angle around a fixed known axis in 3D space, the movement of a point on that axis is a movement on a plane. Shown in the following Figures is only the development on that plane. The error vector $\mathbf{r}(\theta) = (\mathbf{m}(\theta, \mathbf{p}) - \tilde{\mathbf{p}})$ is not necessarily within that plane. However, for optimization only the projection onto the plane is of importance, because the resulting minimizing point lies in that plane.

The problem is shown in Figure 14.5, where the vector $\mathbf{v} = \boldsymbol{\omega} \times (\mathbf{p} - \mathbf{q})$ is the velocity vector of the rotational movement of the model point $\mathbf{p}$, if $\mathbf{p}$ rotates with an angular velocity of one.

Consider now the projection of $\mathbf{r}$ onto $\mathbf{v}$, which reads:

$$\mathbf{r}'_{\mathbf{v}}(\boldsymbol{\theta}) = \frac{\mathbf{r}^T \mathbf{v}}{\mathbf{v}^T \mathbf{v}} \mathbf{v} \tag{14.12}$$

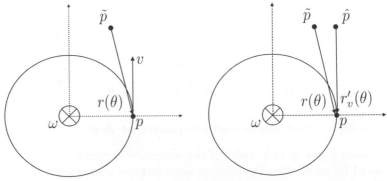

Fig. 14.5. Estimating the rotation angle for a 3D-point–3D-point correspondence.

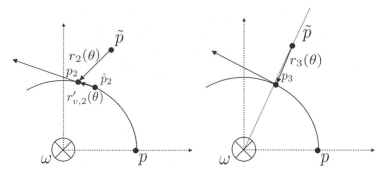

Fig. 14.6. Estimating the rotation angle for a 3D-point–3D-point correspondence. Iterations 2 and 3.

One iteration of the Gauss-Newton results in the parameter change:

$$\Delta\theta = -(J^T J)^{-1} J^T \, \mathbf{r}(\theta) = -(\mathbf{v}^T \mathbf{v})^{-1} \mathbf{v}^T \mathbf{r}(\theta) = -\frac{\mathbf{r}^T \mathbf{v}}{\mathbf{v}^T \mathbf{v}} \qquad (14.13)$$

Hence $\Delta\theta$ is the scale value of $\mathbf{v}$ that gives the projection of the error $\mathbf{r}$ onto $\mathbf{v}$ and $\hat{\mathbf{p}} = \mathbf{p} - \mathbf{r}'_\mathbf{v}(\theta) = \mathbf{p} + \Delta\theta \, \mathbf{v}$.

Inserting $\Delta\theta$ in the movement Equation (14.5) gives the starting point $\mathbf{p_2}$ for the next iteration as shown in Figure 14.6 left. In the example shown $\Delta\theta_1$ is 1.36 *rad* or 78 *deg*.

In the second iteration the projection of $\mathbf{r_2}$ onto $\mathbf{v}$ leads to the minimizer $\hat{\mathbf{p}}_2$ that is almost on the circle. After insertion of $\Delta\theta_2$, which is -0.27 *rad* or -15.6 *deg*, the next starting point $\mathbf{p_3}$ for iteration 3 is found as shown in Figure 14.6 right. This point is so near to the final correct solution, that a drawing of further iterations is not done.

14.3.6 2D–3D Pose Estimation

In the following sections rigid movement is estimated with Nonlinear Least Squares from 2D-point–3D-point and 2D-line–3D-point correspondences. Additionally a comparison with estimation from 3D-line–3D-point correspondences and optical flow is given.

Assume the point $\mathbf{p} \in \mathbb{R}^3$ is observed by a pinhole camera and its projection onto the 2D image plane is $\mathbf{p}' \in \mathbb{R}^2$. If the camera is positioned at the origin and aligned with the world, such that the optical axis is the world z-axis, the combination of camera projection and 3D rigid movement of the point can be written as:

$$\mathbf{m}'(\mathbf{p}, \boldsymbol{\theta}) = \begin{pmatrix} s_x \frac{m_x(\mathbf{p},\boldsymbol{\theta})}{m_z(\mathbf{p},\boldsymbol{\theta})} + c_x \\ s_y \frac{m_y(\mathbf{p},\boldsymbol{\theta})}{m_z(\mathbf{p},\boldsymbol{\theta})} + c_y \end{pmatrix} \qquad (14.14)$$

where s_x, s_y are the focal length of the camera in x- and y-direction expressed in pixel units and $(c_x, c_y)^T$ is the principal point of the camera. The additional

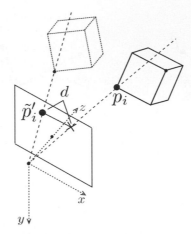

Fig. 14.7. Object movement and 2D–3D correspondences.

assumption here is a possible skew to be zero, i.e., the rows and columns of the image sensor are orthogonal to each other.

If an object with known geometry moves in space from a known pose to a new pose and its known 3D points $\mathbf{p_i}$ are observed in an image at $\tilde{\mathbf{p}_i}'$, its relative movement can be estimated by Nonlinear Least Squares using the 2D–3D correspondence $(\tilde{\mathbf{p}_i}', \mathbf{p_i})$ with the following residual. See also Figure 14.7.

$$d = \mathbf{r_i}(\boldsymbol{\theta}) = (\mathbf{m}'(\mathbf{p_i}, \boldsymbol{\theta}) - \tilde{\mathbf{p}_i}') \tag{14.15}$$

The necessary Jacobian for the optimization is given now for a single correspondence. For m correspondences the Jacobian is simply extended by additional rows. With $\boldsymbol{\theta} = (s_x, s_y, c_x, c_y, \theta_x, \theta_y, \theta_z, \theta_\alpha, \theta_\beta, \theta_\gamma)^T$ the Jacobian reads

$$J = \left[\frac{\partial m'}{\partial s_x}\ \frac{\partial m'}{\partial s_y}\ \frac{\partial m'}{\partial c_x}\ \frac{\partial m'}{\partial c_y}\ \frac{\partial m'}{\partial \theta_x}\ \frac{\partial m'}{\partial \theta_y}\ \ \frac{\partial m'}{\partial \theta_z}\ \ \frac{\partial m'}{\partial \theta_\alpha}\ \frac{\partial m'}{\partial \theta_\beta}\ \frac{\partial m'}{\partial \theta_\gamma} \right]$$

$$= \begin{bmatrix} \frac{m_x}{m_z} & 0 & 1 & 0 & \frac{s_x}{m_z} & 0 & s_x\frac{-m_x}{m_z^2} & \frac{\partial m'_x}{\partial \theta_\alpha} & \frac{\partial m'_x}{\partial \theta_\beta} & \frac{\partial m'_x}{\partial \theta_\gamma} \\ 0 & \frac{m_y}{m_z} & 0 & 1 & 0 & \frac{s_y}{m_z} & s_y\frac{-m_y}{m_z^2} & \frac{\partial m'_y}{\partial \theta_\alpha} & \frac{\partial m'_y}{\partial \theta_\beta} & \frac{\partial m'_y}{\partial \theta_\gamma} \end{bmatrix} \tag{14.16}$$

and

$$\frac{\partial \mathbf{m}'}{\partial \theta_j} = \begin{pmatrix} \frac{\partial \left(s_x \frac{m_x}{m_z} \right)}{\partial \theta_j} \\ \frac{\partial \left(s_y \frac{m_y}{m_z} \right)}{\partial \theta_j} \end{pmatrix} = \begin{pmatrix} \dfrac{s_x \left(\frac{\partial m_x}{\partial \theta_j} m_z - m_x \frac{\partial m_z}{\partial \theta_j} \right)}{m_z^2} \\ \dfrac{s_y \left(\frac{\partial m_y}{\partial \theta_j} m_z - m_y \frac{\partial m_z}{\partial \theta_j} \right)}{m_z^2} \end{pmatrix} \tag{14.17}$$

The partial derivatives $\frac{\partial \mathbf{m}}{\partial \theta_j}, j \in \{\alpha, \beta, \gamma\}$ are given in Equation (14.8). The Jacobian above does not only allow estimation of the pose of an object, but also the estimation of the internal camera parameters. Additionally the formulation allows to estimate only a subset of the parameters, if some of them are known, e.g., rotation only or fixed principal point etc. As the Jacobian above requires derivatives at zero, the estimation should be done relative to the previous estimate iteratively as described in Section 14.3.4.

This analytically derived Jacobian for the problem of camera calibration was already published 1996 as an extension of Lowe's pose estimation algorithm [2]. In the next sections it is extended to articulated objects.

The optimization above minimizes the distance between the projected 3D model point with its corresponding 2D image point, while in [21] the 3D-difference of the viewing ray and its corresponding 3D point is minimized. The minimization in 3D space is not optimal, if the observed image positions are disturbed by noise, as shown in [26], because for 3D points, which are further away from the camera, the error in the optimization will be larger as for points nearer to the camera, which leads to a biased pose estimate due to the least squares solution. In [26] a scaling value was introduced, which downweights correspondences according to their distance to the camera, which is in fact very close to Equation (14.17).

Approximation by 3D-point-3D-line correspondences

An alternative approach to estimate pose from 2D-3D correspondences is a minimization of 3D point–3D line distances. If the inverse projection function is known, as e.g., for a calibrated pinhole camera, it is possible to calculate the 3D viewing ray for the known 2D image point. Assume the viewing ray is described by its normalized direction vector $\boldsymbol{\omega}$ and a point $\mathbf{q}_\omega$ on the ray. The distance d between a point $\mathbf{p}$ and the viewing ray is

$$d = |\, \boldsymbol{\omega} \times \mathbf{p} - \boldsymbol{\omega} \times \mathbf{q}_\omega |\qquad(14.18)$$

Using this distance the pose estimation problem by *Nonlinear Least Squares* with correspondences $((\tilde{\boldsymbol{\omega}}_\mathbf{i}, \tilde{\mathbf{q}}_{\omega,\mathbf{i}}), \mathbf{p}_\mathbf{i})$ is:

$$\min_{\boldsymbol{\theta}} \sum_i |\, \tilde{\boldsymbol{\omega}}_\mathbf{i} \times \mathbf{m}(\boldsymbol{\theta}, \mathbf{p}_\mathbf{i}) - \tilde{\boldsymbol{\omega}}_\mathbf{i} \times \tilde{\mathbf{q}}_{\omega,\mathbf{i}}|^2\qquad(14.19)$$

Because the cross product $\tilde{\boldsymbol{\omega}}_\mathbf{i} \times \mathbf{m}(\boldsymbol{\theta}, \mathbf{p}_\mathbf{i})$ can also be described by a matrix multiplication, it is obvious that this minimization problem is very close to that of Equation (14.4).

Let $[\tilde{\omega} \times]^{3x3}$ be the cross product matrix of one correspondence. The necessary Jacobian J for the Gauss-Newton Method then is:

$$J = [\tilde{\omega} \times]^{3x3} J_{14.7}\qquad(14.20)$$

where $J_{14.7}$ is the Jacobian for 3D point–3D point correspondences from Equation (14.7). For additional correspondences the Jacobian is extended by 3 additional rows for each correspondence.

The minimization in 3D space is inaccurate, if the 2D points are disturbed by noise [26]. Therefore this method should only be applied if the projection function is difficult to derive, e.g., for fisheye cameras. In case of pinhole cameras the minimization in the image plane, where the error is observed, is more accurate. However, the inaccuracies are only significant in the presence of inaccurate correspondences and if the extension in depth of the object is large compared with its distance to the projection center.

Pose estimation from 3D-point–2D-line correspondences

Let $\tilde{\mathbf{n}}_\omega \in \mathbb{R}^2$ be the normal vector of a 2D line and $\tilde{\mathbf{q}}'_\omega \in \mathbb{R}^2$ one point on this line. If the distance of a projected point $\mathbf{p}'_i = \mathbf{m}'(0, \mathbf{p_i})$ to this line is to be minimized as shown in Figure 14.8, then the objective function for minimization is

$$\min_{\boldsymbol{\theta}} \sum_i \left| \tilde{\mathbf{n}}_\omega \left(\mathbf{m}'(\boldsymbol{\theta}, \mathbf{p_i}) - \tilde{\mathbf{q}}'_\omega \right) \right|^2 \tag{14.21}$$

with correspondences $\left((\tilde{\mathbf{n}}_{\omega,i}, \tilde{\mathbf{q}}'_{\omega,i}) , \mathbf{p_i} \right)$.

The residual function is the distance of $\mathbf{p_i}$ to the line in the direction of the normal and therefore may be negative:

$$r_i(\boldsymbol{\theta}) = \tilde{\mathbf{n}}_\omega \left(\mathbf{m}'(\boldsymbol{\theta}, \mathbf{p_i}) - \tilde{\mathbf{q}}'_\omega \right) \Leftrightarrow \tag{14.22}$$

$$= \tilde{\mathbf{n}}_\omega \, \mathbf{m}'(\boldsymbol{\theta}, \mathbf{p_i}) - \tilde{\mathbf{n}}_\omega \, \tilde{\mathbf{q}}'_\omega \Leftrightarrow \tag{14.23}$$

$$= (\tilde{\mathbf{n}}_\omega)_x \, m'_x(\boldsymbol{\theta}, \mathbf{p_i}) + (\tilde{\mathbf{n}}_\omega)_y \, m'_y(\boldsymbol{\theta}, \mathbf{p_i}) - d_\omega \tag{14.24}$$

Applying the Gauss-Newton Method requires the first derivative of the residual function, which reads:

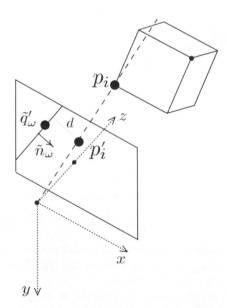

Fig. 14.8. Pose estimation from corresponding 3D-point and 2D-line.

$$\frac{\partial r_i(\boldsymbol{\theta})}{\partial \theta_j} = (\tilde{\mathbf{n}}_\omega)_x \, \frac{\partial m'_x(\boldsymbol{\theta}, \mathbf{p_i})}{\partial \theta_j} + (\tilde{\mathbf{n}}_\omega)_y \, \frac{\partial m'_y(\boldsymbol{\theta}, \mathbf{p_i})}{\partial \theta_j} \tag{14.25}$$

$$= \tilde{\mathbf{n}}_\omega \, \frac{\partial \mathbf{m}'(\boldsymbol{\theta}, \mathbf{p_i})}{\partial \theta_j} \tag{14.26}$$

The objective function requires the absolut value of the residual. However it can be ignored for the derivation as explained now. With the Jacobian $J_{ij} = \frac{\partial r_i(\boldsymbol{\theta})}{\partial \theta_j}$ the gradient of the objective function for a single correspondence is:

$$\nabla f = J^T r(\boldsymbol{\theta}) \tag{14.27}$$

$$= \left(\tilde{\mathbf{n}}_\omega \, \frac{\partial \mathbf{m}'(\boldsymbol{\theta}, \mathbf{p_i})}{\partial \theta_j} \right) \left(\tilde{\mathbf{n}}_\omega \, (\mathbf{m}'(\boldsymbol{\theta}, \mathbf{p_i}) - \tilde{\mathbf{q}}'_\omega) \right) \tag{14.28}$$

Because the first and the last term in the last equation both change signs, if the normal is negated, it is not of importance wether the normal $\tilde{\mathbf{n}}_\omega$ points towards or away from the origin. Therefore the absolut value function in the objective can be ignored and it is valid to derive $r_i(\boldsymbol{\theta})$ instead of $|r_i(\boldsymbol{\theta})|$.

In other words: The distance of the point to the line is given in the direction of the normal in the residual function. The derivative of the point movement is also with respect to that normal. Therefore the minimization makes steps in the correct direction, regardless in which direction the normal is pointing.

Pose estimation from optical flow

In some applications the texture or color of the object is also known in addition to its geometry. In that case, there is the possibility to use the texture information of the model to establish correspondences with the current image. It is not necessary to have a complete textured model, it is sufficient to have colored points.

The idea is to move the object such that the projection of it is most similar to the current image of the object. To achieve this, an assumption is made, that the development of grey values in the near vicinity of the correct point is linear. This assumption is the same as for the KLT tracker [25]. However, this assumption only holds, if the distance between the projected model point and its position in the current image is small. If the displacement is large a multi scale approach is necessary. In that case the current image is smoothed, e.g., with a Gaussian kernel. Increasing sizes of the kernel are applied to get different levels of smoothing. This results in faster image processing for the lower resolution images. The estimation is then applied iteratively to the smallest image first, giving a rough estimate of the pose. Higher resolution images give more accurate pose estimates down to the original image, which yields the most accurate pose. In Figure 14.9 three different levels of smoothing are shown. Let $\mathbf{p}' \in \mathbb{R}^2$ be the projection of the model point for the current pose. And let $\tilde{\mathbf{g}} \in \mathbb{R}^2$ be the grey value gradient vector at position $\mathbf{p}'$ and

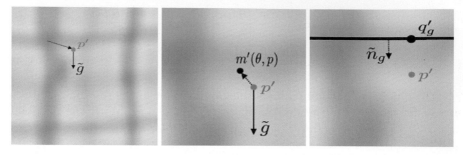

Fig. 14.9. Left: Projection of model point and gradient. Middle: closeup of the movement. Right: Formulation as point line correspondence.

let $d = c_i(\mathbf{p'}) - c_m(\mathbf{p'})$ be the grey value difference between the color of the projected model $c_m(\mathbf{p'})$ point and the pixel value $c_i(\mathbf{p'})$ at the same position in the current image. Then the objective function for minimization is

$$\min_{\boldsymbol{\theta}} \sum_i |\tilde{\mathbf{g}} \ (\mathbf{m'}(\boldsymbol{\theta}, \mathbf{p_i}) - \mathbf{p'}) - d|^2 \qquad (14.29)$$

The geometric interpretation of the objective is given in the following. Assume the object moved to the left and a little upwards resulting in a displacement vector as shown in Figure 14.9 left (the dotted line). This displacement is unknown and depends on the pose parameters of the object. Shown in the left is the projected model point $\mathbf{p'}$ and the image gradient at that position. The goal of minimization is to move the object model point such that the image color at that position equals the object model point's color. If the development of color values would be linear the necessary displacement is $\frac{d}{|\tilde{\mathbf{g}}|}$ in direction of the gradient vector.

This is equal to a 3D-point-2D-line correspondence as in the previous section. The corresponding line has the normal $\tilde{\mathbf{n}}_\mathbf{g} = \frac{1}{|\tilde{\mathbf{g}}|}\tilde{\mathbf{g}}$ and a point $\mathbf{q'_g}$ on the line is found by starting from $\mathbf{p'}$ and taking the point with distance $\frac{d}{|\tilde{\mathbf{g}}|}$ in direction of the gradient vector resulting in $\mathbf{q'_g} = \mathbf{p'} + \frac{d}{|\tilde{\mathbf{g}}|}\tilde{\mathbf{n}}_\mathbf{g}$ as shown in Figure 14.9. The point $\mathbf{p'}$ equals the projected model point at the current pose $\mathbf{m'}(\mathbf{0}, \mathbf{p_i})$. The objective function reads with this formulation:

$$\min_{\boldsymbol{\theta}} \sum_i \left| \tilde{\mathbf{n}}_\mathbf{g} \left(\mathbf{m'}(\boldsymbol{\theta}, \mathbf{p_i}) - \tilde{\mathbf{q}}'_\mathbf{g} \right) \right|^2 \qquad (14.30)$$

with correspondences $((\tilde{\mathbf{n}}_{\mathbf{g},i}, \tilde{\mathbf{q}}'_{\mathbf{g},i}), \ \mathbf{p_i})$.

KLT equations for rigid and articulated objects

An alternative derivation for estimation from model color, leading to the same equations as with (14.30), is obtained by applying the KLT equations [25].

Similar to above, the idea is, that the projected model point has the same image brightness for different poses, also known as *image brightness constancy*.

If the point $\mathbf{p}'$ in the image I_t corresponds to a known 3D object, whose pose is known, the corresponding 3D point $\mathbf{p}$ for that image point can be calculated by intersection of the viewing ray with the 3D object. Now it is possible to find the image displacement to the next image I_{t+1}, which minimizes the brightness difference under the constraint, that the point belongs to a rigid object, which performed a rigid motion from one image to the next. The objective function for multiple points belonging to the object is now with respect to the pose parameters. If the relative motion to the next image is found, the corresponding image displacements can be easily calculated.

Let $I_{t+1}(\mathbf{m}'(\boldsymbol{\theta}, \mathbf{p_i}))$ be the image value at the projected 3D point position, such that $\mathbf{m}'(\mathbf{0}, \mathbf{p_i}) = \mathbf{p}_i'$. The objective function is then

$$\min_{\boldsymbol{\theta}} \sum_i \left(I_{t+1}(\mathbf{m}'(\boldsymbol{\theta}, \mathbf{p_i})) - I_t(\mathbf{p}_i') \right) \tag{14.31}$$

The necessary Jacobian for solving this Nonlinear Least Squares problem consists of the partial derivatives of the residuals:

$$J_{ij} = \frac{\partial I_{t+1}(\mathbf{m}'(\boldsymbol{\theta}, \mathbf{p_i}))}{\partial \theta_j} \tag{14.32}$$

Important is now, that the derivatives are taken at a specific parameter position $\boldsymbol{\theta}_t$, which, after application of the cain rule, leads to:

$$\frac{\partial I_{t+1}(\mathbf{m}'(\boldsymbol{\theta}, \mathbf{p_i}))}{\partial \theta_j}\bigg|_{\boldsymbol{\theta}_t} = \frac{\partial I_{t+1}(\mathbf{q}')}{\partial \mathbf{q}'}\bigg|_{\boldsymbol{\theta}_t} \frac{\partial \mathbf{m}'(\boldsymbol{\theta}, \mathbf{p_i})}{\partial \theta_j}\bigg|_{\boldsymbol{\theta}_t} \tag{14.33}$$

where $\frac{\partial I_{t+1}(\mathbf{q}')}{\partial \mathbf{q}'} = (g_x, g_y)^T$ is the spatial image gradient evaluated at $\mathbf{m}'(\boldsymbol{\theta}_t, \mathbf{p_i})$, which equals $\mathbf{p}_i'$, if $\boldsymbol{\theta}_t$ is zero. As visible, the Jacobian equals those of (14.26) for 3D-point-2D-line correspondences.

14.3.7 Articulated Objects

In this section the estimation of pose or rigid movement as explained in the previous section is extended to kinematic chains. A kinematic chain is a concatenation of transformations. Here the transformations are rotations around arbitrary axes in 3D space. An articulated object consists of one or multiple kinematic chains and has a shape or geometry in addition. Like the arm of the human model in Figure 14.2.

Movement in a kinematic chain

In this work kinematic chains consist only of rotational transformations around known axes at arbitrary positions. The axes for the arm are shown as arrows in Figure 14.2. There is a fixed hierarchy defined for the rotations, e.g., a rotation around the shoulder abduction axis (the one pointing forward) changes the position and orientation of all consecutive axes. In addition to the rotational transformations, there is the position and orientation of the base, which undergoes rigid motions, e.g., the upper body in Figure 14.2.

Consider now the movement of a point $\mathbf{p}$ on the model of the arm with p rotational axes, e.g., a point on the hand. Its movement can be described using the rotation description from Equation (14.1) and the rigid movement from Equation (14.5) resulting in:

$$\mathbf{m}(\boldsymbol{\theta}, \mathbf{p}) = (\theta_x, \theta_y, \theta_z)^T + \tag{14.34}$$
$$\left(R_{\omega_x}(\theta_\alpha) \circ R_{\omega_y}(\theta_\beta) \circ R_{\omega_z}(\theta_\gamma) \circ R_{\omega_1, q_1}(\theta_1) \circ \cdots \circ R_{\omega_p, q_p}(\theta_p) \right) \mathbf{p}$$

The coordinates of the axes $(\mathbf{w_j}, \mathbf{q_j})$ may be given relative to their parent axis (the next one upwards in the chain) or all coordinates of all axes may be given in the same coordinate system, i.e., the same where the base transform is defined, e.g., the world coordinate system. If not stated otherwise $(\mathbf{w_j}, \mathbf{q_j})$ are assumed to be in world coordinates.

If $\boldsymbol{\theta}$ equals $\mathbf{0}$ the position and orientation of the axes define the kinematic chain and also the relative transformations between the axes. To obtain the world coordinates of the axes for a specific $\boldsymbol{\theta}_t$ the direction vector $\boldsymbol{\omega_j}$ and the point on the axis $\mathbf{q_j}$ are transformed by applying the chain transform up to axis $(\mathbf{w_{j-1}}, \mathbf{q_{j-1}})$ to them. Therefore $(\boldsymbol{\omega}_j, \mathbf{q_j})$ are different depending on the current $\boldsymbol{\theta}_t$, if a description in world coordinates is used.

For a specific pose $\boldsymbol{\theta}_t$ a relative movement $\Delta\boldsymbol{\theta}$ can be imagined in the following way: The point $\mathbf{p}$, e.g., on the hand, is rotated at first around the closest axis $(\boldsymbol{\omega}_p, \mathbf{q}_p)$ in the hierarchy with angle $\Delta\theta_p$, then around the fixed second axes $(\boldsymbol{\omega}_{p-1}p, \mathbf{q}_{p-1})$ and so on up to the base, which adds the rigid movement.

Estimating the pose of a kinematic chain from given 3D–3D correspondences $(\mathbf{p_i}, \tilde{\mathbf{p}}_i)$ may be done with *Newton's method*, which is here estimating the joint angles and the global orientation and position. The minimization problem is the same as in Equation (14.4), while the movement function $\mathbf{m}(\boldsymbol{\theta}, \mathbf{p})$ includes here the joint angles as well:

$$\min_{\boldsymbol{\theta}} \sum_i |\mathbf{m}(\boldsymbol{\theta}, \mathbf{p}) - \tilde{\mathbf{p}}|^2 \tag{14.35}$$

To find the minimizer with the iterative *Gauss-Newton* method the Jacobian of the residual functions is necessary.

Jacobian of articulated movement

The partial derivatives of the movement function, which gives the Jacobian, can be derived in the same way as for the rigid movement using the description of rotations around known axes in 3D space from Equation (14.1). If the current pose is $\boldsymbol{\theta}_t$ and only relative movement is estimated the Jacobian is:

$$J = \begin{bmatrix} 1 & 0 & 0 & \frac{\partial m_x}{\partial \theta_\alpha} & \frac{\partial m_x}{\partial \theta_\beta} & \frac{\partial m_x}{\partial \theta_\gamma} & \frac{\partial m_x}{\partial \theta_1} & .. & \frac{\partial m_x}{\partial \theta_p} \\ 0 & 1 & 0 & \frac{\partial m_y}{\partial \theta_\alpha} & \frac{\partial m_y}{\partial \theta_\beta} & \frac{\partial m_y}{\partial \theta_\gamma} & \frac{\partial m_y}{\partial \theta_1} & .. & \frac{\partial m_y}{\partial \theta_p} \\ 0 & 0 & 1 & \frac{\partial m_z}{\partial \theta_\alpha} & \frac{\partial m_z}{\partial \theta_\beta} & \frac{\partial m_z}{\partial \theta_\gamma} & \frac{\partial m_z}{\partial \theta_1} & .. & \frac{\partial m_z}{\partial \theta_p} \end{bmatrix} \qquad (14.36)$$

The derivatives at zero are:

$$\left. \frac{\partial \mathbf{m}(\boldsymbol{\theta}, \mathbf{p})}{\partial \theta_\alpha} \right|_{\boldsymbol{\theta}=\mathbf{0}} = \boldsymbol{\omega}_\mathbf{x} \times (\mathbf{p} - \mathbf{q}_\mathbf{x})$$

$$\left. \frac{\partial \mathbf{m}(\boldsymbol{\theta}, \mathbf{p})}{\partial \theta_\beta} \right|_{\boldsymbol{\theta}=\mathbf{0}} = \boldsymbol{\omega}_\mathbf{y} \times (\mathbf{p} - \mathbf{q}_\mathbf{y})$$

$$\left. \frac{\partial \mathbf{m}(\boldsymbol{\theta}, \mathbf{p})}{\partial \theta_\gamma} \right|_{\boldsymbol{\theta}=\mathbf{0}} = \boldsymbol{\omega}_\mathbf{z} \times (\mathbf{p} - \mathbf{q}_\mathbf{z})$$ $$(14.37)$$

$$\left. \frac{\partial \mathbf{m}(\boldsymbol{\theta}, \mathbf{p})}{\partial \theta_j} \right|_{\boldsymbol{\theta}=\mathbf{0}} = \boldsymbol{\omega}_\mathbf{j} \times (\mathbf{p} - \mathbf{q}_\mathbf{j})$$

where $j \in \{1, .., p\}$ and $\boldsymbol{\omega}_\mathbf{x}, \mathbf{q}_x$ denote rotation around an arbitrary point in space and $\boldsymbol{\omega}_\mathbf{x}, \boldsymbol{\omega}_\mathbf{y}, \boldsymbol{\omega}_\mathbf{z}$ are assumed orthogonal.

Given here is the special Jacobian for $\boldsymbol{\theta} = \mathbf{0}$, which is valid if all axes are given in world coordinates and only relative movement is estimated.

Jacobian of articulated movement under projection

If the movement of the articulated object is observed by a pinhole camera, and 3D-point–2D-point correspondences $(\mathbf{p}, \tilde{\mathbf{p}}')$ are given, the estimation of joint angles and global pose can be done in the same way as for rigid objects by Nonlinear Least Squares (solved with Gauss-Newton). The optimization problem reads:

$$\min_{\boldsymbol{\theta}} \sum_i |\mathbf{m}'(\boldsymbol{\theta}, \mathbf{p}) - \tilde{\mathbf{p}}'|^2 \qquad (14.38)$$

The necessary Jacobian is similar to the partial derivatives of Equation (14.15). Here the additional partial derivatives for the rotation around the joint axes give additional columns in the Jacobian.

To estimate the pose of the articulated object from other kinds of correspondences, the same optimization with the according objective function as for the rigid movement can be applied. The only difference is the different movement function, that now also includes the joint angles of the kinematic chain. The necessary partial derivatives can be looked up in the two previous sections.

Important to note is, that for each model point it is necessary to know, to which joint of the articulated object the point belongs, e.g., a point between the wrist and elbow can give no information about the wrist joint angles. Therefore the Jacobian entries for that point will be zero in the column with the wrist angles.

14.4 Comparison with Gradient Descent

Throughout the literature on body pose estimation from 3D-point–3D-point correspondences there are various optimization techniques that try to minimize the difference between observed 3D points and the surface of the model. Very often this optimization problem is solved with a Gradient Descent method and numerical derivatives, e.g., numerical derivatives are used in [3, 7, 14]. Gradient Descent is still very popular, e.g., in [13] Euler-Lagrange equations are integrated over time, which is similar to standard Gradient Descent with a fixed step size.

Because Gradient Descent (GD) converges very slowly in flat regions, there are various approaches to increase the convergence rate. In the work of Bray [3, 14] a new Gradient Descent method is proposed, called "Stochastic Meta Descent" (GDSMD). The GDSMD method is applied to estimate the motion and pose of a human. The motion function is modeled similar as in this work by rotations around known axes, however optimization relies on numerical derivatives. Additionally in both publications a comparison with other common optimization methods was made, these were the Levenberg-Marquardt extension of Gauss-Newton in [14] and BFGS and Powell's method in [3]. The proposed GDSMSD was reported to work faster and more accurate. Therefore a detailed comparison with Gauss-Newton is made here.

14.4.1 Stochastic Meta Descent

We describe here briefly the SMD method. The stochasticity of this approach is due to random sub sampling of the complete data in each iteration.

A Gradient Descent method updates parameters in one iteration using the update formula

$$\boldsymbol{\theta_{t+1}} = \boldsymbol{\theta_t} - \boldsymbol{\alpha} \cdot \nabla f_i(\boldsymbol{\theta}) = \boldsymbol{\theta_t} - \boldsymbol{\alpha} \cdot J^T \mathbf{r}(\boldsymbol{\theta_t}) \tag{14.39}$$

where t is the current time step or iteration number, the operator "·" indicates a componentwise product, such that each parameter may have a different step size.

According to [14] the idea of SMD is to update the step sizes automatically depending on the steepness and direction of the gradient in last time steps. Especially if the error function exhibits long narrow valleys, a gradient descent might alternate from one side of the valley to the other. To model these behavior, the step size is increased for parameters, whose gradient had the same sign in the last steps and is decreased, if the sign changed. In detail this is:

$$\boldsymbol{\alpha_t} = \boldsymbol{\alpha_{t-1}} \cdot \max\left(\frac{1}{2}, \boldsymbol{\mu} \cdot \mathbf{v_t} \cdot \nabla f_t(\boldsymbol{\theta_t})\right) \tag{14.40}$$

where $\boldsymbol{\mu}$ is a vector of meta step sizes, which control the rate of change of the original step sizes and $\mathbf{v}$ is the exponential average of the effect of all past step sizes. The update of $\mathbf{v}$ is simplified in the following as in [14]:

$$\mathbf{v}_{t+1} = \mathbf{a}_t \cdot \nabla f_t(\boldsymbol{\theta}_t) \qquad (14.41)$$

Constraints on possible motion of the human are imposed by mapping back each parameter to its allowed interval in each iteration. Because the resulting parameter change is smaller in that case, the meta step sizes are updated accordingly.

Global transform

In the first experiment the global transform is estimated, while only the translation in x-direction and the rotation around the height axis are changed. Figure 14.10 shows on the left the starting pose for estimation together with the observed data point cloud, which are generated from the pose shown on the right with added noise (deviation 5 cm). The difference between start and target pose is rotation around the y-axis with -40 degrees and translation in x-direction with 0.35 m.

Contour levels of the error surface are shown in Figure 14.11 with the minimum at $(0,0)$. The error surface of the global transform is calculated by changing the rotation from -40 to $+40$ degrees by 0.5 degrees and the translation in x-direction from -0.4 m to $+0.2$ m by 0.02 cm. For each model pose the error is calculated setting up correspondences by Closest Point using all 6,500 observed points. The plotted error is the average distance between model and observed points. As visible, the error surface is almost quadratic, which can be expected, because the translation in x-direction is linear giving a quadratic error for the translation and the quadratic approximation of the rotation is good for small angles. Visible here is also, that the quadratic approximation is valid even for changing correspondences (due to nearest neighbour).

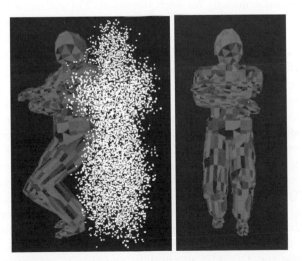

Fig. 14.10. Starting pose on the left with the disturbed point cloud of the target pose (noise deviation 0.05 m) and target pose on the right.

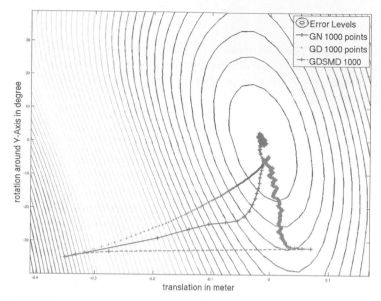

Fig. 14.11. Estimation of the global transform for 1,000 correspondences. Plotted is the error surface and the iterations for Gauss-Newton (GN), Gradient Descent (GD) and Meta Descent (GDSMD).

We compare *Gauss-Newton* (GN), Gradient Descent (GD) and the *Stochastic Meta Descent* (GDSMD) by estimating the 6 parameters of the global transform. The x position and the rotation angle around the y-axis are plotted for each three methods. From the total set of 6,500 observed depth points 1,000 points were chosen randomly for each iteration. The estimation is stopped, if the parameter change is below a certain threshold or at a maximum of 200 iterations. The result for GN is shown in Figure 14.11 as a green solid line, for GD as blue crosses, and for GDSMD as a dashed red line. The initial stepsizes for GD and GDSMD are 0.1 for translation and 1 for rotation. The additional parameters for GDSMD are $\mu = 1,000$ and $\lambda = 0$.

The Gradient Descent methods do not reach the minimum at $(0, 0)$ within 200 iterations. The main reason for the bad performance of the GD methods is their characteristic to make small steps on flat surfaces, because the parameter change in each iteration is proportional to the size of the Jacobian (the gradient), while the GN assumes an quadratic error function resulting in larger steps on smooth surfaces.

The GDSMD approach tries to overcome the slow convergence of the GD in flat regions by adjusting the stepsizes according to previous parameter changes. However, in the case shown, this feature has also the effect to increase the step size dramatically in steep regions. This dramatic increase in the step sizes made the GDSMD method in the shown case even slower than the standard GD without varying step sizes. Possibly better chosen values of

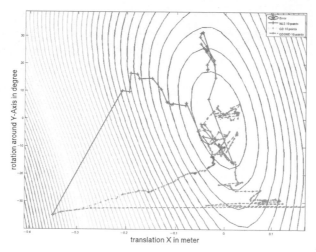

Fig. 14.12. Estimation of the global transform for 10 correspondences. Initial step size of GD and GDSMD was 0.1 for translation and 1 for rotation.

$\mu = 1{,}000$ and $\lambda = 0$ and better initially step sizes could overcome the effect. However, the GN approach does not need any kind of parameters, making it much easier to use.

Using the the same starting point and the same parameters for GD and GDSMD the estimation is repeated using only 10 random correspondences. Results are shown in Figure 14.12. As visible the GN starts to jump very erratic, because of the small amount of data. The GD methods are able to reach a point near the minimum, but the larger effect of noise due to the small data amount is also visible.

Shoulder elbow movement

The error surface of the previous experiment is nearly quadratic. To have a more difficult and less smooth error surface another experiment is carried out. The global position and rotation of the model is unchanged, but the shoulder twist and the elbow flexion are changed. Due to the Closest Point correspondences, which are changing significantly for the lower arm, when it is moving near to the body, the error surface is less smooth. Also the amount of correspondences is lower (around 650), because correspondences were only established for observed points, whose closest corresponding model point is on the upper or lower right arm. The smaller amount of correspondences makes the error surface less smooth.

The estimation is performed using an initial step size of 20 for GD and GDSMD. The additional meta step sizes for GDSMD were $\mu = 1{,}000$ and $\lambda = 0$. Figure 14.13 shows the development of shoulder twist and elbow flexion together with error contour levels. Within 200 iterations all methods reach the correct minimum while performing similarly.

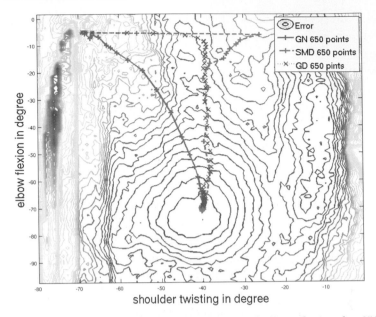

Fig. 14.13. Estimation of the shoulder twisting and elbow flexion for 650 correspondences. Initial stepsizes are 20 for both angles.

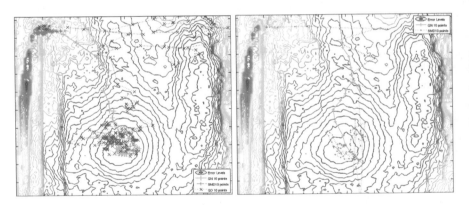

Fig. 14.14. Estimation of the shoulder twisting and elbow flexion for 10 correspondences. Two runs with the same parameters.

Shown in Figure 14.14 are two runs of the same minimization with 10 random correspondences in each iteration. As visible, the noise and the nearest neighbor correspondences lead to an erratic development of estimated angles. Smaller initial step sizes for the GD methods smoothes the development of values, but also increases the possibility to get stuck at a local minimum as in the left plot, where the GDSMD method got stuck in the beginning. On the other hand the small amount of data can lead to large changes in the wrong direction for the GN method. A dampening regularizer for GN is suggested here.

14.4.2 Summary of Comparison

Comparing the Gradient Descent (GD) and the Stochastic Meta Descent (GDSMD) directly, the GDSMD performs not necessarily better than standard GD with fixed step sizes. However, with changes in the step sizes and additional meta step sizes it is possible to tune the parameters, such that GDSMD needs less iterations than standard GD. The GD methods have the advantage that even a small amount of data is sufficient to find a minimum. However, in the presence of noisy measurements averaging smoothes the effect of noise, such that it is beneficial to use more data points at once.

The Gauss-Newton method is not in the need of additional parameters and converges faster in flat regions. The drawback of the GN approach is the need for many data points. There should be at least as much data than parameters to estimate, otherwise it is not possible to calculate a solution in the Least Squares iteration step, because the $J^T J$ matrix is not invertible. Important to note here is the possibility of *dampening* by introducing a single dampening value, which is added as regularizer to the objective function. The *Levenberg-Marquardt* extension of GN is an automatic method to calculate the dampening value. However, if the fraction of data amount to parameters is small, a manual dampening value is more appropriate.

The experiments showed (as can be expected), that optimization in the presence of noise is more accurate and stable (with respect to local minima), if more data is used simultaneously. If only a subset of the whole data is used in each iteration, processing time is decreased, but additional iterations are necessary, because step sizes have to be small.

An argument in [3], which explains the reported worse behaviour of GN like methods, is their characteristic to make large steps in each iteration. If the range of possible motion for each joint angle is constrained by min-max values, it is possible, that the minimization does not improve significantly in one iteration, because a large step may lead far outside the feasible region. However for GN it is possible to reduce the step size by increasing the dampening value, which shifts the GN more to a GD. For GD methods it is also not guaranteed, that the small steps overcome the problem of getting stuck at the border of the feasible region. In our experiments in this and previous work, we did not encounter the problem of getting stuck on the border with GN so far.

14.5 Estimation from Stereo Depth Maps

In this section motion of a person is estimated by Nonlinear Least Squares as explained in the previous sections by use of depth data from a stereo camera setup. At first the algorithm to find 3D-point–3D-point correspondences efficiently using openGL is explained and followed in the end by results.

14.5.1 Stereo

Our motion estimation is based on dense depth information which could be estimated directly from correspondences between images. Traditionally, pairwise rectified stereo images were analyzed exploiting geometrical constraints along the epipolar lines. More recently, generalized approaches were introduced that can handle multiple images and higher order constraints. See [23] for an overview. Achieving realtime performance on standard hardware has become reality with the availability of free programmable graphics hardware (GPU) and the additional benefit of keeping the CPU free for other tasks like our pose estimation [27]. The results presented here are calculated from depth images generated by a dynamic programming-based disparity estimator with a pyramidal scheme for dense correspondence matching along the epipolar lines [10].

14.5.2 Building of 3D–3D Correspondences

To apply the pose estimation algorithm described in the previous chapter, correspondences between observed 3D points and 3D model points are necessary. As the observed points are calculated from depth maps, these correspondences are not known. Equal to [3, 4, 8, 22] an Iterative Closest Point (ICP) approach is taken. For each observed point the nearest point on the model is assumed to be the corresponding one. The body pose of the model is calculated with these correspondences. The following steps are repeated multiple times for each frame of a sequence. The larger the relative motion is, the more correspondences will be incorrect. The principle idea of ICP is, that an estimated relative motion decreases the difference between model and observation, such that more correspondences are correct if build again for the smaller difference.

The steps in one ICP iteration are:

1. Render Model in current pose (OpenSG)
2. Find visible model points by analyzing color of rendered image pixels
3. Build correspondences from depth map by finding nearest model point
4. Estimate relative movement by Nonlinear Least Squares

Between frames it is common to predict the pose for the next image by applying some kind of motion model. In this work the displacement between the last two frames is taken for prediction. Assumed is a constant velocity for each joint angle between frames.

In the following the single steps are explained in detail.

Rendering the model

Because it is assumed, that the pose was calculated correctly in the last frame and the displacement is small between frames, the rendering of the model with the last calculated pose gives approximately the visible body parts. To get a correct synthesized view the internal parameters of the calibrated camera have to be applied to the virtual camera. Because the used rendering library OpenSG, did not support cameras, whose principal point is not in the image center, OpenSG was extended by an "OffCenterCamera", which principal point is given in normalized device coordinates (−1 to 1).

Find visible model points

The rendered image can then be analyzed to get the visible 3D points of the model. A possible solution is to calculate for each image pixel, which belongs to the model (is not zero), the viewing ray and intersect it with the 3D model. However, this is very time consuming.

Therefore it is assumed that the corners of the model's triangles represent the surface sufficiently dense. Then a lookup table of model points can be established, that allows fast computation of the visible subset of model points.

Each triangle of the model is rendered in a unique color, such that the red value indexes the body part (corresponds to a BAP id) and the green and blue values index the vertex in the body part's geometry. The resulting image looks like that shown in Figure 14.15. The set of visible model points is an associative array, which uses a binary search tree (C++ std::map), where the key value is the height value of the point. The height is appropriate, because the extension of the visible point cloud is usually largest in height. A principal component analysis (PCA) could be conducted to find the coordinate with largest variance. However for the experiments in this work the height coordinate has the largest variance.

Fig. 14.15. The 3D point cloud computed from the depth image.

Build correspondences by nearest neighbor

The observations are depth images calculated from disparity maps. Together with the internal camera parameters a 3D point cloud of observed points can be computed. For these observed points it is now necessary to find the closest visible point model point.

An example for the observed point set is shown in Figure 14.15 together with the body model. The green boxes represent the observed points.

The computation time for one frame depends largely on the amount of observed points, therefore the depth image is randomly subsampled to decrease the number of correspondences and reduce computation time. To calculate the 3D observed points from the known focal length and principal point of the camera efficiently, the camera coordinate system is assumed to be aligned with the world coordinate system and the body model is positioned initially, such that it is close to the observed point set. This simplification is only possible as long as a single stereo view is used for pose estimation. For multiple views, the complete projection matrix has to be used to calculate viewing rays for each pixel in the depth image.

For each observed point the closest visible model point is found by searching in the associative array for the nearest point with regard to height. Starting from the point with the smallest height difference $\mathbf{p}_1$, the search is alternated in both directions. and is stopped, if the distance in height to a point is larger than the Euclidean distance d_E to the nearest point $\mathbf{p}_N$ found so far. Figure 14.16 illustrates this. The last tested points are $\mathbf{p}_{top}$ and $\mathbf{p}_{bot}$.

Estimation of relative movement

To fit the body model to the observed point set, a segmentation of the person from the background is necessary. In [3] this is done by using skin color. We assume here only that there are no scene objects (or only negligible parts) within a certain distance to the person by using reweighted least-squares in the pose estimation step. Therefore no explicit segmentation is done, but correspondences, whose error value is larger than a predefined mininmum distance, are not included in the estimation, because the corresponding observed point belongs probably to the background scene.

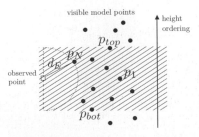

Fig. 14.16. Search for the nearest model point is bound by height distance.

The estimation of body pose for one set of correspondences involves multiple iterations within the dampened Gauss-Newton optimization. There is an upper bound on the number of iterations to have an upper bound on the estimation time. Iterations are also stopped if the parameter change drops below a certain threshold.

14.5.3 Results

The implementation is tested with depth images from a real sequence. The efficiency of our implementation was already shown in [12] where complex arm movement is estimated with 5fps on a 3Ghz Pentium 4. Here experiments are carried out here with 24 DOF motion neglecting computation time. The estimation time per frame depends mainly on the number of used data points. In the experiment shown all points from the depth image are taken, the estimation time with 100,000 points was about 15 seconds per frame, while in [12] 800 correspondences were used. Two depth images from the sequence are shown in 14.17. The initial pose was defined manually. The resulting poses are shown in Figure 14.18, where the model is superimposed on the original image. The 24 DOF are in detail: 4 for each arm (3 shoulder and 1 elbow),

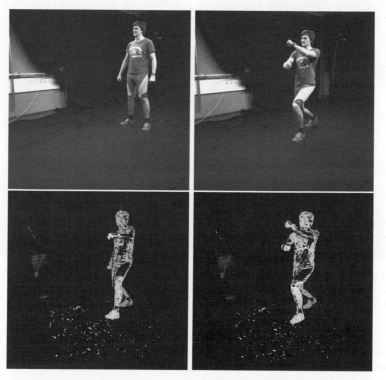

Fig. 14.17. Original undistorted image and calculated depth image below it.

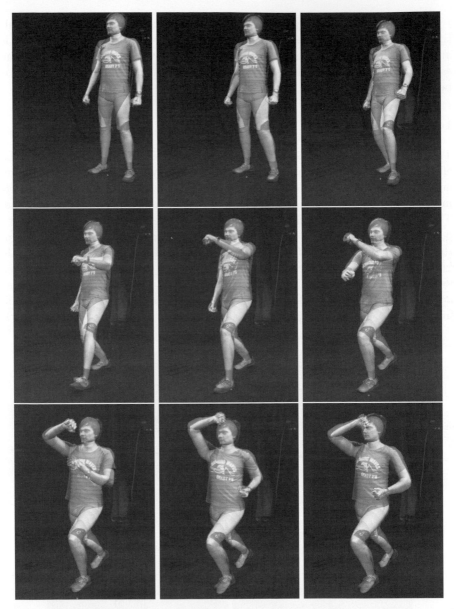

Fig. 14.18. Estimated motion with 24 DOF. Model superimposed on the original image. (COLOR).

6 for the global transform and 5 for each leg (3 hip, 1 knee and 1 ankle). The original sequence was recorded with 25fps, after 130 frames tracking of one arm angle is lost. This is due to the single stereo view.

All available stereo data of about 100,000 points were used for this sequence, though not all are shown in the figure, because the floating point depth image was scaled and shifted, such that the interesting depth range is clearly visible.

14.6 Conclusions

We showed how body pose estimation by nonlinear optimization methods can be improved using the correct derivatives within a *Nonlinear Least Squares framework*. Different kind of correspondences can be handled at once. The resulting equations are similar to previous approaches (12) and even increase the accuracy for 2D-3D correspondences. A comparison of different optimization methods is performed, showing the advantages of the *Gauss-Newton* over *Gradient Descent* methods. The proposed estimation method for articulated objects is applied to motion sequences with complex 24 DOF motion showing the possibilities of motion capture form a single stereo view.

Thanks to Bodo Rosenhahn, Edilson de Aguiar and Prof. H.-P. Seidel (Max-Planck-Institut, Saarbrücken) for recording of the test sequences.

References

1. ISO/IEC 14496. Part 2: Generic coding of audio-visual objects, 1999.
2. Araujo H., Carceroni R. and Brown C. A fully projective formulation to improve the accuracy of lowe's pose estimation algorithm. *Computer Vision and Image Understanding*, 70(2), 1998.
3. Bray M., Koller-Meier E., Mueller P., Van Gool L. and Schraudolph N.N. 3D hand tracking by rapid stochastic gradient descent using a skinning model. In *CVMP*. IEE, March 2004.
4. Bregler C. and Malik J. Tracking people with twists and exponential maps. In *Proc. IEEE CVPR*, pages 8–15, 1998.
5. Chong E. K.P. and Zak S. *An Introduction to Optimization, Second Edition*. Wiley, 2001.
6. Open Source 3DGraphics Creation. www.blender.org.
7. deAguiar E., Theobalt C., Magnor M. and Seidel H.-P. Reconstructing human shape and motion from multi-view video. In *2nd European Conference on Visual Media Production (CVMP)*, London, UK, 2005.
8. Demirdjian D., Ko T. and Darrell T. Constraining human body tracking. In *Proc. ICCV*, Nice, France, October 2003.
9. Mündermann L. Validation of a markerless motion capture system for the calculation of lower extremity kinematics. In *Proc. American Society of Biomechanics*, Cleveland, USA, 2005.
10. Falkenhagen L. Hierarchical block-based disparity estimation considering neighbourhood constraints, 1997.
11. Grest D. and Koch R. Multi-camera person tracking in a cluttered interaction environment. In *CAIP*, Paris, France, 2005.

12. Grest D., Woetzel J. and Koch R. Nonlinear body pose estimation from depth images. In *Proc. DAGM*, Vienna, September 2005.

13. Kakadiaris I. and Metaxas D. Model-based estimation of 3D human motion. *IEEE Trans. on Pattern Analysis and Machine Intelligence (PAMI)*, 22(12), 2000.

14. Roland K., Bray M. and Van Gool L. Full body tracking from multiple views using stochastic sampling. In *Proc. CVPR*, pages 129–136, 2005.

15. Autodesk Maya. www.autodesk.com/maya, 2006.

16. Mohr A. and Gleicher M. Building efficient, accurate character skins from examples. In *ACM SIGGRAPH*, pages 562–568, 2003.

17. Sales Motion CaptureDistribution and Consulting. www.metamotion.com.

18. Murray R.M., Li Z. and Sastry S.S. *A Mathematical Introduction to Robotic Manipulation*. CRC Press, 1994.

19. Niskanen M., Boyer E. and Horaud R. Articulated motion capture from 3-D points and normals. In *CVMP*, London, 2005.

20. Plänkers R. and Fua P. Model-based silhouette extraction for accurate people tracking. In *Proc. ECCV*, pp. 325–339, Springer, 2002.

21. Rosenhahn B., Kersting U., Smith D., Gurney J., Brox T. and Klette R. A system for markerless human motion estimation. In W. Kropatsch, editor, *DAGM*, Wien, Austria, September 2005.

22. Rosenhahn B. and Sommer G. Adaptive pose estimation for different corresponding entities. In *Proc. DAGM*, pages 265–273, Springer, 2002.

23. Scharstein D., Szeliski R. and Zabih R. A taxonomy and evaluation of dense two-frame stereo correspondence algorithms. In *Proc. IEEE Workshop on Stereo and Multi-Baseline Vision*, Kauai, HI, December 2001.

24. ViconMotion Systems and PeakPerformance Inc. www.vicon.com, 2006.

25. Tomasi C. and Kanade T. Detection and tracking of point features. *Technical report*. Carnegie Mellon University, Pittsburg, PA, 1991.

26. Wettegren B., Christensen L.B., Rosenhahn B., Granert O. and Krüger N. Image uncertainty and pose estimation in 3D Euclidian space. *Proc. DSAGM*, pages 76–84, 2005.

27. Yang R. and Pollefeys M. Multi-resolution real-time stereo on commodity graphics hardware. In *Conference of Computer Vision and Pattern Recognition CVPR03*, Madison, WISC., USA, June 2003.

15

Markerless Motion Capture for Biomechanical Applications

Lars Mündermann[1], Stefano Corazza[1], and Thomas P. Andriacchi[1,2,3]

[1] Stanford University, Biomechanical Engineering
   Stanford, CA 94305-4038, USA

[2] Bone and Joint Research Center, VA Palo Alto
   Palo Alto, CA 94304, USA

[3] Department of Orthopedic Surgery, Stanford University Medical Center
   Stanford, CA 94305, USA

Summary. Most common methods for accurate capture of three-dimensional human movement require a laboratory environment and the attachment of markers or fixtures to the bodys segments. These laboratory conditions can cause unknown experimental artifacts. Thus, our understanding of normal and pathological human movement would be enhanced by a method that allows the capture of human movement without the constraint of markers or fixtures placed on the body. The need for markerless human motion capture methods is discussed and the advancement of markerless approaches is considered in view of accurate capture of three-dimensional human movement for biomechanical applications. The role of choosing appropriate technical equipment and algorithms for accurate markerless motion capture is critical. The implementation of this new methodology offers the promise for simple, timeefficient, and potentially more meaningful assessments of human movement in research and clinical practice.

15.1 Introduction

Human motion capture is a well-established paradigm for the diagnosis of the patho- mechanics related to musculoskeletal diseases, the development and evaluation of rehabilitative treatments and preventive interventions for musculoskeletal diseases The use of methods for accurate human motion capture, including kinematics and kinetics, in biomechanical and clinical environment is motivated by the need to understand normal and pathological movement [9]. For example, for the investigation of osteoarthritis (OA) initiation [61,64,65], gait analysis performed through marker based motion capture techniques provides an effective research tool to identify underlying mechanical factors that influence the disease progression. A next critical advancement in human motion capture is the development of a noninvasive and markerless

B. Rosenhahn et al. (eds.), Human Motion – Understanding, Modelling, Capture, and Animation, 377–398.

system. A technique for accurately measuring human body kinematics that does not require markers or fixtures placed on the body would greatly expand the applicability of human motion capture.

15.1.1 Current State of the Art

At present, the most common methods for accurate capture of three-dimensional human movement require a laboratory environment and the attachment of markers, fixtures or sensors to the body segments. These laboratory conditions can cause experimental artifacts. For example, it has been shown that attaching straps to the thigh or shank alters joint kinematics and kinetics [34]. In general, the primary technical factors limiting the advancement of the study of human movement is the measurement of skeletal movement from a finite number of markers or sensors placed on the skin. The movement of the markers is typically used to infer the underlying relative movement between two adjacent segments (e.g., knee joint) with the goal of precisely defining the movement of the joint. Skin movement relative to the underlying bone is a primary factor limiting the resolution of detailed joint movement using skin-based systems [22, 41, 50, 77, 80].

Skeletal movement can also be measured directly using alternative approaches to a skin-based marker system. These approaches include stereoradiography [44], bone pins [47, 77], external fixation devices [41] or single plane fluoroscopic techniques [12, 85]. While these methods provide direct measurement of skeletal movement, they are invasive or expose the test subject to radiation. More recently, real-time magnetic resonance imaging (MRI) using open-access MRI provides noninvasive and harmless in vivo measurement of bones, ligaments, muscle, etc. [79]. However, all these methods also impede natural patterns of movements and care must be taken when attempting to extrapolate these types of measurements to natural patterns of locomotion. With skin-based marker systems, in most cases, only large motions such as flexion–extension have acceptable error limits. Cappozzo et al. [21] have examined five subjects with external fixator devices and compared the estimates of bone location and orientation between coordinate systems embedded in the bone and coordinate systems determined from skin-based marker systems for walking, cycling and flexion–extension activities. Comparisons of bone orientation from true bone embedded markers versus clusters of three skin-based markers indicate a worst-case root mean square artifact of $7°$.

The most frequently used method for measuring human movement involves placing markers or fixtures on the skin's surface of the segment being analyzed [15]. The vast majority of current analysis techniques model the limb segment as a rigid body, then apply various estimation algorithms to obtain an optimal estimate of the rigid body motion. One such rigid body model formulation is given by Spoor and Veldpas [84]; they have described a rigid body model technique using a minimum mean square error approach that lessens the effect of deformation between any two time steps. This assumption limits the scope of application for this method, since markers placed directly on skin will experience nonrigid body movement. Lu and O'Connor [55] expanded the rigid body model approach; rather than seeking the optimal rigid body transformation on each segment individually, multiple, constrained rigid body transforms are sought, modelling the hip, knee, and ankle as ball and socket joints. The difficulty with this approach is modelling the joints as ball and sockets where all joint translations are treated as artifact, which is clearly a limitation for knee motion. Lucchetti et al. [56] presented an entirely different approach, using artifact assessment exercise to determine the correlation between flexion–extension angles and apparent skin marker artifact trajectories. A limitation of this approach is the assumption that the skin motion during the quasi-static artifact assessment movements is the same as during dynamic activities.

A recently described [7,8] point cluster technique (PCT) employs an over-abundance of markers (a cluster) placed on each segment to minimize the effects of skin movement artifact. The basic PCT [2] can be extended to minimize skin movement artifact by optimal weighting of the markers according to their degree of deformation. Another extension of the basic PCT corrects for error induced by segment deformation associated with skin marker movement relative to the underlying bone. This is accomplished by extending the transformation equations to the general deformation case, modelling the deformation by an activity-dependent function, and smoothing the deformation over a specified interval to the functional form. A limitation of this approach is the time-consuming placement of additional markers.

In addition to skin movement artifact, many of the previously described methods can introduce an artificial stimulus to the neurosensory system while measuring human movement yielding motion patterns that do not reflect natural patterns of movement. For example, even walking on a treadmill can produce changes in the stride length–walking speed relationships [13]. Insertion of bone pins, the strapping of tight fixtures around limb segments or constraints to normal movement patterns (such as required for fluoroscopic or other radiographic imaging measurements) can introduce artifacts into the observation of human movement due to local anesthesia and/or interference with musculoskeletal structures. In some cases, these artifacts can lead to incorrect interpretations of movement data.

15.1.2 Need for Unencumbered Motion Capture

The potential for measurement-induced artifact is particularly relevant to studies where subtle gait changes are associated with pathology. For example, the success of newer methods for the treatment and prevention of diseases such as osteoarthritis [10] is influenced by subtle changes in the patterns of locomotion. Thus, the ability to accurately measure patterns of locomotion without the risk of an artificial stimulus producing unwanted artifacts that could mask the natural patterns of motion is an important need for emerging health care applications.

Motion capture is an important method for studies in biomechanics and has traditionally been used for the diagnosis of the patho-mechanics related to musculoskeletal diseases [9, 39]. Recently it has also been used in the development and evaluation of rehabilitative treatments and preventive interventions for musculoskeletal diseases [64]. Although motion analysis has been recognized as clinically useful, the routine clinical use of gait analysis has seen very limited growth. The issue of its clinical value is related to many factors, including the applicability of existing technology to addressing clinical problems and the length of time and costs required for data collection, processing and interpretation [83]. A next critical advancement in human motion capture is the development of a noninvasive and markerless system. Eliminating the need for markers would considerably reduce patient preparatory time and enable simple, time-efficient, and potentially more meaningful assessments of human movement in research and clinical practice. Ideally, the measurement system should be neither invasive nor harmful and only minimally encumber the subject. Furthermore, it should allow measuring subjects in their natural environment such as their work place, home, or on sport fields and be capable of measuring natural activities over a sufficiently large field of view. To date, markerless methods are not widely available because the accurate capture of human movement without markers is technically challenging yet recent technical developments in computer vision provide the potential for markerless human motion capture for biomechanical and clinical applications.

15.2 Previous Work

In contrast to marker-based systems motivated by the expanded need for improved knowledge of locomotion, the development of markerless motion capture systems originated from the fields of computer vision and machine learning, where the analysis of human actions by a computer is gaining increasing interest. Potential applications of human motion capture are the driving force of system development, and the major application areas include smart surveillance, identification, control, perceptual interface, character animation, virtual reality, view interpolation, and motion analysis [62, 87]. Over the last two decades, the field of registering human body motion using computer vision has grown substantially, and a great variety of vision-based systems have been proposed for tracking human motion. These systems vary in the number of cameras used (camera configuration), the representation of captured data, types of algorithms, use of various models, and the application to specific body regions and whole body. Employed configurations typically range from using a single camera [40, 51, 86] to multiple cameras [30, 35, 45, 46, 72].

An even greater variety of algorithms has been proposed for estimating human motion including constraint propagation [73], optical flow [20, 90], medial axis transformation [17], stochastic propagation [43], search space decomposition based on cues [35], statistical models of background and foreground [89], silhouette contours [52], annealed particle filtering [32], silhouette-based techniques [18, 24], shape-encoded particle propagation [63], and fuzzy clustering process [60]. These algorithms typically derive features either directly in the single or multiple 2D image planes [20, 43] or, in the case of multiple cameras, at times utilize a 3D representation [24, 35] for estimating human body kinematics, and are often classified into model-based and model-free approaches. The majority of approaches is model-based in which an a priori model with relevant anatomic and kinematic information is tracked or matched to 2D image planes or 3D representations. Different model types have been proposed including stick-figure [51], cylinders [40], super-quadrics [35], and CAD model [90]. Model-free approaches attempt to capture skeleton features in the absence of an a priori model. These include the representation of motion in form of simple bounding boxes [31] or stick-figure through medial axis transformation [17], and the use of Isomaps [25] and Laplacian Eigenmaps [28] for transforming a 3D representation into a pose-invariant graph for extracting kinematics. Several surveys concerned with computer-vision approaches have been published in recent years, each classifying existing methods into different categories [1, 23, 36, 62, 87]. For instance, Moeslund et al. [62] reviewed more than 130 human motion capture papers published between 1980 and 2000 and categorized motion capture approaches by the stages necessary to solve the general problem of motion capture. Wang et al. [87] provided a similar survey of human motion capture approaches in the field of computer vision ranging mainly from 1997 to 2001 with a greater emphasize on categorizing the framework of human motion analysis in low-level vision, intermediate-level vision, and high-level vision systems.

15.2.1 Utilizing an Articulated Model

The use of a model for identifying a pose and/or individual body segments for subsequently extracting kinematic information highlights a fundamental shift in paradigm between marker-based and markerless systems. A marker-based system typically provides three-dimensional positions of markers attached to a subject. Markers are typically placed upon the area of interest and subsequently processed to gain information. This process includes labeling markers and establishing a correspondence between markers and the object of interest. Numerous marker protocols have been introduced ranging from simple link models to point cluster techniques providing a local coordinate frame for individual body segments. Biomechanical and clinical studies usually utilize marker-based technology for collecting marker positions and adapt marker protocols and analysis techniques to the complexity necessary for providing a valid model for the underlying research question. In comparison, passive model-based markerless approaches typically match a predefined articulated model to multiple image sequences. This eliminates the need for labeling and whole-body kinematics could be obtained instantly. However, since the labeling is part of the tracking/matching process, the model needs to contain enough detailed description of the area of interest for providing useful information for biomechanical and clinical studies. Models introduced in previous markerless approaches typically suffer accurate detail necessary. This is one of the main limitations to utilize previous approaches in biomechanical and clinical practice.

In general, the human body is a very complex system. The skeleton of an adult human is comprised of 206 bones and two systems, the axial skeleton (the trunk of our body) and the appendicular skeleton (our limbs) [37]. The bones themselves are divided up into four classes: long bones (which make up the limbs), short bones (which are grouped together to strengthen our skeleton), flat bones (which protect our body and provide a place for muscles to attach), and irregular bones. Studies on human locomotion are primarily concerned with the interaction among major body segments such as hip, thigh, shin, foot, trunk, shoulder, upper arm, forearm and hand. There are basically three types of limb joint in animals and humans. These are the ball and socket joint (e.g., hip), the pivot joint (e.g., elbow) and the condylar joint (e.g., knee). The hip joint is located between the pelvis and the upper end of the femur (thighbone). The hip is an extremely stable ball and socket joint. The smooth, rounded head of the femur fits securely into the acetabulum, a deep, cuplike cavity in the pelvis. The shoulder joint complex is in fact made up by four joints: the glenohumeral joint (the 'ball-and-socket' joint between the upper arm or humerus and the shoulder blade or scapula, that most sketchy descriptions consider to be the shoulder joint), the acromio-clavicular joint (the joint between the lateral end of the collar bone or clavicle and the scapula), the sternoclavicular joint (the joint between the medial end of the clavicle and the breast bone or sternum) and the scapulo-thoracic joint (the 'virtual' joint between the undersurface of the scapula and the chest wall). There is more movement possible at the shoulder joint than at any other joint in the body. The elbow joint is a ginglymus or hinge joint. The elbow joint is a very complex joint that is created by the junction of three different bones (the humerus of the upper arm, and the paired radius and ulna of the forearm). Normally these bones fit and function together with very close tolerances. Two main movements are possible at the elbow. The hinge-like bending and straightening of the elbow (flexion and extension) happens at the articulation ("joint") between the humerus and the ulna. The complex action of turning the forearm over (pronation or supination) happens at the articulation between the radius and the ulna (this movement also occurs at the wrist joint). The wrist is a complex joint composed of 15 bones, four joint compartments and multiple interconnected ligaments. Working together, these components allow three-dimensional motion, transmitting forces from the hand to the forearm. The knee is a complex joint, which is made up of the distal end of the femur (the femoral condyles), and the proximal end of the tibia (the tibial plateau). The femoral condyles usually roll and glide smoothly on the tibial plateau, allowing for smooth, painless motion of the lower leg. Accurate knee movement is quantified using six degrees of freedom (three rotational and three translational). The ankle joint complex has two major functional axes, the subtalar joint axis, determined by talus and calcaneus, and the ankle joint axis, determined by talus and tibia. Orientation of and movement about the ankle and the subtalar joint axis are difficult to determine. Foot movement is often quantified about an anterior–posterior (in- eversion), a medio–lateral (plantar-dorsiflexion), and an inferior–superior axis (ab-adduction); i.e., axes that do not correspond to an anatomical joint.

15.2.2 Suitability for Biomechanical Applications

While many existing computer vision approaches offer a great potential for markerless motion capture for biomechanical applications, these approaches have not been developed or tested for this applications. To date, qualitative tests and visual inspections are most frequently used for assessing approaches introduced in the field of computer vision and machine learning. Evaluating existing approaches within a framework focused on addressing biomechanical applications is critical. The majority of research on human motion capture in the field of computer vision and machine learning has concentrated on track-ing, estimation and recognition of human motion for surveillance purposes. Moreover, much of the work reported in the literature on the above has been developed for the use of a single camera (monocular approach). Single image stream based methods suffer from poor performance for accurate movement analysis due to the severe ill-posed nature of motion recovery. Furthermore, simplistic or generic models of a human body with either fewer joints or re-duced number of degrees of freedom are often utilized for enhancing compu-tational performance. For instance, existing methods for gait-based human identification in surveillance applications use mostly 2D appearance models and measurements such as height, extracted from the side view. Generic mod-els typically lack accurate joint information, typically assume all joints to have fixed centers of rotation for ease of calculation, and thus lack accuracy for ac-curate movement analysis. However, biomechanical and, in particular, clinical applications typically require knowledge of detailed and accurate representa-tion of 3D joint mechanics. Some of the most challenging issues in whole-body movement capture are due to the complexity and variability of the appearance of the human body, the nonlinear and nonrigid nature of human motion, a lack of sufficient image cues about 3D body pose, including self-occlusion as well as the presence of other occluding objects, and exploitation of multiple image streams. Human body self-occlusion is a major cause of ambiguities in body part tracking using a single camera. The self-occlusion problem is addressed when multiple cameras are used, since the appearance of a human body from multiple viewpoints is available.

Approaches from the field of computer vision have previously been ex-plored for biomechanical applications. These include the use of a model-based simulated annealing approach for improving posture prediction from marker positions [91] and marker-free systems for the estimation of joint centers [48], tracking of lower limb segments [74], analysis of movement disabilities [52,60], and estimation of working postures [75]. In particular, Persson [74] proposed a marker-free method for tracking the human lower limb segments. Only movement in the sagittal plane was considered. Pinzke and Kopp [75] tested the usability of different markerless approaches for automatic tracking and assessing identifying and evaluating potentially harmful working postures from video film. Legrand et al. [52] proposed a system composed of one camera. The human boundary was extracted in each image and a two-dimensional model of the human body, based on tapered super-quadrics, was matched. Marzani et al. [60] extended this approach to a system consisting of three cameras. A 3D model based on a set of articulated 2D super-quadrics, each of them describing a part of the human body, was positioned by a fuzzy clustering process.

These studies demonstrate the applicability of techniques in computer vision for automatic human movement analysis, but the approaches were not validated against marker-based data. To date, the detailed analysis of 3D joint kinematics through a markerless system is still lacking. Quantitative measurements of movement and continuous tracking of humans using multiple image streams is crucial for 3D gait studies. Previous work in the field of computer vision was an inspiration for our work on tracking an articulated model in visual hull sequences. A markerless motion capture system based on visual hulls from multiple image streams and the use of detailed subject-specific 3D articulated models with soft-joint constraints is demonstrated in the following section. Articulated models have been used for popular tasks such as character animation [3, 53], and object tracking in video [20, 40] and in 3D data streams [24, 54]. Our soft-joint constraints approach allows small movement at the joint, which is penalized in least-squares terms. This extends previous approaches [20, 24] for tracking articulated models that enforced hard constraints on the kinematic structure (joints of the skeleton must be preserved). To critically analyze the effectiveness of markerless motion capture in the biomechanical/clinical environment, data obtained from this new system was quantitatively compared with data obtained from marker-based motion capture.

15.3 Markerless Human Movement Analysis Through Visual Hull and Articulated ICP with Soft-joint Constraints

Our approach employs an articulated iterative closest point (ICP) algorithm with soft-joint constraints [11] for tracking human body segments in visual hull sequences (a standard 3D representation of dynamic sequences from multiple images). The soft-joint constraints approach extends previous approaches [20,24] for tracking articulated models that enforced hard constraints on the joints of the articulated body. Small movements at the joint are allowed and penalized in least-squares terms. As a result a more anatomically correct matching suitable for biomechanical applications is obtained with an objective function that can be optimized in an efficient and straightforward manner.

15.3.1 Articulated ICP with Soft-joint Constraints

The articulated ICP algorithm is a generalization of the standard ICP algorithm [16, 78] to articulated models. The objective is to track an articulated model in a sequence of visual hulls. The articulated model M is represented as a discrete sampling of points $p_1, \ldots, p_P$ on the surface, a set of rigid segments $s_1, \ldots, s_S$, and a set of joints $q_{-1}, \ldots, q_Q$ connecting the segments. Each visual hull is represented as a set of points $V = v_1, \ldots, v_N$, which describes the appearance of the person at that time. For each frame of the sequence, an alignment T is computed, which brings the surfaces of M and V into correspondence, while respecting the model joints q. The alignment T consists of a set of rigid transformations T_j, one for each rigid part s_j. Similar to ICP, this algorithm iterates between two steps. In the first step, each point p_i on the model is associated to its nearest neighbor $v_{s(i)}$ among the visual hull points V, where $s(i)$ defines the mapping from the index of a surface point p_i to its rigid part index. In the second step, given a set of corresponding pairs $(p_i, v_{s(i)})$, a set of transformations T is computed, which brings them into alignment. The second step is defined by an objective function of the transformation variables given as $F(T) = H(T) + G(T)$. The term $H(T)$ ensures that corresponding points (found in the first step) are aligned.

$$H(r, t) = w_H \sum_{i=1}^{P} \| R(r_{s(i)}) p_i + t_{s(i)} - v_i \|^2 \qquad (15.1)$$

The transformation T_j of each rigid part s_j is parameterized by a 3x1 translation vector t_j and a 3x1 twist coordinates vector r_j (twists are standard representations of rotation (Ma, Soatto et al. 2004)), and $R(r_{s(i)})$ denotes the rotation matrix induced by the twist parameters $r_{s(i)}$. The term $G(T)$ ensures that joints are approximately preserved, where each joint $q_{i,j}$ can be viewed as a point belonging to parts s_i and s_j simultaneously. The transformations T_i and T_j are forced to predict the joint consistently.

$$G(r, t) = w_G \sum_{(i,j) \in Q} \| R(r_i) q_{i,j} + t_i - R(r_j) q_{i,j} - t_j \|^2 \qquad (15.2)$$

Figure 1: (a) Point-to-point associations used to define the energy H(T). (b) Illustration of the joint mismatch penalty $G(T)$.

Linearizing the rotations around their current estimate in each iteration resulted in a standard least-squares function over the transformation parameters (r, t)

$$argmin = \| A[\tfrac{r}{t}] - b \| \Rightarrow [\tfrac{r}{t}] = (A^T A)^{-1} A^T b$$

where A is a matrix, and b is a vector whose values are dictated by Equations 15.1 and 15.2. Decreasing the value of w_G allows greater movement at the joint, which potentially improves the matching of body segments to the visual hull. The center of the predicted joint locations (belonging to adjacent segments) provides an accurate approximation of the functional joint center. As a result, the underlying kinematic model can be refined and a more anatomically correct matching is obtained.

15.3.2 Methods

The algorithm was evaluated in a theoretical and experimental environment [68, 69]. The accuracy of human body kinematics was evaluated by tracking articulated models in visual hull sequences. Most favorable camera arrangements for a $3 \times 1.5 \times 2$ m viewing volume were used [67]. This viewing volume is sufficiently large enough to capture an entire gait cycle. The settings $w_H = 1$, $w_G = 5000$ (Equations 15.1 and 15.2) were used to underscore the relative importance of the joints. The theoretical analysis was conducted in a virtual environment using a realistic human 3D model. The virtual environment permitted the evaluation of the quality of visual hulls on extracting kinematics while excluding errors due to camera calibration and fore-/background separation. To simulate a human form walking, 120 poses were created using Poser (Curious Labs, CA) mimicking one gait cycle. The poses of the human form consisted of 3D surfaces and had an average volume of 68.01 ± 0.06 liters. Visual hulls of different quality using 4, 8, 16, 32 and 64 cameras with a resolution of 640x480 pixels and an 80-degree horizontal view were constructed of the Poser sequence. In the experimental environment, full body movement was captured using a marker-based and a markerless motion capture system simultaneously. The marker-based system consisted of an eight-Qualisys camera optoelectronic system monitoring 3D marker positions for the hip, knees and ankles at 120 fps. The markerless motion capture system consisted of eight Basler CCD color cameras (656x494 pixels; 80-degree horizontal view) synchronously capturing images at 75 fps. Internal and external camera parameters and a common global frame of reference were obtained through offline calibration. Images from all cameras were streamed in their uncompressed form to several computers during acquisition.

The subject was separated from the background in the image sequence of all cameras using intensity and color thresholding [38] compared to background images (Figure 15.1). The 3D representation was achieved through visual hull construction from multiple 2D camera views [26, 49, 59]. Visual hulls were created with voxel edges of $\lambda = 10$ mm, which is sufficiently small enough for these camera configurations [66]. The number of cameras used for visual hull construction greatly affects the accuracy of visual hulls [67]. The accuracy of visual hulls also depends on the human subjects position and pose within an observed viewing volume [67]. Simultaneous changes in position and pose result in decreased accuracy of visual hull construction. Increasing the

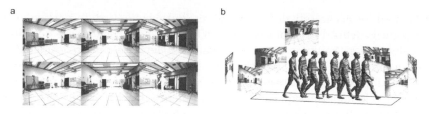

Fig. 15.1. (a) Separated subject data in selected video sequences. (b) Camera configuration, video sequences with separated subject data, and selected visual hulls.

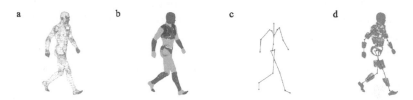

Fig. 15.2. (a) 3D point surface. (b) Body segments. (c) Joint centers. (d) Articulated model.

number of cameras leads to decreased variations across the viewing volume and a better approximation of the true volume value.

A subject-specific 3D articulated model was tracked in the 3D representations constructed from the image sequences. An articulated model is typically derived from a morphological description of the human bodys anatomy plus a set of information regarding the kinematic chain and joint centers. The morphological information of the human body can be a general approximation (cylinders, super-quadrics, etc.) or an estimation of the actual subjects outer surface. Ideally, an articulated model is subject-specific and created from a direct measurement of the subjects outer surface. The kinematic chain underneath an anatomic model can be manually set or estimated through either functional [18, 27] or anthropometric methods [4, 14]. The more complex the kinematic description of the body the more information can be obtained from the 3D representation matched by the model. As previously discussed, while in marker-based systems the anatomic reference frame of a segment is acquired from anatomical landmarks tracked consistently through the motion path, in the markerless system the anatomical reference frames are defined by the model joint centers and reference pose. During the tracking process, the reference frames remain rigidly attached to their appropriate model anatomic segment, thus describing the estimated position and orientation in the subject's anatomic segments. In this study, an articulated body was created from a detailed full body laser scan with markers affixed to the subjects joints (Figure 15.2). The articulated body consisted at least of 15 body segments (head, trunk, pelvis, and left and right arm, forearm, hand, thigh, shank and foot) and 14 joints connecting these segments. These articulated models are well-suited for accurately describing lower limb kinematics. In particular, the knee joint is defined as a joint with 6 DOF. However, the current description lacks details for the upper body. For example, the shoulder is considered as a single joint of 6 DOF and not as a complex structure of 3 individual joints.

The subjects pose was roughly matched based on a motion trajectory to a frame in the motion sequence and subsequently tracked automatically over the gait cycle. The motion trajectory was calculated as a trajectory of center of volumes obtained from the volumetric 3D representations for each frame throughout the captured motion. Several motion sequences typically performed by subjects and patients in a gait laboratory were processed. In addition, more complex sport sequences such as a cricket bowl, handball throw and car wheel performance were analyzed. Joint center locations were extracted for all joints and compared to joint centers from the theoretical sequence. Accuracy of joint center location was calculated as the average Euclidian distance between corresponding joints. Joint centers of adjacent segments were used to define segment coordinate axes. Joint angles for the lower limbs for the sagittal and frontal planes were calculated as angles between corresponding axes of neighboring segments projected into the corresponding planes. Accuracy of human body kinematics was calculated as the average deviation of the deviation of joint angles derived from visual hulls compared to joint angles derived from the theoretical sequence and marker-based system over the gait cycle, respectively. The joint angles (sagittal and frontal plane) for the knee calculated as angles between corresponding axes of neighboring segments are used as preliminary basis of comparison between the marker-based and markerless systems.

Motion sequences were primarily tracked with a subject-specific articulated body consisting of subject-specific morphology and joint center locations. However, the creation of articulated bodies lead to a database of subject-specific articulated bodies. Additional, motion sequences were processed with a model from this database that would match the subject closest based on a height measurement. The settings $w_H = 1$ and $w_G = 1$ were used to grant the lack of detailed knowledge of the morphology and kinematic chain.

15.4 Results

The number of cameras used for visual hull construction greatly affects the accuracy of visual hulls [67]. Surface comparison between visual hulls and the original human form revealed under-approximated and over-approximated regions. Under-approximated regions result from discretization errors in the image plane, which can be reduced with higher imager resolution [66]. However, greater error arises from over-approximated regions, which are characteristic to visual hull construction. The size of over-approximated regions and the maximum surface deviations decreases with increasing number of cameras (Figure 15.3).

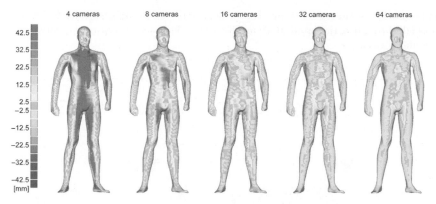

Fig. 15.3. Surface deviations between visual hulls and the original human form for different circular camera arrangements with 4 (a), 8 (b), 16 (c), 32 (d), and 64 (e) camera. Colors indicate the deviation calculated as shortest distance. Colors ranging from cyan to blue indicate areas that are under-approximated and colors ranging from yellow to red indicate areas that are over-approximated.

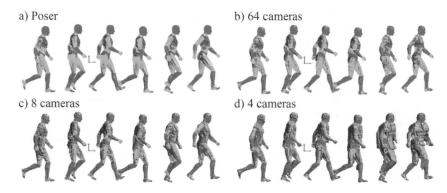

Fig. 15.4. Tracking an articulated body in (a) the Poser and (b–d) visual hull sequences constructed with 64, 8 and 4 cameras.

The accuracy of visual hulls also depends on the human subjects position and pose within the investigated viewing volume [67]. Simultaneous changes in position and pose result in decreasing the accuracy of visual hulls. Increasing the number of cameras leads to decreasing variations across the viewing volume and in a better approximation of the true volume value.

Body segments were tracked accurately in the original Poser sequence (Figure 15.4a). The accuracy of tracking increased with increasing the number of cameras. While all body segments were positioned correctly with 8 to 64 cameras (Figures 15.4b, c), several body segments such as upper limbs were positioned incorrectly with 4 cameras (Figure 15.4d).

Furthermore, the employed articulated ICP algorithm estimated joints centers in the original sequence very accurately (Table 15.1). Joint centers in

Table 15.1. Accuracy of joint center locations [mm] and standard deviation obtained through articulated ICP in the visual hull sequences constructed using 64, 8 and 4 cameras.

	Poser	64	8	4
Full body	1.9 ± 3.7	10.6 ± 7.8	11.3 ± 6.3	34.6 ± 67.0
Lower limbs	0.9 ± 1.2	8.7 ± 2.2	10.8 ± 3.4	14.3 ± 7.6

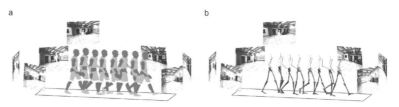

a   b

Fig. 15.5. Articulated body matched to visual hulls. (a) Human body segments. (b) Kinematic chain.

the visual hull sequences constructed using 8 and 64 cameras were predicted with an accuracy that matches the in-plane camera accuracy of magnitude of approximately 1 cm. Joint centers in the visual hull sequences constructed using 8 and 64 cameras were predicted with similar accuracy, while the overall joint center accuracy drastically declines with visual constructed using 4 cameras due to the inaccurate tracking of upper limbs (Figure 15.4). Joint centers for the lower limbs, in particular hip, thighs, shanks and feet were predicted more accurately than for the overall articulated model. Joint center locations for visual hulls constructed with 8 and 64 cameras were estimated within the in-plane camera resolution (Table 15.1). Joint center estimation for lower limbs in visual hull constructed using 4 cameras were estimated less accurate than using higher number of cameras, but still resulted in comparable results (Table 15.1).

Body segments were tracked accurately with the occasional exception of hands and feet due to inaccuracies in the visual hulls (Figure 15.5). Comparable results were obtained for knee joint angles in the sagittal and frontal plane using marker-based and markerless motion capture (Figure 15.6). The accuracy of sagittal and frontal plane knee joint angles calculated from experiments was within the scope of the accuracy estimated from the theoretical calculations ($accuracy_{experimental}$: 2.3 ± 1.0° (sagittal); 1.6 ± 0.9° (frontal); $accuracy_{theoretical}$: 2.1±0.9° (sagittal); 0.4±0.7° (frontal); [68,69]). A similar method, with different model matching formulation and limited to hard-joint constraints, was recently explored by the authors [29]. This method utilized simulated annealing and exponential maps to extract subjects kinematics, and resulted in comparable accuracy.

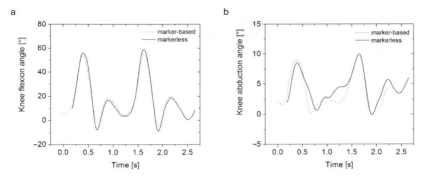

Fig. 15.6. Motion graphs for (a) knee flexion and (b) knee abduction angles.

15.5 Summary of Chapter

The development of markerless motion capture methods for biomechanical and clinical applications is motivated by the need to address contemporary needs to understand normal and pathological human movement without the encumbrance of markers or fixtures placed on the subject, while achieving the quantitative accuracy of marker-based systems. Markerless motion capture has been widely used for a range of applications in the surveillance, film and game industries. However, the biomechanical, medical, and sports applications of markerless capture have been limited by the accuracy of current methods for markerless motions capture.

Previous experience has demonstrated that minor changes in patterns of locomotion can have a profound impact on the outcome of treatment or progression of musculoskeletal pathology. The ability to address emerging clinical questions on problems that influence normal patterns of locomotion requires new methods that would limit the risk of producing artifact due to markers or the constraints of the testing methods. For example, the constraints of the laboratory environment as well as the markers placed on the subjects can mask subtle but important changes to the patterns of locomotion. It has been shown that the mechanics of walking was changed in patients with anterior cruciate ligament deficiency of the knee [6, 10]; functional loading influenced the outcome of high tibial osteotomy [76]; functional performance of patients with total knee replacement was influenced by the design of the implant [5], and the mechanics of walking influenced the disease severity of osteoarthritis of the knee [10, 64, 65, 82]. It should be noted that each of the clinical examples referenced above were associated with subtle but important changes to the mechanics of walking.

The work cited above indicates several necessary requirements for the next significant advancement in our understanding of normal and pathological human movement. First, we need to capture the kinematics and kinetics of human movement without the constraints of the laboratory or the encumbrance of placing markers on the limb segments. Second, we need to relate the external features of human movement to the internal anatomical structures (e.g., muscle, bone, cartilage, and ligaments) to further our knowledge of musculoskeletal function and pathology.

This chapter demonstrates the feasibility of accurately and precisely measuring 3D human body kinematics for biomechanical applications using a markerless motion capture system on the basis of visual hulls. Passive vision systems are advantageous as they only rely on capturing images and thus provide an ideal framework for capturing subjects in their natural environment. The 3D representation of the subject in the form of visual hulls can also be utilized further for anthropometric measurements such as body segment volumes. The ultimate goal for biomechanical applications is to measure all functional degrees of freedom that describe the mechanics of a particular joint (e.g., flexion/extension, ab/adduction, internal/external rotation and anterior/posterior translation at the knee). Accuracy of markerless methods based on visual hulls is dependent on the number of cameras. In general, configurations with fewer than 8 cameras yielded several drawbacks. Volume estimations greatly deviated from original values and fluctuated enormously for different poses and positions across the viewing volume.

The employed algorithm yields great potential for accurately tracking human body segments. The algorithm does not enforce hard constraints for tracking articulated models. The employed cost function consists of two terms, which ensure that corresponding points align and joint are approximately preserved. The emphasis on either term can be chosen globally and/or individually, and thus yields more anatomically correct models. Moreover, the presented algorithm can be employed by either fitting the articulated model to the visual hull or the visual hull to the articulated model. Both scenarios will provide identical results in an ideal case. However, fitting data to the model is likely to be more robust in an experimental environment where visual hull only provide partial information due to calibration and/or segmentation errors.

Limitations of this study are the use of an articulated model consisting of rigid segments. However, inaccuracies in visual hull construction, in particular using 4 and 8 cameras, outweigh segment deformations in the living human. Moreover, the articulated model is given, but could potentially be constructed automatically and accurately with an appropriate number of cameras for the specific viewing volume. Hence, rather than using two separate systems for capturing human shape and human motion, one multi-camera system can be used for both tasks. Furthermore, a well-known problem with ICP methods is that they are prone to local minima, and depend on having a reasonable initial estimate of the transformations T. We are currently extending the approach with a search method which has the capability to get out of local minima. Also, our current model does not explicitly limit the amount of joint rotation. Thus, symmetrical body parts such as the forearm may be rotated around their axis of symmetry. There was enough surface detail in our surface models to prevent this from happening.

This chapter systematically points out that choosing appropriate technical equipment and approaches for accurate markerless motion capture is critical. The processing modules used in this study including background separation, visual hull, iterative closest point methods, etc. yielded results that were comparable to a marker-based system for motion at the knee. While additional evaluation of the system is needed, the results demonstrate the feasibility of calculating meaningful joint kinematics from subjects walking without any markers attached to the limb. This study also demonstrated the potential for markerless methods of human motion capture to address important clinical problems. Future improvements will allow the extraction of not only kinematic but also kinetic variables from the markerless motion capture system. Our research currently focuses primarily on accurately capturing the 6 DOF for the knee. The current model description provides sufficient degrees of freedom for the lower limbs, but needs to be extended for accurate analysis of the upper body. Moreover, utilizing a subject-specific morphology and kinematic chain provides an ideal starting point for tracking a reference model in a dynamic sequence. However, the creation of subject-specific models seems to be infeasible for clinical environments. Our future work will address utilizing accurate articulated models from a database and/or the use of functional methods to accurately determine the underlying kinematic chain.

The markerless framework introduced in this chapter can serve as a basis for developing the broader application of markerless motion capture. Each of the modules can be independently evaluated and modified as newer methods become available, thus making markerless tracking a feasible and practical alternative to marker based systems. Markerless motion capture systems offer the promise of expanding the applicability of human movement capture, minimizing patient preparation time, and reducing experimental errors caused by, for instance, inter-observer variability. In addition, gait patterns can not only be visualized using traces of joint angles but sequences of snapshots can be easily obtained that allow the researcher or clinician to combine the qualitative and quantitative evaluation of a patients gait pattern. Thus, the implementation of this new technology will allow for simple, time-efficient, and potentially more meaningful assessments of gait in research and clinical practice.

Acknowledgment

Funding provided by NSF #03225715 and VA #ADR0001129.

References

1. Aggarwal, JK, Cai, Q *Human motion analysis: a review*, Computer Vision and Image Understanding, 1999, 73(3):295–304.
2. Alexander EJ, Andriacchi TP *Correcting for deformation in skin-based marker systems* Journal of Biomechanics 2001, 34(3):355–361.

3. Allen B, Curless B, Popović A *Articulated body deformation from range scan data* Siggraph, Computer Graphics Proceedings, 2002.
4. Andriacchi TP, Andersson GBJ, Fermier RW, Stern D, Galante JO *A study of lower-limb mechanics during stair-climbing*, Journal of Bone and Joint Surgery 1980, 62(A):749–757.
5. Andriacchi TP, Galante JO, Fermier RW *The influence of total knee-replacement design on walking and stair-climbing*, Journal of Bone and Joint Surgery American volume 1982, 64(9):1328–1335.
6. Andriacchi TP, Birac, D. *Functional testing in the anterior cruciate ligament-deficient knee*, Clinical Orthopaedics and Related Research 1993, 288:40–47.
7. Andriacchi T, Sen K, Toney M, Yoder D *New developments in musculoskeletal testing*, Canadian Society of Biomechanics: 1994, Calgary, Canada; 1994.
8. Andriacchi TP, Alexander EJ, Toney MK, Dyrby C, Sum J *A point cluster method for in vivo motion analysis: applied to a study of knee kinematics*, Journal of Biomechanical Engineering 1998, 120(6):743–749.
9. Andriacchi TP, Alexander EJ *Studies of human locomotion: Past, present and future*, Journal of Biomechanics 2000, 33(10):1217–1224.
10. Andriacchi TP, Mündermann A, Smith RL, Alexander EJ, Dyrby CO, Koo S *A framework for the in vivo pathomechanics of osteoarthritis at the knee*, Annals of Biomedical Engineering 2004, 32(3):447–457.
11. Anguelov D, Mündermann L, Corazza S *An Iterative Closest Point Algorithm for Tracking Articulated Models in 3D Range Scans*, Summer Bioengineering Conference: 2005, Vail, CO, 2005.
12. Banks S, Hodge W *Accurate measurement of three dimensional knee replacement kinematics using single-plane flouroscopy*, IEEE Transactions on Biomedical Engineering 1996, 46(6):638–649.
13. Banks S, Otis J, Backus S, Laskin R, Campbell D, Lenhoff M, Furman G, Haas S *Integrated analysis of knee arthroplasty mechanics using simultaneous fluoroscopy, force-plates, and motion analysis*, 45th Annual Meeting of the Orthopedic Research Society: 1999, Anaheim, CA, 1999.
14. Bell AL, Pedersen DR, Brand RA *A comparison of the accuracy of several hip center location prediction methods*, Journal of Biomechanics 1990, 23:617–621.
15. Benedetti M, Cappozzo A *Anatomical landmark definition and identification in computer aided movement analysis in a rehabilitation context*, Internal Report. Universita Degli Studi La Sapienza, 1994.
16. Besl P, McKay N *A method for registration of 3D shapes* Transactions on Pattern Analysis and Machine Intelligence 1992, 14(2):239–256.
17. Bharatkumar AG, Daigle KE, Pandy MG, Cai Q, Aggarwal JK *Lower limb kinematics of human walking with the medial axis transformation* Workshop on Motion of Non-Rigid and Articulated Objects: 1994, Austin, TX, 1994.
18. Bottino A, Laurentini A *A silhouette based technique for the reconstruction of human movement* Computer Vision and Image Understanding 2001, 83:79–95.
19. Braune W. and Fischer, O. *Determination of the moments of inertia of the human body and its limbs* BerlSpringer, 1988.
20. Bregler C, Malik J *Tracking people with twists and exponential maps*, Computer Vision and Pattern Recognition: 1997; 1997: 568–574.
21. Cappozzo A, Catani F, Leardini A, Benedetti M, Della Croce U *Position and orientation in space of bones during movement: experimental artifacts*, Clinical Biomechanics 1996, 11:90–100.

22. Cappozzo A, Capello A, Della Croce U, Pensalfini F *Surface marker cluster design criteria for 3-D bone movement reconstruction*, IEEE Transactions on Biomedical Engineering, 44:1165–1174, 1997.
23. Cedras C, Shah M *Motion-based recognition: a survey* Image and Vision Computing 1995, 13(2):129–155.
24. Cheung G, Baker S, Kanade T *Shape-from-silhouette of articulated objects and its use for human body kinematics estimation and motion capture*, IEEE Conference on Computer Vision and Pattern Recognition: 2003, Madison, WI, IEEE, 2003, 77–84.
25. Chu CW, Jenkins OC, Matari MJ *Towards model-free markerless motion capture*, Computer Vision and Pattern Recognition: 2003, Madison, WI, 2003.
26. Cheung K, Baker S, Kanade T *Shape-from-silhouette across time part I: theory and algorithm*, International Journal of Computer Vision 2005, 62(3):221–247.
27. Cheung G, Baker S, Kanade T *Shape-From-Silhouette Across Time Part II: Applications to Human modelling and Markerless Motion Tracking*, International Journal of Computer Vision 2005, 63(3):225–245.
28. Corazza S, Mündermann L, Andriacchi TP *Model-free markerless motion capture through visual hull and laplacian eigenmaps*, Summer Bioengineering Conference: 2005, Vail, CO, 2005.
29. Corazza S, Mündermann L, Chaudhari AM, Demattio T, Cobelli C, Andriacchi TP *A markerless motion capture system to study musculoskeletal biomechanics: visual hull and simulated annealing approach*, Annals of Biomedical Engineering, conditionally accepted.
30. Cutler RG, Duraiswami R, Qian JH, Davis LS *Design and implementation of the University of Maryland Keck Laboratory for the analysis of visual movement* Technical Report, UMIACS. University of Maryland, 2000.
31. Darrel T, Maes P, Blumberg B, Pentland AP *A novel environment for situated vision and behavior* Workshop for Visual Behaviors at CVPR: 1994.
32. Deutscher J, Blake A, Reid I *Articulated body motion capture by annealed particle filtering*, Computer Vision and Pattern Recognition: 2000; Hilton Head, SC, 2000.
33. Eberhart H, Inman V. *Fundamental studies of human locomotion and other information relating to design of artificial limbs*, Report to the National Research Council. University of California, Berkeley, 1947.
34. Fisher D, Williams M, Draper C, Andriacchi TP *The therapeutic potential for changing patterns of locomotion: An application to the ACL deficient knee*, Summer Bioengineering Conference, pp. 840–850, Key Biscayne, FL, 2003.
35. Gavrila D, Davis L *3-D model-based tracking of humans in action: a multi-view approach*, Conference on Computer Vision and Pattern Recognition: 1996, San Francisco, CA, 1996.
36. Gavrila D *The visual analysis of human movement: a survey*, Computer Vision and Image Understanding 1999, 73(3):82–98.
37. Gray H. *Gray's Anatomy*, New York City: C.V. Mosby, 2005.
38. Haritaoglu I, Davis L *W4: real-time surveillance of people and their activities*, IEEE Transactions on Pattern Analysis and Machine Intelligence 2000, 22(8):809–830.
39. Harris GF, Smith PA *Human motion analysis*, Current Applications and Future Directions. New York: IEEE Press, 1996.
40. Hogg D *Model-based vision: A program to see a walking person* Image and Vision Computing 1983, 1(1):5–20.

41. Holden J, Orsini J, Siegel K, Kepple T, Gerber L, Stanhope S *Surface movements errors in shank kinematics and knee kinematics during gait*, Gait and Posture 1997, 3:217–227.

42. Inman V, Ralston H, Todd F *Human walking*, Baltimore: Williams & Wilkins, 1981.

43. Isard M, Blake A *Visual tracking by stochastic propagation of conditional density*, 4th European Conference on Computer Vision: 1996, Cambridge, UK, 1996: 343–356.

44. Jonsson H, Karrholm J *Three-dimensional knee joint movements during a step-up: evaluation after cruciate ligament rupture*, Journal of Orthopedic Research 1994, 12(6):769–779.

45. Kanade T, Collins R, Lipton A, Burt P, Wixson L *Advances in co-operative multi-sensor video surveillance*, DARPA Image Understanding Workshop: 1998, 1998: 3–24.

46. Kakadiaris IA, Metaxes D *3D human body model acquisiton from multiple views*, International Journal Computer Vision 1998, 30:191–218.

47. Lafortune MA, Cavanagh PR, Sommer HJ, Kalenak A *Three-dimensional kinematics of the human knee during walking*, Journal of Biomechanics 1992, 25(4):347–357.

48. Lanshammar H, Persson T, Medved V *Comparison between a marker-based and a marker-free method to estimate centre of rotation using video image analysis* Second World Congress of Biomechanics: 1994.

49. Laurentini A *The visual hull concept for silhouette base image understanding*, IEEE Transactions on Pattern Analysis and Machine Intelligence 1994, 16:150–162.

50. Leardini A, Chiari L, Della Croce U, Capozzo A *Human movement analysis using stereophotogrammetry Part 3: Soft tissue artifact assessment and compensation*, Gait and Posture 2005, 21:221–225.

51. Lee HJ, Chen Z *Determination of 3D human body posture from a single view* Comp Vision, Graphics, Image Process 1985, 30:148–168.

52. Legrand L, Marzani F, Dusserre L *A marker-free system for the analysis of movement disabilities* Medinfo 1998, 9(2):1066–1070.

53. Lewis JP, Cordner M, Fong N *Pose space deformation: a unified approach to shape interpolation and skeleton-driven deformation* Siggraph, Computer Graphics Proceedings, 2000.

54. Lin M *Tracking articulated objects in real-time range image sequences*, ICCV, 1999, 1:648–653.

55. Lu T, O'Connor J *Bone position estimation from skin marker coordinates using global optimization with joint constraints*, Journal of Biomechanics 1999, 32:129–134.

56. Lucchetti L, Cappozzo A, Capello A, Della Croce U *Skin movement artefact assessment and compensation in the estimation of knee-joint kinematics* Journal of Biomechanics 1998, 31:977–984.

57. Ma Y, Soatto S, Kosecka Y, Sastry S *An invitation to 3D vision*, Springer, 2004.

58. Marey E. *Animal mechanism: a treatise on terrestrial and aerial locomotion* London: Henry S. King & Co, 1874.

59. Martin W, Aggarwal J *Volumetric description of objects from multiple views*, IEEE Transactions on Pattern Analysis and Machine Intelligence 1983, 5(2):150–158.

60. Marzani F, Calais E, Legrand L *A 3-D marker-free system for the analysis of movement disabilities – an application to the legs* IEEE Trans Inf Technol Biomed 2001, 5(1):18–26.
61. Miyazaki T, Wada M, Kawahara H, Sato M, Baba H, Shimada S *Dynamic load at baseline can predict radiographic disease progression in medial compartment knee osteoarthritis*, Annals of the Rheumatic Diseases 2002, 61(7):617–622.
62. Moeslund G, Granum E *A survey of computer vision-based human motion capture*, Computer Vision and Image Understanding 2001, 81(3):231–268.
63. Moon H, Chellappa R, Rosenfeld A *3D object tracking using shape-encoded particle propagation*, International Conference on Computer Vision: 2001, Vancouver, BC, 2001.
64. Mündermann A, Dyrby CO, Hurwitz DE, Sharma L, Andriacchi TP *Potential strategies to reduce medial compartment loading in patients with knee OA of varying severity: reduced walking speed*, Arthritis and Rheumatism 2004, 50(4):1172–1178.
65. Mündermann A, Dyrby CO, Andriacchi TP *Secondary gait changes in patients with medial compartment knee osteoarthritis: increased load at the ankle, knee and hip during walking*, Arthritis and Rheumatism 2005, 52(9):2835–2844.
66. Mündermann L, Mündermann A, Chaudhari AM, Andriacchi TP *Conditions that influence the accuracy of anthropometric parameter estimation for human body segments using shape-from-silhouette*, SPIE-IS&T Electronic Imaging 2005, 5665:268–277.
67. Mündermann L, Corazza S, Chaudhari AM, Alexander EJ, Andriacchi TP *Most favorable camera configuration for a shape-from-silhouette markerless motion capture system for biomechanical analysis*, SPIE-IS&T Electronic Imaging 2005, 5665:278–287.
68. Mündermann L, Corazza S, Anguelov D, Andriacchi TP *Estimation of the accuracy and precision of 3D human body kinematics using markerless motion capture and articulated ICP*, Summer Bioengineering Conference: 2005, Vail, CO, 2005.
69. Mündermann L, Anguelov D, Corazza S, Chaudhari AM, Andriacchi TP *Validation of a markerless motion capture system for the calculation of lower extremity kinematics*, International Society of Biomechanics & American Society of Biomechanics: 2005, Cleveland, OH, 2005.
70. Mündermann L, Corazza S, Mündermann A, Lin T, Chaudhari AM, Andriacchi TP *Gait retraining to reduce medial compartment load at the knee assessed using a markerless motion capture*, Transactions of the Orthopaedic Research Society 2006, 52:170.
71. Muybridge E. *Animal locomotion*, Philadelphia: J.P. Lippincott Company, 1887.
72. Narayanan PJ, Rander P, Kanade T *Synchronous capture of image sequences from multiple cameras*, Technical Report CMU-RI-TR-95-25. Robotics Institute Carnegie Mellon University, 1995.
73. O'Rourke J, Badler NI *Model-based image analysis of human motion using constraint propagation*, IEEE Transactions on Pattern Analysis and Machine Intelligence 1980, 2:522–536.
74. Persson T *A marker-free method for tracking human lower limb segments based on model matching*, International Journal Biomedical Computing 1996, 41(2):87–97.
75. Pinzke S, Kopp L *Marker-less systems for tracking working postures - results from two experiments*, Applied Ergonomics 2001, 32(5):461–471.

76. Prodromos CC, Andriacchi TP, Galante JO *A relationship between gait and clinical changes following high tibial osteotomy*, Journal of Bone and Joint Surgery 1985, 67(8):1188–1194.
77. Reinschmidt C, van den Bogert A, Nigg B, Lundberg A, Murphy N *Effect of skin movement on the analysis of skeletal knee joint motion during running*, Journal of Biomechanics 1997, 30:729–732.
78. Rusinkiewicz S, Levoy M *Efficient variants of the ICP algorithm* International Conference on 3-D Digital Imaging and modelling: 2001; 2001.
79. Santos J, Gold G, Besier T, Hargreaves B, Draper C, Beaupre G, Delp S, Pauly J *Full-flexion patellofemoral joint kinematics with real-time mRI at 0.5 T*, ISMRM 13th Scientific Meeting: 2005, Miami, FL, 2005.
80. Sati A, De Giuse J, Larouche S, Drouin G *Quantitative assessment of skin-bone movement at the knee*, The Knee, 3:121–138, 1996.
81. Schipplein OD, Andriacchi TP *Interaction between active and apassive knee stabilizers during level walking*, Journal of Orthopaedic Research 1991, 9:113–119.
82. Sharma L, Hurwitz DE, Thonar EJ, Sum JA, Lenz ME, Dunlop DD, Schnitzer TJ, Kirwan-Mellis G, Andriacchi TP *Knee adduction moment, serum hyaluronan level, and disease severity in medial tibiofemoral osteoarthritis* Arthritis and Rheumatism 1998, 41(7):1233–1240.
83. Simon R *Quantification of human motion: gait analysis benefits and limitations to its application to clinical problems*, Journal of Biomechanics 2004, 37:1869–1880.
84. Spoor C, Veldpas F *Rigid body motion calculated from spatial coordinates of markers*, Journal of Biomechanics 1988, 13:391–393.
85. Stiehl J, Komistek R, Dennis D, Paxson R, Hoff W *Flouroscopic analysis of kinematics after posterior-cruciate retaining knee arthroplasty*, Journal of Bone and Joint Surgery 1995, 77:884–889.
86. Wagg DK, Nixon MS *Automated markerless extraction of walking people using deformable contour models*, Computer Animation and Virtual Worlds 2004, 15(3-4):399–406.
87. Wang L, Hu W, Tan T *Recent developments in human motion analysis*, Pattern Recognition 2003, 36(3):585–601.
88. Weber, W, Weber E *Mechanik der menschlichen Gehwerkzeuge*, Göttingen: Dieterich, 1836.
89. Wren CR, Azarbayejani A, Darrel T, Pentland AP *Pfinder: real-time tracking of the human body*, Trans on Pattern Analysis and Machine Intelligence 1997, 19(7):780–785.
90. Yamamoto M, Koshikawa K *Human motion analysis based on a robot arm model*, Computer Vision and Pattern Recognition: 1991.
91. Zakotnik J, Matheson T, Dürr V *A posture optimization algorithm for model-based motion capture of movement sequences*, Journal of Neuroscience Methods 2004, 135:43–54.

Part IV

Biomechanics and Applications

Qualitative and Quantitative Aspects of Movement: The Discrepancy Between Clinical Gait Analysis and Activities of Daily Life

Dieter Rosenbaum and Mirko Brandes[1]

Funktionsbereich Bewegungsanalytik
Klinik für Allgemeine Orthopädie
Universitätsklinikum Münster
Domagkstr. 3, 48129 Münster, Germany

Summary. Two different approaches for the assessment of an individual's movement capabilities and physical activity are available and have to be clearly distinguished:

1. **Clinical gait analysis** is a laboratory-based, i.e., stationary procedure that enables a **qualitative assessment**, i.e., it describes how well patients are able to move or how much they are limited in their movement capabilities.
2. **Activity assessment in daily life, ADL-monitoring** is not confined to a lab environment and assesses the **quantity of movement** or the **activity level** by describing how much patients are using their individual capabilities and which level of mobility is being used.

Both approaches have their specific advantages and disadvantages, which have to be considered before application of either one. However, they may be complementary for a full description of an individual's movement characteristics. In the future, marker-less motion capturing systems might offer an alternative approach halfway between the existing one, i.e., these systems might provide detailed motion analysis in home-based environments.

16.1 Introduction

The observation and analysis of human and animal motion has always been a central interest of researchers from various disciplines. As early as in the 4th century B.C. the Greek philosopher Aristotle was concerned with various aspects of locomotion of different species of animals and sought reasons for bipedal and quadruped gait patterns. He was the first to realize that locomotion can only take by some mechanical action against the supporting surface, i.e., action and reaction. In the 15th and 16th century, the famous and multi-talented scientist-artist Leonardo da Vinci studied the human body in detail,

B. Rosenhahn et al. (eds.), *Human Motion – Understanding, Modelling, Capture, and Animation*, 401–415.

performed anatomical studies and described the mechanical aspects during various movements and drew lines of action and centers of gravity into his sketches. In the 17th century, Giovanni Alfonso Borelli performed detailed calculations on the mechanics of muscles and their moment arms. Three-dimensional movement analysis is also not an invention of modern times. The first quantitative analysis was already performed in the late 19th century in a collaborative approach of the Prussian army that was interested in the effect of internal loading changes induced by the equipment of military recruits. It was based on the successful cooperation between the German anatomist Wilhelm Braune (1831–1892) and the mathematician Otto Fischer (1861–1917). This first approach was extremely time-consuming with respect to preparing the subject (only a single subject was used; Figure 16.1) and analyzing the measurements which took several months of computing time for just a few steps [1].

Since these times, the methods for investigating human movements have become more sophisticated and have greatly benefited from the rapid improvements in computer technology and software developments [3]. Therefore, it is time to reconsider whether gait analysis has achieved a solid standing in clinical practice and where recent and future developments might lead to new and promising possibilities.

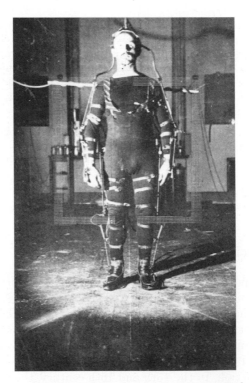

Fig. 16.1. Instrumented subject of the first three-dimensional movement analysis performed by Wilhelm Braune and Otto Fischer in 1895 [1].

16.2 Development and Aims of Clinical Gait Analysis

With an array of cameras and passive or active marker systems, a patient's movement can be captured in the calibrated laboratory environment for performing full-body three-dimensional clinical gait analysis (Figure 16.2). Standard marker sets and biomechanical models (e.g., Helen-Hayes or Cleveland Clinic recommendations) are usually applied to extract parameters of interest which will be used to describe the motion characteristics or identify and understand possible causes of movement disorders [3].

Recent technological developments have helped to reduce the processing times so that clinical gait analysis nowadays can be considered as a standard tool in well-equipped research labs [3]. Nevertheless, there is an ongoing debate about its usefulness and clinical relevance. Certain areas have greatly benefited from the additional information, e.g., the planning of multilevel surgery in children suffering from cerebral palsy [2]. In these patients, the impaired central control of the peripheral body segments can eventually lead to malfunctions and deformities causing severe orthopaedic problems that require surgical interventions concerning soft tissue and/or bony structures. On the other hand, in some less-justified applications the scientists have been accused of having been driven by the availability of instrumentation rather

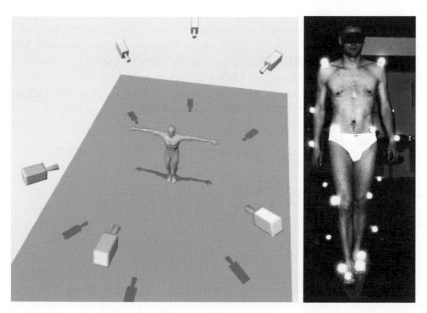

Fig. 16.2. Left: Schematic of the experimental set-up with a 6-camera Motion Analysis Inc. system in the Münster Movement Analysis Lab. Right: Subject equipped with reflective markers attached according to the Helen-Hayes marker set as seen by the cameras, i.e., with the markers clearly visible but no details of the human body distinguishable.

than a clinically relevant approach ("technology searching for application"). A persisting problem is the fact that no "hard data" exists with respect to normative values for comparison with pathological populations that would allow for a distinction between normal and abnormal gait patterns or judgment of the degree of impairment. In general, a complete gait analysis provides a vast amount of data: A spatial resolution of 1 mm results in a data flow of 10^6 bits of information per second. Since the human brain is able to directly process only 10 bits of information per second it is imperative to reduce or condense the data in order to extract the main characteristics of a single patient or a group of subjects [11]. The 3-dimensional observation of the moving object enables a detailed description of the individual's performance during walking. Comparisons between the clinically affected and healthy side or between a patient and a control group are used to detect the cause for movement disorders. In order to achieve this goal, gait analysis of the whole human body is usually performed in the anatomical planes of motion (frontal, sagittal, transverse) and considers different body segments (ankle, knee, and hip joint; pelvis, trunk, arm) as well as kinematic and kinetic parameters (joint motion, joint moment and power). This level of data recording and reduction can be performed with current gait analysis systems in a matter of about a few hours, which has helped greatly to provide a fast feedback in clinical settings. Nevertheless, there is a need for expert knowledge in order to extract meaningful information that is needed for clinical decision-making, i.e., provide information about over-loading or malfunctioning. With recent gait analysis systems patients with specific injuries or diseases can be assessed before and/or after treatment with respect to quality control issues of conservative or surgical treatment options [3]. This way, additional objective information about the individual functional capabilities is available so that the judgment of outcome is not limited to subjective feedback from the patient or conventional clinical and radiographic assessment which are usually based on static measures.

16.3 Applications of Clinical Gait Analysis

As mentioned above, the most common application is in children with cerebral palsy where gait analyses are used for clinical decision-making. A study in 91 CP patients showed that the additional use of gait-analysis data resulted in changes in surgical recommendations in 52% of the patients and reduced the cost of surgery [4]. Another benefit pointed out by the authors was related to preventing inappropriate surgical decisions, which they considered as more likely without gait analysis. Another study supported these findings showing that 39% of procedures recommended before gait analysis were not done when the gait laboratory data were considered. On the other hand, an additional 110 procedures (1.6 per patient) that had not been recommended before the gait study ultimately were performed after addition of the gait laboratory data [10]. A different topic was addressed by an earlier study by Prodromos et al. The authors investigated the clinical outcome in 21 patients with high tibial osteotomy with respect to their knee-joint loading during gait before and after surgery. The results indicate that of preoperative walking characteristics, in particular the knee adduction moment, were associated with postoperative clinical results. The low adduction-moment group had significantly lower postoperative adduction moments and had better clinical results than did patients with high adduction moments. Based on these results the authors saw the ability to predict the clinical outcome based on preoperative gait analysis data [12]. Another important area of application is the evaluation of clinical outcome after surgical or conservative treatment. An example for the effectiveness of physiotherapy and gait retraining was related to the problem of hip dysplasia. In this study, the "Entlastungsgang" was introduced for unloading the hip joint (Figure 16.3). Sixteen hip dysplasia patients received daily gait training combined with intensive physical therapy for 3–4 weeks. The hip abduction torque acting was determined by clinical gait analysis before, during and after the gait training. Gait training resulted in a torque reduction to 77.2% of the initial value [13]. Another study that was performed in our lab

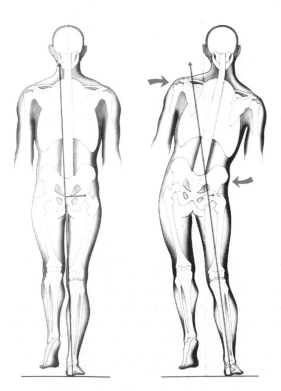

Fig. 16.3. Schematic description of the changes in the hip abduction moment during single stance in normal (left) and modified gait (right; the "Entlastungsgang") [13].

investigated patients before and after rotationplasty [8,9]). This is a surgical treatment option that may be indicated in patients with bone tumors of the lower extremity when the affected bone including the knee joint has to be resected and the remaining shank segment is reattached to the thigh after rotating the foot by 180°. The former ankle joint takes over the function of the former knee joint and the shortened extremity is equipped with a foot prosthesis and an adapter to regain a comparable limb length. These patients have to learn how to load the extremity and to control the "new" knee joint in spite of this specific anatomical situation. Initially, patients are cautious and insecure which is reflected in their gait patterns. After few months, however, they usually learn to adapt to the new situation and regain confidence in their walking abilities so that the gait patterns become more normal and natural. In these cases, gait analysis can be used document the individual progress of single patients (Figure 16.4).

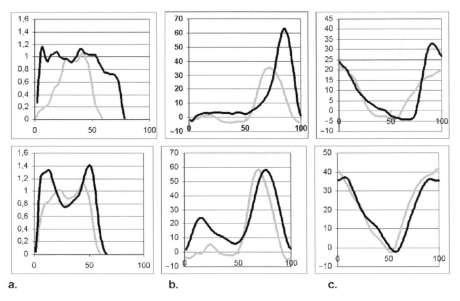

a.   b.   c.

Fig. 16.4. Example of a longitudinal evaluation of a patient with lower leg prosthesis after rotationplasty in the affected leg (grey) and contra-lateral leg (black) approximately three months (top) and six months after surgery (bottom). (a) vertical ground reaction forces [N/BW]; (b) knee joint flexion-extension motion [degree]; (c) hip joint flexion-extension motion [degree]. The changes in the parameters indicate a clear improvement of joint motion and loading with longer follow-up [y-axis = percent of gait cycle]. Other areas of application include pre- and postoperative measurements in patients receiving specific hip prostheses [14], hip revision surgery [7] or knee arthroplasty with a minimally invasive approach [5] or a new prosthetic design [6].

16.4 Potential Limitations of Clinical Gait Analysis

When performing clinical movement analysis certain issues should be considered: Besides the potential limitations caused by instrumentation, i.e., marker and/or soft tissue movements causing a certain artifact, or kinematic crosstalk there are also the procedural limitations caused by investigations of a subject under laboratory conditions. It has been postulated that the measurement tools should not alter the function that is being assessed. It has to be realized however that the patient under observation may be affected or even intimidated by the specific circumstances in the gait lab and may therefore not present his natural movement pattern. Furthermore, the attempt to maintain a constant walking speed or to hit the force plate might further disturb the individuals natural gait pattern. Clinical gait analysis might reveal the best performance under ideal conditions and the patient will probably try to present the best possible gait performance under these "artificial" lab conditions. It is not clear how this compares to the activities that are being performed in daily life, when not under surveillance of several cameras. Therefore, the results obtained in the gait lab might have to be considered as best-case possibilities of **how well** an individual patient is able to move or walk. This may or may not at all be related to how the patient takes advantage of this potential, i.e., **how much** this patient is actually walking. It remains unclear how indicative this is of his or her activities in daily life (ADL).

16.5 Assessment of Daily Life Activities

If both of the above mentioned aspects are of interest the **quality as well as the quantity** of gait have to be assessed. This implies that the methods for clinical gait analysis have to be complemented by methods for the assessment of activities in daily life (ADL). A wide array of tools is available for the assessment of ADL activities. The available systems can be distinguished by their measuring techniques. The most common technique in the last decades are subjective questionnaires, diaries or interviews that have been utilized in a wide range of studies investigating subjects' self-reported activities in daily life. A common and fairly simple method is to ask the patient or subject to fill out a retrospective questionnaire about the activities performed in a certain time frame (i.e., the last week or two). However, these procedures have been tested for reliability and were shown to suffer from bad accuracy because the individual activities are not well remembered or are influenced by subjectivity, i.e. they might reflect how much a subject might have wanted to be active. On the other end of the spectrum are direct observations or video recordings of a subject which are highly accurate and valid but are not very practical or feasible because they require a high effort with respect to recording and analyzing the activities (Figure 16.5).

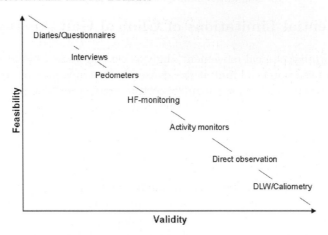

Fig. 16.5. Available methods for activity assessment arranged according to their feasibility and validity. More recently, pedometers have been developed for mechanically detecting impacts during locomotion. Pedometers differ in their construction details as well as their accuracy, although the manufacturers propose also an acceptable estimation of physiologic parameters such as energy consumption.

Recently, body-fixed sensors have been developed with a wide range of shapes and technical specifications. The devices are based on acceleration and/or gyroscopic sensors and can usually store information in a small, lightweight data logger for extended periods (from 24 hour measurements up to several weeks). These activity monitors generally use uni- or triaxial acceleration sensors sensitive to body motion and/or gravity. Other available systems are based on ultrasound sensors. The advantage of body-fixed sensors is that they allow movement free of restrictions for the subject. The measurements are not confined to a calibrated laboratory space and can record the individual activity behavior in the subject's natural home and work environment. Thus, the daily activity level can be assessed under realistic conditions. The only confounding effect might be caused by the patients' awareness about the scope of the measurements so that they might feel especially motivated to reveal a movement pattern that is more active than usual. Therefore, these measurements should be carried out over repeated days (at least one week would be desirable) so that potential differences between normal working days or weekend days with more leisure time would be detected. The Dynaport ADL-monitor developed by McRoberts B.V. (The Hague, The Netherlands) consists of three sensors measuring acceleration induced by body motion and gravity (Figure 16.6). The subject wears two sensors, measuring horizontal and vertical acceleration of the upper body, in a belt around the waist, containing also the battery supply and data logger. The third sensor, worn underneath the trousers and connected to the logger via a cable, is measuring horizontal acceleration of the left thigh. The signals of all sensors are processed by algorithms and automatically classified, for example to distinguish between locomotion, standing, sitting or lying [15, 16].

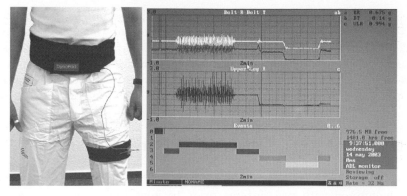

Fig. 16.6. The signals of the Dynaport sensors are automatically classified into locomotion (red), standing (dark blue), sitting (light blue) and lying (yellow).

Data can be stored for longer periods up to several months depending on the activity level of the subjects and the recording intervals. In most cases, the manufacturers offer classification algorithms converting the raw data to activity parameters (walking, bicycling, gait velocity, distances) or posture categories (standing, sitting, and lying). Lab-based methods offer a valid estimation of energy expenditure or can be considered as gold standard in non-lab-based methods, such as direct observation or doubly labeled water (DLW). In the latter technique water is labeled (hydrogen and oxygen are replaced with an uncommon isotope, usually with the heavy non-radioactive forms of the elements deuterium and oxygen-18) for tracking purposes in order to allow for measurement of the metabolic rate or energy expenditure over a period of time. Consequently, the method has to be chosen carefully according to the specific aim of a study regarding the validity of the outcome and feasibility. Detailed overviews of these methods can be found elsewhere [17–25]. The sensor and recording systems are usually less expensive as compared to the more complex and elaborate 3-dimensional gait analysis systems and they do not required large lab space. Therefore, these small, robust, lightweight and less expensive systems are characterized by good cost-effectiveness and wide-range availability. The range of sensors starts with simple and cheap step counters that are worn at the waist and only count the number of steps that the device registers. Due to their simplicity, these devices can be manipulated easily by the subject and may therefore be too unreliable for scientific purposes. Furthermore, the display indicates the number of steps only and no additional timing information is provided so that no further analyses with respect to the actual activity behavior during the day are possible. Finally, the information is visible for the patient, which may be an advantage or disadvantage depending on whether motivational issues are desirable, or not. More sophisticated step counter systems are based on uni-, bi-, or tri-axial accelerometers and store the information about the steps and a time log so that the information can be read out after the recording period. The timing information allows detecting when there are active and less active phases in the daily schedule. This way, differences between work-related and leisure-time activity, between weekdays and weekend-days, or seasonal changes in the activity patterns can be assessed. The basic level of information provides a step count only and more detailed information can be gained when step frequencies or movement intensities are distinguished.

16.6 Clinical Applications of Activity Assessments

The daily activities patients suffering from hip or knee osteoarthritis (OA) were monitored with two systems prior to joint replacement surgery. The subjects wore the DynaPort ADL-monitor for one day and the Step-Activity-Monitor (Cyma Inc., USA) for seven consecutive days. The Step-Activity-Monitor (SAM) is a lightweight and comfortable device which includes an acceleration sensor and interprets acceleration above a specific threshold as steps of the right leg Brandes, 2004 No.172. The SAM is adjustable to individual gait parameters and showed accuracy around 97% [26, 27]. The steps of one foot, in other words, gait cycles are stored in 1-minute-intervals thus enabling a computation of walking intensity in terms of minutes walked with a specific number of gait cycles per minute (Figure 16.7).

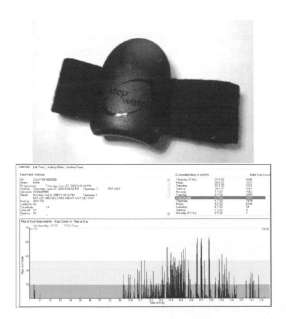

Fig. 16.7. The Step-Activity-Monitor stores the number of gait cycles (y-axis) in 1-minute-intervals (bars at x-axis) up to several months and provides detailed reports for each day.

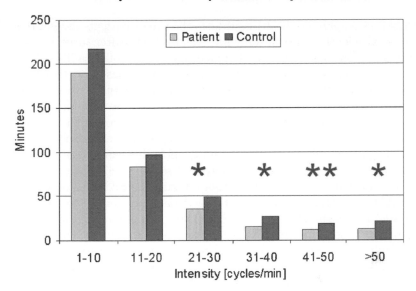

Fig. 16.8. Differences regarding intensity categories between OA patients and healthy subjects (*=p¡0.05, **=p¡0.01).

The time spent for locomotion measured by the ADL-monitor was lower in OA patients (10.5%) compared to healthy adults (11.1%). For the patients, the SAM detected only 4782 (SD 2116) gait cycles per day on average in contrast to 6616 (SD 2387) gait cycles per day in healthy subjects. A closer look at the intensities used in daily life revealed fewer minutes in all intensity categories for the OA patients compared to healthy adults (Figure 16.8). Furthermore, correlation coefficients were computed between the gait parameters assessed with the 3D gait analysis in the lab and the activity parameters measured with the ADL und SAM. Surprisingly, no relevant correlations could be found between lab and non-lab parameters. Thus, the findings of this study show that patients with less movement limitations in lab-based measurements are not inevitably more active in daily life than more severely limited patients. Furthermore, the results show that OA causes the subjects to walk less as well as slower than healthy subjects.

Another study investigated the outcome of limb salvage surgery in 22 patients (mean age 35, SD 18 years) suffering from a malignant tumor in the proximal tibia or distal femur. After resection of the knee joint and the affected parts of the tibia or femur the defects were reconstructed with a modular knee joint prosthesis (MUTARS). To evaluate the activity level of patients after limb salvage surgery, the patients were measured 5.6 (SD 4.1) years after treatment with the DynaPort ADL-monitor and the SAM. The patients spent less time for locomotion (9.9%) as compared to healthy subjects (11.1%). Similar to the previously described study, the patients performed also fewer gait cycles per day (4,786, SD 1,770) than healthy subjects (6,616,

SD 2,387). Compared to healthy subjects the tumor patients did not reach the same activity level as in their daily life even five years after surgery. Compared to amputees, on the other hand, limb salvage surgery offers a life with a moderate activity level taking the complexity of the disease and treatment into account [28]. In laboratory-based assessments, subjects are aware of the test situation and usually try to demonstrate their best locomotion performance. However, activity in daily life underlies many influences, such as job-related activity, leisure-time activities, transportation as well as seasonal variations. Miniature devices allow for an easy and comfortable assessment of these variations. In order to investigate the potential seasonal influence, 13 soldiers of the German Armed Forces were measured with the SAM throughout a year. The SAM was returned by standard mail once per month for downloading the data. The mean gait cycles per day were computed for each month and each soldier to gain insight in seasonal variation of daily activity. The results suggest that daily activity is lower during winter and increases during summer (Figure 16.9). Thus, it could be useful to add a correction factor to activity values obtained in summer or winter.

A possibility to get a more detailed analysis of subjects could be the monitoring of subjects hips by accelerometry with a miniature device attached to the back of the subject. Quite recently, a inverted pendulum model was tested for adults and children under controlled conditions [29, 30]. The accuracy of the method in detecting left and right steps, walking distance and speed under controlled conditions was shown to be 99%. The method was further compared to a manual step counter and global positioning system (GPS) in two walks of a 400 m track with normal (1.5 m/s) and fast (1.9 m/s) gait velocity and a 412 m circuit in the streets with normal velocity (1.5 m/s). For normal velocity, the results of step detection were around 99%, for walking distance and velocity around 95%. For the track with fast speed, the accuracy dropped to 95–90%. However, the method offer great potential

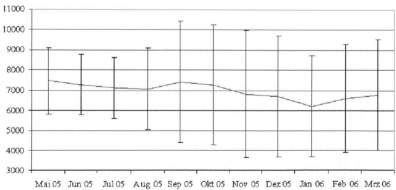

Fig. 16.9. Mean gait cycles/day for all soldiers from May 2005 to March 2006 and mean value for the whole year (blue line, 6,970 gait cycles per day).

to measure subjects activity in daily life after having computed a velocity-sensitive correction factor. Furthermore, a detailed analysis of each gait cycle could add useful information about the variety of gait parameters subjects use in daily life. Finally, it should be mentioned that there are systems available, which are more interested in physiologic parameters. Some of these devices try to determine the energy expenditure from the raw accelerometer signals or from additional skin temperature and impedance or heart rate recordings with appropriate software algorithms.

16.7 Summary and Outlook

While the illustrated activity monitors provide a more or less detailed feedback about the quantity of physical activity in daily life they do not help to determine the reasons for limitations in the individual's mobility. Therefore, the two approaches are complementary because one adds to the other the missing piece of information, i.e., they should be considered as the two different faces of the same coin. With the most recent developments in marker-less motion analyses the link between the two approaches could eventually be added. If the technical problems are solved, marker-less motion analysis might become more widely available because less sophisticated hardware may be used. Furthermore, the systems might be more easily applicable in different environments so that they are less confined to laboratory settings. A future application of these emerging systems might be imaginable in that they can be set-up in the natural environments of patients under observation so that they could allow 3-dimensional movement analysis in daily-life settings.

References

1. C. W. Braune and O. Fischer. Der Gang des Menschen - 1. Theil: Versuche am unbelasteten und belasteten Menschen. *21. Band der Abhandlungen der mathematisch-physischen Classe der Königlich Sächsischen Gesellschaft der Wissenschaften.* Leipzig, S. Hirzel. 21, 1895

2. J. R. Davids, J. R. Quantitative gait analysis in the treatment of children with cerebral palsy. *Journal of Pediatric Orthopaedics* 26(4): 557–559, 2006.

3. Davis, R. B. Reflections on clinical gait analysis. *Journal of Electromyography and Kinesiology* 7(4): 251–257, 1997.

4. P.A. DeLuca, R. B. I. Davis, S. Ounpuu, S. Rose and R. Sirkin (Alterations in surgical decision making in patients with cerebral palsy based on three-dimensional gait analysis. *Journal of Pediatric Orthopaedics* 17: 608–614, 1997.

5. S. Fuchs, B. Rolauffs, T. Plaumann, C. O. Tibesku and D. Rosenbaum Clinical and functional results after the rehabilitation period in minimally-invasive unicondylar knee arthroplasty patients. *Knee Surgery, Sports Traumatology, Arthroscopy* 13(3): 179–186, 2005.

6. S. Fuchs, A. Skwara and D. Rosenbaum Preliminary results after total knee arthroplasty without femoral trochlea: Evaluation of clinical results, quality of life and gait function. *Knee Surgery, Sports Traumatology, Arthroscopy* 13(8): 664–669, 2005.

7. C. Götze, C. Sippel, D. Rosenbaum, L. Hackenberg and J. Steinbeck Ganganalyse bei Patienten nach Hüftpfannen Wechsel [Objective measures of gait following revision hip arthroplasty. First medium-term results 2.6 years after surgery]. *Zeitschrift für Orthopädie und Ihre Grenzgebiete* 141(2): 201–208, 2003.

8. A. Hillmann, D. Rosenbaum, G. Gosheger, C. Hoffmann, R. Rödl and W. Winkelmann Rotationplasty type B IIIa according to Winkelmann: Electromyography and gait analysis. *Clinical Orthopaedics and Related Research* 384: 224–231, 2001.

9. A. Hillmann, A., D. Rosenbaum, J. Schröter, G. Gosheger, C. Hoffmann and W. Winkelmann Clinical and functional results after rotationplasty. Electromyographic and gait analysis in 43 patients. *Journal of Bone and Joint Surgery [Am]* 82(2): 187–196, 2000.

10. R. M. Kay, S. Dennis, S. Rethlefsen, R. A. Reynolds, D. L. Skaggs and V. T. Tolo The effect of preoperative gait analysis on orthopaedic decision making. *Clinical Orthopaedics and Related Research* 372: 217–222, 2000.

11. T. Mittlmeier and D. Rosenbaum Clinical gait analysis. *Unfallchirurg* 108(8): 614–629, 2005.

12. C. C. Prodromos, T. P. Andriacchi and J. O. Galante A relationship between gait and clinical changes following high tibial osteotomy. *Journal of Bone and Joint Surgery [Am]* 67-A(8): 1188–1194, 1985.

13. J. Schröter, V. Güth, M. Overbeck, D. Rosenbaum and W. Winkelmann The "Entlastungsgang. A hip unloading gait as a new conservative therapy for hip dysplasia of the adult. *Gait and Posture* 9(3): 151–157, 1999.

14. W. Steens, D. Rosenbaum, C. Götze, G. G., R. van den Daele and J. Steinbeck Clinical and functional outcome of the thrust plate prosthesis: short- and medium-term results. *Clinical Biomechanics* 18: 647–654, 2003.

15. Brandes, M. and D. Rosenbaum Correlations between the step activity monitor and the DynaPort ADL-monitor. *Clinical Biomechanics* 19(1): 91–94, 2004.

16. Pitta, F., et al. Characteristics of physical activities in daily life in chronic obstructive pulmonary disease. *American Journal of Respir Critical Care Medicine* 171(9): 972–977, 2005.

17. Booth, M. Assessment of physical activity: an international perspective. *Research Quarterly for Exercise and Sport* 71(2 Suppl): S114–120, 2000.

18. Crouter, S.E., et al. Validity of 10 Electronic Pedometers for Measuring Steps, Distance, and Energy Cost. *Medicine and Science in Sports and Exercise* 35(8): 1455–1460, 2003.

19. Ekelund, U., et al. Physical activity assessed by activity monitor and doubly labeled water in children. *Medicine and Science in Sports and Exercise* 33(2): 275–281, 2001.

20. Schutz, Y., R.L. Weinsier, and G.R. Hunter Assessment of free-living physical activity in humans: An overview of currently available and proposed new measures. *Obesity Research* 9(6): 368–379, 2001.

21. Simon, S.R. Quantification of human motion: Gait analysis - Benefits and limitations to its application to clinical problems. *Journal of Biomechanics* 37(12): 1869–1880, 2004.

22. Sirard, J.R. and R.R. Pate Physical activity assessment in children and adolescents. *Sports Medicine* 31(6): 439–454, 2001.

23. Trost, S.G. Objective measurement of physical activity in youth: current issues, future directions. *Exercise and Sport Sciences Reviews* 29(1): 32–36, 2001.

24. Tudor-Locke, C.E. and A.M. Myers Challenges and opportunities for measuring physical activity in sedentary adults. *Sports Medicine* 31(2): 91–100, 2001.

25. Welk, G.J. Principles of design and analyses for the calibration of accelerometry-based activity monitors. *Medicine and Science in Sports and Exercise* 37(11 Suppl): S501–511, 2005.

26. Resnick, B., Nahm, E.S., Orwig, D., Zimmerman, S.S., Magaziner, J. Measurement of activity in older adults: Reliability and validity of the Step Activity Monitor. *Journal of Nursing Measurement* 9(3): 275–290, 2001.

27. Coleman, K.L., Smith, D.G., Boone, D.A., Joseph, A.W., del Aguila, M.A. Step activity monitor: Long-term, continuous recording of ambulatory function. *Journal of Rehabilitation Research and Development* 36(1): 8–18, 1999.

28. Sugiura, H., Katagiri, H., Yonekawa, M., Sato, K., Yamamura, S., Iwata, H. Walking ability and activities of daily living after limb salvage operations for malignant bone and soft-tissue tumors of the lower limbs. *Archives of Orthopaedic and Trauma Surgery* 121(3): 131–134. 2001.

29. Zijlstra, W. and A.L. Hof Assessment of spatio-temporal gait parameters from trunk accelerations during human walking. *Gait and Posture* 18(2): 1–10. 2003.

30. Brandes, M., Zijlstra, W., Heikens, S., van Lummel, R., Rosenbaum, D. Accelerometry based assessment of gait parameters in children. *Gait Posture* (in print) 2006.

Optimization of Human Motion Exemplified with Handbiking by Means of Motion Analysis and Musculoskeletal Models

Harald Böhm and Christian Krämer

Department of Sport Equipment and Materials, TU Munich
Connollystr 32, 80809 Munich, Germany

Summary. This chapter demonstrates the use of computational methods for analyzing human motion to optimize human movements. Motion analysis, inverse dynamics as well as musculoskeletal simulation models are applied to the handcycle, a sport equipment for disabled persons. The optimization problem is hereby to find a arm trajectory that is better adopted to the biomechanical preconditions of the hand–arm–shoulder system, so that propelling is more efficient with respect to metabolic energy consumption of the muscles. The used methods, their results, benefits and limitations are discussed.

17.1 Introduction

Inverse dynamics computation of constraint forces and resultant muscle torques in dynamic movements is a common method to analyze human motion in biomechanics [31] and ergonomics [24]. Visualization of calculated kinematics and kinetics and their presentation by means of three-dimensional human avatars are commonly used. However, the constraint forces and resultant muscle torques obtained from inverse dynamics do not include muscles' physiological properties. This firstly leads to the problem that muscular load is underestimated when the movements include muscular cocontraction [12]. Secondly, important parameters such as metabolic energy consumption [1] or the feeling of discomfort [36] cannot be calculated without determining muscular forces.

Muscular properties such as the force–length and the force–velocity relations [18] have been the subject of detailed measurements in clinical diagnostics and competitive sports. Recently, the knowledge about internal muscle properties has been increased by the application of in vivo ultrasound and MRI techniques [22, 23], and valid parameters are now available, leading to realistic agreement of model calculations and measured quantities such as muscle surface electromyography and joint angles [3] as well as in vivo measured tendon strain [34]. The improved knowledge of muscular parameters

B. Rosenhahn et al. (eds.), Human Motion – Understanding, Modelling, Capture, and
Animation, 417–434.

inverse dynamics

$$m_i \, a_i \xrightleftharpoons{} \Sigma_{ext} \, F_{i,ext}$$

direct dynamics

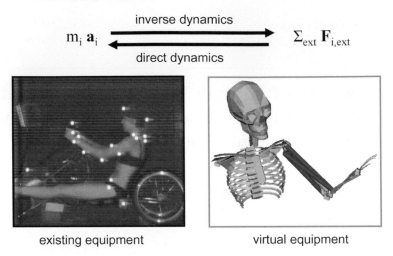

existing equipment   virtual equipment

Fig. 17.1. Inverse and direct dynamic approach explained as by the basic Newton equation formulated for a rigid body.

increased the confidence in musculoskeletal models in order to obtain insight into human motion on a muscular level instead of a purely mechanical level of resultant muscle joint torques [31].

There are currently two methods used to calculate muscular forces: static and dynamic optimization, which use a mechanical inverse or direct dynamic approach respectively. Both inverse and direct dynamic method can best be described by the basic Newton equation for a rigid body (see Figure 17.1): having captured the motion trajectory of the body segment i, e.g., upper arm, as depicted left in Figure 17.1, the acceleration of the upper arm can be computed. Knowing the mass of the upper arm, the right side of the equation contains all external forces on the segment. In case of the upper arm these forces are gravity and the shoulder and elbow joint forces. When using the forward dynamic approach, virtual equipment is tested by applying internal muscle forces to the bones and synthesizing the motion trajectory of the upper arm.

Static optimization has been used extensively for the purpose of estimating in vivo muscle forces [5, 10, 14]. Static models are computationally efficient, allowing full three-dimensional motion and generally incorporate many muscles, e.g., 30 or more muscles per leg. However, static optimization has been criticized on several points. Firstly, the validity of inverse dynamics is intimately related to the accurate collection and processing of body segmental kinematics [11]. Secondly, the fact that static optimization is independent of time makes it relatively difficult to properly incorporate muscle physiology. Finally, analysis based on an inverse dynamics approach may not be appropriate for explaining muscle coordination principles [21, 37].

Dynamic optimization is not subject to these limitations and can provide more realistic estimates of muscle force [1, 3]. Dynamic optimization integrates system dynamics into the solution process. Quantities like muscle forces and the criterion of performance are treated as time-dependent state variables, whose behavior is governed by sets of the following differential equation.

$$\frac{d}{dt}\left(\frac{\partial L}{\partial \dot{x}_i}\right) - \frac{\partial L}{\partial x_i} = F_{x_i}^L + F_{x_i}^M \tag{17.1}$$

L is the lagrangian (kinetic minus potential energy) of a planar rigid multibody system connected by n revolute joints with joint angles $x_1..x_n$ representing the degrees of freedom of the system. $F_{x_i}^L$ and $F_{x_i}^M$ are the passive (ligaments and cartilage) and muscular torques across the respective joint [17]. Ideally, the differential equations accurately represent the system's underlying physiological properties. The time histories of predicted muscle forces are thus consistent with those forces that could naturally arise during movement. Dynamic optimization (forward dynamic approach) unfortunately requires the integration of the equations of motion (Equation 17.1) and muscles' state variables. Consequently a much higher computational expense is the drawback of more detailed modelling.

The aim of this study is firstly to show how muscular properties can be measured and used in order to understand and evaluate human motion and secondly to demonstrate the use of a direct dynamical model for the optimization of sports equipment.

The methods described above will be illustrated by the specific example of handbiking (handcycling). Handbiking has become very popular among disabled athletes and was declared as Paralympic discipline in 2004. As the history of the sport is still very young not many developments have taken place. The drive train in particular has been adopted from conventional bicycles and might not perfectly fit to the upper body's anthropometry. This study compares the circular handbike to a linear rowing movement and derives a new idea of a driving concept from measured muscular properties. This new idea is virtual and therefore a good application for the forward dynamic computer model previously described. The effect of both drive concepts, the conventional circular and the new concept, on metabolic energy consumption were investigated.

17.2 Methods

In this study the following approach was chosen for the purpose of evaluating and optimizing the handbike's propelling movement with respect to increased power output and minimized energy consumption:

1. Collect data of muscular force production ability during the movement (statically and/or dynamically) in Section 17.2.1.
2. Gain knowledge about handbiking and related movements (e.g., rowing, no legs only arms) by means of motion analysis and inverse dynamics in Section 17.2.2.

3. Derive an alternative, improved motion from the preliminary considerations in Sections 17.3 and 17.4.
4. Examine the handbike and the virtual alternative of an elliptical motion by means of a musculoskeletal model and forward dynamics in Section 17.2.3.

17.2.1 Measurement of Muscular Properties

The ability of muscle torque production under static conditions was measured over the complete range of motion [36]. Figure 17.2 displays the setup to measure isometric joint torques of the elbow joint. A similar setup is used to determine muscles torque production of the shoulder joint.

The torques from the force the test person applied to the handle, are measured with strain gauges. The force can be calculated on the basis of the arm of the lever from the hand to the hand grip and from the grip to the torque transducer. The moment generated by the grip is neglected. The force must be perpendicular to the joint distal segment, e.g., forearm, and the body segment must be in line with the lever of the transducer. This can be achieved by moving the device in all directions of rotating as well as up and down. A test series undertaken with n = 7 male subjects, age between 20–26 years, 175–183 cm height and 73–82 kg weight was performed. Different angular positions of the joints in steps of 45 degree and all possible force directions of the shoulder (flexion-extension, abduction-adduction, internal-outward-rotation) and elbow joint (flexion-extension, supination-pronation) have been measured. To avoid muscle fatigue caused by a large number of measurements (three force directions in every angular position) step size of the angular joint positions was limited to 45 degree. The participants had to generate maximum joint torque (in every angular position and force direction), and hold it for 5 seconds with visual feedback.

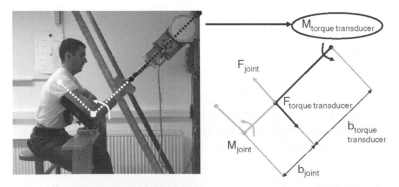

Fig. 17.2. Device to measure isometric joint torques of the elbow joint. The moment is calculated with the formula $\mathbf{M}_{joint} = \mathbf{F}_{joint} b_{joint}$.

17.2.2 Motion Analysis and Inverse Dynamics

Inverse dynamics computation of internal joint forces and torques requires the input of acceleration of the body segments (Figure 17.1), obtained by differentiating motion trajectories. Three different motion capture systems were compared due to their applicability in handbiking: markerless (PCMAN [27], comparable to 12), markerbased with video data and motion analysis software package (SIMI Reality Motion Systems GmbH, Unterschleissheim, Germany) and markerbased infrared filtered image (Vicon, Oxford, UK). All three systems are shown in Figure 17.3. The markerless method was designed for the use of two cameras, which requires manual readjusting of the tracked body segments. To make it run reliably and with high accuracy in the lab, more than eight cameras 15 and an additional laser body scan of the subjects to match the different movement postures would be required. Moreover the predefined definition of the coordinate systems were not in agreement with the conventions of biomechanics which are based on well-defined anatomical landmarks on the body [20]. This problem was solved using a marker-based system with a semi manual tracking of reflecting markers (SIMI) in three video images from different perspectives. This system lead to very time-consuming handling of video sequences and manual correction of the tracking. The infrared system (Vicon) preprocesses the infrared image in the camera, so that only coordinates of the bright marker points are captured. The user finally obtains 3D coordinates of the markers position. This system reliably tracks the markers completely automatically, so that no manual postprocessing is required. The accuracy of tracking the center of a 9 mm diameter marker within a focal volume of 6 cubic meters was less then 0.1 mm. Although the system is very accurate, information about skeletal shoulder and elbow kinematics is limited, particularly about abduction and adduction or internal and external rotation: there is soft tissue between the reflective markers and the bones, which allows for the markers to move in relation to the underlying bones during movement [30]. Based on rigid body mechanics, three-dimensional analysis

Fig. 17.3. Different motion analysis systems: marker-less semiautomatic detection systems (PCMAN [27], left), marker-based semiautomatic tracking with high speed cameras and Simi-Motion (middle) and marker-based automatic tracking systems (Vicon, right).

assumes that markers placed on the body represent the positions of anatomical landmarks for the segment in question [20].

One way to avoid the inaccuracy of surface markers is to use invasive markers and thereby directly measure skeletal motion. This provides the most accurate means for determining bone movements [6]. Nigg [25] reported differences of up to 50% for similar knee angles when comparing knee joint kinematics using external and bone fixed markers. It appears that skin movement artifacts present the most problematic source of error.

Invasive methods are excluded in our study, but the noninvasive marker-based approach can be improved by employing Soederkvist and Wedin's transformation matrices [32]. In this method more than the required three markers are attached to one segment to calculate segments transformation vector (d) and rotation matrix (R) from markers position x_i to markers position y_i in the next timestep. Because of measurement errors the mapping from points x_i to y_i is not exact, the following least squares problem is used to determine R and d.

$$min \sum_{i=1}^{n} \|Rx_i + d - y_i\|^2$$

defining matrix A and B as follows

$$\bar{x} = \frac{1}{n} \sum_{i=1}^{n} x_i$$

$$\bar{y} = \frac{1}{n} \sum_{i=1}^{n} y_i$$

$$A = [x_1 - \bar{x}, ..., x_n - \bar{x}]$$

$$B = [y_1 - \bar{y}, ..., y_n - \bar{y}]$$

Singular value decomposition [2] of the matrix $C = BA^T$ can be used to compute the solution to described in the following algorithm.

$$P\Gamma Q^T = C$$
$$R = Pdiag(1, 1, det(PQ^T))Q^T$$
$$d = \bar{y} - R\bar{x}$$

In this study, elbow and shoulder joint motions are calculated by means of joint coordinate systems (JCS), which are based on anatomical landmarks according to the standard of the International Society of Biomechanics (ISB) [20]. The zero angle configuration of all JCS are calculated from a standing trial as shown in Figure 17.4. The cardan angles are calculated according to the conventions of Grood and Suntay [15], where the first angle is the rotation around one axis defined in the distal segment JCS. The third angle is a rotation around one axis of the proximal JCS. The second rotation axis cannot be freely chosen. It is perpendicular to the first and the third axis.

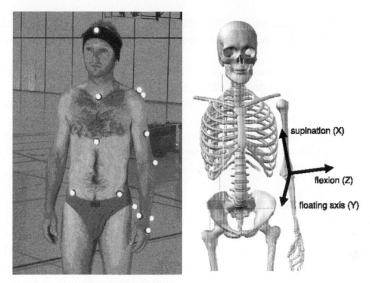

Fig. 17.4. Definition of the joint coordinate system according to ISB [20] from standing trial. Cardan angles are calculated according to Grood and Suntay [15].

Compared to standard cardan rotations this system has the advantage that it can be interpreted clinically. In case of the elbow joint for instance the distal body axis defined with respect to the forearm is selected as the forearm supination-pronation axis (X), in direction from the wrist to the elbow joint centers. The third axis is the elbow flexion-extension axis (Z), which is defined in the upper arm segment from the lateral to the medial epicondylus (see Figure 17.4). The second rotation axis (floating axis, Y) cannot be freely chosen. It is perpendicular to the first and the third axis.

In all approaches, markerless or markerbased, invasive or noninvasive, kinematic crosstalk caused by variability in the experimental determination of the joint axis, e.g., knee joint flexion axis, was shown to be of sufficient magnitude to affect the measurement of joint rotation [28]. Therefore absolute values of segment rotation must be taken with precaution. However, regarding a comparison based on one subject using different types of propelling systems, as it is the case in this study, the error is the same for all systems, since the same anatomical markers were used to determine the joint axis.

The calculated joint angles described above are then used as an input for the Newton Euler equations (see Figure 17.1) to calculate inverse dynamically joint reaction forces and torques and the resultant muscle joint torques. The resultant muscle joint torques are equivalent to the net-muscle joint torques calculated by means of the musculoskeletal model (see Section 17.2.3). It should be noted that inverse computation necessitates the joint velocity and joint acceleration, which are calculated numerically from the joint angles. This generally leads to very rough and unrealistic reaction forces and joint torques.

A marker-driven human model [35] filters motion data according to the rigid body dynamical model. As a result appropriate joint angle functions can be obtained.

This study uses motion capture of handbiking and rowing by means of markerbased tracking systems (SIMI as well as Vicon) and inverse dynamics calculation applying the model described in [35]. Following protocol was carried out on two subjects (male, 32 and 36 years, 175 and 180 cm, 72 and 78 kg): rowing and handcycling for one minute at a cadence of 66 cycles/min and 100 Watt resistance. In case of the comparison between handcycling and rowing, only pulling, i.e., no pushing forces where applied.

17.2.3 Musculoskeletal Model

To simulate a new non existing elliptical handbike drive by means of forward dynamics, a musculoskeletal model has to be implemented, in which the muscles generate the forces to drive the skeletal parts of the model. Segment mass, inertia and joint positions of the rigid skeletal bodies (hand, forearm, upper arm and upper part of the body) are adopted from the Hanavan model [16]. Segment lengths representing individual subjects are determined by using the Vitus 3D body scanner (Tecmat GmbH, Kaiserslautern, Germany). The interface of hand and crank is modeled with constraint forces, to vary between circular and elliptical crank motion the elliptical half axes a and b (see Figure 17.5) are varied. A total of six muscles transfer forces to the skeleton. The muscles were selected with respect to their importance regarding the investigated motion and their functionality (flexor-extensor, wrapping around one or two joints): m. brachialis, m. triceps brachii caput laterale, m. pectoralis

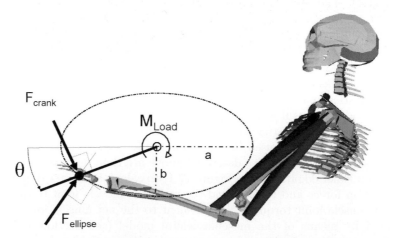

Fig. 17.5. Musculoskeletal model used to simulate handbiking. The human skeleton is modelled as a planar four-segment system linked with frictionless revolute joints at the shoulder and elbow. A total of six muscle groups transfer forces to the skeleton.

major, m. deltoideus pars spinalis, m. biceps brachii caput longum and m. triceps brachii caput longum. The Hill-type muscle model [19], which contains a contractile element (CE) and passive elastic element (PEE) in series with an elastic element (SEE), is applied for the calculation of the muscles' forces (see Figure 17.6). The passive SEE transmits the forces from the muscle CE and PEE to the bones. The force in the contractile element of each muscle depends on the muscle's length, velocity and activation [3,33]. The calculation of the muscle's length from origin to insertion is based on experimentally determined length functions over the joint angle [29]. Muscle activation is defined by a set of control nodes interpolated with sinusoidal functions as depicted in Figure 17.7. The nodes were varied in time and amplitude to make the model drive the handbike. An optimization procedure (Figure 17.8) for the control nodes parameters is used, minimizing the cost function of metabolic energy of all n muscles.

$$min \sum_{i=1}^{n} E_i(l_{CE}, \dot{l}_{CE}, F_{CE}, t)$$

Metabolic energy is calculated

$$E_i = H_i + W_i$$

W is the mechanical work $W = F_{CE}l_{CE}$ and H is the energy required for the contractile process, mostly dissipated as heat. Detailed parameters of calculating H can be found in [1]. The global optimization was performed applying the

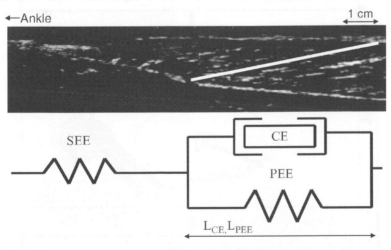

Fig. 17.6. Ultrasound picture of calf muscle and mechanical representation below. One fiber bundle marked in white is associated with the force-producing contractile element (CE). This fiber bundle is attached to the aponeurosis, the white structure surrounding the muscle fibers. When the fiber contract, it pulls at the aponeurosis which transfers the force via the tendon to the bones. The aponeuroses and tendons are associated with the series elastic element SEE.

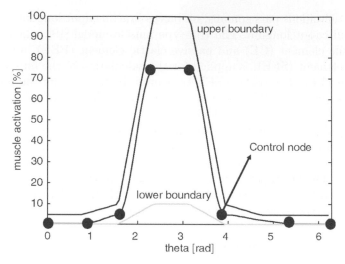

Fig. 17.7. Activation function with 8 control nodes used for m. brachialis. The control nodes can vary in time and amplitude during the optimization process. The upper and lower boundaries are used to reduce optimization time.

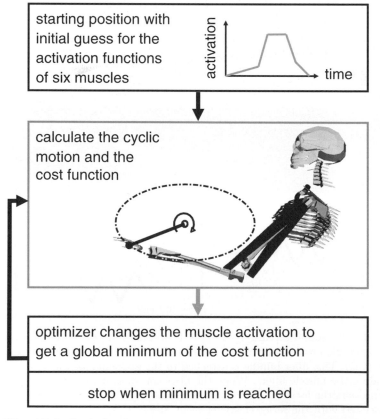

Fig. 17.8. Optimization process to determine muscular activation.

method of simulated annealing [7,13]. The equations of motion were solved and integrated, using the multibody software SIMPACK (Intec GmbH, Wessling, Germany).

17.3 Results

17.3.1 Kinematics of Rowing and Cycling with Respect to the Muscles' Torque Production Ability

The maximum isometric (static) joint torques of 7 subjects as measured with the setup described in Section 17.2.1 are shown in Figure 17.9. The joint torques were normalized and then averaged over the 7 subjects measured.

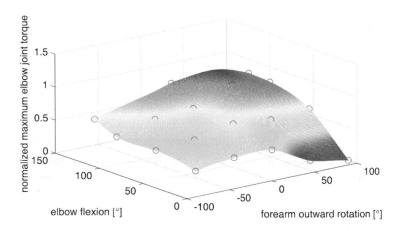

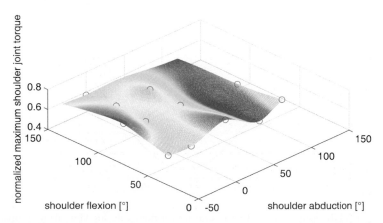

Fig. 17.9. Normalized maximum isometric joint torques of the elbow (top) and shoulder joint (bottom). Dots represent measured data, the surface in-between is interpolated.

In the case of the elbow joint, the flexion-extension angle as well as the supination-pronation angle of the forearm is shown, whereas for the shoulder joint the flexion-extension and abduction-adduction angles are plotted. The inward-outward rotation of the shoulder lies in the fourth dimension and is not displayed here. The joint torques in Figure 17.9 are lower than one, since not all athletes reached their maximum torque at the same joint positions. Figure 17.10 shows a typical example of joint angles derived from motion analysis during one handbike cycle, executed by one subject. The local JCS axis and rotation order are chosen according to Grood and Suntay [15]: the rotation around the z-axis of the proximal body (upper arm) corresponds to flexion and extension of the elbow joint, while the rotation around the y-axis of the distal body (forearm) corresponds to the supination and pronation of the forearm. The "floating axis" (x-axis) is perpendicular to the y- and z-axes and has no anatomical equivalence. The orientation of the axes is shown in Figure 17.4.

These measurements results (maximum isometric joint torque and motion analysis) were used to evaluate the force generation ability of different joint angle trajectories (handcycling and rowing) at the shoulder and elbow joints. This is shown in Figure 17.11 for the handcycle motion and the shoulder joint. This evaluation leads to the conclusion that the muscles' torque potential at the elbow joint is up to 20 percent higher during rowing than during handcycling (see Figure 17.12). This is mainly caused by a higher extend of elbow flexion. Concerning the shoulder joint the muscles' torque potential is the same for both motions.

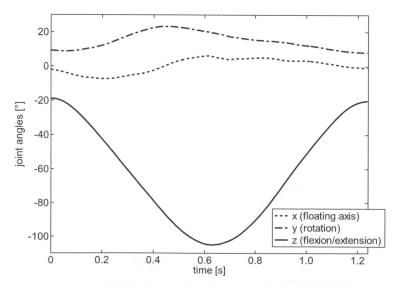

Fig. 17.10. Joint angles during one handbike-cycle for the elbow joint.

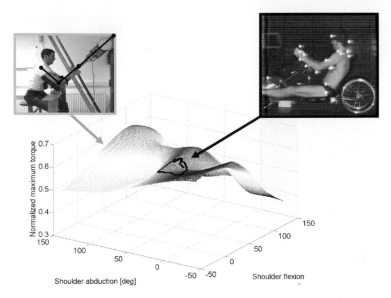

Fig. 17.11. Maximum isometric joint torques at shoulder joint plotted together with the joint angles trajectory of handbiking.

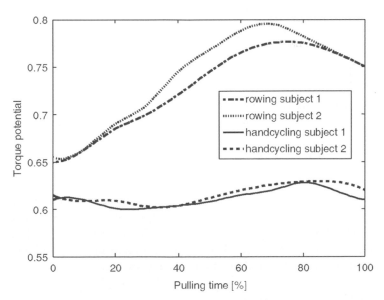

Fig. 17.12. Elbow torque potential of two test persons: in rowing the muscles' torque potential at the elbow joint is up to 20% higher than during handcycling.

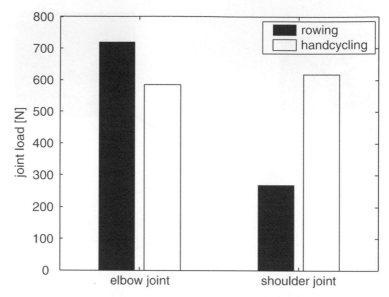

Fig. 17.13. Stress (total force) in elbow and shoulder joint during rowing and handbiking. Elbow joint: rowing: $F_{max} = 720.5N$ at flexion angle of 127.4 deg; handbiking: $F_{max} = 585.2N$ at flexion angle of 111.3 deg. Shoulder joint: rowing: $F_{max} = 269.2N$ at flexion angle of 4.4 deg, abduction angle of 17.1 deg and rotation angle of 14.7 deg. handbiking: $F_{max} = 617.3N$ at flexion angle of 20.6 deg, abduction angle of 9.8 deg and rotation angle of 6.9 deg.

17.3.2 Inverse Dynamics

The forces acting in the elbow and shoulder joint during handbiking and rowing were calculated inverse dynamically as described in Section 17.2.2. The results obtained by motion capturing described in 17.3.1 were used as input for the calculation. The joint forces are determined as the sum of all constraint forces and the force acting around the respective joint, which can be estimated from the muscle joint torques assuming a moment-arm for the ellbow and shoulder of 3.5 and 5.0 cm respectively. Figure 17.13 shows the maximum joint force occuring during the pulling phase of both movements. The maximum force that occurs in the elbow joint during rowing is 135.3 N higher than during handbiking. The maximum force in the shoulder joint is 348.1 N higher during handbiking than during rowing.

17.3.3 Musculoskeletal Model

To test the validity of the model calculations, a conventional circular motion was compared to measurements performed on two subjects as described in Section 17.3.1. The model kinematics and kinetics show close correspondence with the experimental data [4]. The model is therefore valid to compare

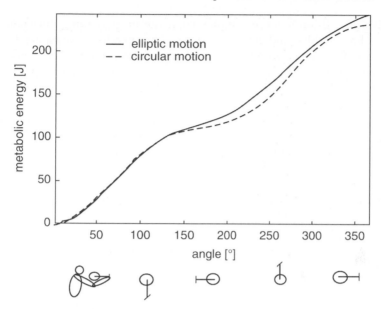

Fig. 17.14. Metabolic energy consumption calculated during one cycle.

metabolic energy in different situations. When handcycling at 100 Watt with a cadence of 66 rpm, the circular drive system requires 3.5% more energy than the elliptic drive system. The progression of calculated metabolic energy consumption is shown in Figure 17.14. A similar result of 3.8% is also obtained for the second subject simulated, having slightly different anthropometric parameters.

17.4 Discussion

The result of the comparison between rowing and handcycling with respect to the muscles' torque production ability (see Section 17.3.1) shows that a linear rowing movement (only arms, no legs) has a higher potential to produce forces than the circular handbike movement. This is caused by a higher extent of elbow flexion. The isometric (static) force is not only responsible for high force production but also more efficient with regards to the muscle energy consumption, i.e., the muscle needs less activation energy to produce the same force in a more suitable joint position (better muscle force–length relation). Limitations of these considerations are that the measured joint torques treat the joints independently. Furthermore the muscle force–velocity relation is not taken into account.

The load (maximum force) in the elbow joint is 135.3 N higher during rowing than during handbiking, while the load in the shoulder joint is 348.1 N higher during handbiking than during rowing (see Section 17.3.2). The circular handbike propulsion is more complex than the linear rowing movement concerning the shoulder and muscle system. This results in a higher shoulder joint force. The linear rowing movement allows for a more effective direction of pulling. Therefore higher forces can be applied by the participating muscles (m. biceps brachii, m. brachialis etc.), which results in higher elbow joint torques. Concerning the task of developing a new drive train for handbikes, this comparison implicates that there is nothing to be said against using a rowing-like motion (only arms) for propulsion: given the high shoulder stress of wheelchair users caused by their everyday activities, it is important that an alternative movement would especially avoid a capacity overload of the shoulder.

The forward dynamical model suggests that the cyclic drive concept is less efficient than the elliptical one. The difference in between circular and elliptic motion of 3.5% and 3.8% on the two simulated subjects is rather small and therefore only relevant for high performance athletes. The next steps will therefore be to optimize elliptic drive parameters for the individual anthropometrical parameters of the subjects in order to obtain a greater difference relevant for practical use in recreational handbiking.

The forward dynamical model successfully synthesizes movements based on a physiological muscle model and a performance criterion of metabolic energy consumption. However, the range of motion in handbiking is restricted by the specific sports equipment. When applying this method to other movements such as walking or entering a car, the cost function might include metabolic energy consumption and additional terms like stability and comfort. This will require multi-objective optimization solved in a paretho sense [26].

References

1. Anderson, F.C., M.G. Pandy: Dynamic optimization of human walking. *Journal of Biomechanical Engineering*, **123**:381–390, 2001.
2. Anderson, E., Z. Bai, C. Bischof, S. Blackford, J. Demmel, J. Dongarra, J. Du Croz, A. Greenbaum, S. Hammarling, A. McKenney, and D. Sorensen: LAPACK User's Guide. *SIAM, Philadelphia*, 3rd edn., 1999.
3. Böhm, H., G.K. Cole, G.P. Brüggemann, and H. Ruder: Contribution of muscle series elasticity to maximum performance in drop jumping. *Journal of Applied Biomechanics*, **22**:3–13, 2006.
4. H. Böhm, C. Krämer, and V. Senner: Optimization of the handbike's drive concept–mathematical approach. *Springer, The Engineering of Sport 6*, **3**:35–40, 2006.
5. Brand, R.A., D.R. Pedersen, and J.A. Friederich: The sensitivity of muscle force predictions to changes in physiologic cross-sectional area. *Journal of Biomechanics*, **19**:589–596, 1986.
6. Cappozzo, A.: Three-dimensional analysis of human walking: experimental methods and associated artifacts. *Human Movement Sciences*, **10**:589–602, 1991.

7. Corana, A., M. Marchesi, C. Martini, and S. Ridella: Minimizing multimodal functions of continuous variables with the simulated annealing algorithm. *ACM Transactions on Mathematical Software*, **13**:262–280, 1987.
8. Chow, J.W., W.G. Darling: The maximum shortening velocity of muscle should be scaled with activation. *Journal of Applied Physiology*, **11**:1–24, 1999.
9. Crowninshield, R.D.: Use of optimization techniques to predict muscle forces. *Journal of Biomechanical Engineering*, **100**:88–92, 1978.
10. Crowninshield, R.D., R.A. Brand: A physiologically based criterion of muscle force prediction in locomotion. *Journal of Biomechanics*, **14**:793–801, 1981.
11. Davy, D.T., M.L. Audu: A dynamic optimization technique for predicting muscle forces in the swing phase of gait. *Journal of Biomechanics*, **20**:187–201, 1987.
12. van Dieen, J.H., I. Kingma, and P. van der Bug: Evidence for a role of antagonistic cocontraction in controlling trunk stiffness during lifting. *Journal of Biomechanics*, **36(12)**:1829–36, 2003.
13. Gall, J.: Learning a dynamic independent pose distribution within a bayesian framework *06241 Dagstuhl Abstracts Collection Human Motion – Understanding, Modelling, Capture and Animation. 13th Workshop "Theoretical Foundations of Computer Vision"*.
14. Glitsch, U.,W. Baumann: The three-dimensional determination of internal loads in the lower extremity. *Journal of Biomechanics*, **30**:1123–1131, 1997.
15. Grood, E.W., W.J. Suntay: A joint coordinate system for the clinical description of three-dimensional motions: applications to the knee. *Journal of Biomedical Engineering*, **105**:97–106, 1983.
16. Hanavan, E.P.: A mathematical model of the human body. *AMRL. Technical Report. Wright-Patterson Air Force Base, Ohio*, 64–102, 1964.
17. Hatze, H.: The complete optimization of human motion. *Mathematical Biosciences*, **28**:99–135, 1976.
18. Epstein, E., W. Herzog: Theoretical models of skeletal muscle. Wiley, New York, 22–57, 1998.
19. Hill, A.V.: The heat of shortening and the dynamic constants of muscle. *Proceedings of the Royal Society London*, **126**:136–195, 1938.
20. Wua, G., F.C.T. van der Helm, H.E.J. Veeger, M. Makhsous, P. Van Roy, C. Anglin, J. Nagels, A.R. Karduna, K. McQuade X. Wang, F.W. Werner, and B. Buchholz: ISB: recommondation on definitions of joint coordinate system of various joints for the reporting of human joint motion - part II: shoulder, elbow, hand and wrist. *Journal of Biomechanics*, **38**:981992, 2005.
21. Kautz, S.A., R.R. Neptune, and F.E. Zajac: General coordination principles elucidated by forward dynamics: minimum fatigue does not explain muscle excitation in dynamic tasks. *Motor Control*, **4**:75–80, 2000.
22. Karamanidis, K., A. Arampatzis: Mechanical and morphological properties of different muscle-tendon units in the lower extremity and running mechanics: effect of aging and physical activity. *Journal of Experimental Biology*, **208**:3907–3923, 2005.
23. Magnusson, S.P., P. Aagaard, S. Rosager, and P.D. Poulsen: Load displacement properties of the human triceps surae aponeuroseis in vivo. *Journal of Physiology*, **531.1**:277–288, 2001.
24. Murray, I.A., G.R. Johnson: A study of the external forces and moments at the shoulder and elbow while performing every day tasks. *Clinical Biomechanics*, **19(6)**:586–594, 2004.

25. Nigg, B.M., G.K. Cole: Optical methods. In: Nigg BM, Herzog W, editors. *Biomechanics of the musculo-skeletal system.* Wiley New York, 86, 254, 1994.
26. Podgaets, A., W. Ockels, K. Seo, V. Stolbov, A. Belyaev, A. Shumihin, and S. Vasilenko: Some problems of pareto-optimization in sports engineering. *Springer, The Engineering of Sport 6,* **3**:35–40, 2006.
27. Seitz, T.: PCMAN Basics, description and manual. *Department of ergonomics, Technical University Munich,* 2002
28. Piazza, S.J., P.R. Cavanagh: Measurement of the screw-home motion of the knee is sensitive to errors in axis alignment. *Journal of Biomechanics,* **33**:1029–1034, 2000.
29. Pigeon, P., L.H. Yahia, and A.G. Feldman: Moment arms and lengths of human upper limb muscles as functions of joint angles. *Journal of Biomechanics,* **29(10)**:1365–1370, 1996.
30. Ramsey, K., P.F. Wretenberg. Biomechanics of the knee: methodological considerations in the in vivo kinematic analysis of the tibiofemoral and patellofemoral joint. *Clinical Biomechanics,* **14**:595–611, 1999.
31. Ren, L., R.K. Jonen, and D. Howard: Dynamic analysis of load carriage biomechanics during level walking. *Journal of Biomechanics,* **38(4)**:853–63, 2005.
32. Soederkvist, I., P.A. Wedin: Determining the movements of the skeleton using well-configured markers. *Journal of Biomechanics,* **26**:1473–1477, 1993
33. van Soest, A.J., P.A. Huijing, and M. Solomonow: The effect of tendon on muscle force in dynamic isometric contractions: a simulation study. *Journal of Biomechanics,* **28**:801–807, 1995.
34. Stafilidis S., K. Karamanidis, G. Morey-Klapsing, G. Demonte, G.P. Bruggemann, and D. Arampatzis: Strain and elongation of the vastus lateralis aponeurosis and tendon in vivo during maximal isometric contraction. *European Journal of Applied Physiology,* **94**:317-322, 2005.
35. Wallrapp, O., T. Grund, and H. Böhm: Human motion analysis and dynamic simulation of rowing. *Multibody Dynamics 2005,* J.M. Goicolea, J. Cuadrado, J.C. García Orden (editors), *in ECCOMAS Advances in Computational Multibody Dynamics, Madrid, Spain [CDROM]* 1–14, 2005.
36. Zacher, I., H. Bubb: Strength based discomfort model of posture and movements. *Digital human modelling for design and engineering, SAE International, Warrendale,* 1, 2139, 2004.
37. Zajac, F.E.: Muscle coordination of movement: a perspective. *Journal of Biomechanics,* **26(1)**:109–124, 1993.

18

Imitation Learning and Transferring of Human Movement and Hand Grasping to Adapt to Environment Changes

Stephan Al-Zubi and Gerald Sommer

Cognitive Systems, Christian Albrechts University, Kiel, Germany

Summary. We propose a model for learning the articulated motion of human arm and hand grasping. The goal is to generate plausible trajectories of joints that mimic the human movement using deformation information. The trajectories are then mapped to a constraint space. These constraints can be the space of start and end configuration of the human body and task-specific constraints such as avoiding an obstacle, picking up and putting down objects. Such a model can be used to develop humanoid robots that move in a human-like way in reaction to diverse changes in their environment and as a priori model for motion tracking. The model proposed to accomplish this uses a combination of principal component analysis (PCA) and a special type of a topological map called the dynamic cell structure (DCS) network. Experiments on arm and hand movements show that this model is able to successfully generalize movement using a few training samples for free movement, obstacle avoidance and grasping objects. We also introduce a method to map the learned human movement to a robot with different geometry using reinforcement learning and show some results.

18.1 Introduction

Human motion is characterized as being smooth, efficient and adaptive to the state of the environment. In recent years a lot of work has been done in the fields of robotics and computer animation to capture, analyze and synthesize this movement with different purposes [1–3]. In robotics there has been a large body of research concerning humanoid robots. These robots are designed to have a one to one mapping to the joints of the human body but are still less flexible. The ultimate goal is to develop a humanoid robot that is able to react and move in its environment like a human being. So far the work that has been done is concerned with learning single gestures like drumming or pole balancing which involves restricted movements primitives in a simple environment or a preprogrammed movement sequence like a dance. An example where more adaptivity is needed would be a humanoid tennis robot which, given its current position and pose and the trajectory of the

B. Rosenhahn et al. (eds.), Human Motion – Understanding, Modelling, Capture, and
Animation, 435–452.

incoming ball, is able to move in a human-like way to intercept it. This idea enables us to categorize human movement learning from simple to complex as follows: (A) imitate a simple gesture, (B) learn a sequence of gestures to form a more complex movement, (C) generalize movement over the range allowed by the human body, and (D) learn different classes of movement specialized for specific tasks (e.g., grasping, pulling, etc.).

This chapter introduces two small applications for learning movement of type (C) and (D). The learning components of the proposed model are not by themselves new. Our contribution is presenting a supervised learning algorithm which learns to imitate human movement that is specifically more adaptive to constraints and tasks than other models. This also has the potential to be used for motion tracking where more diverse changes in movement occur. We will call the state of the environment and the body which affects the movement as constraint space. Constraint space describes any environmental conditions that affect the movement. This may be as simple as object positions which we must reach or avoid, a target body pose or more complex attributes such as the object's orientation and size when grasping it. The first case we present is generating realistic trajectories of a simple kinematic chain representing a human arm. These trajectories are adapted to a constraint space which consists of start and end pose of the arm as shown in Figure 18.4. The second case demonstrates how the learning algorithm can be adapted to the specific task of avoiding an obstacle where the position of the obstacle varies. The third case demonstrates how hand grasping can be adapted to different object sizes and orientations.

The model accomplishes this by aligning trajectories. A trajectory is the sequence of body poses which change in time from the start pose to the end pose of a movement. Aligning trajectories is done by scaling and rotation transforms in angular space which minimizes the distance between similar poses between trajectories. After alignment we can analyze their deformation modes which describe the principal variations of the shape of trajectories. The constraint space is mapped to these deformation modes using a topological map.

In addition to generating adaptive human movement, We also introduce in this chapter a reinforcement learning algorithm which enables us to transfer the learned human movement to a robot with a different embodiment by using reinforcement learning. This is done by learning similar trajectories of the human hand and the robot's end effector. A reward function measures this similarity as a mixture of constraint fitting in the robot's workspace and the similarity of the trajectory shape to the human's. The reinforcement learning algorithm explores trajectory space using a spectral representation of trajectories in order to reduce the state space dimensionality. A special type of DCS networks called QDCS is used for learning.

The Combination of adaptive movement learning and movement transfer enables a complete movement learning and transfer system architecture. This consists of two neural networks (NN) as shown in Figure 18.1. The first

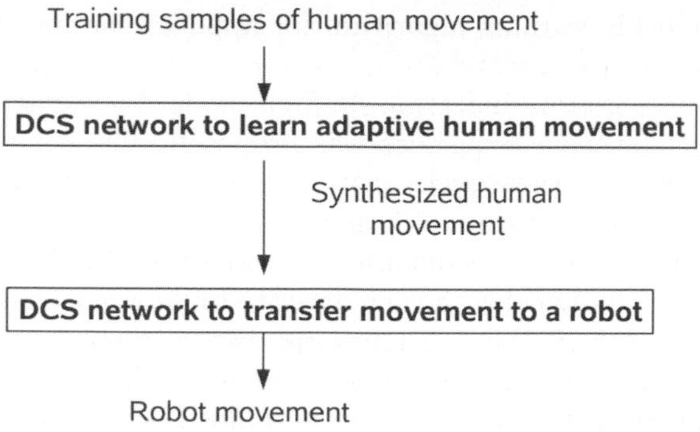

Training samples of human movement

↓

| DCS network to learn adaptive human movement |

Synthesized human
movement

↓

| DCS network to transfer movement to a robot |

↓

Robot movement

Fig. 18.1. Architecture for learning movement and transferring it to a robot.

network reconstructs human movement and the second transforms this movement to the robot space using reinforcement learning.

Next, we describe an overview of the work done related to movement learning and transferring and compare them with the proposed model. After that the adaptive movement algorithm will be presented followed by the transfer algorithm and then experimental results.

18.1.1 State of the Art

There are two representations for movements: pose based and trajectory based. We will describe next pose based methods.

Generative models of motion have been used in [1, 2] in which a nonlinear dimensionality reducing method called Scaled Gaussian Latent Variable Model (SGPLVM) is used on training samples in pose space to learn a nonlinear latent space which represents the probability distribution of each pose. Such a likelihood function was used as a prior for tracking in [1] and finding more natural poses for computer animation in [2] that satisfy constraints such as that the hand has to touch some points in space. Another example of using a generative model for tracking is [4] in which a Bayesian formulation is used to define a probability distribution of a pose in a given time frame as a function of the previous poses and current image measurements. This prior model acts as a constraint which enables a robust tracking algorithm for monocular images of a walking motion. Another approach using Bayesian priors and nonlinear dimension reduction is used in [5] for tracking.

After reviewing pose probabilistic methods, we describe in the following trajectory based methods. Schaal [3] has contributed to the field of learning movement for humanoid robots. He describes complex movements as a set of movement primitives (DMP). From these a nonlinear dynamic system of equations are defined that generate complex movement trajectories. He described a reinforcement learning algorithm that can efficiently optimize the parameters (weights) of DMPs to learn to imitate a human in a high dimensional space.

He demonstrated his learning algorithm for applications like drumming and a tennis swing.

To go beyond a gesture imitation, in [6] a model for segmenting and morphing complex movement sequences was proposed. The complex movement sequence is divided into subsequences at points where one of the joints reaches zero velocity. Dynamic programming is used to match different subsequences in which some of these key movement features are missing. Matched movement segments are then combined with each other to build a morphable motion trajectory by calculating spatial and temporal displacement between them. For example, morphable movements are able to naturally represent movement transitions between different people performing martial arts with different styles.

Another aspect of motion adaptation and morphing with respect to constraints comes from computer graphics on the topic of retargeting. As an example, Gleicher [7] proposed a nonlinear optimization method to retarget a movement sequence from one character to another with an identical structure but different segment lengths. The problem is to satisfy both the physical constraints and the smoothness of movement. Physical constraints are contact with other objects like holding the box.

The closest work to the model presented in this chapter is done by Banarer [8]. He described a method for learning movement adaptive to start and end positions. His idea is to use a topological map called Dynamic Cell Structure (DCS) network [9]. The DCS network learns the space of valid arm configurations. The shortest path of valid configurations between the start and end positions represents the learned movement. He demonstrated his algorithm to learn a single gesture and also obstacle avoidance for a single fixed obstacle.

18.1.2 Contribution

The main difference between pose based methods and our approach is that instead of learning the probability distribution in pose space, we model the variation in trajectory space (each trajectory being a sequence of poses). This representation enables us to generate trajectories that vary as a function of environmental constraints and to find a more compact representation of variations than allowed by pdfs in pose space alone. Pose pdfs would model large variations in trajectories as a widely spread distribution which makes it difficult to trace the sequence of legal poses that satisfy the constraints the human actually makes without some external reference like motion sequence data.

Our approach models movement variation as a function of the constraint space. However, style based inverse kinematics as in [2] selects the most likely poses that satisfy these constraints. This works well as long as the pose constraints do not deviate much from the training data. This may be suitable for animation applications but our goal here is to represent realistic trajectories adapted to constraints without any explicit modelling. Banarer [8] uses also a pose based method and the model he proposed does not generalize well because as new paths are learned between new start and end positions, the DCS network grows very quickly and cannot cope with the curse of dimensionality. Our DCS network generalizes over trajectory space not poses enabling more adaptivity.

Gleicher [7] defines an explicit adaptation model which is suitable to generate a visually appealing movement but requires fine tuning by the animator because it may appear unrealistic. This is because it explicitly morphs movement using a prior model rather than learning how it varies in reality as done in [2].

In the case of Schaal [3], we see that DMPs although flexible are not designed to handle large variations in trajectory space. This is because reinforcement learning adapts to a specific target human trajectory.

Morphable movements [6] define explicitly the transition function between two or more movements without considering the constraint space. Our method can learn the nonlinear mapping between constraint space and movements by training from many samples. The variation of a movement class is learned and not explicitly predefined.

To sum up, we have a trajectory-based learning model which learns the mapping between constraints and movements. The movement can be more adaptive and generalizable over constraint space. It learns movements from samples and avoids explicit modelling which may generate unrealistic trajectories.

18.2 Learning Model

After describing the problem, this section will develop the concept for learning movement and then it describes how this model is implemented.

In order to develop a system which is able to generalize movement, a number of reductions have to be made to the high-dimensional space of start – end configurations or any other environmental constraints. This reduction is done in two steps. The first step is to learn the mechanics of movement itself and the second is to learn how movement changes with start – end configuration. The mechanics of movement are called *intrinsic features*. The changes of intrinsic feature with respect to relative position are called *extrinsic features*. The intrinsic features describe movement primitives that are characteristic for the human being. These features are the following:

1. The acceleration and velocity of joints as they move through space. For example a movement generally begins by a joint accelerating at the start then decelerating as it nears its end position.
2. The nonlinear path taken by the joints to reach their destination. This is what characterizes smooth human movement. Otherwise the movement will look rigid similar to inverse kinematics used by robots. This nonlinearity is not only seen in simple movements like moving from point a

to b but also it can be seen in more complex movements for which it is necessary like obstacle avoidance and walking.

3. The coordination of joints with respect to each other in time. This means that joints co-deform in space and time working together to achieve their goal.

After modelling intrinsic features, extrinsic features can be characterized as the variation of intrinsic feature in the space of all possible start and end positions of the joints and any environmental constraints such as obstacle positions. Extrinsic features describe:

1. The range of freedom of joints.
2. The movement changes with respect to rotation and scale. As an example, we can consider how the movement changes when we draw the letter A in different directions or with different sizes.

The difference between intrinsic and extrinsic features that characterizes movement enables the formulation of a learning model. This model consists of two parts: The first part is responsible for learning intrinsic features which uses principal component analysis (PCA). It is applied on the aligned trajectories of the joints to reduce the dimensionality. The second part models the extrinsic features using a special type of an adaptive topological map called the dynamic cell structure (DCS) network. The DCS learns the nonlinear mapping from the extrinsic features to intrinsic features that are used to construct the correct movement that satisfies these extrinsic features.

In the following subsections we will take a detailed look at these mechanisms.

18.2.1 Intrinsic Features Using PCA

The algorithm which will be used to extract intrinsic features consists of the following steps:

1. Interpolation
2. Sampling
3. Conversion to orientation angles
4. Alignment
5. Principal component analysis

The main step is alignment in which trajectories traced by joints in space are aligned to each other to eliminate differences due to rotation and scale exposing the mechanics of movement which can then be analyzed by PCA. In the following paragraphs each step of the algorithm will be explained in detail as well as the reasons behind it. As an example, we will assume throughout this chapter that there is a kinematic chain of 2 joints: shoulder and elbow. Each joint has 2 degrees of freedom (ϕ, θ) which represent the direction of the corresponding limb in spherical coordinates.

To perform statistical analysis, we record several samples of motion sequences. In each motion sequence the 3D positions of the joints are recorded with their time. Let us define the position measurements of joints of a movement sequence (k) as $\{(x_{i,j,k}, y_{i,j,k}, z_{i,j,k}, t_{i,j,k})\}$ where (x, y, z) is the 3D position of the joint, t is the time in milliseconds from the start of the motion. The index i is the position (frame), j specifies the marker and k specifies the the movement sequence.

The first step is to interpolate between the points of each movement sequence. Usually a 2nd degree B-spline is sufficient to obtain a good interpolation. We end up with a set of parametric curves $\{\mathbf{p}_k(t)\}$ for each motion sequence k where $\mathbf{p}_k(t)$ returns the position vector of all the joints at time t.

The second step is to sample each $\mathbf{p}_k(t)$ at equal time intervals from the start of the sequence $t = 0$ to its end $t = t_{end_k}$. Let n be the number of samples then we form a vector of positions $\mathbf{v}_k = [\mathbf{p}_{1,k}, \mathbf{p}_{2,k} \cdots \mathbf{p}_{n,k}]$ where $\mathbf{p}_{i,k} = \mathbf{p}_k((\frac{i-1}{n-1})t_{end_k})$. This regular sampling at equal time intervals enables us to represent trajectories with a fixed length vector which facilitates statistical analysis on a population of such vectors. This vector form also represents implicitly the acceleration and velocity of these paths through variability of distances between points. This is the reason why time was used as the variable in the parametric curves. In cases where motion is more complex and consists of many curves, this method automatically samples more points of high curvature than points of lower curvature. This places comparable corner points of complex paths near each other and thus does not normally necessitate more complex registration techniques to align the curves as long as the trajectories have up to 4–5 corners.

The third step is to convert the Euclidean coordinates $\mathbf{v}_k$ to direction angles in spherical coordinates $\mathbf{S}_k = [\mathbf{s}_{1,k}, \mathbf{s}_{2,k}, \ldots \mathbf{s}_{n,k}]$ where $\mathbf{s}_{i,k}$ is the vector of direction angles for all the joints. This means that a given joint j in a kinematic chain with some position $\mathbf{p}_j = (x_j, y_j, z_j)$ is attached to its parent with a position $\mathbf{p}_{j-1} = (x_{j-1}, y_{j-1}, z_{j-1})$ and since the distance D between them is constant we need only to represent the direction of joint j with respect to joint $j - 1$. This can be done by taking the relative coordinates $\Delta\mathbf{p} = \mathbf{p}_j - \mathbf{p}_{j-1}$ and then convert $\Delta\mathbf{p}$ to spherical coordinates $(\rho_j, \phi_j, \theta_j)$ where $\rho_j = D$ is constant. After that, we take only the direction angles (ϕ_j, θ_j) as the position of the joint. For a kinematic chain of m joints we will need $m - 1$ such direction angles because the first joint is the basis of the chain therefore has no direction.

The forth step is to align the paths taken by all the joints with respect to each other. This alignment makes paths comparable with each other in the sense that all extrinsic features are eliminated leaving only the deformations of the path set from the mean. In order to accomplish this, we first convert every direction point (ϕ_j, θ_j) of a joint j to a 3D unit vector $\hat{\mathbf{u}}_j = a\mathbf{i} + b\mathbf{j} + c\mathbf{k}$ that corresponds to the direction of the joint at some point in time. With this representation we can imagine that a moving joint traces a path on a unit sphere $\hat{\mathbf{u}}_j(t)$ as the paths shown in Figure 18.2. Given a kinematic chain

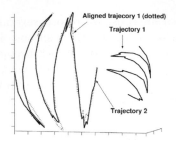

Fig. 18.2. Alignment of two trajectories by scale and rotation. The trajectories are a sequence of direction vectors tracing curves on a unit sphere.

of m joints moving together, we can represent these moving joints as $m-1$ moving unit vectors $\mathbf{U}(t) = [\hat{\mathbf{u}}_1(t), \ldots \hat{\mathbf{u}}_j(t), \ldots \hat{\mathbf{u}}_{m-1}(t)]$. $\mathbf{S}_k$ samples the path at equal time intervals $t_0, \ldots, t_{n_k}$ that correspond to $\mathbf{U}_i = \mathbf{U}(t_i), i = 0 \ldots t_{n_k}$. This enables us to represent the path as a matrix of direction vectors $\mathbf{W} = [\mathbf{U}_0, \ldots \mathbf{U}_{n_k}]'$. The reason why we use direction vector representation $\mathbf{W}$ is because it facilitates alignment of paths with each other by minimizing their distances over rotation and scale transforms. To accomplish this, we define a distance measure between two paths instances $\mathbf{W}_1, \mathbf{W}_2$ as the mean distance between corresponding direction vectors of both paths (i.e., corresponding joints j at the same time ti)

$$pathdist(\mathbf{W}_1, \mathbf{W}_2) = \frac{1}{|\mathbf{W}_1|} \sum_{\forall(\hat{\mathbf{u}}_{i,j}\epsilon\mathbf{W}_1, \hat{\mathbf{v}}_{i,j}\epsilon\mathbf{W}_2)} dist(\hat{\mathbf{u}}_{i,j}, \hat{\mathbf{v}}_{i,j}) \qquad (18.1)$$

where the distance between two direction vectors is simply the angle between

$$dist(\hat{\mathbf{u}}, \hat{\mathbf{v}}) = cos^{-1}(\hat{\mathbf{u}} \cdot \hat{\mathbf{v}}) \qquad (18.2)$$

Two transforms are used to minimize the distances between the paths:

1. Rotation (R): When we multiply each direction vector in the path with a rotation matrix R, we can rotate the whole path as shown in Figure 18.2.
2. Scaling: We can scale the angles between path points. This assumption is made by the observation that movements are scalable. For example, when we draw a letter smaller and then bigger, we basically move in the same way but with larger angles as shown in Figure 18.2. This is of course a simplifying assumption that is not exactly true but it helps us later to fit start-end positions of variable arc-length distances. To scale the path we simply choose a reference direction vector on the path and then shift each direction vector in the direction of the arc length between them by multiplying the arc length θ with a scale factor s as depicted in Figure 18.2.

Minimizing distances can be done by simple gradient descent over the scale parameter s and rotation angles around the three axes $\theta_x, \theta_y, \theta_z$ defining a rotation matrix $R = R_x(\theta_x)R_y(\theta_y)R_z(\theta_z)$. When extending this algorithm to

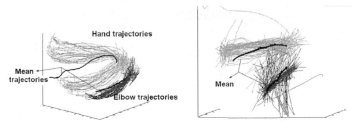

Fig. 18.3. Example of aligning a training set of trajectories represented as direction vectors tracing curves on a unit sphere. Left before alignment and right after. We see how the hand trajectories cluster together and the mean becomes smoother.

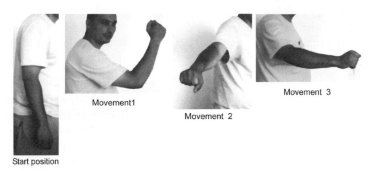

Fig. 18.4. Three movements of the arm that all begin with the same start position (left image), the rest are end positions.

align more than two paths, we can do that by computing the mean path $\overline{\mathbf{W}}$ and then fitting all the sample paths $\{\mathbf{W}_1, \ldots, \mathbf{W}_p\}$ to the the mean. We repeat this cycle until the mean path $\overline{\mathbf{W}}$ converges. The mean path is initialized by an arbitrary sample $\overline{\mathbf{W}} = \mathbf{W}_j$. It is computed from aligned samples in each iteration step by summing the corresponding direction vectors from all the sample paths and then normalizing the sum $\overline{\mathbf{W}} = \dfrac{\sum_i \mathbf{W}_i}{|\sum_i \mathbf{W}_i|}$. An example of aligning is in Figure 18.3.

The fifth step is to convert the aligned trajectories back to the angular representation and form a data matrix of all p aligned motion sequences $X = [\mathbf{S}_1^T \ldots \mathbf{S}_k^T \ldots \mathbf{S}_p^T]^T$. Principal component analysis is applied on X yielding latent vectors $\mathbf{\Psi} = [\psi_1, \psi_2, \ldots, \psi_n]$. Only the first q components are used where q is chosen such that the components cover a large percentage of the data $\mathbf{\Psi}_q = [\psi_1, \psi_2, \ldots, \psi_q]$. Any point in eigenspace can be then converted to the nearest plausible data sample using the following equation

$$\mathbf{S} = \overline{\mathbf{S}} + \mathbf{\Psi}_q \mathbf{b} \tag{18.3}$$

where $\overline{\mathbf{S}} = \frac{1}{p}\sum_{k=1}^{p} \mathbf{S}_k$ and $\mathbf{b}$ is the column vector of an eigenpoint.

The inverse transform from eigenspace to trajectories is approximated by

$$\mathbf{b} = \mathbf{\Psi}'_q(\mathbf{S} - \overline{\mathbf{S}}) \tag{18.4}$$

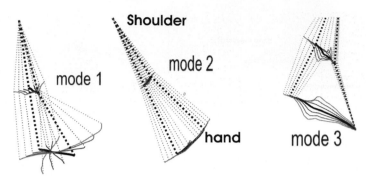

Fig. 18.5. The first three variation modes of a kinematic chain representing the shoulder, elbow and hand constructed in 3D space. The middle thick line is the mean trajectory and the others represent ±1, ±2, ±3 standard deviations along each eigenvector.

The latent coordinates **b** represent the linear combination of deformations from the average paths taken by the joints. An example of that can be seen in Figure 18.5. In this example, the thick lines represent the mean path and the others represent ±3 standard deviations in the direction of each eigenvector which are called modes. The first mode represents the twisting of the hand's path around the elbow and shoulder. The second mode shows the coordination of angles when moving the hand and elbow together. The third mode represent the bulginess of the path taken by the hand and shoulder around the middle. We see that these deformation modes have meaningful mechanical interpretations.

18.2.2 Extrinsic Features Using DCS

PCA performs a linear transform (i.e., rotation and projection in (18.3)) which maps the trajectory space into the eigenspace. The mapping between constraint space and eigenspace is generally nonlinear. To learn this mapping we use a special type of self organizing maps called Dynamic Cell Structure which is a hybrid between radial basis networks and topologically preserving maps [9]. DCS networks have many advantages: They have a simple structure which makes it easy to interpret results, they adapt efficiently to training data and they can cope with changing distributions. They consist of neurons that are connected to each other locally by a graph distributed over the input space. These neurons also have radial basis functions which are Gaussian functions used to interpolate between these neighbors. The DCS network adapts to the nonlinear distribution by growing dynamically to fit the samples until some error measure is minimized. When a DCS network is trained, the output $\mathbf{b}_{DCS}(\mathbf{x})$ which is a point in eigenspace can be computed by summing the activations of the best matching neuron (i.e., closest) to the input vector $\mathbf{x}$ representing a point in constraint space and the local neighbors to which it

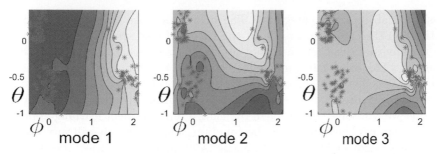

Fig. 18.6. Distribution of eigenvalues (bright regions represent maxima) in the angular space of the end position of the hand.

is connected by an edge which is defined by the function $A_p(\mathbf{x})$. The output is defined as

$$\mathbf{b}_{DCS}(\mathbf{x}) = f_P^{nrbf}(\mathbf{x}) = \frac{\sum_{i \in A_p(\mathbf{x})} \mathbf{b}_i h(\| \mathbf{x} - \mathbf{c}_i \| / \sigma_i)}{\sum_{j \in A_p(\mathbf{x})} h(\| \mathbf{x} - \mathbf{c}_j \| / \sigma_j)}, \qquad (18.5)$$

where $\mathbf{c}_i$ is the receptive center of the neuron i, $\mathbf{b}_i$ represents a point in eigenspace which is the output of neuron i, h is the Gaussian kernel and σ_i is the width of the kernel at neuron i.

The combination of DCS to learn nonlinear mapping and PCA to reduce dimension enables us to reconstruct trajectories from $\mathbf{b}(\mathbf{x})$ using (18.3) which are then fitted to the constraint space by using scale and rotation transformations. For example, a constructed trajectory is fitted to a start and end position.

When using the network to generate new motion paths, the start-end positions Θ are given to the network. It returns the deformation modes $\mathbf{b}$ of the given start-end position. We must use $\mathbf{b}$ to reconstruct the path between the start and positions given by Θ. This is accomplished by converting $\mathbf{b}$ to angular representation $\mathbf{S}$ given by (18.3). $\mathbf{S}$ is converted to direction vector representation $\mathbf{W}$. We take the start and end positions of $\mathbf{W}$ and find the best rotation and scale transform that fits it to Θ using the same method shown in the previous section. The resulting path represents the reconstruction that contains both intrinsic and extrinsic features from the learning model.

18.3 Learning Model for Transferring Movement to a Robot

After building a model which can generate human movement as a function of constraint space, we will introduce in this section a learning algorithm to transfer this movement to the robot. The main problem here is how to transfer human movement to a manipulator with a different geometry and degrees of freedom. There are many solutions proposed in literature which basically fall into two categories: The first class looks only at the effects of actions on the environment [10]. If the effects of goal sequences are the same on the environment then the robot movement is considered equivalent to the humans'. The other category defines some ad hoc function which measures the similarity between the robot and human movement [6]. This function can also be a mixture of degree of goal satisfaction and pose similarity. In this chapter we will solve this problem by using a reinforcement learning approach rather than explicitly defining the function. This approach will mix the two mapping categories. Specifically, we will use similarities of trajectories between the end effector and the hand. This simplifies the problem and enables us to define a meaningful intuitive mapping for any manipulation task involving manipulation of objects. The reward function r will be a weighted mixture of similarities of trajectories and constraint satisfaction as follows

$$r(\mathbf{u}, \mathbf{v}, C) = \alpha_1 f_1(\mathbf{u}, \mathbf{v}) + \alpha_2 f_2(\mathbf{v}, C) \tag{18.6}$$

where $\mathbf{u}$ is the trajectory of the human hand, $\mathbf{v}$ is the trajectory of the robot end effector. C is the constraints to be satisfied in the robots workspace. α_1, α_2 are wieghts, f_1 measures the similarity between the shapes of trajectories and f_2 mesures constraint satisfaction in the robot space.

This approach is similar to programming by demonstration (POD) in [11, 12] where a human teaches the robot to move in a similar way to avoid obstacles for example. The reinforcement learning approach suffers from the drawback that when the action space has a high dimensionality, the solution will not converge in a reasonable time. For this reason we will use a frequency representation of trajectories where only the first few components are used. Specifically, we choose a discrete cosine transform representation of trajectories which uses only real numbers. A trajectory is sampled at N points in 3D space equidistant in time $\mathbf{p}_n, n = 1 \ldots N$. This transform is

$$\mathbf{a}_k = w_k \sum_{n=1}^{N} \mathbf{p}_n cos(\frac{\pi(2n-1)(k-1)}{2N}), k = 1, \ldots, N, \tag{18.7}$$

$$w_k = \{ \begin{matrix} \frac{1}{\sqrt{N}}, k=1 \\ \sqrt{\frac{2}{N}}, 2 \leq k \leq N \end{matrix} \tag{18.8}$$

The reinforcement learning represents trajectories through the parameters $\mathbf{a}_k, k = 1...m$ that where m is some small dimension. The robot trajectory is reconstructed using inverse discrete cosine transform applied on a zeros padded vector of length N: $(\mathbf{a}_1, \mathbf{a}_2, \ldots, \mathbf{a}_m, 0, \ldots, 0)$. This reduces the action space dimension enabling a fast converging solution. This is the advantage of using a spectral representation of trajectories for reinforcement learning. The state is represented as the vector $(\mathbf{a}_1, \mathbf{a}_2, \ldots, \mathbf{a}_m, \mathbf{C}) = (\mathbf{\Theta}, \mathbf{C})$ where $\mathbf{C}$ is the constraint that is satisfied by the trajectory that is reconstructed from $\mathbf{\Theta}$. The action applied on some state is some small increment $\Delta\mathbf{\Theta}$ added to $\mathbf{\Theta}$ to get a new modified trajectory.

Initialize the QDCS network arbitrarily.
For each constraint $\mathbf{C}$ do
  Initialize the start state Θ
  Repeat
    choose an action $\Delta\Theta$, observe $\Theta' = \Theta + \Delta\Theta$ and reward r
    update online $QDCS(\mathbf{C} \cdot \Theta \cdot \Delta\Theta)$ by adding:
      $\alpha[r + \gamma max_{\Delta\Theta'} QDCS(\mathbf{C} \cdot \Theta' \cdot \Delta\Theta') - QDCS(\mathbf{C} \cdot \Theta \cdot \Delta\Theta)]$
    $\Theta \leftarrow \Theta'$
  Until Θ converges to a good solution

Fig. 18.7. Learning optimal trajectories using QDCS.

The reinforcement learning begins exploring the states of trajectories that optimize the reward function by incrementally modifying the trajectory parameters for some initial constraint $\mathbf{C}_0$. Once the optimal trajectory is found, new optimal trajectories for a new constraint $\mathbf{C}_1$ close to $\mathbf{C}_0$ is learned. This will exploit the optimal trajectory learned for $\mathbf{C}_0$ as a prior to learn quickly the optimal trajectory of $\mathbf{C}_1$. The process continues for all constraints in constraint space that are needed to be learned. When fully trained, we can use the QDCS network to find the best robot trajectory that satisfies a given constraint.

A continuous version Q-learning is used to learn the optimal trajectory. In it we replace the discrete Q table in normal Q learning with a DCS network that learns the Q values online as demonstrated in [9]. The QDCS network uses online learning to update its values. The algorithm is depicted in Figure 18.7.

18.4 Experiments

In order to record arm movements, a marker-based stereo tracker was developed in which two cameras track the 3D position of three markers placed at the shoulder, elbow and hand at a rate of 8 frames per second. This was used to record trajectory samples. Two experiments were conducted to show two learning cases: moving between two positions and avoiding an obstacle.

The first experiment demonstrates that our learning model reconstructs the nonlinear trajectories in the space of start-end positions. A set of 100 measurements were made for an arm movement consisting of three joints. The movements had the same start position but different end positions as shown in Figure 18.4.

The first three eigenvalues have a smooth nonlinear almost unimodal distribution with respect to the start-end space as shown in Figure 18.6. The first component explained 72% of the training samples, the second 11% and the third 3%.

The performance of the DCS network was first tested by a k-fold cross validation on randomized 100 samples. This was repeated for $k = 10$ runs. In each run the DCS network was trained and the number of neurons varied between 6 and 11. The average distance between the DCS-trajectory and the data sample was 3.9° and the standard deviation was 2.1°. This shows that the DCS network was able to generalize well using a sample size of about 100.

We can compare with Banarer [8] who fixed the DCS network with an upper bound of 15 neurons to learn a single gesture and not many as in our experiment. He used simulated data of 70 samples with a random noise of up to 5° and the mean error was 4.3° compared to our result of 3.9° on real data. The measurement error of the tracker is estimated to be 4.6° standard deviation which accounts for the similar mean errors. This shows that our model scales well.

Next, we demonstrate the algorithm for obstacle avoidance. In this case 100 measurements were taken for the arm movement with different obstacle positions as shown in Figure 18.8. The black lines show the 3D trajectory of the arm avoiding the obstacle which has a variable position determined by the distance B. We see how the hand backs away from the obstacle and the elbow goes down and then upward to guide the hand to its target. A is the Euclidian distance between the start and end positions of the hand. The grey lines represent a free path without obstacles. In this case we need to only take the first eigenvector from PCA to capture the variation of trajectories due to obstacle position. This deformation mode is shown in Figure 18.9 We define the relative position of the obstacle to the movement as simply $p = \frac{B}{A}$.

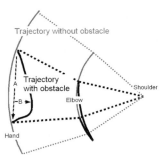

Fig. 18.8. Trajectory for obstacle avoidance in 3D space.

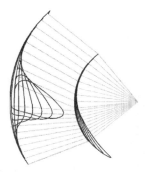

Fig. 18.9. Variation of arm trajectory with respect to the obstacle.

The DCS network learns the mapping between p and the eigenvalue with only 5 neurons. The learned movement can thus be used to avoid any obstacle between the start and end positions regardless of orientation or movement scale. This demonstrates how relatively easy it is to learn new specialized movements that are adaptive to constraints.

Finally, this model was demonstrated on hand grasping. In this case 9 markers were placed on the hand to track the index and thumb fingers using a monocular camera as in Figure 18.10. The 2D positions of the markers were recorded at a rate of 8.5 frames per second from a camera looking over a table. The objects to be grabbed are placed over the table and they vary by both size and orientation. The size ranged from 4 to 12 cm and orientation ranged from 0 to 60 degrees as depicted in Figures 18.11 and 18.12. The tracker recorded 350 grasping samples of which 280 was used for training the DCS and 70 for testing. The DCS learned the variation of movement with 95 neurons and PCA reduced the dimension from 600 to just 23. The first two modes characterize variation of scale and orientation as shown in Figure 18.10. Figures 18.11 and 18.12 depict an example comparison between grasping movement generated by the DCS and an actual sample. Below we

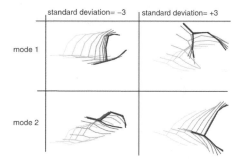

Fig. 18.10. The first two variation modes of grasping.

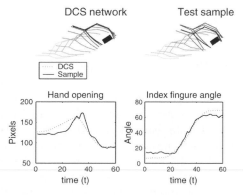

Fig. 18.11. Comparison between DCS and and a grasping movement for a 4 cm object at $60°$ with respect to the hand.

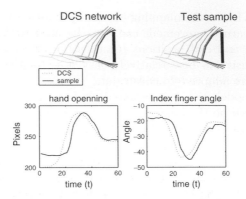

Fig. 18.12. Comparison between DCS and and a grasping movement for a 12 cm object at 0°.

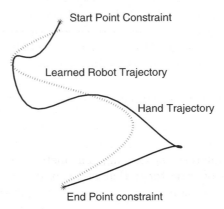

Fig. 18.13. Learned trajectory using start and end position constraints in 2D space.

used two measures that characterize well grasping: distance between the tips of the index finger and the thumb and the direction of the index finger's tip with respect the the direction of the arm. We see that the DCS and sample profiles look very similar. In general, the model's root mean square error for the first measure was 18 pixels for a 800×600 images and 8.5° for the second measure. Training the DCS takes only a few seconds for all the movement classes presented.

After presenting the experiments for movement generation, we demonstrate some results for movement transfer. Figure 18.13 shows a 2D human trajectory that a robot has to learn. The end markers (shown as $(*)$) in the figure are the constraints which represents the start and end position that the robot must maintain. The discrete cosine transform uses 8 coefficients: 4 for the x dimension and 4 for the y dimension. The algorithm was able to learn the nearest trajectory after about 2,000 itarations. The QDCS was set to an upper bound of 200 neurons. The algorithm selects a random action about

10% of the time and the action of maximum Q value the rest of the time. This shows that the algorithm is able to learn in a reasonable time a trajectory because the human motion and the constraints act as strong priors to guide the low dimensional frequency representation of trajectories. Adapting to more constraints is left as future work.

18.5 Conclusion

We proposed a learning model for generation of realistic articulated motion. The model characterizes deformation modes that vary according to constraint space. A combination of DCS network to learn the nonlinear mapping and PCA to reduce dimensionality enables us to find a representation that can adapt to constraint space with a few samples. This trajectory-based method is more suited for movement generation than pose based methods which are concerned with defining priors for good fitting with image data such as tracking. The proposed method models variation of movement with respect to constraints in a more clear way than the previously proposed methods. The potential uses of our method is in developing humanoid robots that are reactive to their environment and also motion tracking algorithms that use prior knowledge of motion to make them robust. Three small applications towards that goal were experimentally validated. We also proposed a trajectory based method that transfers the human movement to a manipulator of a different embodiment using reinforcement leraning. The method uses QDCS to exploit and expore the space of trajectories that fit to constraints specified in robot space. This represents a natural extension of the first algorithm that enables adaptive movement that is retargetable to any robot manipulator system.

Acknowledgments

The work presented here was supported by the the European Union, grant COSPAL (IST-2003-004176). However, this chapter does not necessarily represent the opinion of the European Community, and the European Community is not responsible for any use which may be made off its contents.

References

1. Urtasun, R., Fleet, D.J., Hertzmann, A., Fua, P.: Priors for people tracking from small training sets. In: International Conference on Computer Vision (ICCV). (2005) 403–410
2. Grochow, K., Martin, S.L., Hertzmann, A., Popovic, Z.: Style-based inverse kinematics. ACM Trans. Graph. **23**(3) (2004) 522–531

3. Schaal, S., Peters, J., Nakanishi, J., Ijspeert, A.: Learning movement primitives. In: International Symposium on Robotics Research (ISPR2003), Springer Tracts in Advanced Robotics, Ciena, Italy (2004)
4. Sidenbladh, H., Black, M.J., Fleet, D.J.: Stochastic tracking of 3d human figures using 2d image motion. In: Proceedings of the 6th European Conference on Computer Vision (ECCV '00), London, Springer (2000) 702–718
5. Sminchisescu, C., Jepson, A.: Generative modelling for continuous non-linearly embedded visual inference. In: Proceedings of the twenty-first International Conference on Machine Learning (ICML '04), New York, USA, ACM Press (2004)
6. Ilg, W., Bakir, G.H., Mezger, J., Giese, M.A.: On the repersenation, learning and transfer of spatio-temporal movement characteristics. International Journal of Humanoid Robotics (2004)
7. Gleicher, M.: Retargeting motion to new characters. In: Proceedings of the 25th Annual Conference on Computer Graphics and Interactive Techniques (SIG-GRAPH '98), New York, ACM Press (1998) 33–42
8. Banarer, V.: STRUKTURELLER BIAS IN NEURONALEN NETZEN MITTELS CLIFFORD-ALGEBREN. Technical Report 0501, Technische Fakultät der Christian-Albrechts-Universität zu Kiel, Kiel (2005)
9. Bruske, J., Sommer, G.: Dynamic cell structure learns perfectly topology preserving map. Neural Computation **7**(4) (1995) 845–865
10. Nehaniv, C., Dautenhahn, K.: Of hummingbirds and helicopters: An algebraic framework for interdisciplinary studies of imitation and its applications. In: World Scientific Press (1999)
11. Jacopo Aleotti, S.C.: Trajectory clustering and stochastic approximation for robot programming by demonstration. In: Proceedings of the IEEE/RSJ International Conference on Intelligent Robots and Systems (IROS), Edmonton, Alberta, Canada (2005)
12. J. Aleotti, S. Caselli, M.R.: Toward programming of assembly tasks by demonstration in virtual environments. In: 12th IEEE Int. Workshop on Robot and Human Interactive Communication, Millbrae, California (2003)

19

Accurate and Model-free Pose Estimation of Crash Test Dummies

Stefan K. Gehrig, Hernán Badino, and Jürgen Gall

[1] DaimlerChrysler AG, HPC 050 - G 024
   71059 Sindelfingen, Germany
[2] Frankfurt University, Robert-Mayer-St. 10
   60045 Frankfurt, Germany
[3] Max-Planck-Center Saarbrücken, Stuhlsatzenhausweg 85
   66123 Saarbrücken, Germany

Summary. In this chapter, we present a model-free pose estimation algorithm to estimate the relative pose of a rigid object. In the context of human motion, a rigid object can be either a limb, the head, or the back. In most pose estimation algorithms, the object of interest covers a large image area. We focus on pose estimation of objects covering a field of view of less than 5° by 5° using stereo vision.

With this new algorithm suitable for small objects, we investigate the effect of the object size on the pose accuracy. In addition, we introduce an object tracking technique that is insensitive to partial occlusion. We are particularly interested in human motion in this context focusing on crash test dummies.

The main application for this method is the analysis of crash video sequences. For a human motion capture system, a connection of the various limbs can be done in an additional step. The ultimate goal is to fully obtain the motion of crash test dummies in a vehicle crash. This would give information on which body part is exposed to what kind of forces and rotational forces could be determined as well. Knowing all this, car manufacturers can optimize the passive safety components to reduce forces on the dummy and ultimately on the real vehicle passengers. Since camera images for crash videos contain the whole crash vehicle, the size of the crash test dummies is relatively small in our experiments.

For these experiments, mostly high-speed cameras with high resolution are used. However, the method described here easily extends to real-time robotics applications with smaller VGA-size images, where relative pose estimation is needed, e.g., for manipulator control.

19.1 Introduction

Pose estimation is a problem that has undergone much research in Computer Vision. Most pose estimation algorithms rely on models of the observed object (see, e.g. [1]). For a simple motion capture system it would be nice if a model would not be needed to simplify the handling. We perform model-free pose

453

B. Rosenhahn et al. (eds.), Human Motion – Understanding, Modelling, Capture, and Animation, 453–473.

estimation, i.e., pose estimation with respect to a reference pose, which is the initial pose in our context. So we obtain a relative pose estimate.

We are particularly interested in human motion here with a focus on crash test dummy motion. The motions occurring in vehicle crashes are very different from the typical biomechanical motion models since the occurring accelerations can go beyond $1{,}000\,\mathrm{m/s}^2$. Hence, we do not use any biomechanical motion model. An excellent review on human motion can be found in [2].

The main application for us is the analysis of stereo crash video sequences. Crash test video analysis is a mature technique that uses photogrammetric markers. For optimizing passive safety components in a vehicle, obtaining the full motion of crash test dummies in crashes is of great interest. The sparse measurements performed with photogrammetric markers are not sufficient to obtain all rotational motion parameters. With a stereoscopic sensor setup, it is possible to obtain dense 3D information of the crash scene independent of photogrammetric markers, which are the only source of 3D information in standard crash analysis. In addition, using pose estimation, rotations of rigid parts in the scene can be measured. So for the first time, the maximal torsion of dummy body parts can be obtained. The ultimate goal is to measure all forces exerted on the crash test dummy during a crash. Besides supporting car manufacturers in design of passive safety components, dummy manufacturers can also use this data to put more biological plausibility into their dummies.

A problem closely related to pose estimation is ego-motion estimation. The algorithm proposed here was initially developed for ego-motion estimation by computing the optimal rotation and translation between the tracked static points of multiple frames [3].

The remainder of this chapter is organized as follows: In Section 19.2 we present a short overview of pose estimation using stereo vision. The proposed algorithm is detailed in Section 19.3. Adaptions and input to the algorithm are elaborated in Section 19.4. Simulation results as well as experimental results on real crash image sequences are presented in Section 19.5. In the last section we summarize the chapter.

19.2 Related Work

Pose estimation is a central problem in photogrammetry and Computer Vision. The term is used in several related meanings covering calibration, ego-motion, and object pose estimation. We focus on pose estimation of rigid objects viewed from a static, calibrated stereo camera setup.

There exists a vast literature on pose estimation as can be seen from this book. For a survey that also lists a large number of 3D approaches consult [4]. In this brief literature overview, we limit ourselves to work on pose estimation via stereo vision.

- Grest et al. estimate human poses from stereo images [5]. Here, based on the stereo depth data, human body models are matched. The stereo data itself is only used for segmentation in order to separate the background from the human body which is in the foreground.
- For face tracking, the head pose is estimated using stereo information combined with head models [6]. Again, stereo only aids segmentation.
- Plankers and Fua use stereo and silhouettes to obtain a pose estimate [7]. The stereo data is fitted to the model using a least squares technique. Body model and motion parameters can be estimated from such a stereo sequence, which is computationally expensive.

In above approaches, the object of interest covers a large percentage of the image. Investigations on pose estimation of objects covering only a small field of view are rare in the literature.

Besides utilizing stereo vision to obtain 3D information, many other approaches exist to recover the third dimension for pose estimation. Some other features typically used for pose estimation are:

- **Silhouettes:** Most pose estimation algorithm perform a 3D evaluation of the image, often using several views of the same scene. In [1], the 3D data is obtained by implicitly triangulating multiple views. So, 3D data is obtained in the same step as the pose estimate. Often, silhouettes and edges are used as the main cue to determine pose [8]. Using silhouettes seen from multiple view points, a voxel representation can be obtained [9].
- **Grayscale values:** In [10], the multiple view pose estimation is directly performed on the image grayscale data formulating a large optimization problem on GraphCut basis. The approach is called PoseCut and it produces nice results with large computational effort.
- **Optical flow:** A model-based pose estimation algorithm exploiting optical flow information is presented in [11]. Besides optical flow, also contours are used as a complementary piece of information.

In our approach, we first obtain 3D with standard stereo correspondences and perform pose estimation afterwards. For pose estimation we use solely our 3D points and their tracking information, independent of the image gray values.

Ego-motion and pose estimation are related problems and can be solved using the same mathematical framework (see, e.g. [12]). For an overview of robotic camera pose estimation techniques refer to [13]. We adopt an ego-motion estimation technique for use in pose estimation.

In the next section we present a review of the ego-motion algorithm, first described in [14] and applied to moving object detection in [3]. This chapter is an extended version of a previously published paper [15].

19.3 Robust Ego-motion Estimation

Computing ego-motion from an image sequence means obtaining the change of position and orientation of the observer with respect to a static scene, i.e., the motion is relative to an environment which is considered static. In most approaches this fact is exploited and ego-motion is computed as the inverse of the scene motion [14, 16–18]. In the latter a robust approach for the accurate estimation of the six d.o.f. of motion (three components for translation and three for rotation) in traffic situations is presented. In this approach, stereo is computed at different times and clouds of 3D points are obtained. The optical flow establishes the point-to-point correspondence in time starting with an integer grid position. When reusing the ending position of such a flow for the next correspondence search, we obtain a track, i.e., correspondences over multiple frames. The motion of the camera is computed with a least-squares approach finding the optimal rotation and translation between the clouds. In the next subsections we review the main steps of this approach.

19.3.1 Overview of the Approach

Figure 19.1 shows a block diagram of the method. The inputs are given by the left and right images from a stereo imaging system. We assume that the calibration parameters are known and that the provided images are rectified. Optical flow is computed using the current and previous left image. Disparities between the left and right image are only computed for those image positions where the flow algorithm was successful. Triangulation is performed and a list with the tracked points for the current frame is generated. The list is added to

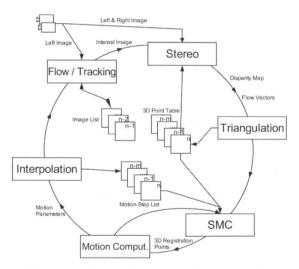

Fig. 19.1. General overview of the approach.

a table where the last m lists of tracked 3D points are stored. This will allow the integration of a multi-frame motion estimation, i.e., motion is not only obtained based on the last observed movement, but also between m frames in the past and the current frame. The six motion parameters (three components for translation and three for rotation) are then computed as the optimal translation and rotation found between the current and previous list of 3D points using a least squares closed-form solution based on rotation quaternions as shown in the next section. In order to avoid the introduction of erroneous data in the least square computation, a smoothness motion constraint is applied, rejecting all pairs of points which represent an incoherent movement with respect to the current ego-motion (Section 19.3.4). These two steps are repeated but using the list of tracked points between the current frame and m frames in the past. The two motion hypotheses are then interpolated obtaining our final ego-motion estimation, and updating the list of motion steps (Section 19.3.5). The whole process is then repeated for every new input data stream.

In our context, tracking a point means associating an image point from the previous frame to the current frame. A track is the connections of such associations over multiple frames. Clearly, these track positions are noninte-gral image positions in general. The flow and stereo algorithms to be used are not constraint to a specific implementation. In fact, our approach was tested with different algorithms obtaining almost identical results. The methods we use are described in Section 19.4.

19.3.2 Obtaining the Absolute Orientation Between Two Frames

Let $X = \{\vec{x}_i : 1 \leq i \leq n\}$ be the set of 3D points of the previous frame and $P = \{\vec{p}_i\}$ the set of 3D points observed at the current frame, where $\vec{x}_i \leftrightarrow \vec{p}_i$, i.e., $\vec{p}_i$ is the version at time t_k, transformed from the point $\vec{x}_i$ at time t_{k-1} (k frame index). In order to obtain the motion of the camera between the current and the previous frame we minimize a function which is expressed as the sum of the weighted residual errors between the rotated and translated data set X with the data set P, i.e.,

$$\sum_{i=1}^{n} w_i \|\vec{p}_i - R_k \vec{x}_i - \vec{d}_k\|^2 \tag{19.1}$$

where n is the amount of points in the sets, R_k is a rotation matrix, $\vec{d}_k$ is a translation vector, and w_i are individual weights representing the expected error in the measurement of the points. This formulation requires a one-to-one mapping of x_i and p_i, which is obtained by the Kanade/Lukas/Tomasi (KLT) tracker. Typically, we track about 1,000 points for an accurate orientation determination. Below $n = 30$, the result becomes error-prone. To solve this least-squares problem we use the method presented by Horn [19], which provides a closed form solution using unit quaternions. In this method the optimal rotation quaternion is obtained as the eigenvector corresponding to the largest positive eigenvalue of a 4×4 matrix. The quaternion is then converted to the rotation matrix. The translation is computed as the difference of the centroid of data set P and the rotated centroid of data set X. The computation of the optimal rotation and translation is not constrained to this specific method. Lorusso et al. [20] shortly describe and compare this method and three more methods for solving the absolute orientation problem in closed form.

19.3.3 Motion over Several Frames

In order to simplify the notation of the following subsections, we represent the motion in homogeneous coordinates. The computed motion of the camera between two consecutive frames, i.e., from frame $k-1$ to frame k, is represented by the matrix M'_k where:

$$M'_k = \begin{bmatrix} R_k & \vec{d}_k \\ 0 & 1 \end{bmatrix} \tag{19.2}$$

The rotation matrix $\hat{R}_k$ and translation vector $\vec{d}_k$, i.e., the object pose are obtained by just inverting M'_k, i.e.,

$$M'^{-1}_k = \begin{bmatrix} \hat{R}_k & \vec{\hat{d}}_k \\ 0 & 1 \end{bmatrix} = \begin{bmatrix} R_k^{-1} & -R_k^{-1}\vec{d}_k \\ 0 & 1 \end{bmatrix} \tag{19.3}$$

The total motion of the camera since initialization can be obtained as the products of the individual motion matrices:

$$M_k = \prod_{i=1}^{k} M'_i \tag{19.4}$$

A sub-chain of movements from time t_n to time t_m is:

$$M_{n,m} = M_n^{-1} M_m = \prod_{i=n+1}^{m} M'_i \tag{19.5}$$

Figure 19.2 shows an example of motion integration with matrices. As we will show later in Section 19.3.5 Equation 19.5 will support the integration of the motion between two nonconsecutive frames (multi-step estimation).

19.3.4 Smoothness Motion Constraint

Optical flow and/or stereo can deliver false information about 3D position or image point correspondence between image frames. Some of the points might also correspond to an independently moving object. A robust method should

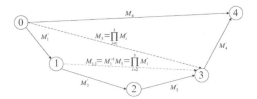

Fig. 19.2. The integration of motion over time can be obtained by just multiplying the individual motion matrices. Every circle denotes the state (position and orientation) of the camera at time t_i. Every vector indicates the motion between two states in 3D-space.

still be able to give accurate results in such situations. If the frame rate is high enough in order to obtain a smooth motion between consecutive frames, then the current motion is similar to the immediate previous motion. This is easily obtained in crash test video analysis since typical frame rates are 500 or 1,000 Hz. Therefore, before including the pair of points $\vec{p}_i$ and $\vec{x}_i$ into their corresponding data sets P and X, we evaluate if the vector $\vec{v}_i = \overrightarrow{p_i x_i}$ indicates a coherent movement. Let us define $\vec{m} = \begin{bmatrix} \dot{x}_{max} & \dot{y}_{max} & \dot{z}_{max} & 1 \end{bmatrix}$ as the maximal accepted error of the position of a 3D point with respect to a predicted position. Based on our previous ego-motion estimation step we evaluate the motion coherence of the vector $\vec{v}_i$ as:

$$\vec{c}_i = M'_{k-1}\vec{x}_i - \vec{p}_i \tag{19.6}$$

i.e., the error of our prediction. If the absolute value of any component of c_i is larger than $\vec{m}$ the pair of points are discarded and not included in the data sets for the posterior computation of relative orientation. Otherwise we weight the pair of points as the ratio of change with respect to the last motion:

$$w_i = 1 - \frac{\|\vec{c}_i\|^2}{\|\vec{m}\|^2} \tag{19.7}$$

which is later used in Equation 19.1. Equations 19.6 and 19.7 define the smoothness motion constraint (SMC). Besides using the prediction error, one could also incorporate the measurement accuracy of the triangulated 3D points. This absolute geometric accuracy decreases quadratically with the measured distance. This geometric stereo accuracy could be included as part of the weight, but this makes little difference for points with similar depth, which is the case for the types of objects we investigate.

19.3.5 Multi-frame Estimation

Single step estimation, i.e., the estimation of the motion parameters from the current and previous frame is the standard case in most approaches. If we are able to track points over m frames, then we can also compute the motion

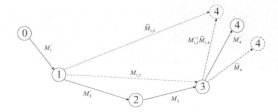

Fig. 19.3. Multi-frame approach. Circles represent the position and orientation of the camera. Vectors indicate motion in 3D-space. $\tilde{M}_4$ (single step estimation) and $M_{1,3}^{-1}\tilde{M}_{1,4}$ (multistep estimation) are interpolated in order to obtain the final estimation M_4'.

between the current and the m previous frames and integrate this motion into the single step estimation (see Figure 19.3). The estimation of motion between frame m and the current frame k ($m < k - 1$) follows exactly the same procedure as explained above. Only when applying the SMC, a small change takes place, since the prediction of the position for $k - m$ frames is not the same as for a single step. In other words, the matrix M_{k-1}' of Equation 19.6 is not valid any more. If the single step estimation for the current frame was already computed as $\tilde{M}_k$ Equation 19.6 becomes:

$$\vec{c}_i = M_{k-m}^{-1} M_{k-1}\tilde{M}_k\vec{x}_i - \vec{p}_i. \tag{19.8}$$

Equation 19.8 represents the estimated motion between times t_{k-m} and t_{k-1} (from Equation 19.5), updated with the current simple step estimation of time t_k. This allows the SMC to be even more precise, since the uncertainty in the movement is now based on an updated prediction. On the contrary in the single step estimation, the uncertainty is based on a position defined by the last motion.

Once the camera motion matrix $\tilde{M}_{m,k}$ between times t_{k-m} and t_k is obtained, it is integrated with the single step estimation. This is performed by an interpolation. The interpolation of motion matrices makes sense if they are estimations of the same motion. This is not the case since the single step motion matrix is referred to as the motion between the last two frames and the multistep motion matrix as the motion between m frames in the past to the current one. Thus, the matrices to be interpolated are $\tilde{M}_k$ and $M_{m,k-1}^{-1}\tilde{M}_{m,k}$ (see Figure 19.3). The corresponding rotation matrices are converted to quaternions in order to apply a spherical linear interpolation. The interpolated quaternion is converted to the final rotation matrix R_k. Translation vectors are linearly interpolated, obtaining the new translation vector $\vec{t}_k$. The factors of the interpolation are given by the weighted sum of the quadratic deviations obtained when computing the relative motion of Equation 19.1.

The multi-frame approach performs better thanks to the integration of more measurements. It also reduces the integration of the errors produced by the single-step estimation between the considered time points. In fact, our experiments have shown that without the multi-frame approach the estimation degenerates quickly and, normally, after a few hundred frames the ego-position diverges dramatically from the true solution. Thus, the multi-frame approach provides additional stability to the estimation process.

19.4 Model-free Pose Estimation

19.4.1 Adaptions to the Ego-motion Estimation Algorithm

In our pose estimation scenario, the stereo camera system remains static and the observed object moves. All equations shown above remain the same since the relative motion is identical. However, the smoothness motion constraint is not related to the motion of the camera anymore but to the observed object. We obtain the initial object position with the initial stereo computation. From there, we transform the motion of the object back into the static reference frame and compare the predicted position via smoothness motion constraint to the measured position. This way, tracks that are not on the observed object can be eliminated for pose estimation.

19.4.2 Stereo and Optical Flow Computation

We describe shortly the stereo and optical flow algorithms used in the experimental results of Section 19.5. The stereo algorithm works based on a coarse-to-fine scheme in which a Gaussian pyramid for left and right images is constructed with a sampling factor of two. The search for the best disparity is only performed at the top level of the pyramid and then a translation of the disparity map is made to the next level, where a correction is done within an interval 1 of the calculated disparity.We use the sum of squared differences as the default correlation function. Different filters and constraints are applied between pyramid translations. The zero-mean normalized cross-correlation (ZNCC) is used in order to check the confidence of the match. A match is considered reliable if the ZNCC coefficient is larger than a predefined threshold (0.7 in our experiments). Dynamic programming can also be applied between pyramid translations in order to eliminate matches which invalidate the ordering constraint. Finally a sub-pixel disparity map is computed as the last step in the pyramid. This is achieved by fitting a second degree curve to the best match and its neighbors and finding the sub-pixel disparity where the slope is zero in the quadratic function. The pyramidal correlation-based stereo computation is described in its initial version in [21]. As a preprocessing step to the stereo algorithm the input images are rectified, i.e., lense distortion and slight orientation deviations from the parallel configuration are removed. We use a planar rectification as described in [22].

The tracking algorithm we use for the computation of optical flow is the Kanade/Lucas/Tomasi (KLT) tracker. An extended description of the algorithm can be found in [23] and therefore we skip the description of this method here. Our experience with different tracker algorithms has shown that the KLT tracker can track feature points with a small error over tens of frames.

Both algorithms can be substituted with other stereo or tracking algorithms. A pure optical flow without tracking cannot be used directly for multi-frame estimation since the flow computation is by definition only performed at integer positions.

19.4.3 Object Tracking and Point Selection

We select the object of interest manually in the scene. For that purpose, we mark a rectangular or circular region in the image. From there on, the object is tracked automatically using the average motion vector of the observed object. To obtain that motion vector, all tracks that originate within the selected region of interest are considered for pose estimation. But only tracks that pass the smoothness motion constraint are used for average motion computation. This way, partial occlusion is handled easily, since these flow vectors are not consistent with the current pose. Therefore, the manual selection does by no means require to be precise as long as the majority of the flow vectors in the selected region corresponds to the object of interest. No precise object boundaries such as silhouettes must be extracted.

19.4.4 The Crash Test Scenario

The algorithm described above is a general technique to obtain a relative pose estimate of a rigid object. We apply this method to the crash test scenario. Compared to standard Human Motion Capture the scenarios differ in several ways:

- The sequences are captured at a frame rate of 500 or 1,000 Hz.
- The image resolution for vehicle crashes is typically 1500 by 1100 pixels.
- The imaging setup covers the whole crash scene, i.e., a crash test dummy might only cover a very small image area. Even worse, the crash test dummies inside the car are occluded by the door and sometimes by the air bag.
- There is no light source within the crash vehicle, so the crash test dummy silhouettes are often not visible due to shading effects.

In Figure 19.4, two typical crash scenarios are shown. On the left side, the standard EuroNCAP (European New Car Assessment Program) offset crash is shown. The car drives onto an aluminum barrier with 40% overlap. Therefore, the car moves laterally during the crash. Another typical scenario is the pedestrian crash is shown at the right side of Figure 19.4. Here, the car approaches at a speed of 10 m/s. The full motion of the pedestrian dummy is of interest here.

Fig. 19.4. Two typical crash test scenarios. EuroNCAP offset crash (left) and pedestrian crash (right).

We also investigated a so-called component crash where only a part of a crash experiment is conducted. The "Free-Motion-Head" experiment simulates the effect of the crash test dummy hitting the engine hood in a pedestrian crash. Other component crashes include lower-leg impact crash, sled crash, pole crash and others.

19.5 Results

19.5.1 Parameters

For our results, we used the parameters listed in Table 19.1 unless otherwise noted. For the initial pose estimation, the algorithm automatically increases the tolerance ranges if the minimum number of points is not reached. Once multi-frame estimation is possible, this situation does not reoccur.

The camera setup for the offset crash and the pedestrian crash was very similar: The baseline was 30cm and the camera resolution was 1504 by 1128 pixels (type RedLake 100 NG) recorded at 1,000 Hz for the offset crash and at 500 Hz for the pedestrian crash. Color cameras were used but for the analysis

Table 19.1. Used parameters for the pose estimation algorithm.

Tolerance in x	0.02 m
Tolerance in y	0.02 m
Tolerance in z	0.03 m
Steps for Multi-frame	20
Minimum number of points for pose estimation	30

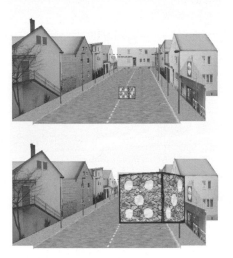

Fig. 19.5. Smallest and largest dice simulation image.

only grayscale values were used. The observation distance for both crashes was between 9 and 10 m for the crash test dummies.

19.5.2 Simulation Results

In order to verify the obtainable accuracy of the algorithm, a simulation scene with a dice of variable size is used. Image dimensions are 1,600 by 1,200 pixels (comparable to high speed/high resolution cameras), and the dice is set at 10 m distance (see Figure 19.5). We varied the size of the dice and let it rotate around the vertical axis to obtain a yaw motion of 2°/frame from the camera perspective. Note, that this motion is much harder to measure than a camera roll motion within the image plane. Stereo and KLT tracks were computed on these images. The obtained errors are listed in the table below (see Table 19.2) and show, that even objects covering a small part of the image allow an accurate pose estimation. The maximum errors, including the integrated pose errors over 100 frames, stay within 10°, the average errors stay well below 1°.

Multi-frame estimation up to 20 frames was used. Other publications rarely show data on pose estimation algorithms of objects covering a small image area, so no comparison data can be provided.

Table 19.2. Accuracy of pose estimation for varying object sizes. Motions are described in the camera coordinate system, pitch motion around the horizontal axis, yaw around the vertical axis and roll in the image plane.

Object size (10 m distance)	0.25 m 70[pix]	0.50 m 140[pix]	0.75 m 210[pix]	1.00 m 280[pix]
Horizontal Angular Resolution	$\approx 2°$	$\approx 4°$	$\approx 6°$	$\approx 8°$
Number of tracks	2,000	8,000	11,000	12,000
Mean error ($\pm$ standard deviation)				
Δ pitch[$°/frame$]	0.04 ± 0.07	0.00 ± 0.02	0.00 ± 0.01	0.00 ± 0.003
Δ yaw[$°/frame$]	0.12 ± 0.25	0.01 ± 0.08	0.01 ± 0.02	0.00 ± 0.02
Δ roll[$°/frame$]	0.01 ± 0.01	0.00 ± 0.003	0.00 ± 0.001	0.00 ± 0.001
integrated pitch[$°$]	0.52 ± 0.19	0.06 ± 0.09	0.06 ± 0.02	0.01 ± 0.01
slope of integrated yaw[$°/frame$]	0.08 ± 0.30	0.00 ± 0.07	0.00 ± 0.05	0.00 ± 0.03
integrated roll[$°$]	0.50 ± 0.18	-0.06 ± 0.09	-0.03 ± 0.02	0.03 ± 0.01
Maximum error				
Δ pitch[$°/frame$]	1.9	0.42	0.19	0.11
Δ yaw[$°/frame$]	4.3	2.9	1.1	0.46
Δ roll[$°/frame$]	0.3	0.13	0.04	0.04
integrated pitch [$°$]	6.9	1.6	0.4	0.2
integrated yaw [$°$]	10.0	2.3	0.8	0.4
integrated roll [$°$]	6.7	0.7	0.38	0.22

19.5.3 Crash Results

We verified the algorithm on several crash video scenes comprising of pedestrian crashes, offset crashes, and component crashes.

We illustrate the performance showing results of an offset crash, where the driver dummy almost disappears in the air bag at one moment in the sequence. The sequence is recorded at 1,000 Hz and consists of 300 images (see Figure 19.6). Here we used multi-frame estimation up to 100 frames in order to cope with the massive occlusion.

In Figure 19.7 the used flows for multi-frame estimation are displayed. No flows outside the dummy head are used for multi-frame estimation since they move differently from the head and are filtered out by the smoothness motion constraint. Most good tracks stay at the five-dot-marker at the right part of the head that stays visible throughout the sequence. The flow vectors computed for this image are shown in Figure 19.8.

A visual inspection of the quality of the algorithm can be obtained when all frames of the sequence are warped into the initial pose using the computed pose estimation. One expects no change for the object of interest whereas the environment might change arbitrarily. A result of such a comparison is shown in Figure 19.9. Note the good agreement to the initial pose shown in the left image.

Fig. 19.6. First and last frame of an A-class offset crash sequence. The middle image shows the moment of deepest air bag penetration.

Fig. 19.7. Used tracks for multi-frame estimation in frame 200 connecting to frame 101 (100 frames multi-frame estimation). Red (dark) denotes large deviation to the predicted position, cyan (light) means good agreement.

Fig. 19.8. KLT tracks on the same scene. Red denotes large flow vectors, green small flows. the circle shows the tracked region of interest.

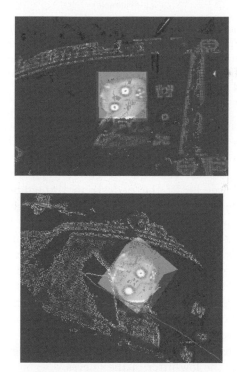

Fig. 19.9. 3D view of the scene. The position of the camera is based on the obtained pose and transforms the head to its initial orientation, viewed from 2.5 m distance. The right image depicts the transformed pose of image 200 after reappearing from the air bag.

To further check the plausibility of the measurements, we also tracked the region of the driver door, specifically the letters. Knowing the vehicle dynamics of such offset crashes we expect a yaw motion away from the barrier and a slight camera roll motion due to jumping. Both behaviors can be clearly seen

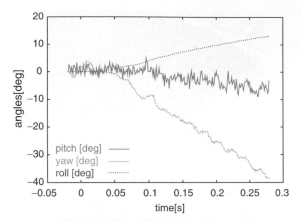

Fig. 19.10. Orientation of the text on the driver door throughout the offset crash. The vehicle pitching up in the air is clearly visible (corresponds to the camera roll angle), and the yaw motion due to the offset barrier is reflected in the yaw angle. The camera pitch angle should remain around 0 in the sequence, but exhibits some noise due to the small size of the object.

in the plot of the integrated angles of the estimated poses in Figure 19.10. The pitch motion is expected to remain around 0° and the deviations from that illustrate the noise level of the measurements.

For the pedestrian crash video (see Figure 19.11), we tracked the head, the back, and the lower leg of the pedestrian throughout the sequence of 250 frames with little integrated rotation error. As an example, we show the results of the back tracking. Here, the pedestrian dummy first rotates mainly in the image plane, then it rotates its own axis, and finally a slight image plane rotation occurs again. In Figure 19.13 the flow and the current region of interest (ROI) is depicted. Note we only adapt the ROI in the image plane. Due to self occlusion, the initial depth determination might be slightly wrong.

Figure 19.12 shows the used flows for multi-frame estimation. No flows are on the arm limbs since they move differently from the back and are filtered out by the smoothness motion constraint.

The obtained integrated orientations and relative orientations are shown in Figure 19.14. Note the low jitter of the curves.

19.5.4 Comparison with a Model-based Approach

We have conducted a crash component experiment called Free-Motion-Head. The goal is to investigate the dynamics of a pedestrian head crashing onto the engine hood. Such a crash has been recorded at 1,000 Hz with two color cameras with 512 by 384 pixel resolution. The cameras were positioned 27 cm apart at an intersecting angle of 10°. The subsequent analysis was done with the described approach and with a model-based approach that incorporates shape and texture of the head as described in [24].

Fig. 19.11. Frame 0, 100, and 200 frame of a pedestrian crash sequence. The sequence was recorded at 500 Hz and the vehicle speed was 36 km/h.

Fig. 19.12. Used tracks for multi-frame estimation in frame 200. All good tracks stay at the back area of the pedestrian that exhibits a rigid motion.

Fig. 19.13. KLT tracks on the same scene. Red (light) denotes large flow vectors, green (dark) small flows. the rotated rectangle shows the tracked region of interest.

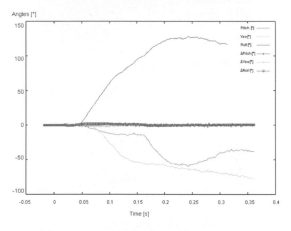

Fig. 19.14. Integrated Orientations and relative frame-to-frame orientations for the pedestrian sequence tracking the back.

In Figure 19.15 some resulting poses of the model-based approach can be seen. There is no noticeable drift in the sequence. Every image pair shows the model projection onto the left and the right image.

We compared the trajectories of the two approaches. In Figure 19.16 the 3D trajectories are compared. Both curves exhibit similar jitter. A quantitative comparison of the rotations is difficult due to the fact that our model-free method only determines rotations with respect to the reference pose of the first frame. The orientations of the model-based approach can be inspected to be consistent with the image views in Figure 19.15. The model-free approach yielded an almost constant view of the head in the stabilized 3D view similar to Figure 19.9.

From this comparison, one cannot draw conclusions about performance against ground truth. However, a prior version of the applied model-based approach has been compared against a marker-based system [25]. There, the

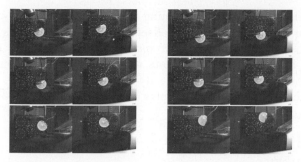

Fig. 19.15. Model-based result of the free-motion-head sequence.

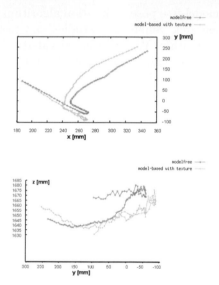

Fig. 19.16. Trajectory curves of the model-based and our model-free approach. The jitter is similar. The first part (starting in the center of the graph) of the y–z-graph is noisier for both approaches due to bad illumination.

finding was that the system has an absolute accuracy of about $3°$ which corresponds to the accuracy statements of commercial marker-based approaches.

19.5.5 Computation Time

The computationally relevant parts of the algorithm are rectification, stereo computation, tracking, and pose estimation. For high resolution images and tracking a 3D cube of 280 by 280 pixel image size, we obtain 160 ms for rectification of the full stereo pair, stereo computation time 850 ms, tracking 750 ms, and the actual pose estimation 750 ms (with 20 frames multi-frame estimation). 12,000 flows were used. Stereo was computed densely and with one pyramid level with a generous margin of 20 pixels to the left, right, top and bottom around the region of interest to make use of stereo post-processing techniques such as dynamic programming and median filter. The same settings applied to the dummy head in the offset crash sequence yields 350 ms for stereo, 250 ms for tracking and 100 ms for pose estimation. 1,000 points were tracked. These time measurements were conducted on a Pentium M 1.7 GHz laptop.

Real-time capable computation times are obtained for the ego-motion version of the algorithm (see [3]).

19.6 Summary

We have proposed a model-free pose estimation algorithm suitable for objects covering a small image area. Simulation studies and test on real image sequences validate the algorithm. We have investigated the accuracy that can be obtained with such a technique on objects covering a small image area. The average error is less than 1° on a 1.5° by 1.5° field-of-view object. In addition, we introduced a simple and efficient tracking technique that is able to handle partial occlusions.

In the future, we also want to apply more thoroughly 3D models to the pose estimation process and we want to investigate the accuracy of such a technique in more detail. Obviously, for human motion capture, a connectivity of the body parts is necessary to obtain a consistent human motion. One could embed our model-free approach into a framework such as presented in [25] to obtain a consistent human motion.

References

1. Rosenhahn, B., Perwass, C., Sommer, G.: Pose estimation of 3d free-form contours. International Journal of Computer Vision **62** (2005) 267–289
2. Aggarwal, J., Cai, Q.: Human motion analysis: a review. Computer Vision Image Understanding (1999) 428–440
3. Badino, H., U. Franke, C. Rabe, S. Gehrig: Stereo-vision based detection of moving objects under strong camera motion. In: International Conference on Computer Vision Theory and Applications, Setubal, Portugal (2006) 253–260
4. Gavrila, D.M.: The visula analysis of human movement: a survey. Computer Vision and Image Understanding **73** (1999) 82–98
5. Grest, D., Woetzel, J., Koch, R.: Nonlinear body pose estimation from depth images. In: DAGM (2005)
6. Seemann, E., Nickel, K., Stiefelhagen, R.: Head pose estimation using stereo vision for human-robot interaction. In: Sixth IEEE International Conference on Automatic Face and Gesture Recognition (2004) 626–631
7. Plankers, R., Fua, P.: Tracking and modelling people in video sequences. Computer Vision and Image Understanding (CVIU) **81** (2001)
8. Deutscher, J., Blake, A., Reid, I.: Articulated body motion capture by annealed particle filtering. In: Proc. Conf. Computer Vision and Pattern Recognition. Volume 2 (2000) 1144–1149

9. Mikic, I., Trivedi, M., Hunter, E., Cosman, P.: Human body model acquisition and tracking using voxel data. International Journal of Computer Vision **53** (2003) 199–223

10. Bray, M., Kohli, P., Torr, P.: Posecut: Simultaneous segmentation and 3d pose estimation of humans using dynamic graph-cuts. In Leonarids, A., Bishof, H., Prinz, A., eds.: Proc. 9th European Conference on Computer Vision, Part II, Volume 3952, Graz, Springer (2006) 642–655

11. T. Brox, B. Rosenhahn, D.C., Seidel, H.P.: High accuracy optical flow serves 3-d pose tracking: exploiting contour and flow based constraints. In: European Conference on Computer Vision (ECCV), Graz, Austria (2006) 98–111

12. Lu, C., Hager, G.D., Mjolsness, E.: Fast and globally convergent pose estimation from video images. IEEE Transactions On Pattern Analysis and Machine Intelligence **22** (2000) 610–622

13. Grinstead, B., Koschan, A., Abidi, M.A.: A comparison of pose estimation techniques: Hardware vs. video. In: SPIE Unmanned Ground Vehicle Technology, Orlando, FL (2005) 166–173

14. Badino, H.: A robust approach for ego-motion estimation using a mobile stereo platform. In: 1st International Workshop on Complex Motion (IWCM'04), Günzburg, Germany, Springer (2004)

15. Gehrig, S.K., Badino, H., Paysan, P.: Accurate and model-free pose estimation of small objects for crash video analysis. In: British Machine Vision Conference BMVC, Edinburgh (2006)

16. Matthies, L., Shafer, S.A.: Error modelling in stereo navigation. In: IEEE Journal of Robotics and Automation. Volume RA-3(3) (1987) 239–248

17. Olson, C.F., Matthies, L.H., Schoppers, M., Maimone, M.W.: Rover navigation using stereo ego-motion. In: Robotics and Autonomous Systems. Volume 43(4) (2003) 215–229

18. van der M., W., Fontijne, D., Dorst, L., Groen, F.C.A.: Vehicle ego-motion estimation with geometric algebra. In: Proc. IEEE Intelligent Vehicle Symposium, Versailles, France, May 18–20 (2002)

19. Horn, B.K.P.: Closed-form solution of absolute orientation using unit quaternions. In: Journal of the Optical Society of America A. Volume 4(4) (1987) 629–642

20. Lorusso, A., Eggert, D., Fisher, R.B.: A comparison of four algorithms for estimating 3-d rigid transformations. In: Proc. British Machine Vision Conference, Birmingham (1995)

21. Franke, U.: Real-time stereo vision for urban traffic scene understanding. In: Proc. Intelligent Vehicles 2000 Symposium (2000)

22. Gehrig, S.K., Klappstein, J., Franke, U.: Active stereo for intersection assistance. In: Vision modelling and Visualization Conference, Stanford, USA (2004) 29–35

23. Shi, J., Tomasi, C.: Good features to track. In: IEEE Conference on Computer Vision and Pattern Recognition (CVPR'94) (1994)

24. Gall, J., Rosenhahn, B., Seidel, H.P.: Robust pose estimation with 3d textured models. In: IEEE Pacific-Rim Symposium on Image and Video Technology (PSIVT06), Springer LNCS (2006) 84–95

25. Rosenhahn, B., Brox, T., Kersting, U., Smith, D., Gurney, J., Klette, R.: A system for marker-less human motion estimation. Kuenstliche Intelligenz (KI) (2006) 45–51

Part V

Modelling and Animation

Modeling and Adaptation

20

A Relational Approach to Content-based Analysis of Motion Capture Data

Meinard Müller and Tido Röder

Universität Bonn, Institut für Informatik III
Römerstr. 164, 53117 Bonn, Germany

Summary. Motion capture or mocap systems allow for tracking and recording of human motions at high spatial and temporal resolutions. The resulting 3D mocap data is used for motion analysis in fields such as sports sciences, biomechanics, or computer vision, and in particular for motion synthesis in data-driven computer animation. In view of a rapidly growing corpus of motion data, automatic retrieval, annotation, and classification of such data has become an important research field. Since logically similar motions may exhibit significant spatio-temporal variations, the notion of similarity is of crucial importance in comparing motion data streams. After reviewing various aspects of motion similarity, we discuss as the main contribution of this chapter a relational approach to content-based motion analysis, which exploits the existence of an explicitly given kinematic model underlying the 3D mocap data. Considering suitable combinations of boolean relations between specified body points allows for capturing the motion content while disregarding motion details. Finally, we sketch how such relational features can be used for automatic and efficient segmentation, indexing, retrieval, classification, and annotation of mocap data.

20.1 Introduction

Historically, the idea of motion capturing originates from the field of gait analysis, where locomotion patterns of humans and animals were investigated using arrays of analog photographic cameras, see Chapter 1. With technological progress, motion capture data or simply *mocap data* became popular in computer animation to create realistic motions for both films and video games. Here, the motions are performed by live actors, captured by a digital mocap system, and finally mapped to an animated character. However, the lifecycle of a motion clip in the production of animations is very short. Typically, a motion clip is captured, incorporated in a single 3D scene, and then never used again. For efficiency and cost reasons, the reuse of mocap data as well as methods for modifying and adapting existing motion clips are gaining in importance. Applying editing, morphing, and blending techniques

B. Rosenhahn et al. (eds.), Human Motion – Understanding, Modelling, Capture, and
Animation, 477–506.

for the creation of new, realistic motions from prerecorded motion clips has become an active field of research [3, 13, 17, 18, 30, 39]. Such techniques depend on motion capture databases covering a broad spectrum of motions in various characteristics. Larger collections of motion material such as [7] have become publicly available in the last few years. However, prior to reusing and processing motion capture material, one has to solve the fundamental problem of identifying and extracting logically related motions scattered in a given database. In this context, automatic and efficient methods for *content-based* motion analysis, comparison, classification, and retrieval are required that only access the raw mocap data itself and do not rely on manually generated annotations. Such methods also play an important role in fields such as sports sciences, biomechanics, and computer vision, see, e. g., Chapters 17, 10, and 11.

One crucial point in content-based motion analysis is the notion of *similarity* that is used to compare different motions. Intuitively, two motions may be regarded as similar if they represent variations of the same action or sequence of actions [18]. Typically, these variations may concern the spatial as well as the temporal domain. For example, the kick sequences shown in Figure 20.1 describe a similar kind of motion even though they differ considerably with respect to motion speed as well as the direction, the height, and the style of the kick. How can a kicking motion be characterized irrespective of style? Or, conversely, how can motion style, the actor's individual characteristics, or emotional expressiveness be measured? Such questions are at the heart of motion analysis and synthesis. We will see that retrieval applications often aim at identifying related motions irrespective of certain motion details, whereas synthesis applications are often interested in just those motion details. Among other aspects of motion similarity, our discussion in Section 20.3 addresses the issue of separating motion details from motion content.

The difficult task of identifying similar motions in the presence of spatio-temporal variations still bears open problems. In this chapter, we will discuss analysis techniques that focus on the rough course of a motion while disregarding motion details. Most of the previous approaches to motion comparison are based on features that are semantically close to the raw data, using 3D positions, 3D point clouds, joint angle representations, or PCA-reduced versions

Fig. 20.1. Top: seven poses from a side kick sequence. **Bottom:** corresponding poses for a frontal kick. Even though the two kicking motions are similar in some logical sense, they exhibit significant spatial and temporal differences.

thereof, see [12, 15, 16, 18, 34, 41]. One problem of such features is their sensitivity towards pose deformations, as may occur in logically related motions. Instead of using numerical, *quantitative* features, we suggest to use relational, *qualitative* features as introduced in [25]. Here, the following observation is of fundamental importance: opposed to other data formats such as images or video, 3D motion capture data is explicitly based on a kinematic chain that models the human skeleton. This underlying model can be exploited by looking for boolean relations between specified body points, where the relations possess explicit semantics. For example, even though there may be large variations between different kicking motions as illustrated by Figure 20.1, all such motions share some common characteristics: first the right knee is stretched, then bent, and finally stretched again, while the right foot is raised during this process. Afterwards, the right knee is once again bent and then stretched, while the right foot drops back to the floor. In other words, by only considering the temporal evolution of the two simple boolean relations "right knee bent or not" and "right foot raised or not", one can capture important characteristics of a kicking motion, which, in retrieval applications, allows for cutting down the search space very efficiently. In Section 20.4, we discuss in detail the concept and design of relational motion features. Then, in Section 20.5, we sketch several applications of relational features, including automatic and efficient motion segmentation, indexing, retrieval, annotation, and classification.

In Section 20.2, for the sake of clarity, we summarize some basic facts about 3D motion capture data as used in this chapter, while describing the data model and introducing some notation. Further references to related work are given in the respective sections.

20.2 Motion Capture Data

There are many ways to generate motion capture data using, e. g., mechanical, magnetic, or optical systems, each technology having its own strengths and weaknesses. For an overview and a discussion of the pros and cons of such systems we refer to [38]. We exemplarily discuss an optical marker-based technology, which yields very clean and detailed motion capture data. Here, the actor is equipped with a set of 40–50 retro-reflective markers attached to a suit. These markers are tracked by an array of six to twelve calibrated high-resolution cameras at a frame rate of up to 240 Hz, see Figure 20.2. From the recorded 2D images of the marker positions, the system can then reconstruct the 3D marker positions with high precision (present systems have a resolution of less than a millimeter). Then, the data is cleaned with the aid of semi-automatic gap filling algorithms exploiting kinematic constraints. Cleaning is necessary to account for missing and defective data, where the defects are due to marker occlusions and tracking errors. For many applications, the 3D marker positions are then converted to a skeletal kinematic chain

Fig. 20.2. Optical motion capture system based on retro-reflective markers attached to the actor's body. The markers are tracked by an array of 6–12 calibrated high-resolution cameras, typically arranged in a circle.

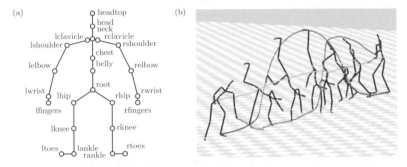

Fig. 20.3. (a) Skeletal kinematic chain model consisting of rigid *bones* that are flexibly connected by *joints*, which are highlighted by circular markers and labeled with joint names. **(b)** Motion capture data stream of a cartwheel represented as a sequence of poses. The figure shows the 3D trajectories of the joints "root", "rfingers", and "lankle".

representation using appropriate fitting algorithms [9, 29]. Such an abstract model has the advantage that it does not depend on the specific number and the positions of the markers used for the recording. However, the mapping process from the marker data onto the abstract model can introduce significant artifacts that are not due to the marker data itself. Here, one major problem is that skeletal models are only approximations of the human body that often do not account for biomechanical issues, see [42].

In this chapter, we assume that the mocap data is modeled using a *kinematic chain*, which may be thought of as a simplified copy of the human skeleton. A kinematic chain consists of *body segments* (the bones) that are connected by *joints* of various types, see Figure 20.3(a). Let J denote the set of joints, where each joint is referenced by an intuitive term such as "root", "lankle" (for "left ankle"), "rankle" (for "right ankle"), "lknee" (for "left knee"), and so on. For simplicity, end effectors such as toes or fingers are also regarded as joints. In the following, a *motion capture data stream* is thought of as a sequence of *frames*, each frame specifying the 3D coordinates of the joints at a certain point in time. Moving from the technical background to an abstract geometric context, we also speak of a *pose* instead of a frame. Mathematically, a pose can be regarded as a matrix $P \in \mathbb{R}^{3 \times |J|}$, where $|J|$ denotes the number of joints. The jth column of P, denoted by P^j, corresponds to the 3D coordinates of joint $j \in J$. A *motion capture data stream* (in information retrieval terminology also referred to as a *document*) can be modeled as a function

$$D : [1 : T] \to \mathcal{P} \subset \mathbb{R}^{3 \times |J|}, \tag{20.1}$$

where $T \in \mathbb{N}$ denotes the number of poses, $[1 : T] := \{1, 2, \ldots, T\}$ corresponds to the time axis (for a fixed sampling rate), and $\mathcal{P}$ denotes the set of poses. A subsequence of consecutive frames is also referred to as a *motion clip*. Finally, the curve described by the 3D coordinates of a single body joint is termed *3D trajectory*. This definition is illustrated by Figure 20.3(b).

20.3 Similarity Aspects

One central task in motion analysis is the design of suitable similarity measures to compare two given motion sequences in a semantically meaningful way. The notion of similarity, however, is an ill-defined term that depends on the respective application or on a person's perception. For example, a user may be interested only in the rough course of the motion, disregarding motion style or other motion details such as the facial expression. In other situations, a user may be particularly interested in certain nuances of motion patterns, which allows him to distinguish, e. g., between a front kick and a side kick, see Figure 20.1. In the following, we discuss some similarity aspects that play an important role in the design of suitable similarity measures or distance functions.

Typically, two motions are regarded as similar if they only differ by certain *global transformations* as illustrated by Figure 20.4(a). For example, one may

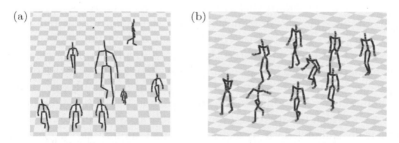

Fig. 20.4. (a) Different global transformations applied to a walking motion. **(b)** Different styles of walking motions.

leave the absolute position in time and space out of consideration by using a similarity measure that is invariant under temporal and spatial translations. Often, two motions are identified when they differ with respect to a global rotation about the vertical axis or with respect to a global reflection. Furthermore, the size of the skeleton or the overall speed of the motions may not be of interest – in such a case, the similarity measure should be invariant to spatial or temporal scalings.

More complex are variations that are due to different motion styles, see Figure 20.4(b). For example, walking motions may differ by performance (e. g., limping, tiptoeing, or marching), by emotional expression or mood (e. g., "cheerful walking", "furious walking", "shy walking"), and by the complex individual characteristics determined by the motion's performer. The abstract concept of *motion style* appears in the literature in various forms and is usually contrasted by some notion of *motion content*, which is related to the semantics of the motion. In the following, we give an overview of how motion style and motion content are treated in the literature.

In the context of gait recognition, Lee and Elgammal, see [21] and Chapter 2, define motion style as the time-invariant, personalized aspects of gait, whereas they view motion content as a time-dependent aspect representing different body poses during the gait cycle. Similarly, Davis and Gao [8] view motions as depending on style, pose, and time. In their experiments, they use PCA on expert-labeled training data to derive those factors (essentially linear combinations of joint trajectories) that best explain differences in style. Rose et al. [32] group several example motions that only differ by style into *verb* classes, each of which corresponds to a certain motion content. They synthesize new motions from these verb classes by suitable interpolation techniques, where the user can control interpolation parameters for each verb. These parameters are referred to as *adverbs* controlling the style of the verbs. To synthesize motions in different styles, Brand and Hertzmann [1] use example motions to train so-called *style machines* that are based on hidden Markov models (HMMs). Here, motion style is captured in certain parameters of the style machine such as average state dwell times and emission probability distributions for each state. On the other hand, motion content is encoded as the most likely state sequence of the style machine. Hsu et al. [15] propose a system for *style translation* that is capable of changing motions performed in a specific input style into new motions with the same content but a different output style. The characteristics of the input and output styles are learned from example data and are abstractly encoded in a linear dynamic system. A physically based approach to grasping the stylistic characteristics of a motion performance is proposed by Liu et al. [23]. They use a complex physical model of the human body including bones, muscles, and tendons, the biomechanical properties of which (elasticity, stiffness, muscle activation preferences) can be learned from training data to achieve different motion styles in a synthesis step. Troje [36] trains linear PCA classifiers to recognize the gender of a person from recorded gait sequences, where the "gender" attribute seems to be located in the first three principal components of a suitable motion representation. Using a Fourier expansion of 3D locomotion data, Unuma et al. [37] identify certain *emotional* or *mood* aspects of locomotion style (for instance, "tired", "brisk", "normal") as gain factors for certain frequency bands.

Pullen and Bregler [30] also use a frequency decomposition of motion data, but their aim is not to pinpoint certain parameters that describe specific styles. Instead, they try to extract those details of the data that account for the natural look of captured motion by means of multiresolution analysis (MRA) on mocap data [3]. These details are found in certain high-frequency bands of the MRA hierarchy and are referred to as *motion texture* in analogy to the texture concept in computer graphics, where photorealistic surfaces are rendered with texture mapping. The term "motion texture" is also used by Li et al. [22] in the context of motion synthesis, but their concept is in no way related to the signal processing approach of Pullen and Bregler [30]. In their parlance, motion textures are generative statistical models describing an entire class of motion clips. Similar to style machines [1], these models consist of a set of *motion textons* together with transition probabilities encoding typical orders in which the motion textons can be traversed. Each motion texton is a linear dynamic system (see also Hsu et al. [15]) that specializes in generating certain subclips of the modeled motion. Parameter tuning at the texton level then allows for manipulating stylistic details.

Inspired by the performing arts literature, Neff and Fiume [27, 28] explore the aspect of *expressiveness* in synthesized motions, see Chapter 24. Their system enables the user to describe motion content in a high-level scripting language. The content can be modified globally and locally by applying procedural *character sketches* and *properties*, which implement expressive aspects such as "energetic", "dejected", or "old man".

Returning to the walking example of Figure 20.4(b), we are faced with the question of how a walking motion can be characterized and recognized irrespective of motion style or motion texture. Video-based motion recognition systems such as [2, 14] tackle this problem by using hierarchical HMMs to model the motion content. The lower levels of the hierarchy comprise certain HMM building blocks representing fundamental components of full-body human motion such as "turning" or "raising an arm". In analogy to *phonemes* in speech recognition, these basic units are called *dynemes* by Green and Guan, see [14] and Chapter 9, or *movemes* by Bregler [2]. Dynemes/movemes and higher-level aggregations of these building blocks are capable of absorbing some of the motion variations that distinguish different executions of a motion.

The focus of this chapter is the automatic analysis of motion content. How can one grasp the gist of a motion? How can logically similar motions be identified even in the presence of significant spatial and temporal variations? How can one determine and encode characteristic aspects that are common to all motions contained in some given motion class? As was mentioned earlier, the main problem in motion comparison is that logically related motions need

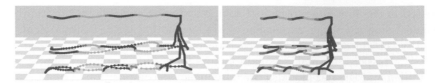

Fig. 20.5. Two walking motions performed in different speeds and styles. The figure shows the 3D trajectories for "headtop", "rfingers", "lfingers", "rankle", and "lankle". Logically corresponding segments in the two motions are indicated by the same colors.

Fig. 20.6. Three repetitions of "rotating both arms forwards". The character on the left is walking while rotating the arms (2.7 s), whereas the character on the right is standing on one spot while rotating the arms (2.3 s). The trajectories of the joints "rankle", "lankle", and "lfingers" are shown.

not be numerically similar as was illustrated by the two kicking motions of Figure 20.1. As another example, the two walking motions shown in Figure 20.5 can be regarded as similar from a logical point of view even though they differ considerably in speed and style. Here, using techniques such as dynamic time warping, one may compensate for spatio-temporal deformations between related motions by suitably warping the time axis to establish frame correspondences, see [18]. Most features and local similarity measures used in this context, however, are based on numerical comparison of spatial or angular coordinates and cannot deal with qualitative variations. Besides spatio-temporal deformations, differences between logical and numerical similarity can also be due to *partial similarity*. For example, the two instances of "rotating both arms forwards" as shown in Figure 20.6 are almost identical as far as the arm movement is concerned, but differ with respect to the movement of the legs. Numerically, the resulting trajectories are very different–compare, for example, the cycloidal and the circular trajectories of the hands. Logically, the two motions could be considered as similar.

Even worse, numerical similarity does not necessarily imply logical similarity. For example, the two actions of picking up an object and placing an object on a shelf are very hard to distinguish numerically, even for a human [18]. Here, the context of the motion or information about interaction with objects would be required, see also [19]. Often, only minor nuances or partial aspects of a motion account for logical differences. Think of the motions "standing on a

Fig. 20.7. A 500-frame ballet motion sampled at 120 Hz, adopted from the CMU mocap database [7]. The motion comprises two 180° right turns, the second of which is jumped. The trajectory of the joint "ltoes" is shown.

spot" compared to "standing accompanied by weak waving with one hand": such inconspicuous, but decisive details are difficult for a full-body similarity measure to pick up unless the focus of the similarity measure is primarily on the motion of the hands. As a further example, consider the difference between walking and running. These motions may of course be distinguished by their absolute speed. Yet, the overall shape of most joints' trajectories is very similar in both motions. A better indicator would be the occurrence of simultaneous air phases for both feet, which is a discriminative feature of running motions.

Last but not least, noise is a further factor that may interfere with a similarity measure for motion clips. Mocap data may contain significant high-frequency noise components as well as undesirable artifacts such as sudden "flips" of a joint or systematic distortions due to wobbling mass or skin shift [20]. For example, consider the toe trajectory shown in the ballet motion of Figure 20.7, where the noise shows as extremely irregular sample spacing. Such noise is usually due to adverse recording conditions, occlusions, improper setup or calibration, or data conversion faults. On the left hand side of the figure, there is a discontinuity in the trajectory, which results from a 3-frame flip of the hip joint. Such flips are either due to confusions of trajectories in the underlying marker data or due to the fitting process. Ren et al. [31] have developed automatic methods for detecting "unnatural" movements in order to find noisy clips or clips containing artifacts within a mocap database. Noise and artifacts are also a problem in markerless, video-based mocap systems, see, e. g., [33] as well as Chapters 11, 12, and 15. In view of such scenarios, it is important to design noise-tolerant similarity measures for the comparison of mocap data.

20.4 Relational Features

Applications of motion retrieval and classification typically aim at identifying related motions by content irrespective of motion style. To cope with significant numerical differences in 3D positions or joint angle configurations that may distinguish logically corresponding poses, we suggest to use qualitative features that are invariant to local deformations and allow for masking out irrelevant or inconsistent motion aspects. Note that mocap data, which is based on an explicit kinematic model, has a much richer semantic content than, for example, pure video data of a motion, since the position and the meaning of all joints is known for every pose. This fact can be exploited by considering features that describe boolean relations between specified points of a pose or short sequences of poses. Summarizing and extending the results of [25], we will introduce in this section several classes of boolean relational features that encode spatial, velocity-based, as well as directional information. The idea of considering relational instead of numerical features is not new and has already been applied by, e.g., Carlsson et al. [4,5,35] in other domains such as visual object recognition in 2D and 3D, or action recognition and tracking.

20.4.1 A Basic Example

As a basic example, we consider a relational feature that expresses whether the right foot lies in front of (feature value one) or behind (feature value zero) the plane spanned by the center of the hip (the root), the left hip joint, and the left foot for a fixed pose, cf. Figure 20.8(a). More generally, let $p_i \in \mathbb{R}^3$, $1 \leq i \leq 4$, be four 3D points, the first three of which are in general position. Let $\langle p_1, p_2, p_3 \rangle$ denote the oriented plane spanned by the first three points, where the orientation is determined by point order. Then define

$$B(p_1, p_2, p_3; p_4) := \begin{cases} 1, & \text{if } p_4 \text{ lies in front of or on } \langle p_1, p_2, p_3 \rangle, \\ 0, & \text{if } p_4 \text{ lies behind } \langle p_1, p_2, p_3 \rangle. \end{cases} \quad (20.2)$$

From this we obtain a feature function $F_{\text{plane}}^{(j_1, j_2, j_3; j_4)} : \mathcal{P} \to \{0, 1\}$ for any four distinct joints $j_i \in J$, $1 \leq i \leq 4$, by defining

$$F_{\text{plane}}^{(j_1, j_2, j_3; j_4)}(P) := B(P^{j_1}, P^{j_2}, P^{j_3}; P^{j_4}). \quad (20.3)$$

The concept of such relational features is simple but powerful, as we will illustrate by continuing the above example. Setting $j_1 = $"root", $j_2 = $"lankle",

(a)   (b)   (c)

Fig. 20.8. Relational features describing geometric relations between the body points of a pose that are indicated by circular markers. The respective features express whether **(a)** the right foot lies in front of or behind the body, **(b)** the left hand is reaching out to the front of the body or not, **(c)** the left hand is raised above neck height or not.

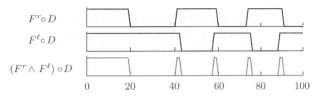

Fig. 20.9. Boolean features F^r, F^ℓ, and the conjunction $F^r \wedge F^\ell$ applied to the 100-frame walking motion $D = D_{\text{walk}}$ of Figure 20.15.

j_3="lhip", and j_4="rtoes", we denote the resulting feature by $F^r :=$ $F_{\text{plane}}^{(j_1,j_2,j_3;j_4)}$. The plane determined by j_1, j_2, and j_3 is indicated in Figure 20.8(a) as a green disc. Obviously, the feature $F^r(P)$ is 1 for a pose P corresponding to a person standing upright. It assumes the value 0 when the right foot moves to the back or the left foot to the front, which is typical for locomotion such as walking or running. Interchanging corresponding left and right joints in the definition of F^r and flipping the orientation of the resulting plane, we obtain another feature function denoted by F^ℓ. Let us have a closer look at the feature function $F := F^r \wedge F^\ell$, which is 1 if and only if both, the right as well as the left toes, are in front of the respective planes. It turns out that F is very well suited to characterize any kind of walking or running movement. If a data stream $D : [1 : T] \to \mathcal{P}$ describes such a locomotion, then $F \circ D$ exhibits exactly two peaks for any locomotion cycle, from which one can easily read off the speed of the motion (see Figure 20.9). On the other hand, the feature F is invariant under global orientation and position, the size of the skeleton, and various local spatial deviations such as sideways and vertical movements of the legs. Furthermore, F leaves any upper body movements unconsidered.

In the following, we will define feature functions purely in terms of geometric entities that are expressible by joint coordinates. Such relational features are invariant under global transforms (Euclidean motions, scalings) and are very coarse in the sense that they express only a single boolean aspect, masking out all other aspects of the respective pose. This makes relational features robust to variations in the motion capture data stream that are not correlated with the aspect of interest. Using suitable boolean expressions and combinations of several relational features then allows to focus on or to mask out certain aspects of the respective motion.

20.4.2 Generic Features

The four joints in $F_{\text{plane}}^{(j_1,j_2,j_3;j_4)}$ can be picked in various meaningful ways. For example, in the case j_1="root", j_2="lshoulder", j_3="rshoulder", and j_4="lwrist", the feature expresses whether the left hand is in front of or behind the body. Introducing a suitable offset, one can change the semantics of a feature. For the previous example, one can move the plane $\langle P^{j_1}, P^{j_2}, P^{j_3} \rangle$

to the front by one length of the skeleton's humerus. The resulting feature can then distinguish between a pose with a hand reaching out to the front and a pose with a hand kept close to the body, see Figure 20.8(b).

Generally, in the construction of relational features, one can start with some *generic relational feature* that encodes information about relative position, velocity, or direction of certain joints in 3D space. Such a generic feature depends on a set of joint variables, denoted by $j_1, j_2, \ldots$, as well as on a variable θ for a threshold value or threshold range. For example, the generic feature $F_{\text{plane}} = F_{\theta,\text{plane}}^{(j_1,j_2,j_3;j_4)}$ assumes the value one iff joint j_4 has a signed distance greater than $\theta \in \mathbb{R}$ from the oriented plane spanned by the joints j_1, j_2 and j_3. Then each assignment to the joints $j_1, j_2, \ldots$ and the threshold θ leads to a boolean function $F : \mathcal{P} \to \{0, 1\}$. For example, by setting $j_1=$"root", $j_2=$"lhip", $j_3=$"ltoes", $j_4=$"rankle", and $\theta = 0$ one obtains the (boolean) relational feature indicated by Figure 20.8(a).

Similarly, we obtain a generic relational feature $F_{\text{nplane}} = F_{\theta,\text{nplane}}^{(j_1,j_2,j_3;j_4)}$, where we define the plane in terms of a normal vector (given by j_1 and j_2) and fix it at j_3. For example, using the plane that is normal to the vector from the joint $j_1=$"chest" to the joint $j_2=$"neck" fixed at $j_3=$"neck" with threshold $\theta = 0$, one obtains a feature that expresses whether a hand is raised above neck height or not, cf. Figure 20.8(c).

Using another type of relational feature, one may check whether certain parts of the body such as the arms, the legs, or the torso are bent or stretched. To this end, we introduce the generic feature $F_{\text{angle}} = F_{\theta,\text{angle}}^{(j_1,j_2;j_3,j_4)}$, which assumes the value one iff the angle between the directed segments determined by (j_1, j_2) and (j_3, j_4) is within the threshold range $\theta \subset \mathbb{R}$. For example, by setting $j_1=$"rknee", $j_2=$"rankle", $j_3=$"rknee", $j_4=$"rhip", and $\theta = [0, 120]$, one obtains a feature that checks whether the right leg is bent (angle of the knee is below 120 degrees) or stretched (angle is above 120 degrees), see Figure 20.10(a).

Other generic features may operate on velocity data that is approximated from the 3D joint trajectories of the input motion. An easy example is the generic feature $F_{\text{fast}} = F_{\theta,\text{fast}}^{(j_1)}$, which assumes the value one iff joint j_1 has an absolute velocity above θ. Figure 20.10(b) illustrates the derived feature

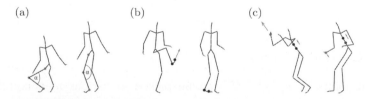

(a)   (b)   (c)

Fig. 20.10. Relational features that express whether **(a)** the right leg is bent or stretched, **(b)** the right foot is fast or not, **(c)** the right hand is moving upwards in the direction of the spine or not.

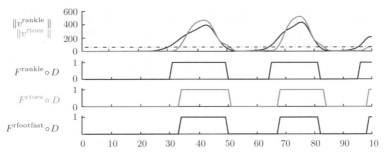

Fig. 20.11. Top: Absolute velocities in cm/s of the joints "rankle" ($\|v^{\mathrm{rankle}}\|$, black) and "rtoes" ($\|v^{\mathrm{rtoes}}\|$, gray) in the walking motion $D = D_{\mathrm{walk}}$ of Figure 20.15. The dashed line at $\theta_{\mathrm{fast}} = 63$ cm/s indicates the velocity threshold. **Middle:** Thresholded velocity signals for "rankle" and "rtoes". **Bottom:** Feature values for $F^{\mathrm{rfootfast}} = F^{\mathrm{rtoes}} \wedge F^{\mathrm{rankle}}$.

$F^{\mathrm{rfootfast}} := F^{\mathrm{rtoes}} \wedge F^{\mathrm{rankle}}$, which is a movement detector for the right foot. $F^{\mathrm{rfootfast}}$ checks whether the absolute velocity of both the right ankle (feature: F^{rankle}) and the right toes (feature: F^{rtoes}) exceeds a certain velocity threshold, θ_{fast}. If so, the feature assumes the value one, otherwise zero, see Figure 20.11. This feature is well suited to detect kinematic constraints such as footplants. The reason why we require both the ankle and the toes to be sufficiently fast is that we only want to consider the foot as being fast if all parts of the foot are moving. For example, during a typical walking motion, there are phases when the ankle is fast while the heel lifts off the ground, but the toes are firmly planted on the ground. Similarly, during heel strike, the ankle has zero velocity, while the toes are still rotating downwards with nonzero velocity. This feature illustrates one of our design principles for relational features: we construct and tune features so as to explicitly grasp the semantics of typical situations such as the occurrence of a footplant, yielding intuitive semantics for our relational features. However, while a footplant always leads to a feature value of zero for $F^{\mathrm{rfootfast}}$, there is a large variety of other motions yielding the feature value zero (think of keeping the right leg lifted without moving). Here, the combination with other relational features is required to further classify the respective motions. In general, suitable combinations of our relational features prove to be very descriptive for full-body motions.

Another velocity-based generic feature is denoted by $F_{\mathrm{move}} = F_{\theta,\mathrm{move}}^{(j_1,j_2;j_3)}$. This feature considers the velocity of joint j_3 relative to joint j_1 and assumes the value one iff the component of this velocity in the direction determined by (j_1, j_2) is above θ. For example, setting j_1="belly", j_2="chest", j_3="rwrist", one obtains a feature that tests whether the right hand is moving upwards or not, see Figure 20.10 (c). The generic feature $F_{\theta,\mathrm{nmove}}^{(j_1,j_2,j_3;j_4)}$ has similar semantics, but the direction is given by the normal vector of the oriented plane spanned by j_1, j_2, and j_3.

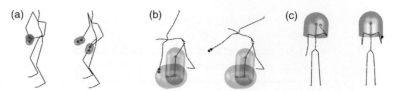

Fig. 20.12. Relational "touch" features that express whether **(a)** the two hands are close together or not, **(b)** the left hand is close to the leg or not, **(c)** the left hand is close to the head or not.

As a final example, we introduce generic features that check whether two joints, two body segments, or a joint and a body segment are within a θ-distance of each other or not. Here one may think of situations such as two hands touching each other, or a hand touching the head or a leg, see Figure 20.12. This leads to a generic feature $F_{\text{touch}}^{(j_1,j_2,\theta)}$, which checks whether the θ-neighborhoods of the joints j_1 and j_2 intersect or not. Similarly, one defines generic touch features for body segments.

20.4.3 Threshold Selection

Besides selecting appropriate generic features and suitable combinations of joints, the crucial point in designing relational features is to choose the respective threshold parameter θ in a semantically meaningful way. This is a delicate issue, since the specific choice of a threshold has a strong influence on the semantics of the resulting relational feature. For example, choosing $\theta = 0$ for the feature indicated by Figure 20.8(b) results in a boolean function that checks whether the left hand is in front of or behind the body. By increasing θ, the resulting feature checks whether the left hand is reaching out to the front of the body. Similarly, a small threshold in a velocity-based feature such as $F_{\theta,\text{fast}}^{(j_1)}$ leads to sensitive features that assume the value 1 even for small movements. Increasing θ results in features that only react for brisk movements. In general, there is no "correct" choice for the threshold θ—the specific choice will depend on the application in mind and is left to the designer of the desired feature set. In Section 20.4.4, we will specify a feature set that is suitable to compare the overall course of a full-body motion disregarding motion details.

To obtain a semantically meaningful value for the threshold θ in some automatic fashion, one can also apply supervised learning strategies. One possible strategy for this task is to use a training set $\mathcal{A}$ of "positive" motions that should yield the feature value one for most of its frames and a training set $\mathcal{B}$ of "negative" motions that should yield the feature value zero for most of its frames. The threshold θ can then be determined by a one-dimensional optimization algorithm, which iteratively maximizes the occurrences of the output one for the set $\mathcal{A}$ while maximizing the occurrences of the output zero for the set $\mathcal{B}$.

To make the relational features invariant under global scalings, the threshold parameter θ is specified relative to the respective skeleton size. For example, the value of θ may by given in terms of the length of the humerus, which scales quite well with the size of the skeleton. Such a choice handles differences in absolute skeleton sizes that are exhibited by different actors but may also result from different file formats for motion capture data.

Another problem arises from the simple quantization strategy based on the threshold θ to produce boolean features from the generic features. Such a strategy is prone to strong output fluctuations if the input value fluctuates slightly around the threshold. To alleviate this problem, we employ a robust quantization strategy using two thresholds: a stronger threshold θ_1 and a weaker threshold θ_2. As an example, consider a feature F^{sw} that checks whether the right leg is stretched sideways, see Figure 20.13. Such a feature can be obtained from the generic feature $F_{\theta,\mathrm{nplane}}^{(j_1,j_2,j_3;j_4)}$, where the plane is given by the normal vector through $j_1=$"lhip" and $j_2=$"rhip" and is fixed at $j_3=$"rhip". Then the feature assumes the value one iff joint $j_4=$"rankle" has a signed distance greater than θ from the oriented plane with a threshold $\theta = \theta_1 = 1.2$ measured in multiples of the hip width. As illustrated by Figure 20.13(a), the feature values may randomly fluctuate, switching between the numbers one and zero, if the right ankle is located on the decision boundary indicated by the dark disc. We therefore introduce a second decision boundary determined by a second, weaker, threshold $\theta_2 = 1.0$ indicated by the brighter disc in Figure 20.13(b). We then define a robust version $F_{\mathrm{robust}}^{\mathrm{sw}}$ of F^{sw} that assumes the value one as soon as the right ankle moves to the right of the stronger decision boundary (as before). But we only let $F_{\mathrm{robust}}^{\mathrm{sw}}$ return to the output value zero if the right ankle moves to the left of the weaker decision boundary. It turns out that this heuristic of *hysteresis thresholding* [11, Chapter 4] suppresses undesirable zero–one fluctuations in relational feature values very effectively, see Figure 20.14.

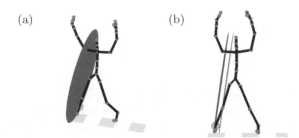

Fig. 20.13. Relational feature that expresses whether the right leg is stretched sideways or not. **(a)** The feature values may randomly fluctuate if the right ankle is located on the decision boundary (dark disc). **(b)** Introducing a second "weaker" decision boundary prevents the feature from fluctuations.

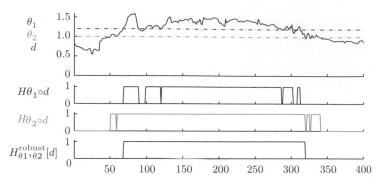

Fig. 20.14. Top: Distance d of the joint "rankle" to the plane that is parallel to the plane shown in Figure 20.13(a) but passes through the joint "rhip", expressed in the relative length unit "hip width" (hw). The underlying motion is a Tai Chi move in which the actor is standing with slightly spread legs. The dashed horizontal lines at $\theta_2 = 1$ hw and $\theta_1 = 1.2$ hw, respectively, indicate the two thresholds, corresponding to the two planes of Figure 20.13(b). **Middle:** Thresholded distance signals using the Heaviside thresholding function, H_θ; black: stronger threshold, θ_1; gray: weaker threshold, θ_2. **Bottom:** Thresholded distance signal using the robust thresholding operator $H_{\theta_1,\theta_2}^{\text{robust}}$.

20.4.4 Example of a Feature Set

Exemplarily, we describe a feature set that comprises $f = 39$ relational features. Note that this feature set has been specifically designed to focus on full-body motions. However, the proposed feature set may be replaced as appropriate for the respective application.

The 39 relational features, given by Table 20.1, are divided into the three sets "upper", "lower", and "mix", which are abbreviated as u, ℓ and m, respectively. The features in the upper set express properties of the upper part of the body, mainly of the arms. Similarly, the features in the lower set express properties of the lower part of the body, mainly of the legs. Finally, the features in the mixed set express interactions of the upper and lower part or refer to the overall position of the body.

Features with two entries in the ID column exist in two versions pertaining to the right/left half of the body but are only described for the right half – the features for the left half can be easily derived by symmetry. The abbreviations "hl", "sw" and "hw" denote the relative length units "humerus length", "shoulder width", and "hip width", respectively, which are used to handle differences in absolute skeleton sizes. Absolute coordinates, as used in the definition of features such as F_{17}, F_{32}, or F_{33}, stand for virtual joints at constant 3D positions w.r.t. an (X, Y, Z) world system in which the Y-axis points upwards. The symbols $Y_{\min}/Y_{\max}$ denote the minimum/maximum Y coordinates assumed by the joints of a pose that are not tested. Features such as F_{22} do not follow the same derivation scheme as the other features and are therefore described in words.

Table 20.1. A feature set consisting of 39 relational features.

ID	set	type	j_1	j_2	j_3	j_4	θ_1	θ_2	description
F_1/F_2	u	F_{nmove}	neck	rhip	lhip	rwrist	1.8 hl/s	1.3 hl/s	rhand moving forwards
F_3/F_4	u	F_{nplane}	chest	neck	neck	rwrist	0.2 hl	0 hl	rhand above neck
F_5/F_6	u	F_{move}	belly	chest	chest	rwrist	1.8 hl/s	1.3 hl/s	rhand moving upwards
F_7/F_8	u	F_{angle}	relbow	rshoulder	relbow	rwrist	$[0°, 110°]$	$[0°, 120°]$	relbow bent
F_9	u	F_{nplane}	lshoulder	rshoulder	lwrist	rwrist	2.5 sw	2 sw	hands far apart, sideways
F_{10}	u	F_{move}	lwrist	rwrist	rwrist	lwrist	1.4 hl/s	1.2 hl/s	hands approaching each other
F_{11}/F_{12}	u	F_{move}	rwrist	root	lwrist	root	1.4 hl/s	1.2 hl/s	rhand moving away from root
F_{13}/F_{14}	u	F_{fast}	rwrist				2.5 hl/s	2 hl/s	rhand fast
F_{15}/F_{16}	ℓ	F_{plane}	root	lhip	ltoes	rankle	0.38 hl	0 hl	rfoot behind lleg
F_{17}/F_{18}	ℓ	F_{nplane}	(0, 0, 0)	(0, 1, 0)	$(0, Y_{\min}, 0)$	rankle	1.2 hl	1 hl	rfoot raised
F_{19}	ℓ	F_{nplane}	lhip	rhip	lankle	rankle	2.1 hw	1.8 hw	feet far apart, sideways
F_{20}/F_{21}	ℓ	F_{angle}	rknee	rhip	rknee	rankle	$[0°, 110°]$	$[0°, 120°]$	rknee bent
F_{22}	ℓ		Plane Π fixed at lhip, normal rhip→lhip. Test: rankle closer to Π than lankle?						feet crossed over
F_{23}	ℓ		Consider velocity v of rankle relative to lankle in rankle→lankle direction. Test: projection of v onto rhip→lhip line large?						feet moving towards each other, sideways
F_{24}	ℓ		Same as above, but use lankle→rankle instead of rankle→lankle direction.						feet moving apart, sideways
F_{25}/F_{26}	ℓ	$F^{\text{rfootfast}}$					2.5 hl/s	2 hl/s	rfoot fast
F_{27}/F_{28}	m	F_{angle}	neck	root	rshoulder	relbow	$[25°, 180°]$	$[20°, 180°]$	rhumerus abducted
F_{29}/F_{30}	m	F_{angle}	neck	root	rhip	rknee	$[50°, 180°]$	$[45°, 180°]$	rfemur abducted
F_{31}	m	F_{plane}	rankle	neck	lankle	root	0.5 hl	0.35 hl	root behind frontal plane
F_{32}	m	F_{angle}	neck	root	(0, 0, 0)	(0, 1, 0)	$[70°, 110°]$	$[60°, 120°]$	spine horizontal
F_{33}/F_{34}	m	F_{nplane}	(0, 0, 0)	(0, −1, 0)	$(0, Y_{\min}, 0)$	rwrist	-1.2 hl	-1.4 hl	rhand lowered
F_{35}/F_{36}	m		Plane Π through rhip, lhip, neck. Test: rshoulder closer to Π than lshoulder?						shoulders rotated right
F_{37}	m		Test: $Y_{\min}$ and $Y_{\max}$ close together?						Y-extents of body small
F_{38}	m		Project all joints onto XZ-plane. Test: diameter of projected point set large?						XZ-extents of body large
F_{39}	m	F_{fast}	root				2.3 hl/s	2 hl/s	root fast

20.5 Applications

In this section, we show how relational features can be used for efficient motion retrieval, classification, and annotation. Fixing a set of boolean relational features, one can label each pose by its resulting feature vector. Such boolean vectors are ideally suited for indexing the mocap data according to these labels. Furthermore, a motion data stream can be segmented simply by grouping adjacent frames with identical labels. Motion comparison can then be performed at the segment level, which accounts for temporal variations, and efficient retrieval is possible by using inverted lists. As a further application, we introduce the concept of motion templates, by which the essence of an entire class of logically related motions can be captured. Such templates, which can be learned from training data, are suited for automatic classification and annotation of unknown mocap data.

20.5.1 Temporal Segmentation

We have seen that relational features exhibit a high degree of invariance against local spatial deformations. In this section, we show how to achieve invariance against local temporal deformations by means of a suitable feature-dependent temporal segmentation. To this end, we fix a list of, say, $f \in \mathbb{N}$ boolean relational features, which define the components of a boolean function $F : \mathcal{P} \to \{0,1\}^f$. From this point forward, F will be referred to as a *feature function* and the vector $F(P)$ as a *feature vector* or simply a *feature* of the pose $P \in \mathcal{P}$. Any feature function can be applied to a motion capture data stream $D : [1 : T] \to \mathcal{P}$ in a pose-wise fashion, which is expressed by the composition $F \circ D$. We say that two poses $P_1, P_2 \in \mathcal{P}$ are F-*equivalent* if the corresponding feature vectors $F(P_1)$ and $F(P_2)$ coincide, i. e., $F(P_1) = F(P_2)$. Then, an F-*run* of D is defined to be a subsequence of D consisting of consecutive F-equivalent poses, and the F-*segments* of D are defined to be the F-runs of maximal length.

We illustrate these definitions by continuing the example from Section 20.4.1. Let $F^2 := (F^r, F^\ell) : \mathcal{P} \to \{0,1\}^2$ be the combined feature formed by F^r and F^ℓ so that the pose set $\mathcal{P}$ is partitioned into four F^2-equivalence classes. Applying F^2 to the walking motion D_{walk} results in the segmentation shown in Figure 20.15, where the trajectories of selected joints have been plotted. F^2-equivalent poses are indicated by the same trajectory color: the color *red* represents the feature vector $(1,1)$, *blue* the vector $(1,0)$, and *green* the vector $(0,1)$. Note that no pose with feature vector $(0,0)$ appears in D_{walk}. Altogether, there are ten runs of maximal length constituting the F^2-segmentation of D_{walk}.

It is this feature-dependent segmentation that accounts for the postulated invariance under temporal deformations. To be more precise, let us start with the sequence of F-segments of a motion capture data stream D. Since each segment corresponds to a unique feature vector, the segments induce a sequence of feature vectors, which we simply refer to as the F-*feature sequence* of D and denote by $F[D]$. If M is the number of F-segments of D and if $D(t_m)$ for $t_m \in [1 : T]$, $0 \le m < M$, is a pose of the m-th segment, then $F[D] = (F(D(t_0)), F(D(t_1)), \dots, F(D(t_{M-1})))$. For example, for the data stream D_{walk} and the feature function F^2 from Figure 20.15, we obtain

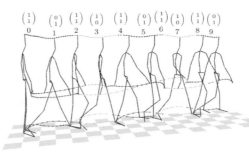

Fig. 20.15. F^2-segmentation of D_{walk}, where F^2-equivalent poses are indicated by uniformly colored trajectory segments. The trajectories of the joints "head-top", "rankle", "rfingers" and "lfingers" are shown.

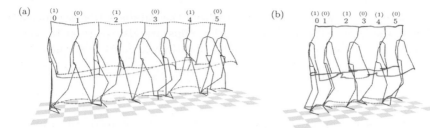

Fig. 20.16. (a) Restricting $F^2 = (F^r, F^\ell)$ to its first component results in an F^r-segmentation, which is coarser than the F^2-segmentation shown in Figure 20.15. (b) Five steps of a slow walking motion performed by an elderly person resulting in exactly the same F^r-feature sequence as the much faster motion of (a).

$$F^2[D_{\text{walk}}] = \left(\begin{pmatrix}1\\1\end{pmatrix}, \begin{pmatrix}0\\1\end{pmatrix}, \begin{pmatrix}1\\1\end{pmatrix}, \begin{pmatrix}1\\0\end{pmatrix}, \begin{pmatrix}1\\1\end{pmatrix}, \begin{pmatrix}0\\1\end{pmatrix}, \begin{pmatrix}1\\1\end{pmatrix}, \begin{pmatrix}1\\0\end{pmatrix}, \begin{pmatrix}1\\1\end{pmatrix}, \begin{pmatrix}0\\1\end{pmatrix}\right). \qquad (20.4)$$

Obviously, any two adjacent vectors of the sequence $F[D]$ are distinct. The crucial point is that time invariance is incorporated into the F-segments: two motions that differ by some deformation of the time axis will yield the same F-feature sequences. This fact is illustrated by Figure 20.16. Another property is that the segmentation automatically adapts to the selected features, as a comparison of Figure 20.15 and Figure 20.16(a) shows. In general, fine features, i. e., feature functions with many components, induce segmentations with many short segments, whereas coarse features lead to a smaller number of long segments.

The main idea is that two motion capture data streams D_1 and D_2 can now be compared via their F-feature sequences $F[D_1]$ and $F[D_2]$ instead of comparing the data streams on a frame-to-frame basis. This has several advantages:

1. One can decide which aspects of the motions to focus on by picking a suitable feature function F.
2. Since spatial and temporal invariance are already incorporated in the features and segments, one can use efficient methods from (fault-tolerant) string matching to compare the data streams instead of applying cost-intensive techniques such as dynamic time warping at the frame level.
3. In general, the number M of segments is much smaller than the number T of frames, which accounts for efficient computations.

Next, we will explain how our concept leads to an efficient way of indexing and searching motion capture data in a semantically meaningful way.

20.5.2 Indexing and Retrieval

In the retrieval context, the *query-by-example* paradigm has attracted a large amount of attention: given a query in form of a short motion clip, the task is to automatically retrieve all motion clips from the database that are logically similar to the query. The retrieved motion clips are also referred to as *hits* with respect to the query. Several general questions arise at this point:

1. How should the data, the database as well as the query, be modeled?
2. How does a user specify a query?
3. What is the precise definition of a hit?
4. How should the data be organized to afford efficient retrieval of all hits with respect to a given query?

In Section 20.5.1, we gave an answer to the first question by introducing the concept of feature sequences, which represent motion capture data streams as coarse sequences of binary vectors. For the moment, we assume that a query is given in form of a short motion clip Q. Furthermore, we assume that the database consists of a collection $\mathcal{D} = (D_1, D_2, \ldots, D_I)$ of mocap data streams or documents D_i, $i \in [1 : I]$. By concatenating the documents $D_1, \ldots, D_I$ while keeping track of document boundaries in a supplemental data structure, we may think of the database $\mathcal{D}$ as consisting of one large document D. Fixing a feature function $F : \mathcal{P} \to \{0,1\}^f$, we use the notation $F[D] = \mathbf{w} = (w_0, w_1, \ldots, w_M)$ and $F[Q] = \mathbf{v} = (v_0, v_1, \ldots, v_N)$ to denote the resulting F-feature sequences of D and Q, respectively. We then simply speak of the database $\mathbf{w}$ and the query $\mathbf{v}$.

Now, the trick is that by incorporating robustness against spatio-temporal variations into the relational features and adaptive segments, we are able to employ standard information retrieval techniques using an index of inverted lists [40]. For each feature vector $v \in \{0,1\}^f$ one stores the *inverted list* $L(v)$ consisting of the indices $m \in [0 : M]$ of the sequence $\mathbf{w} = (w_0, w_1, \ldots, w_M)$ with $v = w_m$. $L(v)$ tells us which of the F-segments of D exhibit the feature vector v. As an example, let us consider the feature function $F^2 = (F^r, F^\ell)$ from Figure 20.9 applied to a walking motion D as indicated by Figure 20.15. From the resulting feature sequence, one obtains the inverted lists $L\left(\binom{1}{1}\right) = \{0,2,4,6,8\}$, $L\left(\binom{0}{1}\right) = \{1,5,9\}$, $L\left(\binom{1}{0}\right) = \{3,7\}$, and $L\left(\binom{0}{0}\right) = \emptyset$. The elements of the inverted lists can then be stored in ascending order, accounting for efficient union and intersection operations in the subsequent query stage. In a preprocessing step, we construct an index $I_F^{\mathcal{D}}$ consisting of the 2^f inverted lists $L(v)$, $v \in \{0,1\}^f$. Since we store segment positions of the F-segmentation rather than individual frame positions in the inverted lists, and since each segment position appears in exactly one inverted list, the index size is proportional to the number M of segments of D. In particular, the time and space required to build and store our index structure is *linear*, opposed to the *quadratic* complexity of strategies based on dynamic time warping, see [18].

Recall that two motion clips are considered as similar (with respect to the selected feature function) if they exhibit the same feature sequence. Adapting concepts from [6], we introduce the following notions. An *exact hit* is an element $k \in [0 : M]$ such that $\mathbf{v}$ is a subsequence of consecutive feature vectors in $\mathbf{w}$ starting from index k. Using the notation $\mathbf{v} \sqsubset_k \mathbf{w}$ for this case, one obtains

$$\mathbf{v} \sqsubset_k \mathbf{w} \; :\Leftrightarrow \; \forall i \in [0:N] : v_i = w_{k+i}. \tag{20.5}$$

The set of all exact hits in the database $\mathcal{D}$ is then given by

$$H_{\mathcal{D}}(\mathbf{v}) := \{ k \in [0:M] \mid \mathbf{v} \sqsubset_k \mathbf{w} \}. \tag{20.6}$$

It is easy to see that $H_{\mathcal{D}}(\mathbf{v})$ can be evaluated very efficiently by intersecting suitably shifted inverted lists:

$$H_{\mathcal{D}}(\mathbf{v}) = \bigcap_{n \in [0:N]} (L(v_n) - n), \tag{20.7}$$

where the substraction of a list and a number is understood component-wise for every element in the list. As an example, we consider $D = D_{\text{walk}}$ and $F = F^2$ and the query sequence $\mathbf{v} = \left(\binom{1}{0}, \binom{1}{1}, \binom{0}{1} \right)$. Then

$$H_{\mathcal{D}}(\mathbf{v}) = \{3,7\} \cap \{-1,1,3,5,7\} \cap \{-1,3,7\} = \{3,7\} \tag{20.8}$$

resulting in two hits starting with the segments 3 and 7, respectively. See also Figure 20.17 for an illustration.

In many situations, the user may be unsure about certain parts of the query and wants to leave certain parts of the query unspecified. Or, the user may want to mask out some of the f components of the feature function F to obtain a less restrictive search leading to more hits. To handle such situations, one can employ the concept of *fuzzy search*. This technique admits at each position in the query sequence a whole set of possible, alternative feature vectors instead of a single one, see [6]. Here, a key idea is that the concept of temporal segmentation can be extended in such a way that segment lengths within a match not only adapt to the granularity of the feature function, but also to the fuzziness of the query. The resulting *adaptive fuzzy hits* can be computed very efficiently using the same index structure as for the case of exact hits. For further details on this strategy we refer to [25, 26].

We now describe how these techniques can be employed in an efficient motion retrieval system based on the query-by-example paradigm, which allows for intuitive and interactive browsing in a purely content-based fashion without relying on textual annotations, see Figure 20.18 for an overview. In the

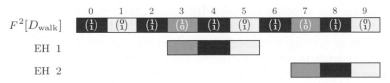

Fig. 20.17. Upper row: feature sequence $F^2[D_{\text{walk}}]$. **Below:** two exact hits (EH) for $\mathbf{v}_{\text{walk},1}$ in $F^2[D_{\text{walk}}]$, indicated by copies of $\mathbf{v}_{\text{walk},1}$ that are horizontally aligned with $F^2[D_{\text{walk}}]$ at the matching positions.

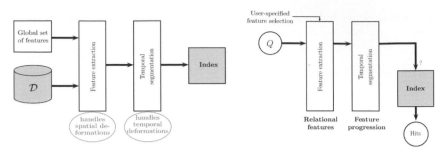

Fig. 20.18. Left: The preprocessing stage. **Right:** The query stage.

preprocessing step, a global feature function F has to be designed that covers all possible query requirements and provides the user with an extensive set of semantically rich features. In other words, it is not imposed upon the user to construct such features (even though this is also possible). Having fixed a feature function F, an index $I_F^{\mathcal{D}}$ is constructed for a given database $\mathcal{D}$ and stored on disk. (In practice, we split up the index into several smaller indices to reduce the number of inverted lists, see [25].) As an example, one may use the feature set comprising 39 relational features as described in Section 20.4.4. Note that this feature set has been specifically designed to focus on full-body motions. However, the described indexing and retrieval methods are generic, and the proposed test feature set may be replaced as appropriate for the respective application. Various query mechanisms of such a content-based retrieval system can be useful in practice, ranging from isolated pose-based queries, over query-by-example based on entire motion clips, up to manually specified geometric progressions. Here, we only consider the case that the input consists of a short query motion clip. Furthermore, the user should be able to incorporate additional knowledge about the query, e.g., by selecting or masking out certain body parts in the query. This is important to find, for example, all instances of "clapping one's hands" irrespective of any concurrent locomotion (recall the problem of partial similarity from Section 20.3.) To this end, the user selects relevant features from the given global feature set (i.e., components of F), where each feature expresses a certain relational aspect and refers to specific parts of the body. The query-dependent specification of motion aspects then determines the desired notion of similarity. In addition, parameters such as fault tolerance and the choice of a ranking or post-processing strategy can be adjusted. In the retrieval procedure, the query motion is translated into a feature sequence, which can be thought of as a progression of geometric constellations. The user-specified feature selection has to be encoded by a suitable fuzzy query, where the irrelevant features correspond to alternatives in the corresponding feature values. In the next step, the adaptive fuzzy hits are efficiently computed using the index. Finally, the hits may be post-processed by means of suitable ranking strategies. For further details we refer to [10, 25].

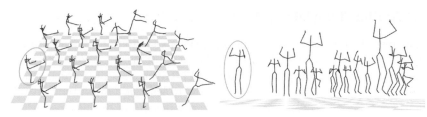

Fig. 20.19. Left: Selected frames from 19 adaptive fuzzy hits for a right foot kick. The query clip is highlighted. Query features: F_{17}, F_{18}, F_{20}, and F_{21}.; see Table 20.1. **Right:** Selected frames from 15 adaptive fuzzy hits for a jump query. Query features: F_3, F_4, F_{25}, and F_{26}.

We implemented our indexing and retrieval algorithms in Matlab 6 and tested them on a database comprising roughly 180 minutes of motion data drawn from the CMU database [7]. The indexing time for $f = 31$ features (similar to the one of Table 20.1) was roughly 6 minutes. The storage requirement was reduced from 370 MB (for the entire database) to 7.5 MB (for the index). The running time to process a query very much depends on the query length (the number of segments), the respective index, as well as the number of resulting hits. For example, Figure 20.19 (left) shows 19 adaptive fuzzy hits for a "kicking" motion (retrieval time: 5 ms), 13 of which are actual martial arts kicks. The remaining six motions (right hand side) are ballet moves containing a kicking component. A manual inspection of the database showed that there were no more than the 13 reported kicks in the database. Similarly, Figure 20.19 (right) shows the top 15 out of 133 hits for a very coarse adaptive fuzzy "jumping" query, which basically required the arms to move up above the shoulders and back down, while forcing the feet to lift off. The hits were ranked according to a simple strategy based on a comparison of segment lengths. This example demonstrates how such coarse queries can be applied to efficiently reduce the search space while retaining a superset of the desired hits.

One major limitation of this retrieval approach is that using all features at the same time in the retrieval process is far too restrictive – even in combination with fault tolerance strategies such as fuzzy or mismatch search – possibly leading to a large number of false negatives. Therefore, the user has to specify for each query a small subset of suitable features that reflect the characteristic properties of the respective query motion. Not only can this be a tedious manual process, but it also prohibits batch processing as needed in morphing and blending applications, where it may be required to identify similarities in a large database for many different motion clips without manual intervention. In the following, we introduce methods for automatic motion classification, annotation, and retrieval that overcome this limitation – however, at the expense of efficiency.

20.5.3 Motion Templates (MTs)

We now introduce a method for capturing the spatio-temporal characteristics of an entire motion class of logically related motions in a compact matrix representation called a *motion template* (MT). Given a set of training motions representing a motion class, we describe how to learn a motion template that explicitly encodes the consistent and the variable aspects of this class. Motion templates have a direct, semantic interpretation: an MT can easily be edited, manually constructed from scratch, combined with other MTs, extended, and restricted, thus providing a great deal of flexibility. One key property of MTs is that the variable aspects of a motion class can be automatically masked out in the comparison with unknown motion data. This strategy can also be viewed as an automatic way of selecting appropriate features for the comparison in a locally adaptive fashion.

In the following, we explain the main idea of motion templates and refer to [24] for details. Given a set of $\gamma \in \mathbb{N}$ example motion clips for a specific motion class, such as the four cartwheels shown in Figure 20.20, the goal is to automatically learn an MT representation that grasps the essence of the class. Based on a fixed set of f relational features, we start by computing the relational feature vectors for each of the γ motions. Denoting the length of a given motion by K, we think of the resulting sequence of feature vectors as a *feature matrix* $X \in \{0, 1\}^{f \times K}$ as shown in Figure 20.20, where, for the sake of clarity, we only display a subset comprising ten features from the feature set of Table 20.1. Now, we want to compute a semantically meaningful average over the γ feature matrices, which would simply be their arithmetic mean if all of the motions agreed in length and temporal structure. However, our matrices typically differ in length and reflect the temporal variations that were present in the original motions. This fact necessitates some kind of temporal alignment prior to averaging, which is done by an iterative, reference-based time warping procedure, see [24] for details. Once the matrices have the same length, their average is computed, yielding as output a matrix with f rows,

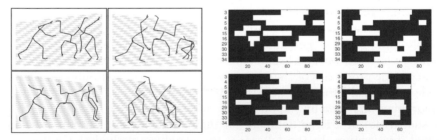

Fig. 20.20. Left: Selected frames from four different cartwheel motions. **Right:** Corresponding relational feature matrices for selected features. The columns represent time in frames, whereas the rows correspond to boolean features encoded as black (0) and white (1). They are numbered in accordance with the features defined in Table 20.1.

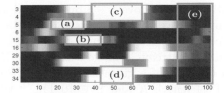 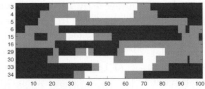

Fig. 20.21. Left: Class MT for "CartwheelLeft" based on $\gamma = 11$ training motions. The framed regions are discussed in Section 20.5.3. **Right:** Corresponding quantized class MT.

referred to as a *motion template*. The matrix entries are real values between zero and one. Figure 20.21 shows a motion template obtained from $\gamma = 11$ cartwheel motions (including the four cartwheels indicated by Figure 20.20), which constitutes a combined representation of all 11 input motions. An MT learned from training motions belonging to a specific motion class $\mathcal{C}$ is referred to as the *class template* for $\mathcal{C}$. Black/white regions in a class MT, see Figure 20.21, indicate periods in time (horizontal axis) where certain features (vertical axis) consistently assume the same values zero/one in all training motions, respectively. By contrast, different shades of gray indicate inconsistencies mainly resulting from variations in the training motions (and partly from inappropriate alignments).

To illustrate the power of the MT concept, which grasps the essence of a specific type of motion even in the presence of large variations, we discuss the class template for the class "CartwheelLeft", which consists of cartwheel motions starting with the left hand, see Figure 20.21. Considering the regions marked by boxes in Figure 20.21, the white region (a) reflects that during the initial phase of a cartwheel, the right hand moves to the top (feature F_5 in Table 20.1). Furthermore, region (b) shows that the right foot moves behind the left leg (F_{15}). This can also be observed in the first poses of Figure 20.20. Then, both hands are above the shoulders (F_3, F_4), as indicated by region (c), and the actor's body is upside down (F_{33}, F_{34}), see region (d) and the second poses in Figure 20.20. The landing phase, encoded in region (e), exhibits large variations between different realizations, leading to the gray/colored regions. Note that some actors lost their balance in this phase, resulting in rather chaotic movements, compare the third poses in Figure 20.20.

20.5.4 MT-based Motion Annotation and Retrieval

Given a class $\mathcal{C}$ of logically related motions, we have derived a class MT $X_{\mathcal{C}}$ that captures the consistent as well as the inconsistent aspects of all motions in $\mathcal{C}$. Our application of MTs to automatic annotation and retrieval are based on the following interpretation: the consistent aspects represent the class characteristics that are shared by all motions, whereas the inconsistent aspects represent the class variations that are due to different realizations. For a given class MT $X_{\mathcal{C}}$, we introduce a *quantized MT* by replacing each entry of $X_\mathcal{C}$ that is below δ by zero, each entry that is above $1 - \delta$ by one, and all remaining entries by 0.5. (In our experiments, we used the threshold $\delta = 0.1$.) Figure 20.21 (right) shows the quantized MT for the cartwheel class.

Now, let D be an unknown motion data stream. The goal is to identify subsegments of D that are similar to motions of a given class $\mathcal{C}$. Let $X \in \{0, 1, 0.5\}^{f \times K}$ be a quantized class MT of length K and $Y \in \{0, 1\}^{f \times L}$ the feature matrix of D of length L. We define for $k \in [1 : K]$ and $\ell \in [1 : L]$ a local cost measure $c^Q(k, \ell)$ between the k-th column $X(k)$ of X and the ℓ-th column $Y(\ell)$ of Y. Let $I(k) := \{i \in [1 : f] \mid X(k)_i \neq 0.5\}$, where $X(k)_i$ denotes a matrix entry of X for $k \in [1 : K]$, $i \in [1 : f]$. Then, if $|I(k)| > 0$, we set

$$c^Q(k, \ell) = \frac{1}{|I(k)|} \sum_{i \in I(k)} |X(k)_i - Y(\ell)_i|, \qquad (20.9)$$

otherwise we set $c^Q(k, \ell) = 0$. In other words, $c^Q(k, \ell)$ only accounts for the consistent entries of X with $X(k)_i \in \{0, 1\}$ and leaves the other entries unconsidered. Based on this local distance measure and a subsequence variant of dynamic time warping, one obtains a distance function $\Delta_\mathcal{C} : [1 : L] \to \mathbb{R} \cup \{\infty\}$ as described in [24] with the following interpretation: a small value $\Delta_\mathcal{C}(\ell)$ for some $\ell \in [1 : L]$ indicates the presence of a motion subsegment of D starting at a suitable frame $a_\ell < \ell$ and ending at frame ℓ that is similar to the motions in $\mathcal{C}$. Note that using the local cost function c^Q of (20.9) based on the quantized MT (instead of simply using the Manhattan distance) is of crucial importance, as illustrated by Figure 20.22.

In the annotation scenario, we are given an unknown motion data stream D for which the presence of certain motion classes $\mathcal{C}_1, \ldots, \mathcal{C}_P$ at certain times is to be detected. These motion classes are identified with their respective class MTs $X_1, \ldots, X_P$, which are assumed to have been precomputed from suitable training data. Now, the idea is to match the input motion D with each of the X_p, $p = 1, \ldots, P$, yielding the distance functions $\Delta_p := \Delta_{\mathcal{C}_p}$. Then, every local minimum of Δ_p close to zero indicates a motion subsegment of D that is

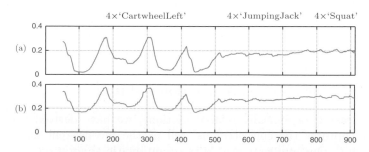

Fig. 20.22. (a) Distance function $\Delta_\mathcal{C}$ based on c^Q of (20.9) for the quantized class MT "CartwheelLeft" and a motion sequence D consisting of four cartwheels (reflected by the four local minima close to zero), four jumping jacks, and four squats. The sampling rate is 30 Hz. (b) Corresponding distance function based on the Manhattan distance without MT quantization, leading to a much poorer result.

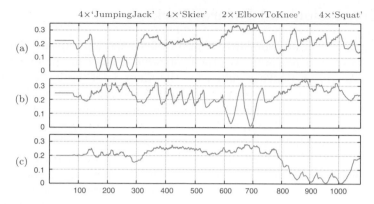

Fig. 20.23. Resulting distance functions for a 35-s gymnastics sequence (30 Hz) consisting of four jumping jacks, four repetitions of a skiing coordination exercise, two repetitions of an alternating elbow-to-knee motion, and four squats with respect to the quantized class MTs for **(a)** "JumpingJack", **(b)** "ElbowToKnee", and **(c)** "Squat".

similar to the motions in C_p. As an example, we consider the distance functions for a 35-s gymnastics motion sequence with respect to the motion classes C_1="JumpingJack", C_2="ElbowToKnee", and C_3="Squat", see Figure 20.23. For C_1, there are four local minima with a cost of nearly zero between frames 100 and 300, which exactly correspond to the four jumping jacks contained in D, see Figure 20.23(a). Note that the remaining portion of D is clearly separated by Δ_1, yielding a value far above 0.1. Analogously, the two local minima in Figure 20.23(b) correspond to the two repetitions of the elbow-to-knee exercise and the four local minima in Figure 20.23(c) correspond to the four squats.

Similarly, motion templates can be used for content-based motion retrieval, where the goal is to automatically extract all motion clips from a database that belong to a specified motion class C. To this end, we compute a distance function Δ_C with respect to the precomputed quantized class MT and the database documents. Then, each local minimum of Δ_C below some quality threshold $\tau > 0$ indicates a hit, see [24] for details. As it turns out, the MT-based retrieval strategy works with high precision and recall for complex motions (such as a cartwheel) even in the presence of significant variations, whereas for short motions with few characteristic aspects it may produce a large number of false positives. Another drawback of the proposed MT-based retrieval strategy is its computational complexity, which is linear in the size of the database. For the future, we plan to combine the MT-based retrieval strategy with index-based retrieval techniques as proposed in Section 20.5.2. First experiments have shown that the use of suitably defined keyframes is a promising concept to cut down the set of candidate motions in an index-based preprocessing step. Such a preselection may also be suitable to eliminate a large number of false positives.

20.6 Conclusion and Future Work

In this chapter, various similarity aspects of 3D motion capture data have been discussed and reviewed. We then introduced the concept of relational features that are particularly suited for the analysis of motion content and that facilitate logical (in contrast to numerical) comparison of motions. Once the features have been specified, they can be used for motion segmentation, efficient indexing, and fast content-based retrieval. As a further application, we introduced the concept of a motion template, which encodes the characteristic and the variable aspects of an entire motion class. By automatically masking out the variable aspects of a motion class in the annotation and retrieval process, logically related motions can be identified even in the presence of large variations and without any user intervention. We will investigate how to automatically learn characteristic keyframes in our template representation, which can then be used to cut down the search space efficiently. As a further promising application in the field of computer vision, we plan to use motion templates and related motion representations as a-priori knowledge to stabilize and control markerless tracking of human motions in video data, see also Chapters 11, 12, and 15.

Acknowledgment

We would like to thank Bernd Eberhardt from HDM school of media sciences (Stuttgart) for providing us with extensive motion capture material. Furthermore, we thank Michael Clausen for constructive and valuable comments.

References

1. Brand M. and Hertzmann A. Style machines. In *Proc. ACM SIGGRAPH 2000*, Computer Graphics Proc., pp. 183–192. ACM Press, 2000.
2. Bregler C. Learning and recognizing human dynamics in video sequences. In *Proc. CVPR 1997*, pp. 568, Washington, DC, USA, 1997. IEEE Computer Society.
3. Bruderlin A. and Williams L. Motion signal processing. In *Proc. ACM SIGGRAPH 1995*, Computer Graphics Proc., pp. 97–104. ACM Press, 1995.
4. Carlsson S. Combinatorial geometry for shape representation and indexing. In *Object Representation in Computer Vision*, pp. 53–78, 1996.
5. Carlsson S. Order structure, correspondence, and shape based categories. In *Shape, Contour and Grouping in Computer Vision*, pp. 58–71. Springer, 1999.
6. Clausen M. and Kurth F. A unified approach to content-based and fault tolerant music recognition. *IEEE Trans. Multimedia*, 6(5):717–731, 2004.
7. CMU. Carnegie-Mellon Mocap Database. `http://mocap.cs.cmu.edu`, March, 2007.
8. Davis J. W. and Gao H. An expressive three-mode principal components model of human action style. *Image Vision Comput.*, 21(11):1001–1016, 2003.

9. de Aguiar E., Theobalt C., and Seidel H.-P. Automatic learning of articulated skeletons from 3D marker trajectories. In *Proc. Intl. Symposium on Visual Computing (ISVC 2006)*, to appear, 2006.

10. Demuth B., Röder T., Müller M., and Eberhardt B. An information retrieval system for motion capture data. In *Proc. 28th European Conference on Information Retrieval (ECIR 2006)*, volume 3936 of *LNCS*, pp. 373–384. Springer, 2006.

11. Faugeras O. *Three-Dimensional Computer Vision: A Geometric Viewpoint*, Chapter 9, pp. 341–400. MIT Press, Cambridge, MA, 1993.

12. Forbes K. and Fiume E. An efficient search algorithm for motion data using weighted PCA. In *Proc. 2005 ACM SIGGRAPH/Eurographics Symposium on Computer Animation*, pp. 67–76. ACM Press, 2005.

13. Giese M. and Poggio T. Morphable models for the analysis and synthesis of complex motion patterns. *IJCV*, 38(1):59–73, 2000.

14. Green R. D. and Guan L. Quantifying and recognizing human movement patterns from monocular video images: Part I. *IEEE Trans. Circuits and Systems for Video Technology*, 14(2):179–190, February 2004.

15. Hsu E., Pulli K., and Popović J. Style translation for human motion. *ACM Trans. Graph.*, 24(3):1082–1089, 2005.

16. Keogh E. J., Palpanas T., Zordan V. B., Gunopulos D., and Cardle M. Indexing large human-motion databases. In *Proc. 30th VLDB Conf., Toronto*, pp. 780–791, 2004.

17. Kovar L. and Gleicher M. Flexible automatic motion blending with registration curves. In *Proc. 2003 ACM SIGGRAPH/Eurographics Symposium on Computer Animation*, pp. 214–224. Eurographics Association, 2003.

18. Kovar L. and Gleicher M. Automated extraction and parameterization of motions in large data sets. *ACM Trans. Graph.*, 23(3):559–568, 2004.

19. Kry P. G. and Pai D. K. Interaction capture and synthesis. *ACM Trans. Graph.*, 25(3):872–880, 2006.

20. Lafortune M. A., Lambert C., and Lake M. Skin marker displacement at the knee joint. In *Proc. 2nd North American Congress on Biomechanics*, Chicago, 1992.

21. Lee C.-S. and Elgammal A. Gait style and gait content: Bilinear models for gait recognition using gait re-sampling. In *Proc. IEEE Intl. Conf. Automatic Face and Gesture Recognition (FGR 2004)*, pp. 147–152. IEEE Computer Society, 2004.

22. Li Y., Wang T., and Shum H.-Y. Motion texture: a two-level statistical model for character motion synthesis. In *Proc. ACM SIGGRAPH 2002*, pp. 465–472. ACM Press, 2002.

23. Liu C. K., Hertzmann A., and Popović Z. Learning physics-based motion style with nonlinear inverse optimization. *ACM Trans. Graph.*, 24(3):1071–1081, 2005.

24. Müller M. and Röder T. Motion templates for automatic classification and retrieval of motion capture data. In *Proc. 2006 ACM SIGGRAPH/Eurographics Symposium on Computer Animation (SCA 2006)*. Eurographics Association, 2006.

25. Müller M., Röder T., and Clausen M. Efficient content-based retrieval of motion capture data. *ACM Trans. Graph.*, 24(3):677–685, 2005.

26. Müller M., Röder T., and Clausen M. Efficient indexing and retrieval of motion capture data based on adaptive segmentation. In *Proc. Fourth International Workshop on Content-Based Multimedia Indexing (CBMI)*, 2005.

27. Neff M. and Fiume E. Methods for exploring expressive stance. In *Proc. 2004 ACM SIGGRAPH/Eurographics Symposium on Computer Animation (SCA 2004)*, pp. 49–58. ACM Press, 2004.

28. Neff M. and Fiume E. AER: aesthetic exploration and refinement for expressive character animation. In *Proc. 2005 ACM SIGGRAPH/Eurographics Symposium on Computer Animation (SCA 2005)*, pp. 161–170. ACM Press, 2005.

29. O'Brien J. F., Bodenheimer R., Brostow G., and Hodgins J. K. Automatic joint parameter estimation from magnetic motion capture data. In *Graphics Interface*, pp. 53–60, 2000.

30. Pullen K. and Bregler C. Motion capture assisted animation: texturing and synthesis. In *Proc. SIGGRAPH 2002*, pp. 501–508. ACM Press, 2002.

31. Ren L., Patrick A., Efros A. A., Hodgins J. K., and Rehg J. M. A data-driven approach to quantifying natural human motion. *ACM Trans. Graph.*, 24(3):1090–1097, 2005.

32. Rose C., Cohen M. F., and Bodenheimer B. Verbs and adverbs: multidimensional motion interpolation. *IEEE Comput. Graph. Appl.*, 18(5):32–40, 1998.

33. Rosenhahn B., Kersting U. G., Smith A. W., Gurney J. K., Brox T., and Klette R. A system for marker-less human motion estimation. In *DAGM-Symposium*, pp. 230–237, 2005.

34. Sakamoto Y., Kuriyama S., and Kaneko T. Motion map: image-based retrieval and segmentation of motion data. In *Proc. 2004 ACM SIGGRAPH/Eurographics Symposium on Computer Animation*, pp. 259–266. ACM Press, 2004.

35. Sullivan J. and Carlsson S. Recognizing and tracking human action. In *Proc. ECCV '02, Part I*, pp. 629–644. Springer, 2002.

36. Troje N. F. Decomposing biological motion: A framework for analysis and synthesis of human gait patterns. *J. Vis.*, 2(5):371–387, 9 2002.

37. Unuma M., Anjyo K., and Takeuchi R. Fourier principles for emotion-based human figure animation. In *Proc. ACM SIGGRAPH 1995*, pp. 91–96. ACM Press, 1995.

38. Wikipedia. http://en.wikipedia.org/wiki/Motion_capture, March, 2007.

39. Witkin A. and Popović Z. Motion warping. In *Proc. ACM SIGGRAPH 95*, Computer Graphics Proc., pp. 105–108. ACM Press/ACM SIGGRAPH, 1995.

40. Witten I. H., Moffat A., and Bell T. C. *Managing Gigabytes*. Morgan Kaufmann Publishers, 1999.

41. Wu M.-Y., Chao S.-P., Yang S.-N., and Lin H.-C. Content-based retrieval for human motion data. In *16th IPPR Conf. on Computer Vision, Graphics, and Image Processing*, pp. 605–612, 2003.

42. Zatsiorsky V. M. *Kinematics of Human Motion*. Human Kinetics, 1998.

21

The Representation of Rigid Body Motions in the Conformal Model of Geometric Algebra

Leo Dorst

Intelligent Systems Laboratory
University of Amsterdam, The Netherlands

Summary. In geometric algebra, the conformal model provides a much more powerful framework to represent Euclidean motions than the customary "homogeneous coordinate" methods. It permits elementary universal operators that can act to displace not only points, but also lines and planes, and even circles and spheres. We briefly explain this model, focusing on its essential structure. We show its potential use for motion capture by presenting a closed form for the bivector logarithm of a rigid body motion, permitting interpolation of motions and combination of motion estimates.

21.1 Algebra for Geometry

In the view of Felix Klein (1849–1925), a geometry is characterized by its motions, and its primitive "objects" are merely identified parts of space that transform well under these motions. Geometric algebra elevates this abstract principle to a constructive representation.

In this chapter, we focus on Euclidean geometry. The motions that define this geometry are the *isometries*, the group of motions that preserve the Euclidean distance between points. This includes the "proper" motions of translation and rotation, and the "improper" reflections. We will construct a representation in which such motions are represented as *versors*, which are algebraically universal operators that can be applied to arbitrary elements of the geometry. To understand the power of such a representation compared to the more common matrices in homogeneous coordinates, we first give an abstract description of the representational principles.

Geometric algebra represents both "objects" (points, lines, planes, circles, spheres) and "operators" (rotations, translations, reflections) on the same representational level. All are basic elements of computation. An operator V is represented as a *versor*, which acts on another element X (object or operator) by a *sandwiching product* between V and its inverse V^{-1}:

$$X \mapsto (-1)^{\xi v} V X V^{-1},$$

507

B. Rosenhahn et al. (eds.), Human Motion – Understanding, Modelling, Capture, and Animation, 507–529.
© 2008 *Springer.*

where the sign depends on the "grades" ξ and v of X and V. Grades are representative dimensionalities, and somewhat technical; let us ignore the sign they contribute for now, and focus on the essential sandwiching structure of the operator representation.

The product involved in the two multiplications in the sandwiching product is the fundamental *geometric product* (denoted by a blank space). It is defined on any vector space $\mathbb{R}^n$ over the scalars $\mathbb{R}$ with bilinear form Q (which is a symmetrical scalar-valued function of two vectors defining the *metric* of the vector space). Such a space contains vectors (denoted in lower case) and a well-defined addition and scalar multiplication among them. The geometric product acts on it to produce a more extended structure from the vector space, its *geometric algebra* $\mathbb{R}_n$, of which the general elements are denoted by upper case. They include vectors and scalars (which can be more specifically denoted by lower case and lower case Greek, respectively). The geometric product is completely defined by the following properties:

1. It is *distributive over addition*: $A(B+C) = AB + AC$.
2. It is *associative*: $A(BC) = (AB)C$.
3. Any *vector squares to a scalar* determined by the metric: $a\,a = Q(a,a)$.
4. *Scalars commute*: $\alpha X = X\alpha$ (but general elements not necessarily).

These simple properties define a surprisingly powerful mathematical structure, according to some (such as [8]) quite sufficient to encode all of geometry (when augmented with a natural concept of differentiation).

The principle of the representation of geometry used in geometric algebra is to encode all geometric constructions as linear combinations of geometric products, and the geometric motions as versors. Then the *structure preservation* of the constructions under the motions, as demanded by Klein, is automatically guaranteed. For example, let $A \circ B$ be the intersection of two elements A and B (such as a line and a sphere), in geometric algebra encoded as a linear combination of geometric products, and V a motion versor. Applying the properties of the geometric product, we find:

$$V(A \circ B)V^{-1} = (V\,A\,V^{-1}) \circ (V\,B\,V^{-1}).$$

This is geometric structure-preservation of the intersection: "the motion of the intersection is identical to the intersection of the moved elements". This property is clearly automatic in the versor representation.

When we construct new objects from the basic vectors of the vector space $\mathbb{R}^n$ by such structurally preserved construction operators (for instance, a line as connecting two points, or as the intersection of two planes), we therefore get universal motions for all elements through the sandwiching operator with V. This makes for a very pleasant algebra of geometry, in which you never have to think about which 'method' to use to apply a motion to a newly constructed object. This is in stark contrast to the usual homogeneous coordinates, in which the 4×4 motion matrices for points and planes are different, and the direct motion of lines involves 6×6 matrices on Plücker coordinates which are even considered too advanced to teach in most introductions to computer graphics or robotics, despite the elementary importance of lines in geometric constructions. In homogeneous coordinates and the matrices based on them, the representation gets in the way of the geometry, sensible geometric operators become dependent on their arguments, and the software becomes more involved than it needs to be. By contrast, versors are universally applicable.

A reform program is currently underway to redo the classical geometries in geometric algebra. This involves defining an interface between the space to be represented, and a well-chosen geometric algebra (i.e., selecting a representational space and a metric Q). For motion capture, we are particularly interested in Euclidean geometry. The geometric algebra for Euclidean geometry is called the *conformal model* [6]. It represents Euclidean motions by versors.

21.2 The Structure of Geometric Algebra

Before we explain the specific use of the conformal model to encode Euclidean geometry, we need to give some more insight in the structure of geometric algebra in general. The explanation is necessarily brief, much more detail may be found in other sources such as [3].

21.2.1 Products

The geometric product is fundamental to geometric algebra, but several derived products are convenient to have. All these derived products can be defined in terms of linear combinations of geometric products, so that their structure is preserved when versors are applied to them.

1. The classical *inner product* of two vectors p and q is defined in terms of the bilinear form as $p \cdot q \equiv Q(p, q)$. It is found in geometric algebra as the symmetric part of the geometric product of p and q:

$$p \cdot q = \tfrac{1}{2}(p\,q + q\,p).$$

   This follows from developing the scalar $(p+q)\,(p+q)$ in terms of both the geometric product and the inner product: $(p+q)\,(p+q) = p\,p+p\,q+q\,p+q\,q$, and also $(p+q)\,(p+q) = (p+q) \cdot (p+q) = p \cdot p + 2p \cdot q + q \cdot q$; the identity follows.

   The inner product can be extended to other elements of the geometric algebra as a *contraction*, denoted $\rfloor$, see [3] and [4].
2. The anti-symmetric part of the geometric product of two vectors p and q is also a product in its own right, called the *outer product*:

$$p \wedge q = \tfrac{1}{2}(p\,q - q\,p).$$

It results in an element called a *2-blade* (which is neither a vector, nor a scalar). A 2-blade represents the 2-dimensional vector space spanned by p and q. This outer product is less familiar than it deserves to be: it spans subspaces as elements of computation. (It is the same outer product as used in Grassmann algebra, for readers familiar with that.) The outer product can be extended to more factors by associativity. An outer product of k terms is called a k-blade, and represents a k-dimensional subspace of the representative space.

3. You can *divide* by a vector, for the inverse of a vector v under the geometric product is $v^{-1} = v/(v \cdot v)$. This follows simply from $v\,v = v \cdot v + v \wedge v = v \cdot v$ (due to the antisymmetry of the outer product) and the commutation of the scalar $v \cdot v$.

Division extends to higher order elements like blades and versors. Elements with norm 0 (such as the null vectors we encounter later) do not have a division.

21.2.2 Objects and Relationships

The elementary objects of geometric algebra are the k-dimensional linear subspaces of the original vector space. These become elements of computation through the outer product, which generates them as k-blades. When using geometric algebra in an application, one chooses a specific *model*, i.e., a relationship between aspects of reality and elements of the geometric algebra. The proper subspaces of the representational algebra can then actually represent more involved elements of reality, such as offset subspaces, or general circles. We enumerate some representational principles for the relationships of such objects.

1. The points of the real space are modeled by vectors in the representation space.
2. In the *direct representation* of a geometric object by a k-blade A, solving the equation $x \wedge A = 0$ for the vector x results in the points contained in the object.
3. In the *dual representation* of a geometric object by a k-blade D, solving the equation $x \rfloor D = 0$ for the vector x results in the points contained in the object.
4. The relationship between the two representatives of the object is through *dualization*, denoted as $D = A^*$. In geometric algebra, dualization is achieved through division by the highest order blade in the representation space. Such a blade is called a *pseudoscalar*, it is of grade n in an n-dimensional space and denoted I_n.
5. The direct representation specifies an object by 'containment' in a representative blade A, the dual representation specifies an object by giving a perpendicular blade D. Dualization is therefore like "taking the orthogonal complement". The direct representation does not require a metric and

is mathematically more fundamental, but in many metric models the dual representation is more compact and convenient.

6. The representation of the object $A \cap B$ that is the intersection of two objects directly represented as the blades A and B can be computed through

$$(A \cap B)^* = B^* \wedge A^*.$$

The dual representation of this meet is therefore the outer product of the duals. (There is some small print here, the blades A and B have to be "in general position", for details see [3].)

7. The element that intersects the elements A and B perpendicularly has as direct representation $B^* \wedge A^*$. It is called the "plunge" of A and B, see [3].

21.2.3 Versors are Orthogonal Transformations

We will represent operators by versors; they produce linear transformations, though not of the most general kind.

1. The versor product implements an *orthogonal transformation*, for the vector mapping $x \mapsto (-1)^v V x V^{-1}$ preserves the inner product of vectors. This is easily shown:

$$x \cdot y = V V^{-1} (x \cdot y) = V (x \cdot y) V^{-1} = \tfrac{1}{2} V (x y + y x) V^{-1} = (V x V^{-1}) \cdot (V y V^{-1}).$$

2. Versors representing consecutive transformations are formed by the geometric product of the versors of the constituent transformations. This follows from

$$(-1)^{\xi \omega} W \left((-1)^{\xi v} V X V^{-1} \right) W^{-1} = (-1)^{\xi(\omega + v)} (W V) X (W V)^{-1},$$

and the fact that the (even or odd) parity of the grades ω and v of W and V is additive.

3. Any versor can be rescaled to a unit versor (for which $V \rfloor V = 1$). The basic versor transformation is provided by a unit vector. A general unit versor can be written as the geometric product of unit vectors (though this factorization is usually not unique).
   The application of a unit vector versor u to a vector x gives:

$$x \mapsto -u x u^{-1} = -(-x u + 2 (x \cdot u)) u^{-1} = x - 2 (x \cdot u) u^{-1}.$$

Realizing that $u^{-1} = u/(u \cdot u) = u$, you see that this reverses the u-component of x. It can therefore be interpreted geometrically as the reflection of x in the plane of which u is the normal vector. Then the geometrical significance of writing a general versor as the product of such operations is that any orthogonal transformation can be written as multiple planar reflections.

4. Proper orthogonal transformations are represented by versors of even parity, improper orthogonal transformations by versors of odd parity.
   This can be derived by defining the *determinant* of a linear transformation as the change of scale of the pseudoscalar (which is the highest order blade) of the algebra. Every vector x of the space is in the subspace spanned by the pseudoscalar I_n, and this can be shown to be equivalent to the commutation statement $x\,I_n = (-1)^{n+1}I_n\,x$. Then transforming I_n by a versor V composed of k vector factors gives: $(-1)^{nk}\,V\,I_n\,V^{-1} = (-1)^k I_n$, so that the determinant of this orthogonal transformation equals $(-1)^k$, and the result follows.

5. *Unit versors are the exponentials of bivectors* (i.e., general linear elements of grade 2). This is not quite true in general (see [10], [3]) , but the mathematical subtleties do not occur in the spaces we will encounter in this chapter (which have a Euclidean metric or a Minkowski metric, and are of sufficiently high dimension). The fact that the bivectors form a linear space makes this logarithmic representation of operators attractive for interpolation and estimation.

21.3 The Conformal Model of Euclidean Geometry

We now move from the general properties of geometric algebras to a specific model, suitable for computations in Euclidean geometry.

21.3.1 Specification

Euclidean motions on points in an n-dimensional space $\mathbb{E}^n$ have two important properties:

- They preserve Euclidean distances between points.
- They preserve the point at infinity.

Both these properties of the motions should be built into their representation, and into the representation of elements they act on. The conformal model does this by selecting a specific representational space of $n + 2$ dimensions, and relating its vectors and metric to Euclidean geometry.

1. The *weighted points* of n-dimensional Euclidean space are represented in an $(n + 2)$-dimensional representation space as *vectors*.
2. The *point at infinity* is represented by a *special vector* ∞ in this representational space.
3. This *inner product* of the representational space is chosen to represent the *squared Euclidean distance* by setting:

$$\frac{p \cdot q}{(-\infty \cdot p)\,(-\infty \cdot q)} = -\tfrac{1}{2}\,d_E^2(p, q)$$

Here $-\infty \cdot p$ is the weight of the point represented by the vector p, for a normalized point this equals 1. The scaling ensures that Euclidean distance is a geometrical property defined among normalized points.

4. The versors of the geometric algebra of the representational space represent orthogonal transformations of that space. These preserve the inner product, and therefore Euclidean distances in $\mathbb{E}^n$; therefore they can represent Euclidean motions in $\mathbb{E}^n$. But they also need to preserve the point at infinity. Algebraically, this implies that the unit vectors that can be used as factors of the versors should satisfy: $\infty = -n \infty n^{-1}$, and therefore $\infty \cdot n = 0$. The geometrical meaning is that Euclidean motions can be represented as successive reflections in general planes.

21.3.2 The Vectors of the Conformal Model

The vectors of the representational space that represent points are somewhat unusual: they square to zero, which makes them *null vectors*. This is clear from the relationship between the Euclidean metric and the inner product: a Euclidean point should have distance zero to itself. The vector ∞ representing the point at infinity is also a null vector. The existence of null vectors implies that the representational space is not a Euclidean space; instead, it is an $(n + 2)$-dimensional Minkowski space, denoted $\mathbb{R}^{n+1,1}$. One can define an orthonormal basis for it with $(n + 1)$ basis vectors squaring to $+1$, and one basis vector squaring to -1.

Being a vector space, the representational space also contains all linear combinations of null vectors. Vectors of the form $c - \frac{1}{2}\rho^2\infty$, with c a null vector, are of specific interest. Let us "probe" them with a unit point, to see what they dually represent, i.e. let us solve $x \cdot (c - \frac{1}{2}\rho^2\infty) = 0$ for a null vector x. The equation can be rewritten as $d_E^2(x, c) = \rho^2$. This is the equation for x to lie on a sphere, so the vector $\sigma = c - \frac{1}{2}\rho^2\infty$ is the dual representation of that sphere. The vector σ is not a null vector: $\sigma^2 = \rho^2$, as you may verify. Imaginary dual spheres are represented for $\rho^2 < 0$. A point is then a dual sphere of zero radius.

The $(n + 2)$-dimensional representational space also contains purely Euclidean vectors, for the n-dimensional Euclidean space is a subspace of the representational space. We will denote those purely Euclidean vectors in bold font. Such a vector $\mathbf{n}$ dually represents a plane through the origin (i.e., it is what we would classically call a normal vector). General planes (not necessarily passing through the origin) are dually represented as the difference of two normalized null vectors: the midplane between points a and b satisfies $d_E(x, a) = d_E(x, b)$, which gives $x \cdot a = x \cdot b$, so that $x \cdot (a - b) = 0$ and therefore $(a - b)$ dually represents the midplane. Note that $\infty \cdot (a - b) = 0$, so that such a plane indeed contains the point at infinity.

21.3.3 Introducing a Basis

The operations of geometric algebra can be specified without reference to a basis. Yet you may prefer to have a standard basis to see how a point is represented explicitly, in order to understand how one might implement the operations, and to relate the conformal model to more classical approaches. This requires appointing an arbitrary point as the origin, and denoting locations by direction vectors relative to that. Let us denote the vector representing the arbitrary origin point by o, a basis for the Euclidean 3D direction by $\{\mathbf{e}_1, \mathbf{e}_2, \mathbf{e}_3\}$, and of course we have the point at infinity denoted by the vector ∞. An arbitrary vector of the representational space is a general linear combination of these components, but a null vector representing a point at location $\mathbf{x}$ should have the specific form:

$$x = \alpha \left(o + \mathbf{x} + \tfrac{1}{2}\,\mathbf{x}^2\,\infty \right). \tag{21.1}$$

Please verify that $x^2 = 0$.

In this expression, the purely Euclidean vector $\mathbf{x}$ is the location of the point x relative to o, and the scalar constant α is its weight. The point x at location $\xi_1 \mathbf{e}_1 + \xi_2 \mathbf{e}_2 + \xi_3 \mathbf{e}_3$ therefore has the coordinates $\left(\alpha, \alpha\,\xi_1, \alpha\,\xi_2, \alpha\,\xi_3, \tfrac{1}{2}\alpha(\xi_1^2 + \xi_2^2 + \xi_3^2) \right)$ on the basis $\{o, \mathbf{e}_1, \mathbf{e}_2, \mathbf{e}_3, \infty\}$. The first four components are reminiscent of the usual homogeneous coordinates, where an extra o-dimension is added to n-dimensional space to represent the point at the origin. The extra ∞-dimension of the conformal model provides the basis vectors have a very specific inner product, defining the metric through the inner product relationships in the following table:

	o	$\mathbf{e}_1$	$\mathbf{e}_2$	$\mathbf{e}_3$	∞
o	0	0	0	0	-1
$\mathbf{e}_1$	0	1	0	0	0
$\mathbf{e}_2$	0	0	1	0	0
$\mathbf{e}_3$	0	0	0	1	0
∞	-1	0	0	0	0

Let us verify that the inner product of two normalized points indeed gives their squared Euclidean distance:

$$
\begin{aligned}
p \cdot q &= (o + \mathbf{p} + \tfrac{1}{2}\mathbf{p}^2\infty) \cdot (o + \mathbf{q} + \tfrac{1}{2}\mathbf{q}^2\infty) \\
&= -\tfrac{1}{2}\mathbf{q}^2 + \mathbf{p} \cdot \mathbf{q} - \tfrac{1}{2}\mathbf{p}^2 \\
&= -\tfrac{1}{2}(\mathbf{q} - \mathbf{p})^2.
\end{aligned}
$$

This shows explicitly how the conformal model works, and why the strange metric is essential. That metric gives the conformal model quantitative properties that the homogeneous coordinates lack, and that are precisely right to permit its encoding Euclidean motions as orthogonal transformations, with manifest structure preservation when represented in versor form.

21.3.4 Elementary Euclidean Objects as Blades

In the conformal model, where Euclidean points are represented as representative vectors, the outer product has a pleasing semantics. We only state the results; a detailed derivation may be found in [3].

1. A *line* is constructed from two points represented by vectors a and b as the 3-blade $L = a \wedge b \wedge \infty$. This represents the line directly, in the sense that solving the equation $x \wedge L = 0$ for vectors x representing Euclidean points results exactly in points on the line through a and b. (Note that the properties of associativity and anti-symmetry imply that the point ∞ is on the line: it passes through infinity.) Yet L has more properties than merely being a point set: it represents a weighted line, with a specific orientation. This permits encoding its velocity or density, and other such physical properties, in its algebraic structure. Alternatively, a line through a with a direction vector $\mathbf{u}$ may be encoded as $L = a \wedge \mathbf{u} \wedge \infty$, which is essentially a **plunge** into the dual plane $\mathbf{u}$.
2. A *plane* is constructed from three points a, b and c as the 4-blade $\Pi = a \wedge b \wedge c \wedge \infty$. Besides its spatial attitude and location, such a plane has an orientation, and a weight that can be used to denote density. Alternatively, a plane in dual representation can be constructed from a normal vector $\mathbf{n}$ and a point p on it as $\pi = p \rfloor (\mathbf{n} \wedge \infty)$.
3. A *flat point* is a 2-blade of the form $a \wedge \infty$. It is the kind of element in common between a line and a plane.
4. A *circle* through the points a, b, c is represented by the 3-blade $K = a \wedge b \wedge c$. This circle is oriented, and has a weight measure proportional to the area of the triangle formed by a, b, c. In the conformal model, a line is merely a circle that passes through infinity.
5. A *sphere* through the points a, b, c, d is represented by the 4-blade $\Sigma = a \wedge b \wedge c \wedge d$. This sphere is oriented, and has a weight measure proportional to the volume of the tetrahedron formed by a, b, c, d. Alternatively, a sphere in dual representation can be constructed from its center q and a point p on it as $\sigma = p \rfloor (q \wedge \infty)$.
6. A *point pair* is a 2-blade $a \wedge b$. Just like a circle is "a sphere on a plane", a point pair is "a sphere on a line".

You should realize that all of these are basic elements of computation, and that they can be combined using the products of geometric algebra to produce new elements. For instance, an alternative characterization of the circle is by writing its dual representation as $\kappa = \sigma \wedge \pi$, where σ is a dual sphere and π is a dual plane. Or one could specify the circle that orthogonally intersects three dual spheres $\sigma_1, \sigma_2, \sigma_3$ as $K = \sigma_3 \wedge \sigma_2 \wedge \sigma_1$, a **plunge**. Even tangent elements can be made in this way: the intersection of two touching circles results in a blade that represents their common tangent vector. A direct representation of the tangent vector at a point a in direction $\mathbf{n}$ is $a \rfloor (a \wedge \mathbf{n} \wedge \infty)$. For more details, see [3].

21.4 Euclidean Motions as Versors

The planar reflection is the basic Euclidean transformation, all others can be constructed as multiple planar reflections. A translation can be made as a reflection in two parallel planes, a rotation by a reflection in two intersecting planes (where the intersection line is the rotation axis). You see that *proper* Euclidean motions are represented by an even number of reflections, and therefore by a versor consisting of an even number of vector terms (appropriately called an *even versor*). Improper motions involve odd versors. All preserve Euclidean distances, though not necessarily handedness.

21.4.1 Planar Reflections

Reflections in general planes are the geometrical basis of our representation of general motions. We therefore need to find out how to represent a general plane. To be specific, let us introduce an origin o, and aim to encode the plane with unit normal vector $\mathbf{n}$ at distance δ from this origin. In the usual Euclidean geometry the location $\mathbf{x}$ of a general point on this plane would satisfy the "Hesse normal equation"

$$\mathbf{x} \cdot \mathbf{n} = \delta.$$

Using the conformal model, this can be written in terms of the conformal point x as the homogeneous equation:

$$x \cdot (\mathbf{n} + \delta\infty) = 0.$$

Therefore the vector $\pi = \alpha(\mathbf{n} + \delta\infty)$, or any multiple of it, is the dual representation π of the plane Π. The Euclidean parameters of the plane can be retrieved from its dual representation π by the equations:

$$\alpha\,\delta = -o \cdot \pi, \quad \text{and} \quad \alpha\,\mathbf{n} = o\rfloor(\pi \wedge \infty).$$

(The latter requires a property of the contraction $\rfloor$ which reads in general $a\rfloor(b \wedge c) = (a \cdot b)\,c - b\,(a \cdot c)$, see [3].) If required, the scaling factor α can be eliminated by the demand for unit norm on $\mathbf{n}$. These equations are algebraic expressions that amount to straightforward coordinate manipulation in an implementation.

The sandwiching product with π produces a reflection of a point x. This may be verified by using the commutation properties of the various vector components involved. Let us work out the action on the components (using a unit plane since α is eliminated in the sandwiching anyway):

$$-\pi\,o\,\pi^{-1} = -(\mathbf{n} + \delta\infty)\,o\,(\mathbf{n} + \delta\infty)$$
$$= o - \delta\mathbf{n}\,(\infty\,o + o\,\infty) - \delta^2\,\infty\,o\,\infty$$
$$= o + 2\delta\,\mathbf{n} + 2\delta^2\,\infty$$

$$-\pi\,\mathbf{x}\,\pi^{-1} = -(\mathbf{n} + \delta\infty)\,\mathbf{x}\,(\mathbf{n} + \delta\infty)$$
$$= -\mathbf{n}\,\mathbf{x}\,\mathbf{n} - 2\delta\infty\,(\mathbf{x}\cdot\mathbf{n})$$
$$= \mathbf{x} - 2(\mathbf{n}\cdot\mathbf{x})\,\mathbf{n} - 2\delta\infty\,(\mathbf{x}\cdot\mathbf{n})$$
$$-\pi\,\infty\,\pi^{-1} = -(\mathbf{n} + \delta\infty)\infty(\mathbf{n} + \delta\infty) = \infty.$$

The invariance of ∞ confirms that planar reflection is a Euclidean transformation. Putting these elements together, we find that the result is the vector representing the point at location:

$$\mathbf{x} + 2(\delta - \mathbf{n}\cdot\mathbf{x})\,\mathbf{n},$$

which is the correct answer for a reflection in the dual plane π. If you follow the computations above in detail, you find that the geometry of reflection is completely represented by the algebra of commutation relationships. But you would not compute all this in practice; the above is just a proof of the correctness of the simple conformal model statement

$$x \;\mapsto\; -\pi\,x\,\pi^{-1}$$

to specify this operation in a program based on the conformal model. You learn to read this as "the dual plane π acting on the point x". Because of the structure preservation of the outer product, the equation is simply extended to the reflection formula for a k-blade X (which can represent a line, plane, circle, and so on):

$$X \;\mapsto\; (-1)^k\,\pi\,X\,\pi^{-1}.$$

That is how easy general planar reflections are in the conformal model.

21.4.2 Translations

A translation over $\mathbf{t}$ can be made by a successive reflection in two planes π_1 and π_2 with well chosen normal (parallel to $\mathbf{t}$) and separation (proportional to $\|\mathbf{t}\|/2$). That successive reflection leads to an element that is the composite operation. The composition product for the versors π_1 and π_2 is the geometric product $\pi_2\,\pi_1$ (in that order!):

$$\left(-\pi_2\left(-\pi_1\,x\,\pi_1^{-1}\right)\pi_2^{-1}\right) = (\pi_2\,\pi_1)\,x\,(\pi_2\,\pi_1)^{-1}.$$

Therefore, the translation versor over $\mathbf{t}$ can be made by any multiple of:

$$(\mathbf{t} + \tfrac{1}{2}\mathbf{t}^2\infty)\,\mathbf{t} = \mathbf{t}^2(1 - \tfrac{1}{2}\mathbf{t}\infty).$$

We define the *translation versor*

$$T_{\mathbf{t}} \equiv (1 - \mathbf{t}\,\infty/2) = e^{-\mathbf{t}\infty/2}.$$

The final expression in terms of an exponential prepares later usage; it is a bit of overkill here, since the power series defining the exponential has all terms beyond the second equal to zero (because ∞ is a null vector).

Applying this translation versor $T_{\mathbf{t}}$ to the point at the (arbitrary) origin o gives a familiar result:

$$
\begin{aligned}
T_{\mathbf{t}}\, o\, T_{\mathbf{t}}^{-1} &= (1 - \mathbf{t}\infty/2)\, o\, (1 + \mathbf{t}\infty/2) \\
&= o - \mathbf{t}\infty o/2 + o\mathbf{t}\infty/2 - \mathbf{t}\infty o\mathbf{t}\infty/4 \\
&= o + \mathbf{t} + \tfrac{1}{2}\mathbf{t}^2\,\infty,
\end{aligned}
$$

in agreement with our representation (21.1) of a point at location $\mathbf{t}$ relative to the origin.

21.4.3 Rotations at the Origin

A rotation over ϕ around an axis with direction $\mathbf{a}$ through the origin can be made by reflection in two planes containing the normalized axis $L = o \wedge \mathbf{a} \wedge \infty$, and making an angle $\phi/2$ with each other, see Figure 21.1. Such planes through the origin can dually represented by purely Euclidean unit vectors $\mathbf{u}$ and $\mathbf{v}$, so the total rotation versor is $\mathbf{v}\,\mathbf{u}$. We need to establish the relationship to the desired rotation parameters.

The 2-blade $\mathbf{u} \wedge \mathbf{v}$ is the purely Euclidean attitude of the rotation plane, and as a dual meet is proportional to the dual axis L^* by $\mathbf{u}\wedge\mathbf{v} = -\sin(\phi/2)\,L^*$. We can use this 2-blade to encode the rotation, for the geometric product of the two unit vectors $\mathbf{v}$ and $\mathbf{u}$ can be written as:

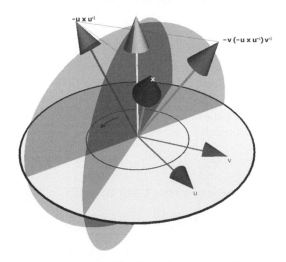

Fig. 21.1. A rotation over ϕ as a double reflection in the planes with normals $\mathbf{u}$ and $\mathbf{v}$, making a relative angle $\phi/2$. The rotation plane is the common plane of $\mathbf{u}$ and $\mathbf{v}$; the rotation axis is the dual of the plane.

$$\mathbf{v}\,\mathbf{u} = \mathbf{v} \cdot \mathbf{u} + \mathbf{v} \wedge \mathbf{u} = \cos(\phi/2) + L^{*}\,\sin(\phi/2)$$

For $n = 3$, the square $(L^{*})^{2}$ is evaluated as:

$$(L^{*})^{2} = L\,I_{n+2}^{-1}\,L\,I_{n+2}^{-1} = (-1)^{\mathrm{grade}(L)(n+3)}(I_{n+2}^{-1})^{2}L^{2} = -L^{2} = -1.$$

Therefore L^{*} is a normalized 3-component element (parametrized by the axis vector $\mathbf{a}$), squaring to -1. Algebraically, this makes the element $\mathbf{v}\,\mathbf{u}$ isomorphic to a unit quaternion representation, even though it is completely expressed in terms of real geometric elements. Because of the negative square of L^{*}, we can write this rotation operator in terms of an exponential (defined in terms of the usual power series):

$$\mathbf{v}\,\mathbf{u} = e^{L^{*}\phi/2}.$$

The knowledge of the unit rotation axis L and the rotation angle ϕ therefore immediately produces the rotation operator in versor form.

21.4.4 General Rotations

A general rotation takes place around an arbitrary axis. We can make such a rotation by first performing an inverse translation on the rotation argument, then rotating around the origin, and moving the result back to the original location of the rotation axis. This produces the operator $T_{\mathbf{t}}\,\exp(L^{*}\phi/2)\,T_{\mathbf{t}}^{-1}$. By developing the exponent in this expression in a power series, and rearranging, one can simply demonstrate that this is equivalent to $\exp(T_{\mathbf{t}}(L^{*}\phi/2)T_{\mathbf{t}}^{-1})$, the exponential of the translated dual axis with its rotation angle. This is effectively due to the structure preserving properties of the versor representation. As a consequence, a general unit axis L and angle ϕ can be used to produce the corresponding rotation operator immediately:

3D rotation around line L over angle ϕ: $R = e^{L^{*}\phi/2}.$

That capability of the conformal model greatly extends the geometric power of quaternion-like expressions and techniques.

21.4.5 General Rigid Body Motions

When we focus on the proper Euclidean motions, we can represent those by an even number of reflections and therefore represent them algebraically by an even number of the unit vectors representing Euclidean planes. These are operators with useful properties of their own, and they are called *rotors*.[1] Rotors

---

[1] If you use all available vectors in the conformal model as versors, you would also get reflections in spheres, and reflections in imaginary spheres. Not all those are rotors, which should moreover have their inverse equal to their reverse, see [3] or [10] for details.

can be used to transform general elements X through $X \mapsto V X V^{-1}$. They preserve handedness, and are therefore abstract rotations in the representative space. But more importantly, these transformations are "continuously connected to the identity", which means that we can do them in arbitrarily small amounts. This makes them true *motions*, for we can do them "a little at a time".

Their exponential representation shows this most clearly. By choosing a small bivector δB for the exponent, we compute:

$$X \mapsto e^{-\delta B/2} \, X \, e^{\delta B/2} \approx (1 - \delta B/2) X (1 + \delta B/2) = X + \tfrac{1}{2}\delta\,(X\,B - B\,X),$$

which is a small additive disturbance of X. The second term involves the commutator product of X with the bivector B, which can be shown to be grade-preserving. (It also explains why small motions form a Lie algebra, see [2]). We can even interpolate motions represented exponentially, for by dividing the exponent we obtain the Nth root of a rotor V as $r = V^{1/N}$. Then the total motion generated by V could be done in N steps by performing r a total of N times.

So we have a desire to *express a rotor as the exponential of a bivector*. In the Euclidean and Minkowski spaces we consider in this chapter, that can be done for almost all rotors [10]. (We will comment on the single geometrically relevant exception below.)

21.4.6 Screws

When we treated the translations and rotations, we showed that the rotors of each can be represented as the exponential of a 2-blade, a special case of a bivector. When the motions are composed, their rotors multiply. The result can always still be written as the exponential of a bivector, but this bivector is not simply found as the sum of the 2-blades of the contributing rotors:

$$e^{\mathbf{B}} \, e^{\mathbf{A}} \neq e^{\mathbf{A}+\mathbf{B}} \quad \text{in general.}$$

The reason is simply that the right-hand side is symmetric in $\mathbf{A}$ and $\mathbf{B}$, whereas the left-hand side is not. However, it can be shown that we are indeed allowed to add the exponents if the two rotors commute, which happens precisely when their bivectors commute.

There is a way of constructing general rigid body motions with commuting translations and rotations. Geometrically, one follows a rotation around a certain axis L with a translation of $\mathbf{w}$ along that axis. The bivectors of the corresponding rotors are $-\mathbf{w}\,\infty/2$ and $L^*\phi/2$. If rotors are to commute, so should their bivectors, and we must have $(\mathbf{w}\,\infty)\,L^* = L^*\,(\mathbf{w}\,\infty)$. That implies $\mathbf{w} \wedge L = 0$, confirming that $\mathbf{w}$ is parallel to the axis. With this condition, the full rigid body motion rotor is:

$$e^{-\mathbf{w}\,\infty/2} \, e^{L^*\phi/2} = e^{-\mathbf{w}\,\infty/2 + L^*\phi/2}.$$

Fig. 21.2. A screw motion and its parameters $\mathbf{I}\phi$ (the rotation plane and angle), $\mathbf{v}$ (the location of the screw axis in the plane) and $\mathbf{w}$ (the translation along the screw axis).

If we divide the exponent by N, we have the rotor r for the Nth root of the motion. The total motion can be performed in small steps of this rotor r, and they trace out a screw, see Figure 21.2. That is why this is called the *screw representation* of a rigid body motion. It is important because of its simple form: it relates a motion versor to a characterizing bivector through its exponential. This can be reversed: the bivector of the screw is then the *logarithm* of the rigid body motion.

21.4.7 General Rigid Body Motions as Screws

The screw characterization of a rigid body motion has been known since the 19th century, and survives as a specialized representation for time-critical robotics dynamics algorithms. In computer vision and computer graphics, it is more common to have a general motion given as "a rotation in the origin followed by a translation". Perhaps the main reason is that each of those operations is clearly recognizable in the homogeneous coordinate matrix $[\![M]\!]$ that represents such a rigid body motion (e.g., in OpenGL):

$$[\![M]\!] = \left[\!\!\left[\begin{array}{cc} [\![R]\!] & [\![t]\!] \\ [\![0]\!]^T & 1 \end{array} \right]\!\!\right] \tag{21.2}$$

Though this representation is straightforward for the composition of motions, it is awkward in almost all other uses. The main reason is that it represents the motion as a discrete pose transformation rather than in a continuous parametrization that can be naturally interpolated.

To comply with this common usage, let us suppose we have been given the rotation plane and angle $\mathbf{I}\phi$, and a translation $\mathbf{t}$, together specifying the rigid body motion versor as:

$$V = T_\mathbf{t}\, R_{\mathbf{I}\phi} = e^{-\mathbf{t}\infty/2} e^{-\mathbf{I}\phi/2}.$$

If we want to interpolate this motion, we need to write it as a pure exponential, so that we can interpolate its bivector logarithm. Since the translation rotor and rotation rotor do not commute, this is not trivial. We take the detour of rewriting the versor in its screw representation, to revert to the results of the previous section. Our technique of rewriting follows [7].

The rotation of the screw representing this motion will be around an axis perpendicular to the $\mathbf{I}$-plane, but translated to a location $\mathbf{v}$ yet to be determined. We may of course take $\mathbf{v}$ to be in the $\mathbf{I}$-plane, without loss of generality: so we demand $\mathbf{v} \wedge \mathbf{I} = 0$, which implies $\mathbf{v}\,\mathbf{I} = -\mathbf{I}\mathbf{v}$. The translation $\mathbf{w}$ of the screw needs to be perpendicular to the $\mathbf{I}$-plane, so that $\mathbf{w}\,\mathbf{I} = \mathbf{I}\,\mathbf{w}$. Under these conditions, we need to solve for $\mathbf{v}$ and $\mathbf{w}$ in:

$$T_\mathbf{t}\, R_{\mathbf{I}\phi} = T_\mathbf{w}\left(T_\mathbf{v}\, R_{\mathbf{I}\phi} T_{-\mathbf{v}}\right). \tag{21.3}$$

Since the only translation perpendicular to $\mathbf{I}$ is performed by $\mathbf{w}$, this must be the component of the translation vector $\mathbf{t}$ perpendicular to $\mathbf{I}$. This gives

$$\mathbf{w} = (\mathbf{t} \wedge \mathbf{I})/\mathbf{I}.$$

That leaves the other component of the translation $\mathbf{u} \equiv (\mathbf{t}\rfloor\mathbf{I})/\mathbf{I}$, which is the projection onto $\mathbf{I}$. Multiplying both sides of Equation (21.3) by $T_{-\mathbf{w}}$, we should solve the equation

$$T_\mathbf{u}\, R_{\mathbf{I}\phi} = T_\mathbf{v}\, R_{\mathbf{I}\phi}\, T_{-\mathbf{v}}, \tag{21.4}$$

with all quantities residing in the $\mathbf{I}$-plane. We can swap the rightmost translation with $R_{\mathbf{I}\phi}$, according to the following swapping rule

$$R_{\mathbf{I}\phi}\, T_{-\mathbf{v}} = R_{\mathbf{I}\phi}\left(1 + \mathbf{v}\infty/2\right) = \left(1 + R_{\mathbf{I}\phi}\mathbf{v}R_{\mathbf{I}\phi}^{-1}\infty/2\right) R_{\mathbf{I}\phi} = T_{-R_{\mathbf{I}\phi}\mathbf{v}R_{\mathbf{I}\phi}^{-1}}\, R_{\mathbf{I}\phi}.$$

We also observe that since $\mathbf{v}$ is a vector in the plane of $R_{\mathbf{I}\phi}$, we can represent the versor product with $R_{\mathbf{I}\phi}$ in the one-sided form:

$$R_{\mathbf{I}\phi}\, \mathbf{v}\, R_{\mathbf{I}\phi}^{-1} = R_{\mathbf{I}\phi}^2\, \mathbf{v}.$$

Collating these results, Equation (21.4) can be rewritten as

$$T_\mathbf{u}\, R_{\mathbf{I}\phi} = T_{(1-R_{\mathbf{I}\phi}^2)\mathbf{v}}\, R_{\mathbf{I}\phi}.$$

When $R_{\mathbf{I}\phi}^2$ is not equal to 1, the solution for $\mathbf{v}$ is therefore

$$\mathbf{v} = \left(1 - R_{\mathbf{I}\phi}^2\right)^{-1}\mathbf{u} = \left(1 - R_{\mathbf{I}\phi}^2\right)^{-1}(\mathbf{t}\rfloor\mathbf{I})/\mathbf{I}. \tag{21.5}$$

The feasibility of this solution proves the possibility of the screw decomposition of the rigid body motion, and also computes its parameters in terms of $\mathbf{t}$ and $\mathbf{I}$. But we may only have been given the rigid body motion versor V, so we need to show that $\mathbf{t}$ and $\mathbf{I}$ can be extracted from that. This is done through the equations:

$$R_{\mathbf{I}\phi} = -o\rfloor(V\infty), \quad \mathbf{t} = -2\,(o\rfloor V)R_{\mathbf{I}\phi}^{-1},$$

and realizing that $\mathbf{I}$ is just the normalized grade-2 part of $R_{\mathbf{I}\phi}$.

21.4.8 Logarithm of a Rigid Body Motion

Given the versor of a rigid body motion, the above indicates how we should find its bivector logarithm: just compute the corresponding screw parameters from the translational and rotational parts. When doing this, one has the need for the logarithm of a pure rotation around the origin, which generally has the form:

$$R_{\mathbf{I}\phi} = e^{-\phi/2} = \cos(\phi/2) - \mathbf{I}\sin(\phi//2).$$

This is easy enough: $\mathbf{I}$ is obtained from normalizing the ratio of the grade-2 and grade-0 parts, and the angle ϕ by the atan2 function (which is robust against the scalar part being zero). When the grade-2 part is zero, one needs to be more careful. In that case, $R_{\mathbf{I}\phi}$ must be equal to ± 1. Then $R_{\mathbf{I}\phi}^2 = 1$, which is precisely the exceptional condition for solving Equation (21.5). Even though we cannot continue the standard computation, a screw decomposition still exists.

- The case $R_{\mathbf{I}\phi} = 1$ is the identity rotation. The rigid body motion is then a pure translation, and has the simpler form $\exp(-\mathbf{t}\infty/2) = 1 - \mathbf{t}\infty/2$ which presents no difficulties for the logarithm.
- The case $R_{\mathbf{I}\phi} = -1$ is a rotor with a rotation angle $\phi = 2\pi$, but in an undetermined plane. Such a rotation over 2π is the identity rotation on a vector, but the rotor is nevertheless useful in the representation of rotations of composite elements.[2] We can in principle choose any rotation plane $\mathbf{J}$ to write this rotor of -1 in exponential form as the rotor $e^{-\mathbf{J}\pi}$, but in the screw representation we should make sure that $\mathbf{J}$ is perpendicular to $\mathbf{t}$ (for we must preserve the commutativity of the rotational and translational rotors permitting the addition of their bivectors). In spaces with 3 or more dimensions, we can simply choose $\mathbf{J}$ proportional to $\mathbf{t}^*$. But in 2-D this is impossible, and therefore we cannot make a screw representation of a rigid body motion with rotational rotor -1 in 2-D. For this one exceptional situation a logarithm does not exist, and interpolation is impossible.

---

[2] Remember the "plate trick": rotating a horizontally held plate in your hand leads to an elbow-up position after a rotation of 2π, only to revert to the elbow-down starting position after a 4π rotation. The parity of the elbow is encoded in the sign of the rotor corresponding to the plate rotation.

Close to these exceptional situations, the determination of the rotation plane **I** is still possible, but numerically unstable. Around the identity, this is tempered in all outcomes by the near-zero value of ϕ; around -1, the numerical instability cannot be resolved.

Combining these results and those of the previous section gives the following pseudocode for the logarithm of a rigid body motion in 3-D.

```
log(V) {
        R = -o⌋(V∞)
        t = -2(o⌋V)/R
        if (‖⟨R⟩₂‖ < ε)
                if (R > 0)                        // the case R = 1
                        log = -t∞/2
                else                              // the case R = -1
                        J = t*/‖t*‖
                        log = -t∞/2 - Jπ
                endif
        else                                      // the regular case
                I = ⟨R⟩₂/‖⟨R⟩₂‖
                φ = -2 atan2(‖⟨R⟩₂‖, ⟨R⟩₀)
                log = (-(t ∧ I)/I + 1/(1 - R²) t⌋Iφ) ∞/2 - Iφ/2
        endif
}
```

In this pseudocode, the notation $\langle R \rangle_k$ denotes taking the kth grade part of the versor R. The variable ϵ is determined by demands on numerical stability. One may improve subsequent numerics by making sure to return a bivector through taking the grade-2 part of the log.

We should mention that the classical homogeneous coordinates matrix representation of rigid body motion given in Equation (21.2) also has a closed form logarithm; it may for example be found in [9].

21.5 Applications of the Rigid Body Motion Logarithm

21.5.1 Interpolation of a Rigid Body Motions

According to Chasles' theorem, any rigid body motion can be viewed as a screw motion. It is then natural to interpolate the original motion by performing this screw gradually.

Figure 21.3 shows how simple this has become in the conformal model of geometric algebra: the ratio of two unit lines L_2 and L_1 defines the square of the versor that transforms one into the other. Performing this motion in N steps implies using the versor

$$V^{1/N} = e^{\log(L_2/L_1)/(2N)}.$$

Fig. 21.3. The interpolation and extrapolation of the rigid body motion transforming a spatial line L_1 into a line L_2 is done by repeated application of the versor
$\exp\left(\log(L_2/L_1)/(2N)\right)$. The screw nature of the motion is apparent.

Applying this versor repeatedly to the line L_1 gives the figure. It interpolates
the transformation of L_1 into L_2, and extrapolates naturally. Note how all
can be defined completely, and simply, in terms of the geometric elements
involved. You do not need coordinates to specify how things move around!
The same rotor $V^{1/N}$ can of course be applied to any element that should
move similarly to L_1.

21.5.2 Interpolation of a Sequence of Rigid Body Motions

Since the resulting logarithms are bivectors, they can be interpolated naturally themselves. Any linear technique from a linear space, such as the usual
point based splines can be used to produce sensible results, for the exponential
of any of the interpolated bivectors gives a valid intermediate pose transformation. As an example, Figure 21.4 shows a cardinal spline interpolation of
the bivectors characterizing 4 frames. These cardinal splines are independent
of the parametrization of the bivectors since they work "per dimension".

It is truly the bivector representation of the intermediate poses that is
being interpolated, and this allows for interesting phenomena. In Figure 21.5,
the two solidly drawn frames look like being related by a simple translation,
but the bivectors of their versors are actually 0 and $\mathbf{e}_1 \wedge \mathbf{e}_2\,\pi - \mathbf{e}_3 \wedge \infty$. Therefore, the second frame is both translated over $\mathbf{e}_3$ and rotated over 2π around
the $\mathbf{e}_3$ axis. This does not show in their rendering, but when the versors are
interpolated the extra twist is properly interpolated. Clearly, other multiples
of 2π rotation angles in other planes would be equally feasible. Figure 21.5(b)
shows the interpolation between two seemingly identical frames.

Fig. 21.4. Cardinal spline interpolation of rigid body motion. A sequence of four frame poses is given, drawn solid. A cardinal spline with tension 0.75 is fitted through the bivectors of their versors, and the exponentials of the results are used as intermediate poses. The frames are drawn by performing the versors on the tangent vectors $o \wedge \mathbf{e}_i$ of the standard frame at the origin.

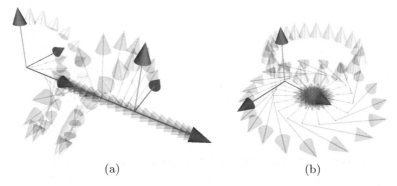

(a)   (b)

Fig. 21.5. Extra twists can appear between seemingly purely translated frames, depending on the actual values of their bivectors. This can be considered either a feature or a bug, depending on the user. Figure 21.5(a) shows the interpolation of the versors with bivectors 0 and $\mathbf{e}_1 \wedge \mathbf{e}_2 \, \pi - \mathbf{e}_3 \wedge \infty$. Figure (b) shows the interpolation of versors with bivectors 0 and $\mathbf{e}_1 \wedge \mathbf{e}_3 \, \pi - \mathbf{e}_3 \wedge \infty$. The latter is the versor of a rotation of 2π around the $\mathbf{e}_3$ axis at $2/\pi \, \mathbf{e}_1$, as the interpolation between the apparently identical initial and final frames shows.

On the one hand, this can be used to compactly encode such rotations; on the other hand, if one wants to interpolate measured data, it shows that one should take care in the assignment of bivectors to frames, making sure that they belong to the same turn of the screw if this phenomenon of extra twists is to be avoided.

The resulting curves, though valid as motions, are not designed to minimize physical quantities like overall torque, or some trade-off between rotational and translational energy (for which one would need to know mass and inertia). Such measures should first be related to the bivector characterization, and since a formulation of classical physics has been given in geometric algebra [2, 5], this should be relatively easy.

More seriously, the simple scheme of interpolating the bivectors leads to a motion that is *not covariant* under a change of the interpolants: moving the target frames and then interpolating is not the same as moving the interpolated sequence. The fundamental mathematical reason is that the group of rigid body motions $SE(3)$ cannot be endowed with a "bi-invariant Riemannian metric" [9]. This means that the concept of straight line (or geodesic) becomes more involved, and that interpolations based on concepts of minimization of length (De Casteljau) and/or energy (splines) are compromised: they cannot be covariant simultaneously under changes of body frame and of world frame. Two practical solutions to this dilemma are compared in [1], using the classical homogeneous coordinate representation. We should study whether the more direct bivector representation could suggest other methods (but the fundamental mathematical impossibility will of course remain).

21.5.3 Combination of Estimates

In applications in computer vision and robotics, one often is able to estimate a pose in various ways. The problem then arises how to combine such estimates. The linearity of the bivector representation is very helpful here. For example, if the poses are known to have a normal distribution in their bivector parameters, the best estimate of the pose versor is the exponential of the average bivector. This is illustrated in Figure 21.6. Here it is obviously required that the bivector characterizations of the frames do belong to the same turn (2π-interval) of the screw.

However, the mere averaging of bivectors combines their rotational and translational part in equal measure, as it were comparing angles and distances by counting the angles as the corresponding arc lengths on the unit circle (since they are radians). In applications, one typically has different accuracies for these angles and positions, or even for certain orientations or directions (as for instance in stereo vision, where the radial distance from the observer is less accurate than the tangential distances).

Fig. 21.6. From ten Gaussian distributed estimated poses of a frame, the "best" frame (denoted solidly) is estimated as the exponent of the average of the bivectors of their versors.

An advantage of the linearity of the bivector space is that it permits one to incorporate such differences in the second order statistics in a simple and structural manner, using covariance ellipsoids. This also permits a direct borrowing of techniques like time-dependent Kalman filtering, and classification and segmentation of poses using Mahalanobis distances. The first applications of such conformal estimation methods are beginning to appear (see e.g. [11]).

21.6 Conclusion

Our goal was to demonstrate the geometrical coherence of the conformal model for the treatment of Euclidean geometry. We focused on the representation and interpolation of the rigid body motions, and hope to have convinced the reader that this unfamiliar representation deserves to be explored for applications in which those are central. The advantage of the completely linear representation by bivectors permits one to adapt familiar techniques from the positional interpolation and linear estimation without structural change. The fact that the Euclidean properties are baked into the algebra at the fundamental level of its inner product means that all computations will result in directly usable elements, so that no computational effort needs to be wasted in projecting back to the motion manifold, as happens so often in techniques based on approximate linearizations of homogeneous coordinate matrices. This could lead to faster motion processing, since it has been shown that by employing the proper implementational techniques, computations in the conformal model can be as cheap as those using traditional homogeneous coordinates [2]. The field is beginning to take off, and we hope to report on more results soon.

Acknowledgment and US Patent Alert

I am grateful to Daniel Fontijne for spotting and helping to fix the exceptional case $R = -1$ in the logarithm, and for his whipping up a cardinal spline interpolation program to generate Figures 21.4 and 21.5.

Commercial applications of conformal geometric algebra to model statics, kinematics and dynamics of particles and linked rigid bodies are protected by US Patent 6,853,964 (System for encoding and manipulating models of objects). Anyone contemplating such applications should contact Alyn Rockwood (alynrock@yahoo.com) or David Hestenes (hestenes@asu.edu). The patent does not restrict use of conformal geometric algebra for academic research and education or other worthy purposes that are not commercial.

References

1. Altafini, A.: The De Casteljau Algorithm on SE(3). In: *Non-Linear Control in the Year 2000*, Springer, pp. 23–34, 2000.

2. Doran, C. and A. Lasenby: *Geometric Algebra for Physicists*. Cambridge University Press, 2000.
3. Dorst, L., D. Fontijne and S. Mann: *Geometric Algebra for Computer Science, An Object-Oriented Approach to Geometry*. Morgan Kaufman, 2007. See also www.geometricalgebra.net.
4. Dorst, L.: The Inner Products of Geometric Algebra. In: *Applications of Geometric Algebra in Computer Science and Engineering*, L. Dorst, C. Doran, and J. Lasenby (Eds.), Birkhäuser, Boston, pp. 35–46, 2002.
5. Hestenes, D.: *New Foundations for Classical Mechanics* (2nd edn). Reidel, Dordrecht, 2000.
6. Hestenes, D., A. Rockwood and H. Li: *System for encoding and manipulating models of objects*. U.S. Patent 6,853,964, granted February 8, 2005.
7. Hestenes, D. and E. D. Fasse: Homogeneous Rigid Body Mechanics with Elastic Coupling. In: *Applications of Geometric Algebra in Computer Science and Engineering*, L. Dorst, C. Doran, and J. Lasenby (Eds.), Birkhäuser, Boston, pp. 197–212, 2002.
8. Lasenby, J., A. N. Lasenby and C. J. L. Doran: A Unified Mathematical Language for Physics and Engineering in the 21st Century. *Phil. Trans. R. Soc. Lond. A*, vol. 358, pp. 21–39, 2000.
9. Park, F.C.: Distance Metrics on the Rigid-Body Motions with Applications to Mechanism Design. *Transactions of the ASME*, vol. 117, pp. 48–54, 1995.
10. Riesz, M.: *Clifford Numbers and Spinors*. E. F. Bolinder and P. Lounesto (Eds.), Kluwer Academic, 1993.
11. Zhao, Y., R. J. Valkenburg, R. Klette and B. Rosenhahn: Target Calibration and Tracking using Conformal Geometric Algebra. In: *IEEE Pacific-Rim Symposium on Image and Video Technology (PSIVT)*, Chang L.-W., Lie W.-N. (Eds.), Springer, Berlin Heidelberg, LNCS 4319, pp. 74–83, Taiwan 2006.

22

Video-based Capturing and Rendering of People

Christian Theobalt[1], Marcus Magnor[2], and Hans-Peter Seidel[1]

[1] MPI Informatik, Saarbrücken, Germany
[2] Technical University of Braunschweig, Germany

Summary. The interesting combination of chapters in this book gives a nice overview of the importance of the analysis of human motion in three different scientific disciplines, computer vision, biomechanics, and computer graphics. In the computer vision field the main goal has been to develop robust and efficient techniques for capturing and measuring human motion. Researchers from biomechanics, on the other hand, try to capitalize on such measurement methods to analyze the characteristics of human motion. In contrast, the focus of computer graphics research has been to realistically animate and render moving humans. In this chapter, we will outline that a joint solution to the acquisition, the reconstruction, and the rendering problem paves the trail for video-based capturing and rendering of people. We present a model-based system to create realistic 3D videos of human actors that puts this idea into practice. Our method enables us to capture human actors with multiple video cameras and to render their performances in real-time from arbitrary novel viewpoints and under arbitrary novel lighting conditions.

22.1 Introduction

In recent years, an increasing research interest in the field of 3D Video Processing has been observed. This young and challenging discipline lives on the boundary between computer graphics and computer vision. The goal of 3D Video Processing is the extraction of spatio-temporal models of dynamic scenes from multiple 2D video streams. These scene models comprise of descriptions of the scene's shape, the scene's appearance, as well as the scene's motion. Having these dynamic representations at hand, one can display the captured real world events from novel synthetic camera perspectives. In order to put this idea into practice, algorithmic solutions to three major problems have to be found: the problem of multi-view acquisition, the problem of scene reconstruction from image data, and the problem of scene display from novel viewpoints. While the first two problems have been widely investigated in computer vision, the third question is a core problem in computer graphics.

531

B. Rosenhahn et al. (eds.), Human Motion – Understanding, Modelling, Capture, and Animation, 531–559.
© 2008 *Springer.*

Human actors are presumably the most important elements of many real-world scenes. Unfortunately, it is well known to researchers in computer graphics and computer vision that both the analysis of shape and motion of humans from video, as well as their convincing graphical rendition are very challenging problems. To tackle the difficulties of the involved inverse problems, we propose in this chapter a model-based approach to capture and render free-viewpoint videos of human actors. Human performances are recorded with a handful of synchronized video cameras. A template model is deformed to match the shape and proportions of the captured human actor and it is made to follow the motion of the person by means of a marker-free optical motion capture approach. In a first algorithmic variant, the human performances can be rendered in real-time from arbitrary synthetic viewpoints. Time-varying surface appearance is generated by means of a dynamic multi-view texturing from the input video streams. In a second variant of the method, we do not only capture time-varying surface appearance but reconstruct sophisticated dynamic surface reflectance properties. This enables real-time free-viewpoint video rendering not only from novel viewpoints but also under novel virtual lighting conditions. Finally, we also give an outlook on future directions in model-based free-viewpoint video of human actors, like on how to incorporate high-quality laser-scanned shape priors into the overall workflow. This chapter is a summary of several algorithms that we have developed in the last couple of years and methods that we continue to work on. It also provides a comprehensive set of references to other papers where the individual method are described in greater detail than in this overview chapter.

The chapter is organized as follows. We begin with a review of important related work in Section 22.2. Section 22.3 gives details about the multi-view video studio, the camera system and the lighting setup that we employ for acquisition. Our template body model is described in Section 22.4. A central ingredient of all our free-viewpoint video methods, the silhouette-based analysis-through-synthesis method, is described in Section 22.5. We employ this algorithm to adapt the shape of our template model such that it matches the subject currently recorded, and also employ it to capture the motion of the human actor. In Section 22.6 we describe how to combine the silhouette-based analysis-through-synthesis method with a dynamic texture-generation approach to produce free-viewpoint videos of human actors. We also present and discuss visual results obtained with this method. Our extension of the algorithm which enables us to generate dynamically textured free-viewpoint videos is described in Section 22.7. We show several example sequences that we have captured and rendered under different virtual lighting conditions. We end the chapter with an outlook to future directions in model-based free-viewpoint video that we have started to work on in Section 22.8 and conclude in Section 22.9.

22.2 Related Work

Since the work presented in this chapter jointly solves a variety of algorithmic subproblems, we can capitalize on a huge body of previous work in the fields human motion capture, image-based rendering, 3D video image-based reflectance estimation, and mesh-based animation processing.

For a detailed review on human motion capture, we would like to refer the reader to the Chapters 12 and 13, and constrain our review to relevant research from the other fields.

22.2.1 Free-Viewpoint Video

Research in free-viewpoint video aims at developing methods for photo-realistic, real-time rendering of previously captured real-world scenes. The goal is to give the user the freedom to interactively navigate his or her viewpoint freely through the rendered scene. Early research that paved the way for free-viewpoint video was presented in the field of image-based rendering (IBR). Shape-from-silhouette methods reconstruct geometry models of a scene from multi-view silhouette images or video streams. Examples are image-based [28, 52] or polyhedral visual hull methods [27], as well as approaches performing point-based reconstruction [17]. The combination of stereo reconstruction with visual hull rendering leads to a more faithful reconstruction of surface concavities [26]. Stereo methods have also been applied to reconstruct and render dynamic scenes [21, 58], some of them employing active illumination [50]. On the other hand, light field rendering [25] is employed in the 3D TV system [30] to enable simultaneous scene acquisition and rendering in real-time.

In contrast, we employ a complete parameterized geometry model to pursue a model-based approach towards free-viewpoint video [6, 43–46]. Through commitment to an adaptable body model whose shape is made consistent with the actor in multiple video streams, we can capture a human's motion and his dynamic surface texture [7]. We can also apply our method to capture personalized human avatars [1].

IBR methods can visualize a recorded scene only for the same illumination conditions that it was captured in. For correct relighting, it is inevitable to recover complete surface reflectance characteristics as well.

22.2.2 Image-based Reflectance Estimation

The estimation of reflection properties from still images has been addressed in many different ways. Typically, a single point light source is used to illuminate an object of known 3D geometry consisting of only one material. One common approach is to take HDR (High Dynamic Range) images of a curved object, yielding a different incident and outgoing directions per pixel and thus capturing a vast number of reflectance samples in parallel. Often, the parameters of an analytic BRDF (Bidirectional Reflectance Distribution Function) model are fitted to the measured data [24, 37] or a data-driven model is used [29]. Zickler et al. [57] proposed a scattered data interpolation method to reconstruct a reflectance model. Reflectance measurements of scenes with more complex incident illumination can be derived by either a full-blown inverse global illumination approach [4, 15, 53] or by representing the incident light field as an environment map and solving for the direct illumination component only [31, 35, 54]. In a method that we will briefly outline in Section 22.7, we approximate the incident illumination by multiple point light sources and estimate BRDF model parameters taking only direct illumination into account.

Reflection properties together with measured photometric data can also be used to derive geometric information of the original object [56]. Rushmeier et al. [36] estimate diffuse albedo (i.e., diffuse reflectance) and normal maps from photographs with varied incident light directions [3, 36]. A linear light source is employed by Gardner et al. [13] to estimate BRDF properties and surface normal. In [14, 16], reflectance and shape of static scenes are simultaneously refined using a single light source in each photograph.

Instead of explicitly reconstructing a mathematical reflectance model, it has also been tried to take an image-based approach to relighting. In [18] a method to generate animatable and relightable face models from images taken with a special light stage is described. Wenger et al. [51] extend the light stage device such that it enables capturing of dynamic reflectance fields. Their results are impressive, however it is not possible to change the viewpoint in the scene. Einarsson et al. [12] extends it further by using a large light stage, a tread-mill where the person walks on, and light field rendering for display. This way, human performances can be rendered from novel perspectives and relit. Unfortunately the method can only handle periodic motion, such as walking, and is only suitable for low frequency relighting. For our 3D video scenario, we prefer a more compact scene description based on parametric BRDFs that can be reconstructed in a fairly simple acquisition facility, that allows for arbitrary viewpoint changes as well as high-frequency relighting.

Carceroni and Kutulakos [5] present a volumetric method for simultaneous motion and reflectance capture for nonrigid objects [5]. They have shown nice results for spatially confined 3D scenes where they used a coarse set of surfels as shape primitives.

In Section 22.7, we briefly talk about an extension of our original model-based free-viewpoint video pipeline that enables us to capture shape, motion parameters and dynamic surface reflectance of the whole human body at high accuracy [42, 48]. As input, we only expect a handful of synchronized video streams showing arbitrary human motion. Our reconstructed dynamic scene description enables us to render virtual people in real-time from arbitrary viewpoints and under arbitrary lighting conditions.

22.2.3 Mesh-based Deformation and Animation

In Section 22.8, we identify some of the conceptual restrictions that are imposed by using an adaptable kinematic template model. Following these insights, we propose a way to modify our framework to use more general models, e.g., high-quality scanned geometry, for model-based free-viewpoint video. The method we present couples our marker-less motion capture with a mesh-based deformation scheme. To learn more about these mesh-deformation schemes, we would like to suggest to the reader to look into the references in the following paragraph. We have developed a novel simple and fast procedure that overcomes many limitations of the traditional animation pipeline by capitalizing on and extending ideas from mesh deformation.

A variety of concepts have been proposed in the literature to efficiently and intuitively deform 3D triangle meshes. Among the most promising approaches have been algorithms that use differential coordinates, see [2,39] for reviews on this subject. The potential of these methods for animation has already been stated in previous publications, however, the focus always lay on deformation transfer between moving meshes [55]. Using a complete set of correspondences between different synthetic models, [40] enables transferring the motion of one model to the other. Following a similar line of thinking, [10,41] propose a mesh-based inverse kinematics framework based on pose examples with potential application to mesh animation. Recently, [38] presents a multigrid technique for efficient deformation of large meshes and [20] presents a framework for performing constrained mesh deformation using gradient domain techniques. Both methods are conceptually related to our algorithm and could also be used for animating human models. However, none of the papers provides a complete integration of the surface deformation approach with a motion acquisition system, nor does any of them provide a comprehensive user interface.

We developed a novel skeleton-less deformation method that allows us to easily make a high-quality scan of a person move the same way as the template model used for motion capture [9]. To this end, only a handful of correspondences between the two models has to be specified once. Realistic motion of the scan as well as nonrigid surface deformations are catered for automatically.

22.3 Acquisition: A Studio for Multi-view Video Recording

The input to our system are multiple synchronized video streams of a moving person, so-called MVV sequences, that we capture in our free-viewpoint video studio. The studio features a multi-camera system that enables us to capture a volume of approximately $4 \times 4 \times 3$ m with eight externally synchronized video cameras [47]. The imaging sensors can be placed in arbitrary positions, but typically we resort to an approximately circular arrangement around the center of the scene. Optionally, one of the cameras is placed in an overhead position. Over time, we enhanced and upgraded the studio, such that we have two 8-camera systems at our disposition. Camera system I, the older one of the two setups, features 8 IEEE-1394 cameras which we run at 320x240 image resolution and 15 fps. This setup has been used for our original work on non-relightable 3D video, Section 22.6. Camera setup II, features 8 Imperx MDC 1004 video cameras providing 1004x1004 pixel frame resolution, 12-bit color depth and a frame rate of 25 fps. Camera system II has been used in our recent work on 3D Video relighting, Section 22.7.

Prior to recording, the cameras are calibrated, and inter-camera color consistency is ensured by applying a color-space transformation to each video stream. The lighting conditions in the studio are fully-controllable and the scene background can optionally be draped with black molleton. We have a set of different light setups at our disposition. While for the free-viewpoint video with dynamics textures we prefer a diffuse illumination, our work on relighting requires spot-light illumination. The specific requirements of the lighting setup used for dynamic reflectance estimation are explained in Section 22.7.1.

22.4 The Adaptable Human Body Model

While 3D object geometry can be represented in different ways, we employ a triangle mesh representation because it offers a closed and detailed surface description, and, even more importantly, it can be rendered very fast on graphics hardware. Since the model must be able to perform the same complex motion as its real-world counterpart, it is composed of multiple rigid-body parts that are linked by a hierarchical kinematic chain. The joints between segments are suitably parameterized to reflect the object's kinematic degrees of freedom. Besides object pose, also the dimensions of the separate body parts must be kept adaptable as to be able to match the model to the object's individual stature.

As geometry model, a publicly available VRML (Virtual Reality Modelling Language) geometry model of a human body is used, Figure 22.1a. The model consists of 16 rigid body segments, one for the upper and lower torso, neck, and head, and pairs for the upper arms, lower arms, hands, upper legs, lower legs and feet. In total, the human body model comprises more than 21000 triangles. A hierarchical kinematic chain connects all body segments, resembling the anatomy of the human skeleton. 17 joints with a total of 35 joint parameters define the pose of the virtual character. Different joints in the body model provide different numbers of rotational degrees of freedom the same way as the corresponding joints in an anatomical skeleton do. For global positioning, the model provides three translational degrees of freedom which influence the position of the skeleton root. The root of the model is located at the pelvis. The

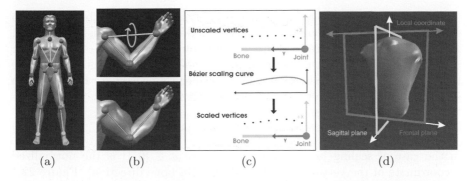

(a)   (b)   (c)   (d)

Fig. 22.1. (a) Surface model and the underlying skeletal structure. Spheres indicate joints and the different parameterizations used; blue sphere – 3 DOF ball joint, green sphere – 1 DOF hinge joint, red spheres (two per limb) – 4 DOF. The black/blue sphere indicates the location of three joints, the root of the model and joints for the upper and lower half of the body. (b) The upper figure shows the parameterization of a limb, consisting of 3 DOF for the wrist position in local shoulder coordinates (shown in blue) and 1 DOF for rotation around the blue axis. The lower right figure demonstrates an exaggerated deformation of the arm that is achieved by appropriately tweaking the Bézier parameters. (c) Schematic illustration of local vertex coordinate scaling by means of a Bézier scaling curve for the local $+x$ direction. (d) The two planes in the torso illustrate the local $+x,-x$ and $+z,-z$ scaling directions, respectively.

kinematic chain is functionally separated in an upper body half and a lower body half. The initial joints of both kinematic sub-chains spatially coincide with the model root.

In Figure 22.1a individual joints in the body model's kinematic chain are drawn and the respective joint color indicates if it is a 1-DOF hinge joint, a 3-DOF ball joint, or a joint being part of our custom limb parameterization. Each limb, i.e., complete arm or leg, is parameterized via four degrees of freedom. These are the position of the tip, i.e., wrist or ankle, and the rotation around an axis connecting root and tip, Figure 22.1b. This limb parameterization was chosen because it is particularly well-suited for an efficient grid search of its parameter space, Section 22.5. The head and neck articulation is specified via a combination of a 3-DOF ball joint and a 1-DOF hinge joint. The wrist provides three degrees of freedom and the foot motion is limited to a 1-DOF hinge rotation around the ankle. In total, 35 pose parameters fully specify a body pose.

In addition to the pose parameters, the model provides anthropomorphic shape parameters that control the bone lengths as well as the structure of the triangle meshes defining the body surface.

Each of the 16 body segments features a scaling parameter that scales the bone as well as the surface mesh uniformly in all three coordinate directions (in the local coordinate frame of the segment).

In order to match the geometry more closely to the shape of the real human each segment features four one-dimensional Bézier curves $B_{+x}(u)$, $B_{-x}(u)$, $B_{+z}(u)$, $B_{-z}(u)$ which are used to scale individual coordinates of each vertex $v = (v_x, v_y, v_z)$ in the local triangle mesh. The scaling is performed in the local $+x$,$-x$,$+z$, and $-z$ directions of the segment's coordinate frame, which are orthogonal to the direction of the bone axis. For instance

$$v_{x\_scaled} = B_{+x}(u) \cdot v_x \qquad (22.1)$$

in the case of scaling in $+x$ -direction, where $u \in [0,1]$ is the normalized y-coordinate of the vertex in the local frame (in bone direction). Figure 22.1 shows the effect of changing the Bézier scaling values using the arm segment as an example. Intuitively, the four scaling directions lie on two orthogonal planes in the local frame. For illustration, we show these two planes in the torso segment in Figure 22.1d.

22.5 Silhouette-based Analysis-Through-Synthesis

The challenge in applying model-based analysis for free-viewpoint video reconstruction is to find a way how to automatically and robustly adapt the geometry model to the subject's appearance as it was recorded by the video cameras. Since the geometry model is suitably parameterized to alter its shape and pose, the problem reduces to determining the parameter values that achieve the best match between the model and the video images. This task is regarded as an optimization problem. The subject's silhouettes, as seen from the different camera viewpoints, are used to match the model to the video images (an idea used in similar form in [23]): The model is rendered from all camera viewpoints, and the rendered images are thresholded to yield binary masks of the model's silhouettes. The rendered model silhouettes are then compared to the corresponding image silhouettes [6, 43–46]. As comparison measure, the number of silhouette pixels that do not overlap is determined. Conveniently, the exclusive-or (XOR) operation between the rendered model silhouette and the segmented video-image silhouette yields those pixels that are not overlapping. The energy function thus evaluates to:

$$E_{XOR}(\mu) = \sum_{i=0}^{N} \sum_{x=0}^{X} \sum_{y}^{Y} (P_s(x,y) \wedge !P_m(x,y)) \vee (!P_s(x,y) \wedge P_m(x,y)) \quad (22.2)$$

where μ is the model parameters currently considered, e.g., pose or anthropomorphic parameters, N the number of cameras, and X and Y the dimensions of the image. $P_s(x,y)$ is the 0/1-value of the pixel (x,y) in the capture image silhouette, while $P_m(x,y,\mu)$ is the equivalent in the reprojected model image given that the current model parameters are μ. Fortunately, this XOR energy function can be very efficiently evaluated in graphics hardware, Figure 22.2.

CPU   **GPU**

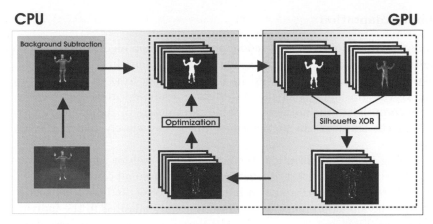

Fig. 22.2. Hardware-based analysis-through-synthesis for free-viewpoint video: To match the geometry model to the multi-video recordings of the actor, the image foreground is segmented and binarized. The model is rendered from all camera viewpoints. The boolean XOR operation is executed between the foreground images and the corresponding model renderings, and the number of remaining pixels in all camera views serves as matching criterion. Model parameter values are varied via numerical optimization until the XOR result is minimal. The numerical minimization algorithm runs on the CPU while the energy function evaluation is implemented on the GPU.

With 8 cameras, a Pentium III with a GeForce 6800 graphics board easily performs more than 250 of such matching function evaluations per second.

To adapt model parameter values such that the mismatch score becomes minimal, a standard numerical optimization algorithm, such as Powell's method [34], runs on the CPU. As a direction set method it always pertains a number of candidate descend directions in parameter space. The optimal descend in one direction is computed using Brent's line search method. For each new set of model parameter values, the optimization routine invokes the matching function evaluation routine on the graphics card.

One valuable benefit of model-based analysis is the low-dimensional parameter space when compared to general reconstruction methods: The parameterized model provides only a few dozen degrees of freedom that need to be determined, which greatly reduces the number of potential local minima. Furthermore, many high-level constraints are implicitly incorporated, and additional constraints can be easily enforced by making sure that all parameter values stay within their anatomically plausible range during optimization. Finally, temporal coherence is straightforwardly maintained by allowing only some maximal rate of change in parameter value from one time step to the next.

The silhouette-based analysis-through-synthesis approach is employed for two purposes, the initialization or shape adaptation of the model's geometry and the computation of the body pose at each time step.

Shape Adaptation

To apply the silhouette-based model pose estimation algorithm to real-world multi-video footage, the generic geometry model must first be initialized, i.e. its proportions must be adapted to the subject in front of the cameras. To achieve this, we apply the silhouette-based analysis-through-synthesis algorithm to optimize the anthropomorphic parameters of the model. This way, all segment surfaces can be deformed until they closely match the actor's stature.

During model initialization, the actor stands still for a brief moment in a predefined pose to have his silhouettes recorded from all cameras. The generic model is rendered for this known initialization pose, and without user intervention, the model proportions are automatically adapted to the individual's silhouettes. First, only the torso is considered. Its position and orientation is determined approximately by maximizing the overlap of the rendered model images with the segmented image silhouettes. Then the pose of arms, legs and head are recovered by rendering each limb in a number of orientations close to the initialization pose and selecting the best match as starting point for refined optimization. This step is identical to the optimization which we perform for pose determination (see Section 22.5). Following the model hierarchy, the optimization itself is split into several suboptimizations in lower-dimensional parameter spaces. After the model has been coarsely adapted in this way, the uniform scaling parameters of all body segments are adjusted. For selected body segments (e.g., arm and leg segments) we found it advantageous to scale their dimension only in the bone direction, and to leave the control of the triangle mesh shape orthogonal to this direction to the Bézier parameters. The algorithm then alternates typically around 5–10 times between optimizing joint parameters and segment scaling parameters until it has converged. Finally, the Bézier control parameters of all body segments are optimized in order to fine-tune each segment's outline such that it complies with the recorded silhouettes. In Figure 22.3, the initial model shape, its shape after five iterations of pose and scene optimization and its shape after Bézier scaling are shown.

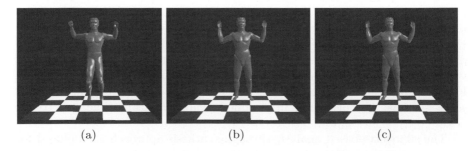

(a)   (b)   (c)

Fig. 22.3. (a) template model geometry; (b) model after 5 iterations of pose and scale refinements; (c) model after adapting the Bézier scaling parameters.

Obviously, an exact match between model outline and image silhouettes is not attainable since the parameterized model has far too few degrees of freedom. Thanks to advanced rendering techniques an exact match is neither needed for convincing dynamic texturing (Section 22.6) nor for reflectance estimation (Section 22.7). The initialization procedure takes only a few seconds. From now on the anthropomorphic shape parameters remain fixed.

Alternatively, not only the static initialization pose but a sequence of postures can be employed to reconstruct a spatio-temporally consistent shape model. After recovery of this spatio-temporally consistent model using the previously mentioned silhouette matching, it is also feasible to reconstruct smaller-scale per-time step surface deformations, i.e. displacements $\mathbf{r_j}$ for each vertex j in a mesh [7]. We achieve this by minimizing an energy functional of the form

$$
\begin{aligned}
E_{disp}(v_j, \mathbf{r_j}) = {} & w_I E_I(v_j + \mathbf{r_j}) + w_S E_S(v_j + \mathbf{r_j}) \\
& + w_D E_D(v_j + \mathbf{r_j}) + E_P(v_j, v_j + \mathbf{r_j})
\end{aligned}
\tag{22.3}
$$

$E_I(v_j + \mathbf{r_j})$ is a multi-view color-consistency measure. The term $E_S(v_j + \mathbf{r_j})$ penalizes vertex positions that project into image plane locations that are very distant from the boundary of the person's silhouette. The term $E_D(v_j + \mathbf{r_j})$ aims at maintaining the segment's mesh quality by measuring the distortion of triangles. We employ a distortion measure which is based on the Frobenius norm [32]. Finally, the term $E_P(v_j, v_j + \mathbf{r_j})$ penalizes visibility changes that are due to moving a vertex j from position v_j to position $v_j + \mathbf{r_j}$. Appropriate weighting coefficients w_I, w_S, w_D, and w_P were found through experiments.

After calculating the optimal displacement for a subset of vertices, these displacements are used to smoothly deform the whole region by means of a Laplacian interpolation method. While this approach can capture time-varying geometry at sufficient detail, its resolution is still limited by the mesh granularity and the implicit smoothness constraint imposed by the Laplacian deformation scheme.

Marker-less Pose Tracking

The individualized geometry model automatically tracks the motion of the human dancer by optimizing the 35 joint parameters for each time step. The analysis-through-synthesis framework enables us to capture these pose parameters without having the actor wear any specialized apparel. This is a necessary precondition for free-viewpoint video reconstruction, since only if motion is captured completely passively can the video imagery be used for texturing. The model silhouettes are matched to the segmented image silhouettes of the actor so that the model performs the same movements as the human in front of the cameras, Figure 22.4.

At each time step t an optimal set of pose parameters, P_t, is found by performing a numerical minimization of the silhouette XOR energy functional, Equation (22.2), in the space of pose parameters.

(a)   (b)   (c)

Fig. 22.4. (a) Silhouette XOR; (b) body model; (c) textured body model from same camera view.

The numerical optimization of the multidimensional, nonconvex matching functional can potentially result in suboptimal fitting. A straightforward approach would be to apply any standard numerical minimization method to optimize all pose-related degrees of freedom in the model simultaneously. This simple strategy, however, exhibits some of the fundamental pitfalls that make global optimization infeasible. In the global case, the energy function reveals many erroneous local minima. Fast movements between consecutive time frames are almost impossible to track since it may happen that no overlap between the model and the image silhouette occurs that guides the optimizer towards the correct solution. A different problem arises if one limb moves very close to the torso. In this case, it is quite common for global minimization method to find a local minimum in which the limb penetrates the torso. Instead, we present a method that enables us to use a standard direction set minimization scheme to robustly estimate pose parameters. We effectively constrain the search space by exploiting structural knowledge about the human body, knowledge about feasible body poses, temporal coherence in motion data and a grid sampling preprocessing step.

To efficiently avoid local minima, the model parameters are not all optimized simultaneously. Instead, the model's hierarchical structure is exploited. Model parameter estimation is performed in descending order with respect to the individual segments' impact on silhouette appearance and their position along the model's kinematic chain, Figure 22.5. First, position and orientation of the torso is varied to find its 3D location. Next, arms, legs and head are considered. Finally, hands and feet are examined.

Temporal coherence is exploited by initializing the optimization for one body part with the pose parameters P_{t-1} found in the previous time step. Optionally, a simple linear prediction based on the two preceding parameter sets is feasible.

Due to the limb parameterization described in Section 22.4, fitting an arm or leg is a four-dimensional optimization problem. In order to cope with

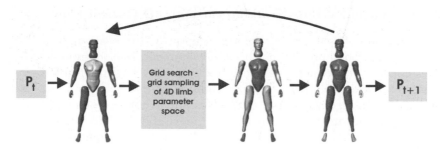

Fig. 22.5. For motion capture, the body parts are matched to the images in hierarchical order: the torso first, then arms, legs and head, finally hands and feet. Local minima are avoided by a limited regular grid search for some parameters prior to optimization initialization.

fast body motion that can easily mislead the optimization search, we precede the numerical minimization step with a regular grid search. The grid search samples the four-dimensional parameter space at regularly spaced values and checks each corresponding limb pose for being a *valid pose*. Using the arm as an example, a valid pose is defined by two criteria. Firstly, the wrist and the elbow must project into the image silhouettes in every camera view. Secondly, the elbow and the wrist must lie outside a bounding box defined around the torso segment of the model. For all valid poses found, the error function is evaluated, and the pose that exhibits the minimal error is used as starting point for a direction set downhill minimization. The result of this numerical minimization specifies the final limb configuration. The parameter range, from which the grid search draws sample values, is adaptively changed based on the difference in pose parameters of the two preceding time steps. The grid sampling step can be computed at virtually no cost and significantly increases the convergence speed of the numerical minimizer.

Optionally, the whole pose estimation pipeline can be iterated for a single time step. Our results show that with the appropriate specification of constraints even a standard downhill minimizer can perform as well in human body tracking as more complicated statistical optimization schemes, such as condensation [11].

The performance of the silhouette-based pose tracker can be further improved by capitalizing on the structural properties of the optimization problem [44]. First, the XOR evaluation can be sped up by restricting the computation to a sub-window in the image plane and excluding stationary body parts from rendering. Second, optimization of independent sub-chains can be performed in parallel. A prototype implementation using 5 PCs and 5 GPUs, as well as the improved XOR evaluation exhibited a speed-up of up to factor 8 (see also Section 22.6.3). For details about these improvements, the interested reader is referred to [44].

22.5.1 Augmented Pose Tracking through 3D Optical Flow

While silhouette-based analysis robustly captures large-scale movements, in some cases the body pose cannot be resolved unambiguously from object silhouettes alone. Especially small-scale motions may be irresolvable because at limited image resolution, small protruding features, such as nose or ears, may not be discernible in the silhouettes. In a free-viewpoint video system, inaccuracies in recovered body pose inevitably lead to rendering artifacts when the 3D videos are displayed. To resolve ambiguities as well as to refine silhouette-estimated model pose, we enhance our original analysis-through-synthesis approach such that the object's surface texture is considered in addition [43,46]. We incorporate texture information into the tracking process by computing corrective motion fields via 3D optical flow or *scene flow*, Figure 22.6b, c. The reconstruction of the 3D motion field from the 2D optical flows is possible using a technique described in [49].

If correspondences in the image plane are known, i.e., it is known to which image coordinates 3D points project in each camera view, the scene flow can be reconstructed by solving a linear system of equations. In our free-viewpoint video approach, the correspondences are known for each vertex because we have an explicit body model, and the projection matrices $\mathbf{P}_i$ for each recording camera i have been determined via calibration. The projection matrix $\mathbf{P}_i$ describes the relationship between a 3D position of a vertex and its projection into the image plane of the camera i, $u_i = (u_i, v_i)^T$.

The differential relationship between the vertex $\mathbf{x}$ with coordinates $(x, y, z)^T$ and u_i is described by the 2×3 Jacobian matrix $J_i = \frac{\partial u_i}{\partial \mathbf{x}_i}$:

$$o_i = \frac{du_i}{dt} = J_i \frac{d\mathbf{x}}{dt} \tag{22.4}$$

In other words, the Jacobian describes the relationship between a small change in 3D position of a vertex, and the change of its projected image in camera

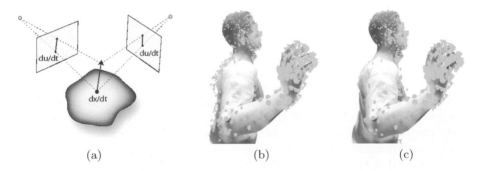

(a)   (b)   (c)

Fig. 22.6. (a) 3D motion (scene flow) of a surface point and the corresponding observed optical flows in two camera views. Body model with corrective motion field (green arrows) before (b) and after (c) pose update.

i. The term $\frac{du_i}{dt}$ is the optical flow o_i observed in camera i, $\frac{d\mathbf{x}}{dt}$ is the corresponding scene flow of the vertex (Figure 22.6). Given a mathematical camera model, the Jacobian can be computed analytically [49]. If a vertex is visible from at least two camera views, an equation system of the form $\mathbf{B}\frac{d\mathbf{x}}{dt} = \mathbf{U}$ can be formulated, whose solution is the scene flow of the vertex.

We incorporate the scene flow information into our silhouette-based framework by developing the following predictor-corrector approach. Considering an arbitrary time step $t+1$, the augmented motion capture algorithm works as follows: Starting with a set of 35 body pose parameters P_t that were found to be optimal for time step t, the system first computes an estimate of the pose parameters $P'_{sil,t+1}$ at time $t+1$ by employing the silhouette-based motion estimation scheme, Section 22.5. In a second step, estimate $P'_{sil,t+1}$ is augmented by computing a 3D corrective motion field from optical flows. The model that is standing in pose $P'_{sil,t+1}$ and that is textured with the video images from time t is rendered into all camera views. The images of the back-projected model form a prediction of the person's appearance at $t+1$. The optical flows are computed for each pair of back-projected model view and corresponding segmented video frame at time $t+1$. From camera calibration, the camera matrix of each recording imaging sensor is known. Since, in addition, the geometric structure of the body model is available, for each model vertex corrective flow vectors in 3D can be computed from the corrective 2D optical flows in all camera views. The end-point of each motion vector is the position at which the respective vertex should be in order for the whole model to be in a stance that is photo-consistent with all camera views. This information has to be translated into pose update parameters for the model's joints that bring the model into the photo-consistent configuration. We compute the differential pose update, $P_{\text{diff},t+1}$, in a least-squares sense [19] and apply it to the model after $P'_{sil,t+1}$ in order to obtain the final pose estimate P_{t+1} for time $t+1$. The final pose parameter estimate serves as a starting point for the pose determination in the next time step.

22.6 Free-viewpoint Video with Dynamic Textures

By combining the silhouette-based analysis-through synthesis method with a dynamic texture generation we can reconstruct and render free-viewpoint videos of human actors that reproduce the omnidirectional appearance of the actor under fixed lighting conditions. First, the shape and the motion of the body model are reconstructed by means of the approach described in Section 22.5.

A high-quality 3D geometry model is now available that closely matches the dynamic object in the scene over the entire length of the sequence. To display the object photo-realistically, the recorded video images are used for texturing the model surface.

Since time-varying video data is available, model texture doesn't have to be static. To create convincing surface appearance we capitalize on the projective texturing capability of modern GPUs.

Prior to display, the geometry model as well as the video cameras' calibration data is transferred to the graphics board. During rendering, the user's viewpoint information, the model's updated pose parameter values, the current video images, as well as the visibility and blending coefficients ν_i, ω_i for all vertices and cameras i are continuously transferred to the graphics card.

The color of each rendered pixel $c(j)$ is determined by blending all l video images I_i according to

$$c(j) = \sum_{i=1}^{l} \nu_i(j) * \rho_i(j) * \omega_i(j) * I_i(j) \qquad (22.5)$$

where $\omega_i(j)$ denotes the blending weight of camera i, $\rho_i(j)$ is the optional view-dependent rescaling factor, and $\nu_i(j) = \{0, 1\}$ is the local visibility. During texture preprocessing, the weight products $\nu_i(j)\rho_i(j)\omega_i(j)$ have been normalized to ensure energy conservation. Technically, Equation (22.5) is evaluated for each fragment by a fragment program on the graphics board. The rasterization engine interpolates the blending values from the triangle vertices.

By this means, time-varying cloth folds and creases, shadows and facial expressions are faithfully reproduced, lending a very natural, dynamic appearance to the rendered object. The computation of the blending weights and the visibility coefficients is explained in the following two subsections.

22.6.1 Blending Weights

The blending weights determine the contribution of each input camera image to the final color of a surface point. Although our model geometry is fairly detailed, it is still an approximation, and therefore the surfaces may locally deviate from the true shape of the human actor. If surface reflectance can be assumed to be approximately Lambertian, view-dependent reflection effects play no significant role, and high-quality, detailed model texture can still be obtained by blending the video images cleverly. Let θ_i denote the angle between a vertex normal and the optical axis of camera i. By emphasizing for each vertex individually the camera view with the smallest angle θ_i, i.e., the camera that views the vertex most head-on, a consistent, detail-preserving texture is obtained. A visually convincing weight assignment has been found to be

$$\omega_i = \frac{1}{(1 + \max_{j}(1/\theta_j) - 1/\theta_i)^{\alpha}} \qquad (22.6)$$

where the weights ω_i are additionally normalized to sum up to unity. The parameter α determines the influence of vertex orientation with respect to

(a) $\alpha = 0$ (b) $\alpha = 3$ (c) $\alpha = 15$

Fig. 22.7. Texturing results for different values of the control factor α.

camera viewing direction and the impact of the most head-on camera view per vertex, Figure 22.7. Singularities are avoided by clamping the value of $1/\theta_i$ to a maximal value.

Although it is fair to assume that everyday apparel has purely Lambertian reflectance, in some cases the reproduction of view-dependent appearance effects may be desired. To serve this purpose, our method provides the possibility to compute view-dependent rescaling factors, ρ_i, for each vertex on-the-fly while the scene is rendered:

$$\rho_i = \frac{1}{\phi_i} \tag{22.7}$$

where ϕ_i is the angle between the direction to the outgoing camera and the direction to input camera i.

22.6.2 Visibility

Projective texturing on graphics hardware has the disadvantage that occlusion is not taken into account, so hidden surfaces also get textured. The z-buffer test, however, allows determining for every time step which object regions are visible from each camera.

Due to the use of a parameterized geometry model, the silhouette outlines in the images do not correspond exactly to the outline of the model. When projecting video images onto the model, a texture seam belonging to some frontal body segment may fall onto another body segment farther back, Figure 22.8a. To avoid such artifacts, *extended soft shadowing* is applied: For each camera, all object regions of zero visibility are determined not only from the camera's actual position, but also from several slightly displaced virtual camera positions, Figure 22.8b. Each vertex is tested whether it is visible from all camera positions, actual as well as virtual. A triangle is textured by a camera image only if all of its three vertices are completely visible from that camera.

While too generously segmented silhouettes do not affect rendering quality, too small outlines can cause annoying non-textured regions. To avoid such rendering artifacts, all image silhouettes are expanded by a couple of pixels

(a)   (b)

Fig. 22.8. (a) Small differences between object silhouette and model outline cause erroneous texture projections (l) that can be removed using soft-shadow visibility computation. (b) Morphologically dilated segmented input video frames that are used for projective texturing.

prior to rendering. Using a morphological filter operation, the object outlines of all video images are dilated to copy the silhouette boundary pixel color values to adjacent background pixel positions, Figure 22.8.

22.6.3 Results

Our free-viewpoint video reconstruction and rendering approach has been tested on a variety of scenes, ranging from simple walking motion over karate performances to complex and expressive ballet dance. The sequences are between 100 and 400 frames long and were recorded from eight camera perspectives.

Ballet dance performances are ideal test cases as they exhibit rapid, complex motion. The motion capture subsystem demonstrates that it is capable of robustly following human motion involving fast arm motion, complex twisted poses of the extremities, and full body turns (Figure 22.9). The comparison to true input images shows that the virtual viewpoint renditions look very lifelike. Certainly, there are extreme body poses such as the fetal position that cannot be reliably tracked due to insufficient visibility. To our knowledge, no nonintrusive system has demonstrated that it is able to track such extreme positions. In combination with our texture generation approach convincing novel viewpoint renditions can be generated, as it is also shown in Figure 22.10. Subtle surface details, such as wrinkles in clothing, are nicely reproduced in the renderings. In Figure 22.10 snapshots from a freeze-and-rotate sequence, in which the body motion is stopped and the camera flies around the scene, are depicted.

The free-viewpoint renderer can easily replay dynamic 3D scenes at the original capture frame rate of 15 fps. The maximal possible frame rate is significantly higher. Standard TV frame rate of 30 fps can easily be attained even on a XEON 1.8 GHz CPU featuring a GeForce 3 GPU. On a Pentium IV with a GeForce 6800 frame rates of more than 100 fps are feasible.

Fig. 22.9. Novel viewpoints are realistically synthesized. Two distinct time instants are shown on the left and right with input images above and novel views below.

Fig. 22.10. Conventional video systems cannot offer moving viewpoints of scenes frozen in time. However, with our free-viewpoint video system *freeze-and-rotate* camera shots of body poses are possible. The pictures show such novel viewpoints of scenes frozen in time for different subjects and different types of motion.

We tested the execution speeds of individual algorithmic components of our reconstruction method. We have 5 PCs at our disposition, each featuring an Intel XEON 1.8 GHz CPU and a GeForce 3 GPU. On a single PC, silhouette fitting for pose capture takes between 3 s and 14 s. If we employ the optimized XOR evaluation and the parallel motion capture system [44], fitting times of below a second are feasible. To get an impression of how much the employed hardware influences the fitting performance, we have performed tests on a Pentium IV 3.0 GHz with a GeForce 6800. On this machine, fitting times of around a second are feasible even with the nonoptimized version of the tracker. We expect that with five such machines, fitting times in the range of 0.2 s per frame are feasible.

The augmented pose tracking can improve the accuracy of the recovered poses, however at the cost of increasing the fitting time [46]. Using eight cameras and 320x240 pixel images, it takes between 10 s and 30 s per time step to compute the 2D optical flows by means of the hierarchical Lukas Kanade approach, depending on what window size and hierarchy depth are chosen [46]. Any faster flow algorithm would be equally applicable. The reconstruction of the corrective flow field and the pose update only takes around 0.3 s even on the old XEON machine. The employment of the corrective motion field step leads to PSNR (Peak Signal-to-Noise Ratio) improvements with respect to the ground truth data in the range of 0.3–0.9 dB, depending on the sequence. Although these improvements are on a small scale, they may well be noticeable if one zooms into the scene.

The method presented in this Section is subject to a few limitations. First, the motion estimation process is done offline, making the system unsuitable for live broadcast applications. However, it is foreseeable that the ongoing performance advancement of graphics hardware will make this feasible in a few years time. The appearance of the actor cannot be faithfully represented if he wears very loose apparel. A further limitation is that we can currently only reproduce the appearance of the actor under the illumination conditions that prevailed at the time of recording. We describe how to overcome this limitation in the following Section.

Even though our approach exhibits these limitations, our results show that our method enables high-quality reconstruction of free-viewpoint videos. Convincing novel viewpoint renditions of human actors can be generated in real-time on off-the-shelf graphics hardware.

22.7 Relightable Free-viewpoint Video

In the previous section, we have introduced an approach to realistically render human actors for all possible synthetic viewpoints. However, this algorithm can only reproduce the appearance of the actor under the lighting conditions that prevailed at the time of acquisition. To implant a real-world actor into surroundings different from the recording environment, her appearance must be adapted to the new illumination situation. To this end, a description of the actor's surface reflectance is required. We have enhanced our original free-viewpoint video pipeline such that we are able to reconstruct such dynamic surface reflectance descriptions. The basic idea behind this process that we call dynamic reflectometry is that when letting a person move in front of a calibrated static setup of spot lights and video cameras the cameras are not only texture sensors but actually reflectance sensors. Due to the motion of the person, each point on the body surface is seen under many different incoming light and outgoing viewing directions. Thus, we can fit a dynamic surface reflectance model to each point on the body surface, which consists of a per-texel parametric bidirectional reflectance distribution function and a per-texel time-varying normal direction. We now briefly review the algorithmic steps needed to generate relightable free-viewpoint videos and refer the interested reader to [42, 48] for technical details.

22.7.1 Modifications to Original Pipeline

During recording, we employ two calibrated spot lights, i.e., we know their positions and photometric properties. For each person and each type of apparel, we record one sequence, henceforth termed reflectance sequence (RS), in which the person performs a rather simple rotation motion in front of the setup. The RS will be used to estimate the surface reflectance properties. The actual relightable free-viewpoint videos, as well as the time-varying normal map are reconstructed from the so-called dynamic sequences (DS), in which the actor can do arbitrary movements.

Reflectance estimation causes more strict requirements to the quality of the employed body model. In order to prevent rendering artifacts at body segment boundaries and to facilitate spatio-temporal texture registration, we transform the shape-adapted segmented body model (Section 22.4) into a single-skin model by means of an interactive procedure.

Prior to reflectance estimation we transform each input video frame into a 2D surface texture. Textural representation of surface attributes facilitates rendering of the relightable free-viewpoint videos and also enables us to take measures to enhance spatio-temporal multi-view texture registration. Incorrect multi-view registration would eventually lead to erroneous reflectance estimates. There are two primary reasons for inaccurate texture registration, first the fact that we use only an approximate model, and second, transversal shift of the apparel while the person is moving. We counter the first problem by warping the multi-view input images such that they comply with the body model. The motion of textiles is identified and compensated by optical flow computation and texture warping. Please refer to [42, 48] for the details of these methods.

22.7.2 Reflectance Estimation and Rendering

As stated previously, our dynamic reflectance model is comprised of two components, namely a parametric BRDF model, and a time-varying surface normal field. The BRDF is the quotient between outgoing radiance in one direction and incoming irradiance from another direction. Parametric representations of the BRDF are very advantageous, because they are able to represent the complex reflectance function in terms of a few parameters of a predefined functional skeleton. The BRDF thus compactly represents surface reflectance in terms of 4 direction parameters, the incoming light direction and the outgoing viewing direction, as well as the model parameters. In our approach, we can employ any arbitrary BRDF representation, but we mainly used the Phong [33] and Lafortune [22] models.

In the dynamic sequence, we collect several samples of the BRDF for each surface point, or in GPU terminology, for each texel. The goal is to find optimal model parameters for each texel that best reproduce the collected reflectance samples. We formulate this as the solution to a least-squares problem in the difference between collected samples and predicted appearance according to the current estimate. The current appearance is predicted by evaluating the illumination equation for each surface point given the current estimate. Once the BRDF model has been estimated for each texel, we can use it to estimate time-varying surface normals from every DS sequence in which the person wears the same clothing. This way, we can correctly capture and relight also time-varying surface details, such as wrinkles in clothing.

The BRDF parameters and normal maps are stored in respective texture images. At rendering time, the body model is displayed in the sequence of captured body poses and the illumination equation is, in graphics hardware, evaluated for each rendered fragment of the model. We can render and relight free-viewpoint videos in real-time on commodity graphics hardware. For illumination, it is feasible to use both normal point or directional light sources, or even captured real-world illumination from HDR environment maps. Example renderings of actors under novel virtual lighting conditions can be seen in Figure 22.12. Even subtle surface details are faithfully reproduced in the synthetic lighting environment, Figure 22.11.

22.7.3 Results and Discussion

We have tested our approach on several motion sequences showing three different actors, different styles of motion, and six different styles of apparel. The sequences were recorded with our new camera setup at 25 fps, thus providing 1004x1004 pixel frame resolution. The typical length of a RS sequence is around 300 frames, the length of the employed motion sequences was in the range of 300–500 frames. For our tests, we resorted to the Phong reflectance model.

Our dynamic scene description allows us to photo-realistically render human actors under both artificial and real world illumination that has been captured in high-dynamic range environment maps, see Figure 22.12a, b. Even with realistically cast shadows, relightable 3D videos can be displayed in real-time on commodity graphics hardware. We can also implant actors into virtual environments as they are commonly used in computer games, such as a little pavilion with mirroring floor, Figure 22.12b (right). Our dynamic reflectometry method faithfully captures time-varying surface details, such as wrinkles in clothing, Figure 22.11a. This way, they can be realistically displayed under varying artificial lighting conditions, Figure 22.11b.

Reflectance estimation typically takes one hour on a Pentium IV 3.0 GHz. Normal estimation takes approximately 50 s per time step, and it can be parallelized to reduce computation time. Optional input frame warping takes

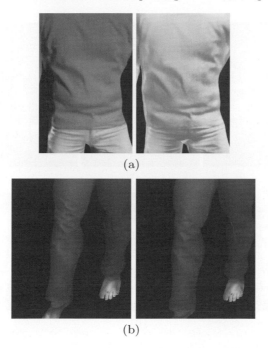

(a)

(b)

Fig. 22.11. (a) Comparison between an input frame (left) and the corresponding normal map that we reconstructed. For rendering, the three components of the surface normals were encoded in the RGB color channels. (Even in a grayscale rendition) one can see that the geometry and not only the texture of subtle surface details has been captured. (b) This level of accuracy also enables us to faithfully reproduce time-varying geometric details, such as the wrinkles in the trousers around the knee.

around 10 s for one pair of reference image and reprojected image. Cloth shift compensation accounts for an additional 35 s of computation time for one time step of video.

We have validated our dynamic reflectometry method both visually and quantitatively via comparison to ground truth image data and reflectance descriptions obtained with laser-scanned geometry. For a detailed elaboration on these evaluations, we would like to refer the reader to [42, 48].

Our approach is subject to a couple of limitations: The single-skin surface model is generated in an interactive procedure. The method presented in Section 22.8 would be one way to overcome these limitations. Furthermore, although we can handle normal everyday apparel, we can not account for loose apparel whose surface can deviate almost arbitrarily from the body model. The approach described in Chapter 12 could be employed to conquer this limitation. Sometimes, we observe small rendering artefacts due to undersampling (e.g., underneath the arms). However, we have verified that the

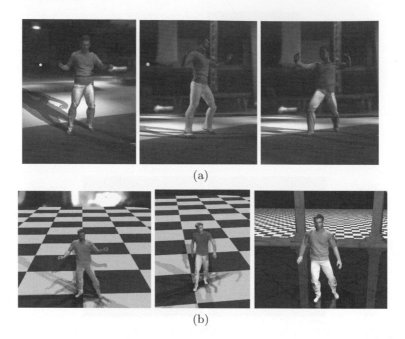

Fig. 22.12. (a) 3D videos rendered under different disco-type artificial lighting conditions. (b) Performances of different subjects rendered in captured real-world lighting conditions (l), (m) and in a game-type lighting environment (r). In either case, the actors appear very lifelike and subtle surface details are faithfully reproduced. Also shadows and mirroring effects can be rendered in real-time.

application of a RS sequence showing several rotation motions with different body postures almost completely solves this problem.

Despite these limitations, our method is an effective combination of algorithmic tools that allows for the creation of realistic relightable dynamic scene descriptions.

22.8 Future Directions

The commitment to a parameterized body model enables us to make the inverse problems of motion estimation and appearance reconstruction tractable. However, a model-based approach also implies a couple of limitations. Firstly, a template model is needed for each type of object that one wants to record. Secondly, we currently cannot handle people wearing very loose apparel. Chapter 12 explains a method to capture clothed people that we could combine with our appearance estimation framework. Furthermore, while a relatively smooth template model enables easy fitting to a wide range of body shapes, more detailed geometry specific to each actor would improve rendering quality even

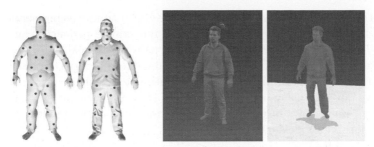

Fig. 22.13. (a) The motion of the template model (l) is mapped onto target (r) by only specifying correspondences between individual triangles. (b) We can now use the moving laser scan instead of the moving template model in our free-viewpoint video pipeline. The image on the left is an input frame, the image on the right the free-viewpoint video with laser-scanned geometry.

more. For instance, it would be intriguing to have a method at hand that enables us to make a high-quality laser scan follow the motion of the actor in each video frame without having to manually design skeleton models or surface skinning parameters.

To achieve this goal, we have developed a method that enables us to make our moving template model drive the motion of a high quality laser scan of the same person without resorting to a kinematic skeleton [9]. The user only needs to mark a handful of corresponding triangles between the moving template and the target mesh, Figure 22.13a. The transformations of the marked triangles on the source are mapped to their counterparts on the high-quality scan. Deformations for all the other triangles are interpolated on the surface by means of a harmonic field. The surface of the appropriately deformed scan at each time step is computed by solving a Poisson system. Our framework is based on the principle of differential mesh editing and only requires the solution of simple linear systems to map poses of the template to the target mesh. As an additional benefit, our algorithm implicitly solves the motion retargeting problem and automatically generates convincing non-rigid surface deformations. Figure 22.13b shows an example where we mapped the motion of our moving template model onto a high-quality static laser scan. This way, we can easily use detailed dynamic scene geometry as our underlying shape representation. For details on the correspondence specification and the mesh-deformation framework, we would like to refer the reader to [8, 9].

22.9 Conclusions

We have presented a compendium of methods from a young and challenging field of research on the boundary between Computer Vision and Computer Graphics. By jointly using an adaptable template body model and a marker-free motion capture approach we can capture both time-varying shape, time-varying appearance and even time-varying reflectance properties of moving actors. These dynamic scene models enable us to realistically display human performances in real-time from novel virtual camera perspectives and under novel virtual lighting conditions. The commitment to a shape prior enables us to achieve our goal using only a handful of cameras. In future, we will further investigate novel ways of incorporating high-quality shape models and more general clothing models into our framework.

Acknowledgments

We would like to thank the current and former coworkers who were cooperating with the authors on many of the projects described in this chapter. In particular, we would like to thank Edilson de Agiar, Naveed Ahmed, Joel Carranza, Hendrik Lensch, and Gernot Ziegler.

References

1. Ahmed N., deAguiar E., Theobalt T., Magnor M. and Seidel H.-P. Automatic generation of personalized human avatars from multi-view video. In *VRST '05: Proc. ACM symposium on Virtual reality software and technology*, pp. 257–260, Monterey, USA, December 2005. Association for Computing Machinery (ACM), ACM.
2. Alexa M., Cani M. and Singh K. Interactive shape modelling. *In Eurographics course notes (2005)*.
3. Bernardini F., Martin I. M. and Rushmeier H. High-quality texture reconstruction from multiple scans. *IEEE TVCG*, 7(4):318–332, 2001.
4. Boivin S. and Gagalowicz A. Image-based rendering of diffuse, specular and glossy surfaces from a single image. In *Proc. ACM SIGGRAPH 2001*, pp. 107–116, 2001.
5. Carceroni R. and Kutulakos K. Multi-view scene capture by surfel sampling: From video streams to non-rigid 3D motion shape & reflectance. In *ICCV*, pp. 60–67, 2001.
6. Carranza J., Theobalt C., Magnor M.A. and Seidel H.-P. Free-viewpoint video of human actors. In *Proc. SIGGRAPH'03*, p. 569–577, 2003.
7. deAguiar E., Theobalt C., Magnor M. and Seidel H.-P. Reconstructing human shape and motion from multi-view video. In *Proc. 2nd European Conference on Visual Media Production (CVMP)*, pp. 42–49, London, UK, December 2005. The IEE.
8. deAguiar E., Zayer R., Theobalt C., Magnor M. and Seidel H.-P. A framework for natural animation of digitized models. Research Report MPI-I-2006-4-003, Max-Planck-Institut fuer Informatik, Saarbruecken, Germany, July 2006.
9. deAguiar E., Zayer R., Theobalt C., Magnor M. and Seidel H.-P. Video-driven animation of human body scans. In *IEEE 3DTV Conference*, pageNN, Kos Island, Greece, to appear 2007.

10. Der K., Sumner R. and Popovic J. Inverse kinematics for reduced deformable models. *ACM Trans. Graph.*, 25(3):1174–1179, 2006.
11. Deutscher B., Blake A. and Reid I. Articulated body motion capture by annealed particle filtering. In *Proc. CVPR'00*, volume 2, 2126ff, 2000.
12. Einarsson P., Chabert C., Jones A., Ma W., Hawkins L., Bolas M., Sylwan S. and Debevec P. Relighting human locomotion with flowed reflectance fields. In *Rendering Techniques*, pp. 183–194, 2006.
13. Gardner A., Tchou C., Hawkins T. and Debevec P. Linear light source reflectometry. *ACM Trans. Graphics. (Proc. SIGGRAPH'03)*, 22(3):749–758, 2003.
14. Georghiades A. Recovering 3-d shape and reflectance from a small number of photographs. In *Eurographics Symposium on Rendering*, pp. 230–240, 2003.
15. Gibson S., Howard T. and Hubbold R. Flexible image-based photometric reconstruction using virtual light sources. *Computer Graphics Forum*, 20(3), 2001.
16. Goldman D., Curless B., Hertzmann A. and Seitz S. Shape and spatially-varying brdfs from photometric stereo. In *Proc. ICCV*, pp. 341–448, 2004.
17. Gross M., Würmlin S., Näf M., Lamboray E., Spagno C., Kunz A., Koller-Meier E., Svoboda T., Van Gool L., Lang S., Strehlke K. Moere A. and Staadt O. blue-c: a spatially immersive display and 3d video portal for telepresence. *ACM Trans. Graph. (Proc. SIGGRAPH'03)*, 22(3):819–827, 2003.
18. Hawkins T., Wenger A., Tchou C., Gardner A., Göransson F. and Debevec P. Animatable facial reflectance fields. In *Proc. Eurographics Symposium on Rendering*, pp. 309–319, 2004.
19. Horn B.K.P. Closed-form solution of absolute orientation using unit quaternions. *Journal of the Optical Sociey of America*, 4(4):629–642, 1987.
20. Huang J., Shi X., Liu X., Zhou K., Wei L., Teng S., Bao H., Guo B. and Shum H. Subspace gradient domain mesh deformation. *ACM Trans. Graph.*, 25(3):1126–1134, 2006.
21. Kanade T., Rander P. and Narayanan P.J. Virtualized reality: Constructing virtual worlds from real scenes. *IEEE MultiMedia*, 4(1):34–47, 1997.
22. Lafortune E., Foo S., Torrance K. and Greenberg D. Non-linear approximation of reflectance functions. In *Proc. SIGGRAPH'97*, pp. 117–126. ACM Press, 1997.
23. H. Lensch, W. Heidrich and H.P. Seidel. A silhouette-based algorithm for texture registration and stitching. *Graphical Models*, 64(3):245–262, 2001.
24. Lensch H., Kautz J., Goesele M., Heidrich W. and Seidel H.-P. Image-based reconstruction of spatial appearance and geometric detail. *ACM Transactions on Graphics*, 22(2):27, 2003.
25. Levoy M. and Hanrahan P. Light field rendering. In *Proc. ACM SIGGRAPH'96*, pp. 31–42, 1996.
26. Li M., Schirmacher H., Magnor M. and Seidel H.-P. Combining stereo and visual hull information for on-line reconstruction and rendering of dynamic scenes. In *Proc. IEEE Multimedia and Signal Processing*, pp. 9–12, 2002.
27. Matsuyama T. and Takai T. Generation, visualization and editing of 3D video. In *Proc. 1st International Symposium on 3D Data Processing Visualization and Transmission (3DPVT'02)*, 234ff, 2002.
28. Matusik W., Buehler C., Raskar R., Gortler S.J. and McMillan L. Image-based visual hulls. In *Proc. ACM SIGGRAPH 00*, pp. 369–374, 2000.
29. Matusik W., Pfister H., Brand M. and McMillan L. A data-driven reflectance model. *ACM Trans. Graph. (Proc. SIGGRAPH'03)*, 22(3):759–769, 2003.

30. Matusik W. and Pfister H. 3d tv: a scalable system for real-time acquisition, transmission and autostereoscopic display of dynamic scenes. *ACM Trans. Graph. (Proc. SIGGRAPH'04)*, 23(3):814–824, 2004.
31. Nishino K., Sato Y. and Ikeuchi K. eigen-texture method: Appearance compression and synthesis based on a 3d model. *IEEE Trans. PAMI*, 23(11):1257–1265, Nov. 2001.
32. Pebay P.P. and Baker T.J. A comparison of triangle quality measures. *In Proc. to the 10th International Meshing Roundtable*, pp. 327–340, 2001.
33. Phong B.-T. Illumnation for computer generated pictures. *Communications of the ACM*, pp. 311–317, 1975.
34. Press W., Teukolsky S., Vetterling W. and Flannery B. *Numerical recipes in C++*. Cambridge University Press, 2002.
35. Ramamoorthi R. and Hanrahan P. A signal-processing framework for inverse rendering. In *Proc. SIGGRAPH 2001*, pp. 117–128. ACM Press, 2001.
36. Rushmeier H., Taubin G. and Guéziec A. Applying Shape from Lighting Variation to Bump Map Capture. In *Eurographics Workshop on Rendering*, pp. 35–44, June 1997.
37. Sato Y., Wheeler M. and Ikeuchi K. Object Shape and Reflectance Modelling from Observation. In *Proc. SIGGRAPH'97*, pp. 379–388, 1997.
38. Shi L., Yu Y., Bell N. and Feng W. A fast multigrid algorithm for mesh deformation. *ACM Trans. Graph.*, 25(3):1108–1117, 2006.
39. Sorkine O. Differential representations for mesh processing. *Computer Graphics Forum*, 25(4), 2006.
40. Sumner R. and Popovic J. Deformation transfer for triangle meshes. *ACM Trans. Graph.*, 23(3):399–405, 2004.
41. Sumner R., Zwicker M., Gotsman C. and Popovic J. Mesh-based inverse kinematics. *ACM Trans. Graph.*, 24(3):488–495, 2005.
42. Theobalt C., Ahmed N., deAguiar E., Ziegler G., Lensch H., Magnor M. and Seidel H.-P. Joint motion and reflectance capture for creating relightable 3d videos. Research Report MPI-I-2005-4-004, Max-Planck-Institut fuer Informatik, Saarbruecken, Germany, April 2005.
43. Theobalt C., Carranza J., Magnor M. and Seidel H.-P. Enhancing silhouette-based human motion capture with 3d motion fields. In Jon Rokne, Reinhard Klein and Wenping Wang, editors, *11th Pacific Conference on Computer Graphics and Applications (PG-03)*, pp. 185–193, Canmore, Canada, October 2003. IEEE, IEEE.
44. Theobalt C., Carranza J., Magnor M. and Seidel H.-P. A parallel framework for silhouette-based human motion capture. In *Proc. Vision, Modelling and Visualization 2003 (VMV-03)*, pp. 207–214, Munich, Germany, November 2003.
45. Theobalt C., Carranza J., Magnor M. and Seidel H.-P. 3d video - being part of the movie. *ACM SIGGRAPH Computer Graphics*, 38(3):18–20, August 2004.
46. Theobalt C., Carranza J., Magnor M. and Seidel H.-P. Combining 3D flow fields with silhouette-based human motion capture for immersive video. *Graphical Models*, 66:333–351, September 2004.
47. Theobalt C., Li M., Magnor M. and Seidel H.-P. A flexible and versatile studio for multi-view video recording. In Peter Hall and Philip Willis, editors, *Vision, Video and Graphics 2003*, pp. 9–16, Bath, UK, July 2003. Eurographics, Eurographics.

48. Theobalt C., Ahmed N., Lensch H., Magnor M. and Seidel H.-P. Enhanced dynamic reflectometry for relightable free-viewpoint video,. Research Report MPI-I-2006-4-006, Max-Planck-Institut fuer Informatik, Saarbrücken, Germany, 2006.
49. Vedula S., Baker S., Rander P., Collins R. and Kanade T. Three-dimensional scene flow. In *Proc. 7th IEEE International Conference on Computer Vision (ICCV-99)*, volume II, pp. 722–729. IEEE.
50. Waschbüsch M., Würmlin S., Cotting D., Sadlo F. and Gross M. Scalable 3D video of dynamic scenes. In *Proc. Pacific Graphics*, pp. 629–638, 2005.
51. Wenger A., Gardner A., Tchou C., Unger J., Hawkins T. and Debevec P. Performance relighting and reflectance transformation with time-multiplexed illumination. In *ACM TOG (Proc. SIGGRAPH'05)*, volume 24(3), pp. 756–764, 2005.
52. Würmlin S., Lamboray E., Staadt O.G. and Gross M. 3d video recorder. In *Proc. IEEE Pacific Graphics*, pp. 325–334, 2002.
53. Yu Y., Debevec P., Malik J. and Hawkins T. Inverse global illumination: Recovering reflectance models of real scenes from photographs. In *Proc. ACM SIGGRAPH'99*, pp. 215–224, August 1999.
54. Yu Y. and Malik J. Recovering photometric properties of architectural scenes from photographs. In *Proc. ACM SIGGRAPH'98*, pp. 207–218, 1998.
55. Zayer R., Rössl C., Karni Z. and Seidel H.-P. Harmonic guidance for surface deformation. In Marc Alexa and Joe Marks, editors, *Proc. Eurographics 2005*, volume 24, pp. 601–609, 2005.
56. Zhang R., Tsai P., Cryer J. and Shah M. Shape from Shading: A Survey. *IEEE Trans. PAMI*, 21(8):690–706, 1999.
57. Zickler T., Enrique S., Ramamoorthi R. and Belhumeur P. Reflectance sharing: Image-based rendering from a sparse set of images. In *Proc. Eurographics Symposium on Rendering*, pp. 253–264, 2005.
58. Zitnick C., Kang S., Uyttendaele M., Winder S. and Szeliski R. High-quality video view interpolation using a layered representation. *ACM TOC (Proc. SIGGRAPH'04)*, 23(3):600–608, 2004.

23

Interacting Deformable Objects

Matthias Teschner[1], Bruno Heidelberger[2], and Matthias Müller-Fischer[2]

[1] University of Freiburg
[2] AGEIA Technologies, Inc.

Summary. This chapter discusses approaches for the efficient simulation of interacting deformable objects. Due to their computing efficiency, these methods might be employed in the model-based analysis of human motion.

The realistic simulation of geometrically complex deformable objects at interactive rates comprises a number of challenging problems, including deformable modelling, collision detection, and collision response. This chapter proposes efficient models and algorithms for these three simulation components. Further, it discusses the interplay of the components in order to implement an interactive system for interacting deformable objects.

A versatile and robust model for geometrically complex solids is employed to compute the dynamic behavior of deformable objects. The model considers elastic and plastic deformation. It handles a large variety of material properties ranging from stiff to fluid-like behavior. Due to the computing efficiency of the approach, complex environments consisting of up to several thousand primitives can be simulated at interactive speed.

Collisions and self-collisions of dynamically deforming objects are detected with a spatial subdivision approach. The presented algorithm employs a hash table for representing a potentially infinite regular spatial grid. Although the hash table does not enable a unique mapping of grid cells, it can be processed very efficiently and complex data structures are avoided.

Collisions are resolved with a penalty approach, i.e., the penetration depth of a colliding primitive is processed to compute a force that resolves the collision. The presented method considers the fact that only sampled collision information is available. In particular, the presented solution avoids non-plausible collision responses in case of large penetrations due to discrete simulation steps. Further, the problem of discontinuous directions of the penalty forces due to coarse surface representations is addressed.

All presented models and algorithms process tetrahedral meshes with triangulated surfaces. Due to the computing efficiency of all simulation components, complex environments consisting of up to several thousand tetrahedrons can be simulated at interactive speed. For visualization purposes, tetrahedral meshes are coupled with high-resolution surface meshes.

B. Rosenhahn et al. (eds.), Human Motion – Understanding, Modelling, Capture, and Animation, 561–596.

23.1 Introduction

There exist a variety of human motion analysis approaches that are based on models [1, 52, 53, 70, 77]. For example, modelling the human body with articulated rigid bodies allows for reconstructing the geometry of a human body based on detected body parts and joints. While some existing solutions consider the entire human body as non-rigid, they are nevertheless restricted to rigid body parts and elastically deformable structures are neglected. In this chapter, we discuss efficient solutions for the simulation of interacting deformable objects that might be employed in realistic human body models including deformable soft tissue. Augmenting articulated rigid bodies with efficient deformable structures would improve the flexibility of model-based tracking methods. Further, the accuracy and the robustness of human motion analysis approaches could be improved. We discuss an efficient model for the simulation of deformable solids and we present approaches to detect and to process colliding deformable objects.

23.2 Related Work

Deformable modelling: Deformable models have been extensively investigated in the last two decades [6, 32, 68, 85, 86]. Approaches based on mass-spring models [13], particle systems [24], or FEM [50, 66] have been used to efficiently represent 3D objects or deformable 2D structures, such as cloth [94] or discrete shells [38]. Typical applications for these approaches can be found in computational surgery [21] and games. Although very efficient algorithms have been proposed [20, 21, 95], these approaches could handle only a few hundred deformable primitives in real-time.

While many approaches are restricted to elastic deformation, models for plastic deformation have been introduced in [69, 87] . However, no real-time approximations of these models have been presented so far. In [38], a method to simulate the dynamic behavior of discrete shells has been described. Very promising results have been shown. However, the approach is computationally expensive. Many approaches focus on solutions to specific problems in deformable modelling. However, there exist no efficient, unified approach to physically plausible simulation of 2D and 3D deformable models with elasticity and plasticity. Further, there exist no framework where complex deformable objects can be handled with integrated collision handling at interactive rates.

Collision detection: Efficient collision detection is an essential component in physically-based simulation or animation [10, 21], including cloth modelling [13, 74, 93]. Further applications can be found in robotics, computer animation, medical simulations, computational biology, and games [90].

Collision detection algorithms based on bounding-volume (BV) hierarchies have proven to be very efficient and many types of BVs have been investigated. Among the acceleration structures we find spheres [48, 75, 83], axis-aligned bounding boxes [12, 49], oriented bounding boxes [35], and discrete-oriented polytopes [57]. In [96], various optimizations to BV hierarchies are presented.

Initially, BV approaches have been designed for rigid objects. In this context, the hierarchy is computed in a pre-processing step. In the case of deforming objects, however, this hierarchy must be updated at run time. While effort has been spent to optimize BV hierarchies for deformable objects [59], they still pose a substantial computational burden and storage overhead for complex objects. As an additional limitation for physically based applications, BV approaches typically detect intersecting surfaces. The computation of the penetration depth for collision response requires an additional step.

As an alternative to object partitioning, other approaches employ discretized distance fields as volumetric object representation for collision detection. The presented results, however, suggest that this data structure is less suitable for real-time processing of geometrically complex objects [44]. In [42], a hybrid approach is presented which uses BVs and distance fields.

Recently, various approaches have been introduced that employ graphics hardware for collision detection. In [3], a multi-pass rendering method is proposed for collision detection. However, this algorithm is restricted to convex objects. In [60], the interaction of a cylindrical tool with deformable tissue is accelerated by graphics hardware. [46] proposes a multi-pass rendering approach for collision detection of 2-D objects, while [54] and [55] perform closest-point queries using bounding-volume hierarchies along with a multi-pass-rendering approach. The aforementioned approaches decompose the objects into convex polytopes.

There exist various approaches that propose spatial subdivision for collision detection. These algorithms employ uniform grids [29,91,97] or BSPs [61]. In [91], spatial hashing for collision detection is mentioned, but no details are given. [64] presents a hierarchical spatial hashing approach as part of a robot motion planning algorithm which is restricted to rigid bodies.

We employ spatial hashing for the collision detection of deformable tetrahedral meshes. $\mathbb{R}^3$ is implicitly subdivided into small grid cells. Information about the global bounding box of our environment is not required and 3D data structures, such as grids or BSPs are avoided. Further, our approach inherently detects collisions and self-collisions. The parameters of the algorithm have been investigated and optimized.

Collision response: Contact models and collision response for rigid and deformable bodies are well investigated. Analytical methods for calculating the forces between dynamically colliding rigid bodies have been presented in [4,5,7,8,26,40,65,71]. These approaches solve inequality-constrained problems which are formulated as linear complementarity problems (LCP). In addition to analytical methods, a second class of collision response schemes is based on so-called penalty forces. These approaches calculate response forces based on penetration depths in order to resolve colliding objects. First solutions have been presented in [72, 85]. Penalty-based approaches have been used in simulations with deformable objects, cloth and rigid bodies [22,62,65]. A third approach which directly computes contact surfaces of colliding deformable objects is presented in [30].

Due to their computing efficiency, penalty-based approaches are very appropriate for interactive simulations of deformable objects. They can consider various elasto-mechanical object properties. Friction and further surface characteristics can also be incorporated. Penalty forces are computed based on penetration depths and there exist many approaches that compute the exact or approximative penetration depth of two colliding objects which is defined as the minimum translation that one object undergoes to resolve the collision. Exact penetration depth computations can be based on Minkowski sums [15,39] or hierarchical object presentations [23], while approximative solutions based on the GJK algorithm [33] and iteratively expanding polytopes have been presented in [14,16]. Further approaches are based on object space discretizations [27], employ graphics hardware [47,82], or introduce incremental optimization steps [56].

While existing approaches very efficiently compute the minimal penetration depth, they do not address inconsistency problems of the result in discrete-time simulations. One solution to this problem is continuous collision detection [76]. However, these approaches are computationally expensive compared to discrete collision detection approaches and not appropriate for deformable objects. This section presents an alternative solution to the inconsistency problems. The approach computes penetration depths which significantly reduce artifacts in the respective collision response scheme.

23.3 Deformable Modelling

There is a growing demand for interactive deformable modelling in computational surgery and entertainment technologies, especially in games and movie special effects. These applications do not necessarily require physically correct deformable models, but efficient deformable models with physically plausible dynamic behavior. Additionally, simulations should be robust and controllable, and they should run at interactive speed.

This section describes a unified method suitable for modelling deformable tetrahedral or triangulated meshes with elasticity and plasticity. The proposed model extends existing deformable modelling techniques by incorporating efficient ways for volume and surface area preservation. The computing efficiency of our approach is similar to simple mass-spring systems. Thus, environments of up to several thousand deforming primitives can be handled at interactive speed.

In order to optimize the dynamics computation various numerical integration schemes have been compared. Comparisons have been performed with respect to robustness and performance in the context of our model. The deformable modelling approach is integrated into a simulation environment that can detect and handle collisions between tetrahedral meshes (see Sections 23.4 and 23.5). For visualization purposes, deformable tetrahedral meshes are coupled with high-resolution surface meshes in the spirit of free-form deformation.

23.3.1 Model

We consider deformable solids that are discretized into tetrahedra and mass points. In order to compute the dynamic behavior of objects we derive forces at mass points from potential energies. These forces preserve distances between mass points, they preserve the surface area of the object, and they preserve the volume of tetrahedra. The material properties of a deformable object are described by weighted stiffness coefficients of all considered potential energies.

Potential Energies

In order to represent deformations of objects, we consider constraints of the form $C(\mathbf{p}_0, \ldots, \mathbf{p}_{n-1})$. These scalar functions depend on mass point positions $\mathbf{p}_i$. They are zero if the object is undeformed. In order to compute forces based on these constraints we consider the potential energy

$$E(\mathbf{p}_0, \ldots, \mathbf{p}_{n-1}) = \frac{1}{2} kC^2 \qquad (23.1)$$

with k denoting a stiffness coefficient. This coefficient has to be defined for each type of potential energy. The potential energy is zero if the object is undeformed. Otherwise, the energy is larger than zero. The potential energies of our model are independent of rigid body modes of the object. The overall potential energy derived from our constraints can be interpreted as deformation energy of the object. Now, forces at mass points $\mathbf{p}_i$ are derived as

$$\mathbf{F}^i(\mathbf{p}_0, \ldots, \mathbf{p}_{n-1}) = -\frac{\partial}{\partial \mathbf{p}_i} E = -kC \frac{\partial C}{\partial \mathbf{p}_i} \qquad (23.2)$$

The overall force at a mass point is given as the sum of all forces based on potential energies that consider this mass point. Damping which significantly improves the robustness of the dynamic simulation can be incorporated as

$$\mathbf{F}^i(\mathbf{p}_0, \ldots, \mathbf{p}_{n-1}, \mathbf{v}_0, \ldots, \mathbf{v}_{n-1}) = \left(-kC - k_d \sum_{0 \leq j < n} \frac{\partial C}{\partial \mathbf{p}_j} \mathbf{v}_j \right) \frac{\partial C}{\partial \mathbf{p}_i} \qquad (23.3)$$

with $\mathbf{v}_i$ denoting the velocity of a mass point and k_d denoting the damping coefficient. We do not consider any additional constraints or boundary conditions for our forces. In contrast to similar approaches [13, 17, 73], we do not explicitly bound potential energies or forces resulting from the energies. The direction of a force $\mathbf{F}$ based on a potential energy E corresponds to the negative gradient of E, i.e., a dynamic simulation resulting from these forces reduces the deformation energy of an object. Further, these forces are orthogonal to rigid body modes, i.e., they conserve linear and angular momentum of the object.

Distance Preservation

The first potential energy E_D considers all pairs of mass points that are connected by tetrahedral edges. E_D represents energy based on the difference of the current distance of two points and the initial or rest distance D_0 with $D_0 \neq 0$:

$$E_D(\mathbf{p}_i, \mathbf{p}_j) = \frac{1}{2} k_D \left(\frac{|\mathbf{p}_j - \mathbf{p}_i| - D_0}{D_0} \right)^2 \tag{23.4}$$

Forces $\mathbf{F}_D$ resulting from this energy are computed as stated in Equation (23.3). While damping of theses forces is very useful to improve the stability of the numerical integration process, experiments have shown no significant improvement of the stability if damping is applied to forces resulting from the other two energies that we consider in our deformation model.

Surface Area Preservation

The second energy E_A considers triples of mass points that build surface triangles. E_A represents energy based on the difference of the current area of a surface triangle and its initial area A_0 with $A_0 \neq 0$:

$$E_A(\mathbf{p}_i, \mathbf{p}_j, \mathbf{p}_k) = \frac{1}{2} k_A \left(\frac{\frac{1}{2} |(\mathbf{p}_j - \mathbf{p}_i) \times (\mathbf{p}_k - \mathbf{p}_i)| - A_0}{A_0} \right)^2 \tag{23.5}$$

Forces $\mathbf{F}_A$ based on this energy are computed as stated in Equation (23.2). Preservation of surface area is considered in the animation of discrete shells and thin plates. In the animation of volumetric tetrahedral meshes the effect of this energy is negligible.

Volume Preservation

Our third potential energy E_V considers sets of four mass points that build tetrahedra. E_V represents energy based on the difference of the current volume of a tetrahedron and its initial volume V_0:

$$E_V(\mathbf{p}_i, \mathbf{p}_j, \mathbf{p}_k, \mathbf{p}_l) = \frac{k_V}{2} \frac{\left(\frac{1}{6} (\mathbf{p}_j - \mathbf{p}_i) \cdot ((\mathbf{p}_k - \mathbf{p}_i) \times (\mathbf{p}_l - \mathbf{p}_i)) - V_0 \right)^2}{\tilde{V}_0^2} \tag{23.6}$$

with $\tilde{V}_0 = V_0$ if our model is applied to volumetric tetrahedral meshes. In this case we assume $V_0 \neq 0$. However, if our model is applied to thin plates or discrete shells, we can not assume $V_0 \neq 0$. In this case we use $\tilde{V}_0 = \frac{\sqrt{2}}{12} \bar{l}^3$ with $\bar{l}$ denoting the average edge length of a tetrahedron. Then, $\tilde{V}_0$ corresponds to the volume of a regular tetrahedron with edge length $\bar{l}$. Based on E_V forces $\mathbf{F}_V$ are computed as stated in Equation (23.2).

The preservation of the signed volume as it is calculated with the mixed product in Equation (23.6) is of major importance to our deformation model.

In contrast to the energies E_D and E_A which are not sensitive to inverted tetrahedra, forces based on E_V preserve the initial orientation of the vectors in the mixed product. If a tetrahedron is inverted and the orientation of these vectors changes, the sign of the volume represented with the mixed product in Equation (23.6) changes accordingly. Thus, inverting a tetrahedron results in forces $\mathbf{F}_V$ that restore the original orientation of the tetrahedron. Due to the normalization of all constraints that are considered in the potential energies the stiffness coefficients k_D, k_A, k_V are scale-invariant. These stiffness coefficients can be used to mimic a wide range of material properties. Refer to Section 23.3.4 for an overview of sets of stiffness coefficients for various materials.

23.3.2 Tetrahedral and Triangulated Meshes

The proposed deformable model is designed to work with tetrahedral meshes. However, the proposed deformation model can also be applied to arbitrary triangle meshes. In this case, distance-preserving forces are considered for all edges and area-preserving forces are considered for all triangles of the mesh. Employing these forces, a dynamic simulation preserves all distances between mass points and the surface area. However, there is no resistance of the model against bending which is essential in cloth and discrete shell simulation [9, 13, 28, 38, 41, 73].

In order to control bending of triangulated surface we generate a tetrahedron for each pair of adjacent triangles with one common edge. If vertices opposite to this common edge are connected with an additional edge, a virtual tetrahedron is generated. Now, preservation of the volume of the virtual tetrahedron, while also preserving the area of the triangles, corresponds to a preservation of the angle between the two adjacent triangles. If this angle is zero, the volume of the virtual tetrahedron is zero. Otherwise, the volume is larger or smaller than zero depending on a concave or convex angle. If the volume-preserving force is considered for a virtual tetrahedron, its volume and the respective angle between adjacent triangles are preserved. In case of thin plates or cloth, the rest angle is zero. In case of discrete shells, arbitrary rest angles can occur.

23.3.3 Numerical Integration

In order to compute the dynamic behavior of our deformable models, Newton's equation of motion is applied to all mass points. Based on initial values for positions and velocities, a time step h, internal forces at mass points resulting from our deformation energies, and external forces, such as gravity, piecewise linear trajectories for all mass points are calculated employing a numerical integration scheme. Based on performance comparisons we have chosen the Verlet scheme for numerical integration [92]. This method has been very popular in molecular dynamics for decades and has recently been proposed in the context of physically-based simulation of cloth, general mass-spring systems, and rigid bodies [28, 41, 51]. The Verlet algorithm uses positions and forces at time t, and positions at the previous time $t - h$ to calculate new positions at time $t + h$:

$$\mathbf{x}(t + h) = 2\mathbf{x}(t) - \mathbf{x}(t - h) + h^2 \frac{\mathbf{F}(t)}{m} + O(h^4)$$

$$\mathbf{v}(t + h) = \frac{\mathbf{x}(t + h) - \mathbf{x}(t - h)}{2h} + O(h) \qquad (23.7)$$

with $\mathbf{F}(t) = \mathbf{F}_D(t) + \mathbf{F}_A(t) + \mathbf{F}_V(t)$. The Verlet method has several advantages in environments with interacting, dynamically deforming objects. First, only one force computation is required per integration step. This is essential, since force computation is the most expensive part in the calculation of an integration step. Second, the integration of positions has a local discretization error of $O(h^4)$. This high accuracy allows for comparatively large time steps. Third, the integration of positions is independent of the integration of velocities if undamped forces are used. Depending on the application this could be employed to omit the integration of velocities which would further improve the performance. However, we do not use this property, since damping of distance-preserving forces is essential for the robustness of our model. Further, collision response, as utilized in some of the experiments, requires velocities of mass points.

Computing Time

In our experiments, we distinguish between computing efficiency of a numerical integration scheme which is discussed in this section, and the performance of an integration method which is discussed in the following section. In a first experiment, we have tested the computing efficiency of the Verlet scheme in environments with dynamically deforming objects of varying geometrical complexity. Table 23.1 shows the computing time for one numerical integration step with various deformable objects that are depicted in Figure 23.1. Our measurements show that 1500 forces can be updated at 1 KHz, while more complex objects with 14000 forces can be updated at 140 Hz. Since we are interested in an interactive behavior of our simulations, it is essential to have update rates of the numerical integration that are above 20 Hz.

Note that the computing time for an integration step does not correspond to the performance of an integration method. In order to assess the performance, the ratio of integration time step and computing time has to be considered. This problem is addressed in the following section.

Comparison to Other Approaches

In order to optimize the performance of our dynamically deforming objects, we have implemented and compared several numerical integration methods that

Table 23.1. Computing time for one Verlet step (Intel Pentium 4, 2.8 GHz). The number of mass points is given for each model. Further, n_D, n_A, n_V denote the numbers of considered distance-preserving, area-preserving, and volume-preserving forces per integration step, respectively. In case of $n_A = 0$, the object surface is not considered in the deformation model.

Setup	Points	n_D	n_A	n_V	Time [ms]
Cube 1	8	18	0	5	0.03
Cuboid	242	981	0	500	1.00
Face	472	2622	874	1277	2.26
Cloth	1301	5478	1826	2739	5.03
Cube 2	1331	6930	0	5000	7.11
Santa	915	7500	2500	3700	7.36
Membrane	20402	90801	0	50000	114.61

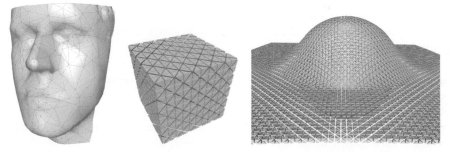

Fig. 23.1. Models used for performance measurements: face, cube, membrane. Geometrical complexity and computing times are given in Table 23.1.

have been proposed in previous approaches to physically-based deformation of mass-spring or particle systems. We are not only interested in maximal time steps or minimal computing time, but in optimal performance. Therefore, we propose to consider the ratio of the numerical integration time step and the computing time for one numerical integration time step as performance measure.

As a test case, we have applied various integration schemes to a cube with 1331 mass points, considering 5000 volume-preserving forces and 6930 distance-preserving forces. We have implemented and compared the following integration schemes: Verlet [92], velocity Verlet [84], Runge-Kutta, Beeman [11], explicit Euler, Leap-Frog [45], Heun, implicit Theta-Scheme, and Gear [31]. Table 23.2 shows measurements of time steps and computing times for various integration methods. Our measurements suggest that Verlet and Leap-frog can be computed very efficiently, while providing a reasonable time step. Although the fourth order Runge-Kutta scheme allows for a larger time step, its computation is significantly more expensive. Two classes of integration schemes, namely predictor-corrector methods (Gear) and implicit

Table 23.2. Maximum time step and computing time for various integration schemes. The last column shows the ratio of time step and computing time for one integration step. This ratio can be interpreted as performance measure of an integration method.

Method	Time step [ms]	comp. Time [ms]	Ratio
Verlet	3.1	7.3	0.427
Leap-frog	3.1	7.3	0.426
RK 2	4.9	14.3	0.342
vel. Verlet	2.5	7.3	0.341
Beeman	2.5	7.4	0.337
Heun	4.2	18.4	0.229
expl. Euler	1.5	7.3	0.205
RK 4	6.3	33.0	0.191

methods (Theta), have been implemented and considered in the comparison, but are not listed in Table 23.2. Both methods suffer from drawbacks that are related to our deformable model and to our application.

Although the Gear scheme is very efficient and robust, we do not consider it in our environment. This is due to the fact that the Gear scheme has to be re-initialized after collision handling which significantly reduces the stability of the method. Since collision handling is an important component in our environment, Gear is not appropriate for our application. Implicit integration schemes have shown to be very robust in physically-based simulations [9, 20, 21, 37, 94]. They are very popular, since they allow for large time steps. On the other hand, they are expensive to compute. Implicit methods require to solve a sparse linear system per integration step. Further, computing and storing the system matrix causes additional costs. In our application with comparatively complex objects with several thousand degrees of freedom, these costs are significant. Although, we use an efficient Conjugate Gradient algorithm with only a few iterations (5–30), we do not achieve computing times faster than 200 ms in our test scenario. This is not appropriate for our application.

A second aspect is the combination of dynamically deforming objects and collision handling. Larger time steps cause larger penetration depths of objects which are difficult to resolve. Robust collision response commonly requires a small intersection volume of two colliding objects. This is difficult to guarantee, if the time step is too large. In some of the experiments, collision handling is employed. In these experiments, the limiting factor for the time step is not the numerical integration, but our collision handling scheme.

From our perspective, the optimal numerical integration scheme does not depend on the size of the time step. Instead, it depends on the underlying model, on the application, and on the complexity of the data structures. Although implicit integration methods are very useful in many applications, we propose to use Verlet or Leap-Frog for our specific problem.

23.3.4 Results

In this section, we describe some applications of our deformable model. Examples with elastically deforming volumetric tetrahedral meshes are given and it is shown how to apply the approach to interactive simulations of cloth and discrete shells. Further, the incorporation of plasticity into the deformable model is described and it is illustrated how to incorporate the variation of stiffness constants in the animation of melting or fluid-like objects. Finally, an overview of all parameters and the performance of the simulations are given.

Elastic Deformation

We have integrated our model into a simulation environment for deformable objects. In this environment, our tetrahedral meshes can be coupled with high-resolution surface meshes in the spirit of FFD [18,63]. Figure 23.2 illustrates this visualization technique. Further, collisions between deformable objects can be handled based on a spatial hashing approach presented in Section 23.4. If more than one deformable tetrahedral mesh is used, this collision handling scheme is employed.

Figure 23.3 shows a sequence of interactive cloth simulation to illustrate that our approach can also be employed to interactively animate thin plates and discrete shells.

Plastic Deformation

In addition to elastic deformation, the proposed model can also handle plastic deformation as introduced in [69,87]. Therefore, the deformation of an object is decomposed into elastic and plastic deformation, whereas the plastic deformation does not contribute to the deformation energy of an object. In our model, this can be represented employing the energy E_D which represents

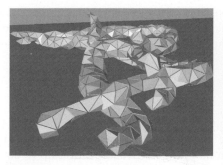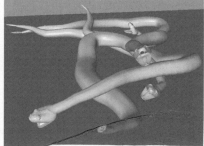

Fig. 23.2. A low-resolution tetrahedral mesh and a high-resolution surface mesh of a snake. Deformation is computed for the tetrahedral mesh, while the high-resolution mesh is visualized.

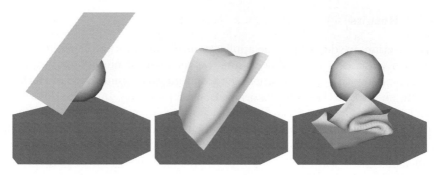

Fig. 23.3. Sequence of a deformable cloth model interacting with a rigid sphere.

distance differences. Therefore, E_D is decomposed into two components for elasticity and plasticity:

$$E_D = E_D^{elastic} + E_D^{plastic} \qquad (23.8)$$

This corresponds to a decomposition of forces $\mathbf{F}_D$ resulting from E_D:

$$\mathbf{F}_D = \mathbf{F}_D^{elastic} + \mathbf{F}_D^{plastic} \qquad (23.9)$$

In the case of elastic deformation, E_D contributes to the deformation energy of an object and F_D is applied to minimize this energy. However, if plasticity is considered, only the elastic part $E_D^{elastic}$ contributes to the deformation energy. Therefore, only $\mathbf{F}_D^{elastic} = \mathbf{F}_D - \mathbf{F}_D^{plastic}$ is considered accordingly in the numerical integration process. In order to model the plastic evolution, O'Brien [69] proposes three parameters that we have implemented in our model. The first parameter specifies a minimum value for E_D that must be met before the decomposition of elasticity and plasticity occurs. This represents the fact that small deformations are only elastic. The second parameter provides a maximum value for $E_D^{plastic}$. This parameter controls the maximum amount of deformation that can be stored by an object. The third parameter specifies the rate of plastic flow. This parameter can be used to model hysteresis of a material. Further details on these three parameters can be found in [69]. Plastic deformation is only considered in the distance-preserving forces. Volume preservation and - if applied - surface area preservation is not affected by plasticity.

Performance and Parameters

The complexity of the simulation scenarios varies from 700 tetrahedrons to 3700 tetrahedrons. Depending on the quality of the tetrahedrons and the additional computing costs for collision handling and visualization up to 5000 tetrahedrons can be handled at interactive rates. Note, that the numerical integration process itself is capable of handling up to 25000 tetrahedrons at interactive rates.

The most relevant parameter for material properties is k_D. The stiffness coefficient k_V is responsible for the volume preservation. This parameter also avoids inverted tetrahedrons. The parameter k_A is only used for discrete shells and thin plates. In the case of deformable triangle meshes it is difficult to map values of individual parameters to certain properties such as resistance against stretch, shearing, or bending. We have not further investigated the correlation between the stiffness coefficients and material properties for triangulated meshes. The damping coefficient k_d improves the stability of dynamic simulations. Damping is only applied to distance-preserving forces, thus reducing internal oscillations of mass points. Experiments have shown that there exist optimal values for k_d. If k_d is too large, energy is unintentionally added to the simulation.

In our experiments, the time step for the integration is usually similar to the computing time required to compute this time step. If the ratio of both values is one, the simulation runs at real-time. The performance of the visualization obviously depends on the complexity of our surface meshes. If the visualization is too expensive, we usually perform several simulation steps until the scene is rendered. However, a rendering rate of more than 20 Hz is always guaranteed.

23.3.5 Conclusion

In this section, a versatile and robust model has been presented that can be used to represent deformable tetrahedral meshes and deformable triangle meshes. The model considers elastic and plastic deformation. It handles a large variety of material properties ranging from stiff to fluid-like behavior. The proposed model extends existing deformable modelling techniques by incorporating efficient ways for volume and surface area preservation. The computing efficiency of the approach is similar to simple mass-spring systems. Thus, environments of up to several thousand deforming primitives can be handled at interactive speed. Experiments have been described to show the capabilities of the simulation system with integrated collision handling which is described in the remaining part of this chapter. Ongoing work focusses on the integration of the presented deformable model into surgery simulators. First projects investigate potential applications in hysteroscopy simulation and simulation of stent placement.

23.4 Collision Detection

In order to realistically process the interaction between deformable objects, efficient collision detection algorithms are required. Further, the information provided by the collision detection approach should allow for an efficient and physically correct collision response (see Section 23.5).

This section describes an algorithm for the detection of collisions and self-collisions of deformable objects based on spatial hashing. The algorithm classifies all object primitives, i.e. vertices and tetrahedrons, with respect to small axis-aligned bounding boxes AABB. Therefore, a hash function maps the 3D boxes (cells) to a 1D hash table index. As a result, each hash table index contains a small number of object primitives that have to be checked against each other for intersection. Since a hash table index can contain more than one primitive from the same object as well as primitives from different objects, self-collisions and collisions of different objects can be detected. The actual collision detection test computes barycentric coordinates of a vertex with respect to a penetrated tetrahedron. This information can be employed to estimate the penetration depth for a pair of colliding tetrahedrons. The penetration depth can be used for further processing, such as collision response.

Using a hash function for spatial subdivision is very efficient. While spatial subdivision usually requires one preprocessing pass through all object primitives to estimate the global bounding box and the cell size, this pass can be omitted in our approach. On the other hand, the hash mechanism does not always provide a unique mapping of grid cells to hash table entries. If different 3D grid cells are mapped to the same index, the performance of the proposed algorithm decreases. In order to reduce the number of index collisions, we have optimized the parameters of the collision detection algorithm that have an impact on this problem, namely the characteristics of the hash function, the hash table size, and the 3D cell size. This section investigates these factors.

Further, this section presents experimental results that have been obtained using physically-based environments for deformable objects with varying geometrical complexity. Environments with up to 20000 tetrahedrons can be tested for collisions and self-collisions in real-time on a PC. The remainder of this section on collision detection is organized as follows. First, the proposed algorithm is explained. The relevant parameters of the algorithm are introduced and their influence on the performance is investigated. Results and experiments are described. Finally, limitations of our approach are discussed, followed by directions for ongoing research.

23.4.1 Method

The collision detection algorithm implicitly subdivides $\mathbb{R}^3$ into small AABBs. In a first pass, all vertices of all objects are classified with respect to these small 3D cells. In a second pass, all tetrahedrons are classified with respect to the same 3D cells. If a tetrahedron intersects with a cell, all vertices that have been associated with this cell in the first pass, are checked for interference with this tetrahedron. The actual intersection test computes barycentric coordinates of a vertex with respect to a tetrahedron in order to estimate, whether a vertex penetrates a tetrahedron.

The consistent processing of all object primitives enables the detection of collisions and self-collisions. If a vertex penetrates a tetrahedron, a collision is detected. If the vertex and the tetrahedron belong to the same object, a self-collision is detected. If a vertex is part of a tetrahedron, the intersection test is omitted.

Spatial Hashing of Vertices: In the first pass, the positions of all vertices are discretized with respect to a user-defined cell size. Therefore, the coordinates of the vertex position (x, y, z) are divided by the given grid cell size l and rounded down to the next integer (i, j, k): $i = \lfloor x/l \rfloor$, $j = \lfloor y/l \rfloor$, $k = \lfloor z/l \rfloor$. The hash function *hash* maps the discretized 3D position (i, j, k) to a 1D index h and the vertex and object information is stored in a hash table at this index h: $h = hash(i, j, k)$. In addition to generating a hash value for each vertex, this first pass also computes the AABBs of all tetrahedrons based on their current deformed state.

Spatial Hashing of Tetrahedrons: While the first pass has considered all vertices to build the hash table and to update the AABBs of the tetrahedrons, the second pass of the algorithm traverses all tetrahedrons. First, the minimum and maximum values describing the AABB of a tetrahedron, are discretized. Again, these values are divided by the user-defined cell size and rounded down to the next integer. Second, hash values are computed for all cells affected by the AABB of a tetrahedron. Therefore, all cells are traversed from the discretized minimum to the discretized maximum of the AABB (see Figure 23.4). All vertices found at the corresponding hash table index are tested for intersection.

Intersection Test: If a vertex $\mathbf{p}$ and a tetrahedron t are mapped to the same hash index and $\mathbf{p}$ is not part of t, a penetration test has to be performed. The actual intersection test consists of two steps. First, $\mathbf{p}$ is checked against the AABB of t which has been updated in the first pass. If $\mathbf{p}$ penetrates the AABB

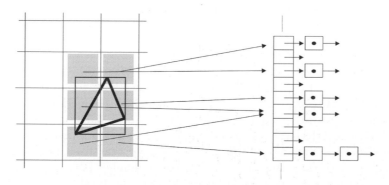

Fig. 23.4. Hash values are computed for all grid cells covered by the AABB of a tetrahedron. The tetrahedron is checked for intersection with all vertices found at these hash indices.

of t, the second step actually tests, whether $\mathbf{p}$ is inside t. This test computes barycentric coordinates of $\mathbf{p}$ with respect to a vertex of t.

23.4.2 Parameters

In this section, we investigate the parameters of the presented algorithm. The characteristics of the hash function, the size of the hash table, the size of a 3D cell for spatial subdivision, and the actual intersection test influence the performance of the algorithm. We have optimized all these aspects of the algorithm.

Hash Function

In the first pass of the algorithm, hash values are computed for all discretized vertex positions. These hash values should be uniformly distributed to guarantee an adequate performance of the algorithm. The hash function has to work with vertices of the same object that are close to each other, and with vertices of different objects that are farther away. We have tested several hash functions in our implementation, basically variants of additive and rotating hash functions. The hash function gets three values, describing a vertex position (x, y, z), and returns a hash value $hash(x,y,z) = (\ x\,p1 \ \mathbf{xor} \ y\,p2 \ \mathbf{xor} \ z\,p3)\ \mathbf{mod}$ n, where $p1, p2, p3$ are large prime numbers, in our case 73856093, 19349663, 83492791, respectively. The value n is the hash table size. The function can be evaluated very efficiently and produces a comparatively small number of hash collisions for small hash tables. The quality of the hash function is less important for larger hash tables.

Hash Table Size

The size of the hash table significantly influences the performance of the collision detection algorithm. Experiments indicate that larger hash tables reduce the risk of mapping different 3D positions to the same hash index. Therefore, the algorithm generally works faster with larger hash tables. On the other hand, the performance slightly decreases for larger hash tables due to memory management. Figure 23.5 and Figure 23.6 show the performance of our algorithm for two test scenarios with a varying hash table size. If the hash table is significantly larger than the number of object primitives, the risk of hash collisions is minimal. Although it is known that hash functions work most efficiently if the hash table size is a prime number [19], Figure 23.5 and Figure 23.6 show performance measurements with hash table sizes of 99, 199, 299 and so on.

Our implementation of the hash table does not require a reinitialization in each simulation step which would reduce the efficiency in case of larger tables. To avoid this problem, each simulation step is labeled with a unique

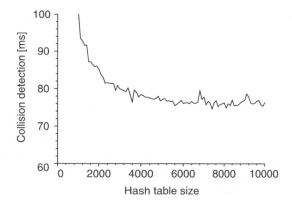

Fig. 23.5. Performance of the collision detection algorithm for two deformable objects with an overall number of 5898 vertices and 20514 tetrahedrons with varying hash table size of 99, 199, 299 and so on. The grid cell size is set to the average edge length of all tetrahedrons.

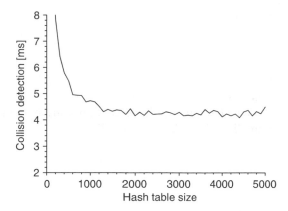

Fig. 23.6. Performance of the collision detection algorithm for 100 deformable objects (see Figure 23.15) with an overall number of 1200 vertices and 1000 tetrahedrons with varying hash table size of 99, 199, 299 and so on. The grid cell size is set to the average edge length of all tetrahedrons.

time stamp. If the first pass stores vertices in a hash table cell with outdated time stamp, the time stamp is updated and the cell is reset before new vertices are inserted. If the time stamp is up to date, new vertices are appended to the hash table cell. When the second pass generates hash indices for the tetrahedrons, the current time stamp is compared to the time stamp found in the hash table entry. If the time stamps differ, the information in the hash table is outdated and no intersection tests have to be performed. Therefore, no reinitialization of the hash table has to be performed during the simulation which would be comparatively costly for larger hash tables.

Grid Cell Size

The grid cell size which is used for spatial hashing, influences the number of object primitives that are mapped to the same hash index. In case of larger cells, the number of primitives per hash index increases and the intersection test slows down. If the cell size is significantly smaller than a tetrahedron, the tetrahedron covers a larger number of cells and has to be checked against vertices in a larger number of hash entries. The measurements in Figures 23.7 and 23.8 indicate that a grid cell should have the size of the average edge length of all tetrahedrons to achieve optimal performance. The graphs illustrate that the grid cell size has a more significant impact on the performance than hash table size or hash function.

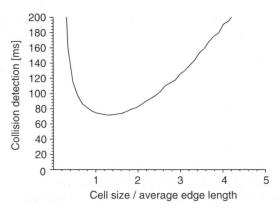

Fig. 23.7. Performance of the collision detection algorithm for two deformable objects with an overall number of 5898 vertices and 20514 tetrahedrons with varying grid cell size. Hash table size is 9973.

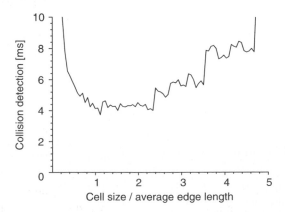

Fig. 23.8. Performance of the collision detection algorithm for 100 deformable objects with an overall number of 1200 vertices and 1000 tetrahedrons with varying grid cell size. Hash table size is 4999.

Intersection Test

Now we have to detect, whether a vertex $\mathbf{p}$ penetrates a tetrahedron t, whose vertices are at positions $\mathbf{x}_0, \mathbf{x}_1, \mathbf{x}_2, \mathbf{x}_3$. *Barycentric coordinates with respect to* $\mathbf{x}_0$: We express $\mathbf{p}$ in new coordinates $\beta = (\beta_1, \beta_2, \beta_3)^T$ with respect to a coordinate frame, whose origin coincides with $\mathbf{x}_0$ and whose axis coincide with the edges of t adjacent to $\mathbf{x}_0$: $\mathbf{p} = \mathbf{x}_0 + \mathbf{A}\beta,$, where $\mathbf{A} = [\mathbf{x}_1 - \mathbf{x}_0, \mathbf{x}_2 - \mathbf{x}_0, \mathbf{x}_3 - \mathbf{x}_0]$ is a 3 by 3 dimensional matrix. The coordinates β of $\mathbf{p}$ in this new coordinate frame are: $\beta = \mathbf{A}^{-1}(\mathbf{p} - \mathbf{x}_0)$. Now, $\mathbf{p}$ lies inside tetrahedron t, if $\beta_1 \geq 0$, $\beta_2 \geq 0$, $\beta_3 \geq 0$ and $\beta_1 + \beta_2 + \beta_3 \leq 1$.

23.4.3 Time Complexity

Let n be the number of primitives (vertices and tetrahedrons). To find all intersecting vertex-tetrahedron pairs a naive approach would test all vertices against all tetrahedrons resulting in a time complexity of the order of $O(n^2)$. The goal of our approach is to reduce this complexity to $O(n)$. Since a deformation algorithm needs to process all the primitives at each time step, linear time complexity for collision detection does not decrease its performance significantly.

During the first pass, all vertices are inserted into the hash table. This pass takes $O(n)$ time. First, the hash table does not need to be initialized, so the time is independent of the hash table size. Second, for each vertex, the hash function can be evaluated and a vertex reference can be added to the hash cell in $O(1)$ time. In the second pass, for all tetrahedrons all vertices in a local neighborhood are tested for collision. The time complexity of this pass is of the order of $O(n \cdot p \cdot q)$ where p is the average number of cells intersected by a tetrahedron and q is the average number of vertices per cell. If the cell size is chosen to be proportional to the average tetrahedron size, p is a constant. If there are no hash collisions, the average number of tetrahedrons per cell is constant too and so is q, since there are at most four times as many vertices as tetrahedra in a cell. With both, p and q being constant, the time complexity of the algorithm turns out to be linearly dependent on the number of primitives.

23.4.4 Results

We have performed experiments with various setups of deformable objects (see Table 23.3). The dynamic behavior of all deformable objects is computed with the constraint-based approach described in Section 23.3. Experiments indicate that the detection of all collisions and self-collisions for dynamically deforming objects can be performed with 15 Hz with up to 20k tetrahedrons and 6k vertices on a PC, Intel Pentium 4, 1.8 GHz. The performance is independent from the number of objects. It only depends on the number of object primitives. The performance varies slightly during simulations due to the changing number of hash collisions and due to a varying distribution of hash table elements.

Table 23.3. Performance of the collision detection algorithm. Average collision detection time, minimum, maximum, and standard deviation for 1000 simulation step are given.

Objects	Tetras	Vertices	Ave [ms]	Min [ms]	Max [ms]	Dev [ms]
100	1000	1200	4.3	4.1	6.5	0.24
8	4000	1936	12.6	11.3	15.0	0.59
20	10000	4840	30.4	28.9	34.4	1.25
2	20514	5898	70.0	68.5	72.1	0.86
100	50000	24200	172.5	170.5	174.6	1.08

23.4.5 Discussion

The presented algorithm performs two passes on the objects, even though it would be sufficient to only perform one pass which computes hash values for all tetrahedrons. In this case, each hash table entry contains tetrahedrons that could intersect. We have implemented this approach, but found it less efficient compared to the two-pass approach. While vertex positions are mapped to the hash table exactly once, tetrahedrons are usually mapped to several hash indices which leads to a larger number of elements in the hash table, thus decreasing the performance of the algorithm.

The comparison of the performance with other existing collision detection approaches is difficult. There exist numerous public domain libraries, such as RAPID [35], PQP [58], and SWIFT [25]. However, these approaches are not optimized for deformable objects. They work with data structures that can be pre-computed for rigid bodies, but have to be updated in case of deformable objects. The collision detection approach has been implemented based on tetrahedrons. However, it is not restricted to tetrahedrons and could handle other object primitives as well by simply replacing the intersection test. Since the algorithm provides the exact position of a vertex inside a penetrated tetrahedron, we can employ this information for collision response which is described in the following section. If we assume that a face or a vertex of the penetrated tetrahedron is part of the object surface, we can easily derive the penetration depth which allows for a correct collision response.

23.4.6 Conclusion

We have presented a method for detecting collisions and self-collisions of dynamically deforming objects. Instead of computing the global bounding box of all objects and explicitly performing a spatial subdivision, we propose to use a hash function that maps 3D cells to a hash table, thus realizing a very efficient, implicit spatial subdivision. The actual vertex-in-tetrahedron test is based on Barycentric coordinates. It provides information that can be used for physically based collision response. We have investigated and optimized the parameters of our approach. Experiments, performed with various test scenarios, show that environment of up to 20k tetrahedrons can be processed in real-time, independent from the number of objects.

23.5 Collision Response

In order to realistically simulate the behavior of colliding objects, an appropriate collision response has to be considered. One idea commonly used in discrete-time simulations is to generate forces which eventually separate colliding objects. These response or penalty forces are computed for penetrating object vertices as a function of their penetration depth which represents the distance and the direction to the surface of the penetrated object. In case of deformable objects, this force computation is intended to reflect the fact that real colliding objects deform each other. The deformation induces forces in the contact area which are approximated with penetration depth approaches in virtual environments. Response forces commonly consider additional features such as friction which is computed as a function of the relative velocity of colliding structures and their penetration depth.

Penetration depth approaches work very well for sufficiently dense sampled surfaces and in case of small penetrations. However, in interactive discrete-time simulations with discretized object representations, these two requirements are rarely met. Depending on the size of the simulation time step, large penetrations can occur which result in the computation of non-plausible penetration depths and directions. Figure 23.9 illustrates this problem. Further, discrete surface representations can result in discontinuous penetration directions. These discontinuities illustrated in Figure 23.10 degrade the stability of the response process.

In this section, we present a method to compute consistent penetration depths and directions for colliding tetrahedral meshes with triangulated surfaces. In contrast to approaches that only consider one closest surface feature,

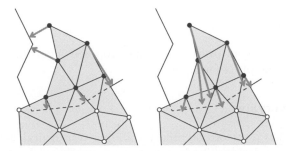

Fig. 23.9. The presented approach addresses the problem of non-plausible penetration depth estimation. Instead of strictly computing minimal distances as illustrated in the left-hand image, the approach computes consistent penetration distances as illustrated in the right-hand image. Therefore, collision response artifacts in discrete-time simulations are significantly reduced.

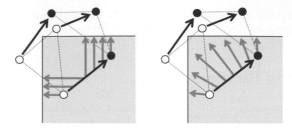

Fig. 23.10. The presented approach also addresses the problem of discontinuous penetration depths for small displacements of penetrating vertices as illustrated in the left-hand image. Instead, smooth and plausible approximations are computed which reduce artifacts in the collision response scheme.

the presented approach considers a set of close surface features to significantly reduce discontinuities of estimated penetration depth directions for small displacements of penetrating vertices. Further, a propagation scheme is introduced to approximate the penetration depth and direction for vertices with deep penetrations. This significantly reduces artifacts of the penetration direction in case of large penetrations.

The method works with any underlying deformation model and any contact model that computes penalty forces based on a given penetration depth. The scheme requires a volumetric collision detection approach. Although the method is primarily intended to work with deformable objects, it can also be applied to rigid bodies. The method has been integrated into a collision response scheme for dynamically deforming tetrahedral meshes. Experiments show that the scheme significantly reduces artifacts compared to standard penetration depth approaches. It provides a plausible collision response for a wide range of simulation time steps even in case of large object penetrations. The scheme works with objects of any geometrical complexity, but is especially advantageous for coarsely sampled objects.

23.5.1 Method

This section provides an overview of the proposed algorithm followed by a detailed description of its four stages.

Algorithm Overview

The method takes a set of potentially colliding tetrahedral meshes as input and computes consistent n-body penetration depths and directions for all colliding mesh points. The method proceeds in four consecutive stages:

Stage 1 detects all *colliding points* in the scene based on a spatial hashing approach. **Stage 2** identifies all colliding points adjacent to one or more non-colliding points as *border points*. Further, it detects all *intersecting edges* that contain one non-colliding point and one border point. The exact *intersection point* and corresponding surface normal of the penetrated surface are computed for each intersection edge. **Stage 3** approximates the penetration depth and direction for each border point based on the adjacent intersection points and surface normals obtained from the second stage. **Stage 4** propagates the penetration depth and direction to all colliding points that are not border points.

As a result of this algorithm, all colliding mesh points in the scene have an appropriate penetration depth and direction. This information can be used as input to any penalty-based collision response scheme. For our experiments, we use linear response forces. Further, surface friction is considered.

Point Collisions

The first stage detects all object points that collide with any tetrahedral mesh in the scene as described in Section 23.4. At the end of the first stage, all mesh points in the scene are either classified as colliding points or non-colliding points (see Figure 23.11).

Edge Intersections

The second stage identifies all colliding points with at least one adjacent non-colliding point as border points. The underlying idea is to classify colliding points with respect to their penetration depth. Based on this information, the second stage finds all intersecting edges that contain one non-colliding point and one border point. Moreover, the exact intersection point of each of those edges with the surface along with the corresponding surface normal of the penetrated mesh is computed. In order to efficiently compute this information, the original spatial hashing approach has been extended to handle collisions between edges and surfaces.

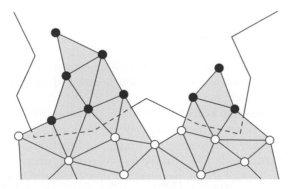

Fig. 23.11. The first stage classifies all mesh points either as colliding points (black) or non-colliding points (white).

584   M. Teschner et al.

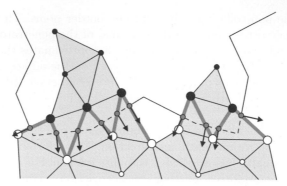

Fig. 23.12. The second stage finds all intersecting edges (red) of the tetrahedral meshes that contain one non-colliding point (white) and one border point (black). Further, the exact intersection point and the corresponding surface normal are computed for each intersection edge.

In a first step, all intersecting edges are classified with respect to the hash grid cells by using an efficient voxel traversal technique, e.g. [2]. In a second step, a simplified box-plane intersection test [36] is performed to classify all mesh faces. If a face intersects with a hash grid cell, all associated edges of the cell are checked for intersection with the respective face. The actual intersection test computes Barycentric coordinates of the intersection point in order to detect, whether the edge intersects the face or not. In addition, the Barycentric coordinates can also be used to interpolate a smooth surface normal based on the three vertex normals of the face. This results in a smooth approximation of the penetration direction. Each edge can possibly intersect with more than one mesh face. Therefore, only the intersection point nearest to the non-colliding point of the edge is considered in further stages. At the end of the second stage, each border point is adjacent to one or more intersection edges. Further, all intersecting edges have an exact intersection point and a corresponding surface normal (see Figure 23.12).

Penetration Depth and Direction

The third stage approximates the penetration depth and direction for all border points based on the adjacent intersection points and surface normals computed in the second stage.

First, the influence on a border point is computed for all adjacent intersection points. This influence is dependent on the distance between an intersection and a border point. The respective weighting function has to be positive for all nonzero distances and increasing for decreasing distances. Further, it has to ensure convergence to the penetration depth information with respect to a intersection point $\mathbf{x}_i$ if a colliding point $\mathbf{p}$ approaches $\mathbf{x}_i$. This leads to the following weighting function for the influence $\omega(\mathbf{x}_i, \mathbf{p})$:

$$\omega(\mathbf{x}_i, \mathbf{p}) = \frac{1}{\|\mathbf{x}_i - \mathbf{p}\|^2} \qquad (23.10)$$

with $\mathbf{x}_i$ denoting an intersection point and $\mathbf{p}$ denoting the border point. The weighting function does not have to be normalized, since this would not avoid any normalization steps in further processing. The weight is undefined for coinciding points. However, the first stage ensures that there is no collision detected in this case. The penetration depth $d(\mathbf{p})$ of a border point $\mathbf{p}$ is now computed based on the influences resulting from Equation (23.10):

$$d(\mathbf{p}) = \frac{\sum_{i=1}^{k}(\omega(\mathbf{x}_i, \mathbf{p}) \cdot (\mathbf{x}_i - \mathbf{p}) \cdot \mathbf{n}_i)}{\sum_{i=1}^{k} \omega(\mathbf{x}_i, \mathbf{p})} \qquad (23.11)$$

with $\mathbf{n}_i$ denoting the unit surface normal of the penetrated object surface at the intersection point. The number of intersection points adjacent to the border point $\mathbf{p}$ is given by k. Finally, the penetration direction $\hat{\mathbf{r}}(\mathbf{p})$ of a border point is computed as a weighted average of the surface normals

$$\hat{\mathbf{r}}(\mathbf{p}) = \frac{\sum_{i=1}^{k}(\omega(\mathbf{x}_i, \mathbf{p}) \cdot \mathbf{n}_i)}{\sum_{i=1}^{k} \omega(\mathbf{x}_i, \mathbf{p})} \qquad (23.12)$$

and the normalized penetration direction $\mathbf{r}(\mathbf{p})$ is obtained as

$$\mathbf{r}(\mathbf{p}) = \frac{\hat{\mathbf{r}}(\mathbf{p})}{\|\hat{\mathbf{r}}(\mathbf{p})\|}. \qquad (23.13)$$

At the end of the third stage, consistent penetration depths and directions have been computed for all border points (see Figure 23.13). In contrast to existing penetration depth approaches that consider only one distance, the weighted averaging of distances and directions provides a continuous behavior of the penetration depth function for small displacements of colliding points and for colliding points that are adjacent to each other. Non-plausible penetration directions due to the surface discretization of the penetrated object are avoided.

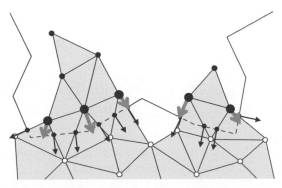

Fig. 23.13. The third stage approximates the penetration depth and direction for all border points based on the adjacent intersection points and surface normals.

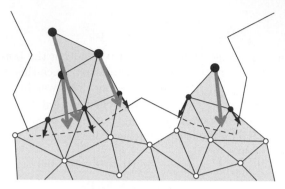

Fig. 23.14. Stage 4 propagates the penetration depth and direction to all colliding points that are not border points.

Propagation

Based on the computed penetration depth information for border points, the fourth stage propagates the information to all other colliding points that are not border points (see Figure 23.14). This is in contrast to existing penetration depth approaches that compute the penetration depth for all points independently. The idea of the propagation scheme is to avoid non-plausible penetration depths in case of large penetrations.

The propagation is an iterative process that consists of the following two steps: First, the current border points are marked as *processed points*. Second, a new set of border points is identified from all colliding points that are adjacent to one or more processed points. The iteration is aborted, if no new border points are found. Otherwise, the penetration depth and direction for the new border points is computed based on the information available from all adjacent processed points. A weighting function is used to compute the influence $\mu(\mathbf{p}_j, \mathbf{p})$ of an adjacent processed point $\mathbf{p}_j$ on a border point $\mathbf{p}$:

$$\mu(\mathbf{p}_j, \mathbf{p}) = \frac{1}{\|\mathbf{p}_j - \mathbf{p}\|^2}. \tag{23.14}$$

Based on the influences $\mu(\mathbf{p}_j, \mathbf{p})$, the penetration depth $d(\mathbf{p})$ of a border point $\mathbf{p}$ is computed as:

$$d(\mathbf{p}) = \frac{\sum_{j=1}^{l}(\mu(\mathbf{p}_j, \mathbf{p}) \cdot ((\mathbf{p}_j - \mathbf{p}) \cdot \mathbf{r}(\mathbf{p}_j) + d(\mathbf{p}_j)))}{\sum_{j=1}^{l} \mu(\mathbf{p}_j, \mathbf{p})}$$

with $\mathbf{r}(\mathbf{p}_j)$ denoting the normalized penetration direction of the processed point $\mathbf{p}_j$ and $d(\mathbf{p}_j)$ denoting its penetration depth. The number of processed points adjacent to the border point $\mathbf{p}$ is given by l.

Finally, the penetration direction $\hat{\mathbf{r}}(\mathbf{p})$ is computed as a weighted average of the penetration direction of the processed points adjacent to the border point as

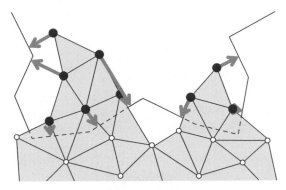

Fig. 23.15. The algorithm computes consistent penetration depths and directions for all colliding points.

$$\hat{\mathbf{r}}(\mathbf{p}) = \frac{\sum_{j=1}^{l} \mu_j \mathbf{r}_j}{\sum_{j=1}^{l} \mu_j}. \tag{23.15}$$

and normalized

$$\mathbf{r}(\mathbf{p}) = \frac{\hat{\mathbf{r}}(\mathbf{p})}{\|\hat{\mathbf{r}}(\mathbf{p})\|}. \tag{23.16}$$

At the end of the fourth stage, all colliding points have a consistent penetration depth and direction assigned (see Figure 23.15).

23.5.2 Results

We have integrated our method in a simulation environment for deformable objects based on [89]. Various experiments have been carried out to compare the quality and performance of the proposed method with the standard closest-feature approach. All test scenarios presented in this section have been performed on a PC Pentium 4, 3 GHz, GeForce FX Ultra 5900 GPU.

In a first test, two deformable cubes consisting of 1250 tetrahedrons are simulated. Large penetrations between the objects occur due to the high relative velocity and the discrete-time simulation. As illustrated in Figure 23.16, the standard approach fails to compute a consistent penetration depth. This results in a non-plausible collision response. Employing our approach to the same scenario results in consistent, plausible penetration depth information.

The second scenario simulates 120 deformable spheres consisting of 2400 tetrahedrons. Starting from a random position, they build a stack of spheres. Computing the penetration depth with the standard approach leads to heavy artifacts. The spheres tend to stick together due to inconsistent handling of penetrated object points. In this case, inconsistent penetration depths and response forces cause non-plausible equilibrium states. By applying our approach, these response artifacts are avoided. Figure 23.17 illustrates this second experiment. Our approach scales linearly with the number of colliding

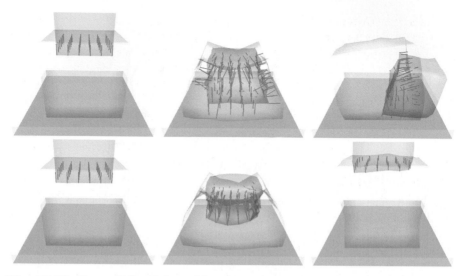

Fig. 23.16. Two colliding deformable cubes. The standard closest-feature approach shown in the first row causes non-plausible penetration depth information in case of large penetrations. This causes artifacts in the collision response scheme which are eliminated with the presented approach illustrated in the second row.

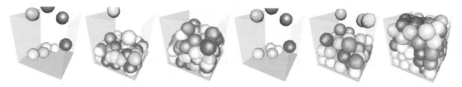

Fig. 23.17. 120 colliding deformable spheres. The first three images illustrate the sticking artifact of the standard penetration depth approach. These non-plausible equilibrium states are avoided with the presented approach as shown in the three images on the right-hand side.

points. In all experiments presented in this section, an average time of 35 μs is needed for resolving a colliding point. Most time is spent for detecting the edge intersections required by the second stage of the method. We experienced similar computing costs to calculate the closest feature in the standard approach.

23.5.3 Discussion

While the presented approach eliminates many collision response artifacts inherent to existing approaches, there still exist configurations where a plausible collision response can not be computed. If a colliding object is entirely enclosed by the penetrated object, the algorithm presented in this section does not compute any penetration depth, since there are no border points. The response scheme would not generate any forces until at least one object point leaves the penetrated object. In contrast, standard approaches would compute penetration depth information for all object points and probably resolve the collision in an arbitrary direction. However, if at least one object point of a colliding object is outside the penetrated object, the presented approach is likely to compute plausible and consistent penetration depth information for all colliding points. Further, there exist cases of objects crossing each other, where neither the existing nor the proposed approach are able to compute useful penetration depth information.

The presented approach does not compute the penetration depth according to its definition. Instead of computing the shortest distance to the surface of the penetrated object, the approach approximates the penetration depth only for points close to the surface. For all colliding non-border points, the depth is propagated from border points without considering the penetrated object. This supports consistency, but leads to results that can differ significantly from the actual penetration depth according to the definition. However, this disregard of the definition eliminates many artifacts in the respective collision response scheme. Further, if colliding points converge to the surface of a penetrated object, the computed penetration depth converges to the exact penetration depth.

23.5.4 Conclusions

We presented an approach to consistent penetration depth estimation. In discrete-time simulations, the method eliminates many collision response artifacts inherent to existing penetration depth approaches. Instead of computing only the closest surface feature for colliding points, a set of surface features is considered to avoid dynamic discontinuities of the penetration depth function. Further, the penetration depth is only computed for colliding points close to the surface, whereas consistent information is propagated to colliding points with larger penetrations. In general, the algorithm is faster than standard penetration depth approaches due to the propagation process. Experiments with dynamically deforming objects have illustrated some advantages of the consistent penetration depth estimation compared to existing methods.

23.6 Interactive Simulation of Interacting Deformable Objects

The combination of the presented approaches for deformable modelling, collision detection and penetration depth estimation is surprisingly straightforward and a first application in the area of computational medicine can be seen in Figure 23.18. All approaches work with tetrahedral meshes. While the deformable model and the collision detection also work with other object primitives, the penetration depth approach is limited to volumetrically sampled objects, since the depth propagation does not work for objects that are

Fig. 23.18. A first prototype of a hysteroscopy simulator is shown. Seven deformable polyps are placed into the cavity of a deformable uterus. With the presented simulation components, the objects and their interaction can be simulated at interactive rates.

represented with surface meshes. All approaches provide information that is necessary for the subsequent component. The mass-point positions are governed by the deformable model. These positions are employed by the collision detection stage. Further, the collision detection method estimates pairs of penetrating mass points and penetrated tetrahedrons. This information is used by the last component that computes the actual penetration depth of a point. Based on the penetration depth, a penalty force is computed that is passed to the deformable model as an external force. All simulation components can handle comparatively large time steps. While the penetration depth approach is particularly appropriate for large time steps, the explicit numerical integration scheme that is employed to compute the object dynamics is only conditionally stable. Nevertheless, the integration method is not the limiting factor for the time step. Instead, the time step is limited by the discrete collision detection scheme that misses collisions in case of large relative object velocities.

23.7 Conclusion

This chapter has illustrated approaches for the efficient processing of dynamically interacting deformable objects. While specific solutions to three simulation components have been considered, there is still extensive potential for future research. In terms of deformable modelling, combinations of efficient geometrically motivated [68] and accurate physically based models [67] might be investigated in order to enable the balancing of performance and accuracy within the simulation. In terms of collision detection, continuous approaches might enable larger simulation time steps and GPU-based approaches might improve the overall simulation performance [90]. Further, it might be useful to employ the deformation model in order to accelerate the collision detection [79]. In the area of collision response, the consistent penetration depth approach might be used to determine contact surface information which would allow for physically correct collision handling [78, 81]. Additionally, further simulation components could be investigated, e.g., mesh generation [80] and geometric constraints [34]. Since human motion is not only governed by deformable objects, rigid object dynamics and the two-way coupling of rigid and deformable solids might be investigated.

References

1. Aggarwal J. K. and Cai Q., Human Motion Analysis: A Review, *Computer Vision and Image Understanding: CVIU*, vol. 73, no. 3, 428–440, 1999.
2. Amanatides J. and Woo A., A Fast Voxel Traversal Algorithm for Ray Tracing, *Proc. Eurographics*, 3–9, 1987.
3. Baciu G., Wong W., Sun H., RECODE: an Image-based Collision Detection Algorithm, *The Journal of Visualization and Computer Animation*, vol. 10, 181–192, 1999.
4. Baraff D., Analytical Methods for Dynamic Simulation of Non-penetrating Rigid Bodies, *ACM SIGGRAPH Computer Graphics, Proc. 16th Annual Conference on Computer Graphics and Interactive Techniques SIGGRAPH '89*, pp. 223–232, 1989.
5. Baraff D., Coping with Friction for Non-Penetrating Rigid Body Simulation, *Computer Graphics*, vol. 25, no. 4, 31–40, 1991.
6. Baraff D., Witkin A., Dynamic Simulation of Non-penetrating Flexible Bodies, *Computer Graphics*, vol. 26, no. 2, 303–308, 1992.
7. Baraff D., Issues in Computing Contact Forces for Non-Penetrating Rigid Bodies, *Algorithmica*, vol. 10, 292–352, 1993.
8. Baraff D., Fast Contact Force Computation for Non-penetrating Rigid Bodies, *ACM SIGGRAPH '94: Proc. 21st Annual Conference on Computer Graphics and Interactive Techniques*, pp. 23–34, 1994.
9. Baraff D., Witkin A., Large Steps in Cloth Simulation, *ACM SIGGRAPH '98: Proc. 25th Annual Conference on Computer Graphics and Interactive Techniques*, pp. 43–54, 1998.
10. Baraff D., *Collision and contact*, ACM SIGGRAPH Course Notes, 2001.
11. Beeman D., Some Multistep Methods for use in Molecular Dynamics Calculations, *Journal of Computational Physics*, vol. 20, 130–139, 1976.
12. van den Bergen G., Efficient Collision Detection of Complex Deformable Models Using AABB Trees, *Journal of Graphics Tools*, vol. 2, no. 4, 1–13, 1997.
13. Bridson R., Fedkiw R., Anderson J., Robust Treatment of Collisions, Contact and Friction for Cloth Animation, *ACM Transactions on Graphics (TOG), Proc. 29th Annual Conference on Computer Graphics and Interactive Techniques SIGGRAPH '02*, vol. 21, no. 3, 594–603, 2002.
14. van den Bergen G., Proximity Queries and Penetration Depth Computation on 3D Game Objects, *Proc. Game Developers Conference*, 2001.

15. Cameron S. A., Culley R. K., Determining the Minimum Translational Distance Between Two Convex Polyhedra, *Proc. International Conference Robotics and Automation*, pp. 591–596, 1986.
16. Cameron S., Enhancing GJK: Computing Minimum and Penetration Distance Between Convex Polyhedra, *Proc. International Conference Robotics and Automation*, pp. 3112–3117, 1997.
17. Caramana E., Burton D., Shashkov M., Whalen P., The Construction of Compatible Hydrodynamics algorithms Utilizing Conservation of Total Energy, *Journal of Computational Physics,* vol. 146, 227–262, 1998.
18. Chadwick J., Haumann D., Parent R., Layered Construction for Deformable Animated Characters, *ACM SIGGRAPH Computer Graphics, Proc. 16th Annual Conference on Computer Graphics and Interactive Techniques SIGGRAPH '89,* pp. 243–252, 1989.
19. Cormen T., Leiserson C., Rivest R., *Introduction to Algorithms,* ISBN 0-262-03141-8, The MIT Press, Cambridge, Massachusetts, 1990.
20. Debunne G., Desbrun M., Cani M.-P., Barr A. Adaptive Simulation of Soft Bodies in Real-Time, *Proc. Symposium on Computer Animation,* pp. 133–144, 2000.
21. Debunne G., Desbrun M., Cani M.-P., Barr A. Dynamic Real-time Deformations Using Space & Time Adaptive Sampling, *Proc. 28th Annual Conference on Computer Graphics and Interactive Techniques SIGGRAPH '01,* pp. 31–36, 2001.
22. Desbrun M., Schröder P., Barr A., Interactive Animation of Structured Deformable Objects, *Proc. Graphics Interface,* pp. 1–8, 1999.
23. Dobkin D., Hershberger J., Kirkpatrick D., Suri S., Computing the Intersection Depth of Polyhedra, *Algorithmica,* vol. 9, 518–533, 1993.
24. Eberhardt B., Weber A., Strasser W., A Fast, Flexible Particle-System Model for Cloth Draping, *IEEE Computer Graphics and Applications,* vol. 16, no. 5, 52–59, 1996.
25. Ehmann S., Lin M., *SWIFT: Accelerated proximity queries between convex polyhedra by multi-level voronoi marching,* Technical Report TR00-026, Univ. of North Carolina at Chapel Hill, 2000.
26. Faure F., An Energy-Based Method for Contact Force Computation, *Proc. Eurographics,* pp. 357–366, 1996.
27. Fisher S., Lin M. C., Deformed Distance Fields for Simulation of Non-Penetrating Flexible Bodies, *Proc. Workshop Computer Animation and Simulation,* pp. 99–111, 2001.
28. Fuhrmann A., Gross C., Luckas V., Interactive Animation of Cloth Including Self Collision Detection, *Proc. WSCG,* University of West Bohemia, Czech Republic, pp. 141–148, 2003.
29. Ganovelli F., Dingliana J., O'Sullivan C., BucketTree: Improving Collision Detection Between Deformable Objects, *Proc. Spring Conference on Computer Graphics,* 2000.
30. Gascuel M.-P., An Implicit Formulation for Precise Contact Modelling Between Flexible Solids, *Proc. 20th Annual Conference on Computer Graphics and Interactive Techniques SIGGRAPH '93,* pp. 313–320, 1993.
31. Gear C., *Numerical Initial Value Problems in Ordinary Differential Equations,* Prentice Hall, Englewood Cliffs, New Jersey, 1971.

32. Gibson S., Mitrich B., *A Survey of Deformable Models in Computer Graphics*, Technical Report TR-97-19, Mitsubishi Electric Research Laboratories MERL, Cambridge, Massachusetts, 1997.

33. Gilbert E. G., Johnson D. W., Keerthi S. S., A Fast Procedure for Computing the Distance Between Objects in Three-Dimensional Space, *IEEE Journal Robotics and Automation*, vol. 4, 193–203, 1988.

34. Gissler M., Becker M., Teschner M., Local Constraint Methods for Deformable Objects, *Proc. virtual reality interactions and physical simulations VriPhys*, Madrid, Spain, pp. 25–32, Nov. 6–7, 2006.

35. Gottschalk S., Lin M., Manocha D., OBBTree: A hierarchrical structure for rapid interference detection, *Proc. 23rd Annual Conference on Computer Graphics and Interactive Techniques SIGGRAPH '96*, pp. 171–180, 1996.

36. Greene N., Detecting Intersection of a Rectangular Solid and a Convex Polyhedron, *Graphics Gems IV*, pp. 74–82, 1994.

37. Grispun E., Krysl P., Schröder P., CHARMS: A Simple Framework for Adaptive Simulation, *ACM Transactions on Graphics (TOG), Proc. 29th Annual Conference on Computer Graphics and Interactive Techniques SIGGRAPH '02*, vol. 21, no. 3, 281–290, 2002.

38. Grispun E., Hirani A., Desbrun M., Schröder P., Discrete Shells, *Proc. ACM SIGGRAPH/Eurographics Symposium on Computer Animation*, pp. 62–67, 2003.

39. Guibas L., Seidel R., Computing Convolutions by Reciprocal Search, *Proc. Symposium on Computational Geometry*, pp. 90–99, 1986.

40. Hahn J. K., Realistic Animation of Rigid Bodies, *ACM SIGGRAPH Computer Graphics, Proc. 15th Annual Conference on Computer Graphics and Interactive Techniques SIGGRAPH '88*, vol. 22, no. 4, 299–308, 1988.

41. Hauth M., Etzmuss O., Eberhardt B., Klein R., Sarlette R., Sattler M., Daubert K., Kautz J., Cloth Animation and Rendering, *Eurographics Tutorials*, 2002.

42. He T., Kaufman A., Collision detection for volumetric objects, *Proc. IEEE Visualization*, pp. 27–34, 1997.

43. Heidelberger B., Teschner M., Keiser R., Mueller M., Gross M., Consistent Penetration Depth Estimation for Deformable Collision Response, *Proc. Vision, Modelling, Visualization*, pp. 339–346, 2004.

44. Hirota G., Fisher S., Lin M., *Simulation of non-penetrating elastic bodies using distance fields*, Technical Report TR00-018, University of North Carolina at Chapel Hill, 2000.

45. Hockney R., The Potential Calculation and Some Applications, Alder B., Fernbach S., Rotenberg M. (eds.): *Methods in Computational Physics*, Plasma Physics, Academic Press, New York, vol. 9, pp. 136–211, 1970.

46. Hoff K., Zaferakis A., Lin M., Manocha D., Fast and Simple 2D Geometric Proximity Queries Using Graphics Hardware, *Proc. Symposium on Interactive 3D Graphics*, pp. 145–148, 2001.

47. Hoff K., Zaferakis A., Lin M., Manocha D., Fast 3D Geometric Proximity Queries Between Rigid and Deformable Models Using Graphics Hardware Acceleration, *Technical Report University of North Carolina, Computer Science*, 2002.

48. Hubbard P., Interactive Collision Detection, *Proc. IEEE Symposium on Research Frontiers in Virtual Reality*, pp. 24–31, 1993.

49. Hughes M., DiMattia C., Lin M., Manocha D., Efficient and Accurate Interference Detection for Polynomial Deformation and Soft Object Animation, *Proc. Computer Animation,* pp. 155–166, 1996.
50. James D., Pai D., Artdefo. Accurate Real-Time Deformable Objects, *Proc. 26th Annual Conference on Computer Graphics and Interactive Techniques SIGGRAPH '99,* pp. 65–72, 1999.
51. Kacic-Alesic Z., Nordenstam M., Bullock D., A Practical Dynamics System, *Proc. ACM SIGGRAPH/Eurographics Symposium on Computer Animation,* pp. 7–16, 2003.
52. Kakadiaris I.A. and Metaxas D., Vision-Based Animation of Digital Humans, *Proc. Computer Animation '98 Conf.,* pp. 144–152, June 1998.
53. Kakadiaris I., Metaxas D., Model-Based Estimation of 3D Human Motion, *IEEE Transactions on Pattern Analysis and Machine Intelligence,* vol. 22, no. 12, pp. 1453–1459, 2000.
54. Kim Y., Otaduy M., Lin M., Manocha D., Fast Penetration Depth Computation for Physically-based Animation, *Proc. ACM SIGGRAPH/Eurographics Symposium on Computer Animation,* pp. 23–31, 2002.
55. Kim Y., Hoff K., Lin M., Manocha D., Closest Point Query Among the Union of Convex Polytopes Using Rasterization Hardware, *Journal of Graphics Tools,* vol. 7, no. 4, 43–51, 2002.
56. Kim Y. J., Lin M. C., Manocha D., Incremental Penetration Depth Estimation Between Convex Polytopes Using Dual-Space Expansion, *IEEE Transactions on Visualization and Computer Graphics,* vol. 10, no. 2, pp. 152–163, 2004.
57. Klosowski J., Held M., Mitchell J., Sowizral H., Zikan K., Efficient Collision Detection Using Bounding Volume Hierarchies of k-DOPs, *IEEE Transactions on Visualization and Computer Graphics,* vol. 4, no. 1, 21–36, 1998.
58. Larsen E., Gottschalk S., Lin M., Manocha D., *Fast proximity queries with swept sphere volumes,* Technical Report TR99-018, University of North Carolina at Chapel Hill, 1999.
59. Larsson T., Akenine-Moeller T., Collision Detection for Continuously Deforming Bodies, *Proc. Eurographics,* pp. 325–333, 2001.
60. Lombardo J., Cani M.-P., Neyret F., Real-time Collision Detection for Virtual Surgery, *Proc. Computer Animation,* pp. 33–39, 1999.
61. Melax S., Dynamic Plane Shifting BSP Traversal, *Proc. Graphics Interface,* pp. 213–220, 2000.
62. McKenna M., Zeltzer D., Dynamic Simulation of Autonomous Legged Locomotion *ACM SIGGRAPH Computer Graphics, Proc. 17th Annual Conference on Computer Graphics and Interactive Techniques SIGGRAPH '90,* vol. 24, no. 4, 29–38, 1990.
63. Milliron T., Jensen R., Barzel R., Finkelstein A., A Framework for Geometric Warps and Deformations, *ACM Transactions on Graphics,* vol. 21, no. 1, 20–51, 2002.
64. Mirtich B., *Efficient algorithms for two-phase collision detection,* Technical Report TR-97-23, Mitsubishi Electric Research Laboratory, 1997.
65. Moore M., Wilhelms J., Collision Detection and Response for Computer Animation, *ACM SIGGRAPH Computer Graphics , Proc. 15th Annual Conference on Computer Graphics and Interactive Techniques SIGGRAPH '88,* vol. 22, no. 4, 289–298, 1988.

66. Müller M., Dorsey J., McMillan L., Jagnow R., Cutler B., Stable Real-Time Deformations, *Proc. ACM SIGGRAPH/Eurographics Symposium on Computer Animation*, pp. 49–54, 2002.
67. Müller M., Gross M., Interactive Virtual Materials, *Proc. Graphics Interface*, pp. 239–246, May 17–19, 2004.
68. Müller M., Heidelberger B., Teschner M., Gross M., Meshless Deformations Based on Shape Matching, *ACM Transactions on Graphics (TOG), ACM SIGGRAPH 2005 Papers SIGGRAPH '05*, vol. 24, no. 3, 471–478, 2005.
69. O'Brien J., Bargteil A., Hodgins J., Graphical Modelling and Animation of Ductile Fracture, *ACM Transactions on Graphics (TOG), Proc. 29th Annual Conference on Computer Graphics and Interactive Techniques SIGGRAPH '02*, vol. 21, no. 3, pp. 291–294, 2002.
70. O'Rourke J. and Badler N.I., Model-Based Image Analysis of Human Motion using Constraint Propagation, *IEEE Trans. Pattern Analysis and Machine Intelligence*, vol. 2, no. 6, 522–536, 1980.
71. Pauly M., Pai D. K., Guibas L. J., Quasi-Rigid Objects in Contact, *Proc. ACM SIGGRAPH/Eurographics Symposium on Computer Animation*, pp. 109–119, 2004.
72. Platt J. C., Barr A. H., Constraint Methods for Flexible Models, *ACM SIGGRAPH Computer Graphics, Proc. 15th Annual Conference on Computer Graphics and Interactive Techniques SIGGRAPH '88*, vol. 22, no. 4, 279–288, 1988.
73. Provot X., Deformation Constraints in a Mass-Spring Model to Describe Rigid Cloth Behavior, *Graphics Interface,* pp. 147–154, 1995.
74. Provot X., Collision and Self-collision Handling in Cloth Model Dedicated to Design Garment, *Proc. Graphics Interface*, pp. 177–189, 1997.
75. Quinlan S., Efficient Distance Computation Between Non-convex Objects, *Proc. IEEE International Conference on Robotics and Automation*, pp. 3324–3329, 1994.
76. Redon S., Kim Y. J., Lin M. C., Manocha D., Fast Continuous Collision Detection for Articulated Models, *Proc. Symposium on Solid Modelling and Applications*, 2004.
77. Rehg J.M. and Kanade T., Model-Based Tracking of Self-Occluding Articulated Objects, *Proc. International Conference Computer Vision*, pp. 612–617, June 1995.
78. Spillmann J., Teschner M., Contact Surface Computation for Coarsely Sampled Deformable Objects, *Proc. Vision, Modelling, Visualization VMV'05*, Erlangen, Germany, pp. 289–296, Nov. 16–18, 2005.
79. Spillmann J., Becker M., Teschner M., Efficient Updates of Bounding Sphere Hierarchies for Geometrically Deformable Models, *Proc. Virtual Reality Interactions and Physical Simulations VriPhys*, Madrid, Spain, pp. 53–60, Nov. 6–7, 2006. Best Paper Award.
80. Spillmann J., Wagner M., Teschner M., Robust Tetrahedral Meshing of Triangle Soups, *Proc. Vision, Modelling, Visualization VMV'06*, Aachen, Germany, pp. 9–16, Nov. 22–24, 2006.
81. Spillmann J., Becker M., Teschner M., Non-iterative Computation of Contact Forces for Deformable Objects, *Journal of WSCG 2007*, vol. 15, no. 1–3, 33–40, Feb. 2007.

82. Sud A., Otaduy M. A., Manocha D., DiFi: Fast 3D Distance Field Computation Using Graphics Hardware, *Computer Graphics Forum*, vol. 23, no. 3, 557–566, 2004.
83. Bradshaw G., O'Sullivan C., Sphere-tree construction using medial-axis approximation, *Proc. ACM SIGGRAPH/Eurographics Symposium on Computer Animation*, pp. 33–40, 2002.
84. Swope W., Andersen H., Berenc P., Wilson K., A Computer Simulation Method for the Calculation of Equilibrium Constants for the Formation of Physical Clusters of Molecules: Application to Small Water Clusters, *Journal of Chemical Physics,* vol. 76, no. 1, 1982.
85. Terzopoulos D., Platt J. C., Barr A. H., Elastically Deformable Models, *ACM SIGGRAPH Computer Graphics, Proc. 14th Annual Conference on Computer Graphics and Interactive Techniques SIGGRAPH '87*, vol. 21, no. 4, 205–214, 1987.
86. Terzopoulos D., Fleischer K., Deformable Models, *The Visual Computer,* vol. 4, pp. 306–331, 1988.
87. Terzopoulos D., Fleischer K., Modelling Inelastic Deformation: Viscoelasticity, Plasticity, Fracture, *ACM SIGGRAPH Computer Graphics, Proc. 15th Annual Conference on Computer Graphics and Interactive Techniques SIGGRAPH '88*, vol. 22, no. 4, 269–278, 1988.
88. Teschner M., Heidelberger B., Müller M., Pomeranets D., Gross M..Optimized Spatial Hashing for Collision Detection of Deformable Objects, *Proc. Vision, Modelling, Visualization*, pp. 47–54, 2003.
89. Teschner M., Heidelberger B., Müller M., Gross M., A Versatile and Robust Model for Geometrically Complex Deformable Solids, *Proc. Computer Graphics International*, pp. 312–319, 2004.
90. Teschner M., Kimmerle S., Heidelberger B., Zachmann G., Raghupathi L., Fuhrmann A., Cani M.-P., Faure F., Magnetat-Thalmann N., Strasser W., Collision Detection for Deformable Objects, *Computer Graphics Forum*, vol. 24, no. 1, 61–81, 2005.
91. Turk G., *Interactive collision detection for molecular graphics*, Technical Report TR90-014, University of North Carolina at Chapel Hill, 1990.
92. Verlet L., Computer Experiments on Classical Fluids. Ii. Equilibrium Correlation Functions, *Physical Review,* vol. 165, 201–204, 1967.
93. Volino P., Courchesne M., Magnenat-Thalmann N., Versatile and Efficient Techniques for Simulating Cloth and Other Deformable objects, *Proc. 22nd Annual Conference on Computer Graphics and Interactive Techniques SIGGRAPH '95*, pp. 137–144, 1995.
94. Volino P., Magnenat-Thalmann N., Comparing Efficiency of Integration Methods for Cloth Animation, *Proc. Computer Graphics International,* pp. 265–274, 2001.
95. Wu X., Downes M., Goktekin T., Tendick F., Adaptive Nonlinear Finite Elements for Deformable Body Simulation Using Dynamic Progressive Meshes, *Proc. Eurographics*, pp. 349–358, 2001.
96. Zachmann G., Minimal Hierarchical Collision Detection, *Proc. Symposium on Virtual Reality Software and Technology*, pp. 121–128, 2002.
97. Zhang D., Yuen M., Collision Detection for Clothed Human Animation, *Proc. Pacific Graphics*, pp. 328–337, 2000.

24

From Performance Theory to Character Animation Tools

Michael Neff[1] and Eugene Fiume[2]

[1] Department of Computer Science and Program for Technocultural Studies
   University of California, Davis, One Shields Ave, Davis, CA, 95616, USA
[2] Department of Computer Science
   University of Toronto, Toronto, Canada

Summary. The artistic performance literature has much to say about expressive human movement. This chapter argues that powerful, expressive animation tools can be derived from the classic lessons and best practices in this field. A supporting software system for creating expressive character animation is presented that takes advantage of these lessons. To this end, an analysis of the literature was conducted and a set of key aspects of expressive movement were synthesized. To validate the approach, a subset of these aspects is represented computationally as *movement properties*, or *properties* for short. These properties provide a set of handles and language for expressing movement. They can be used to write *character sketches*, which are executable descriptions of a particular character's movement style. They can also be used to edit and refine an animation. A semantics is provided that describes how the software framework allows these varied properties to be combined in specifying an animation. Examples of character sketches and edits are shown on movements of a standing character.

24.1 Introduction

By analogy to other fields of scholarly research, we can view researchers interested in understanding and manipulating motion as starting from either primary or secondary sources. Primary approaches begin with raw motion data, while secondary approaches rely on analyses conducted by other researchers that in turn have their ultimate bases in actual motion. Much of the work in this volume relies on primary approaches, and often does not require an exlicit understanding of motion. For instance, Theobalt et al. (Chapter 22) capture motion from video and recreate the motion in novel lighting conditions. Elgammal et al. (Chapter 2) use learning techniques to construct implicit models of motion applied to tasks like generation, recognition and style transfer. In a step closer to secondary techniques, Mueller et al. (Chapter 20) explicitly identify features of movement and use these to identify similar motions in databases.

B. Rosenhahn et al. (eds.), Human Motion – Understanding, Modelling, Capture, and Animation, 597–629.

Our work seeks to create better tools for expressive character animation, and relies on explicit, secondary sources to do this. The process followed was to first find an analysis of key expressive aspects of movement. There are numerous potential sources, including psychology, biomechanics, physics, and direct experience. We focus on the performing arts literature. Based on an analysis from this field, we synthesize computational models. Specifically, we build concrete representations of the ideas presented in the literature. These models are then combined within a software framework and used to generate motion.

The advantage of working from secondary sources is that experts in a relevant domain, namely expressive movement, have already identified the salient aspects of movement, the "handles", as it were, that are necessary for people to manipulate in order to construct expressive movement. We are ultimately interested in modelling motion that reflects personality, mood and style, and this literature identifies the primitives from which these qualities may be constructed. Providing these handles directly to animators offers a motion control workflow that is better aligned to their traditional and effective practice. The disadvantage of secondary approaches is that they are constructive in nature: the motion must be built from scratch, which can be a challenging task.

The remainder of this chapter describes a software system built to model expressive motion, based on ideas from the arts literature. The system can produce either kinematic animation or dynamic animation using a forward simulation/controller framework. In our prototype, we focus on a limited range of movement of a standing character: posture adjustments, gesturing, balance shifts and crouching. This range of motion is both reasonably stable from a dynamics perspective and expressively rich.

The important aspects of movement identified from the arts literature are encapsulated in small pieces of code called *movement properties*, or *properties* for short. These properties see expression through a *base representation* that defines a small set of constructs, and related commands, that are used to generate the motion. Algorithms have been developed that allow an artist to specify various high-level properties for a movement sequence that will then be combined into a single base representation and used to generate the character's motion.

These properties can be seen as defining a language by which the animator, or a high level character system, can express ideas about motion. This language is used to write *character sketches*, brief descriptions of a person's movement style that are applied to all the actions specified for a character. Not only do these character sketches allow an animator to move quickly to a particular portrayal of a character, perhaps more importantly, they allow an animator to *explore* different movement possibilities. By changing character sketches, an animator can very quickly try a different take on a movement sequence. Additional movement properties can then be applied on top of a selected sequence, allowing the animator to *refine* the motion. We believe these twin properties of exploration and refinement are particularly important in an artistic workflow. Examples shown below will demonstrate how a single movement sequence could be performed in considerably different ways solely through the application of character sketches and animator edits: in a neutral fashion; in a languid, sensual style; in a dejected manner; and in the manner of an old, tired man.

The chapter is organized as follows: Section 24.2 describes related technical work. Section 24.3 describes the different levels of information stored in the system and Section 24.4 describes how an animator works with the system. Section 24.5 introduces the performing arts literature and Sections 24.6 through 24.8 present specific ideas from the arts literature followed immediately by details of how they are represented in the system. This serves to illustrate the mapping between movement ideas and their corresponding computational representation. Section 24.9 provides examples of how the system supports both exploration and refinement. System details related to motion specification are formalized in Section 24.10 followed by details on motion generation in Section 24.11 and some discussion in the final section.

24.2 Background

Expressive or stylistic aspects of motion have seen increasing interest in the computer animation field in recent years. These approaches can be grouped based on how they derive their "understanding of motion".

24.2.1 Direct Reuse of Data

Some approaches do not develop a model of motion at all, but directly reuse captured motion by interpolating and extrapolating a set of sample motions. These approaches generally produce high quality results as they reuse real motion, but they provide limited control and flexibility to animators. The range of motions is defined by the convex hull of what is captured plus a small neighbourhood about which extrapolation produces reasonable results. Examples of these approaches include "Verbs and Adverbs" of Rose et al. [26] for locomotion and the reaching system of Wiley and Hahn [36].

24.2.2 Models from Data

A second approach builds models of certain aspects of motion from captured data. These approaches have been used to generate some impressive animations. Animator control is generally better than in the direct reuse approaches, but remains limited. One reason for this is the extracted models are generally not human understandable and editable. For instance, motion classes may be defined by very large vectors whose components do not have intuitive meanings.

Examples of this approach include: motion transformation [2,35] in which the difference between motion done in a neutral style and an expressive style is used to compute a transform that can then be applied to novel motions; style-machines [5], in which different dance styles are learnt from motion capture data allowing the style to be applied to new motion; and style translation [12] in which motion done in one style can be mapped to another style learnt from training data. Pullen and Bregler [25] take a different approach, extracting correlation information and high frequency data from motion capture data and using the correlation data to complete a roughly defined keyframe animation and the high frequency data to add "texture" or nuance to the motion. Liu et al. [18] use a learning approach to extract the parameters for a physical model from motion capture data with a particular style and then use spacetime optimization to synthesize variations on the motion with different physical conditions.

24.2.3 Procedural Approaches

Procedural approaches explicate a list of movement properties from a variety of sources such as direct experience, physics, the arts literature or psychology, and then hand craft computational representations of them. These approaches are synthetic in nature, in that the final motion is built up from the modelled properties, rather than using motion capture data. For this reason, more work is usually required to create appealing motion. The strength of these approaches is that they offer much better control. Because the properties they represent are conceptually well understood, they are more easily refined. They only rarely require complete changes to the style of the motion, and they are represented in an explicit way that may facilitate further customization.

Our work falls into this category. Other related approaches include the EMOTE system [6] which models Laban's Effort-Shape analysis, and the procedural work of Perlin [24] which relies on animator knowledge and useful interpolation primitives consisting of sinusoids and noise functions.

24.3 Levels of Representation

In moving from the conceptual definitions used in the performing arts literature to concrete requirements for software tools, we employ two levels of representation. The first is a low-level *base* representation that defines the core time-series data needed to generate motion. The second is a set of higher-level *actions* and *properties* corresponding to the incorporation of ideas in the arts literature that are used to provide animators with macroscopic controls. The base representation provides the foundation upon which to define the semantics of an open-ended set of such higher level motion concepts. The information flow is shown in Figure 24.1.

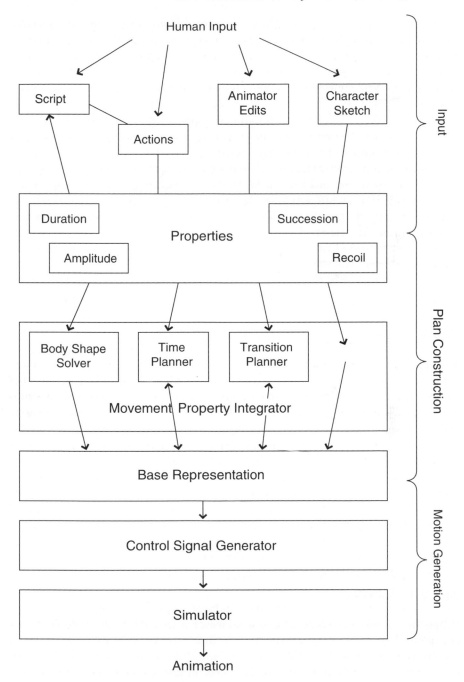

Fig. 24.1. Basic system architecture showing the three main stages of animator input, assembling the motion plan and movement generation.

The *base representation*, or *BRep*, is derived from poses consisting of subsets of a character's Degrees of Freedom (DOFs) at particular points in time. This makes it analogous to keyframe systems or pose-based control structures used in hand-tuned dynamic control. We chose this representation because of its simplicity and representational richness – keyframe representations may be low-level and very painstaking for an animator to use, but they offer excellent control over all aspects of movement. Such a representation also allows for a more compact specification and representation of motion than would be afforded with a frame-by-frame time series.

The base representation has a time ordered track for each DOF in the skeleton. It may also include tracks that define *signal parameters* for real-time controllers that are active in the system. These tracks can be used, for example, to specify the desired look-at location for a gaze tracker or desired balance point. All the tracks are populated with *transition elements* which span a short period of time and specify a transition curve that takes the DOF from its value at the start of the transition to a desired pose value. They specify the duration of the movement and an optional hold time. The BRep will be formalized in Section 24.10.

24.3.1 Actions

Actions define the structure of a motion. They specify how many poses are part of the motion, if these poses are repeated, which DOFs are part of each pose, etc. Actions are defined hierarchically. The lowest level, a *DOFEntry*, corresponds to one transition element. It specifies the behaviour of one DOF for a short period of time. *Poses* consist of a set of DOF Entries. Poses are grouped into possibly repeating *cycles*. An *action* consists of a sequence of cycles.

Actions provide handles on well-defined portions of motion. For instance, they can be used to specify a waving motion and subsequent edits can be applied to affect the entire wave, or parts of the wave. Actions may overlap in time if they use different DOFs. Actions define *what* motion is to be performed, and what DOFs are involved in it, but not *how* is to be performed. The latter is resolved by the use of specific *properties*, which define the value each DOF should take on, the length of a transition, etc.

24.3.2 Properties

Properties define the content or semantics of a motion and are designed to encapsulate particular, aesthetically relevant aspects of movement, such as the amplitude of a motion, the amount of muscular tension, a particular class of posture, etc. They provide a higher level language by which an animator can specify a desired motion. For instance, an animator might want to say "Increase the amplitude of the wave, keep the motion loose and do it with a higher tempo". Each of these edits can be invoked by selecting the appropriate property.

Table 24.1. A few of the properties defined in the system.

Property	Description
SetDOFValue	Specifies the value for a DOF
SetTension	Adjusts the amount of joint tension during a motion
Synchronize	Sets a timing relationship between two actions
SetDuration	Varies the transition time to a pose
SetTransitionCurve	Varies the transition envelope
VarySuccession	Adjusts the relative timing of joints in a motion
VaryExtent	Adjusts the amount of space a pose occupies
VaryAmplitude	Adjusts the angle range covered by a motion
GenerateWeightShifts	Generates idle weight shifting behaviour
SetReachShape	Varies posture during a reaching motion
SetPosture	Varies character posture
AdjustBalance	Alters the balance point of the character
SetRhythm	Coordinates actions to occur with a certain rhythm
CreateRecoil	Varies all aspects of movement to recoil from point

The short pieces of code that define properties generally operate by executing commands defined by the Base Representation. Some properties require more complex calculations and hence parameterize larger routines as discussed below. The set of properties is open ended and extensible. Since property definitions are explicitly represented in a human-comprehensible way, they can also be modified to meet the needs of a particular animator or production. The goal is that over time, this approach will allow a rich library of properties to be developed. Some properties are listed in Table 24.1. Note the wide range of granularity at which properties may act, from *SetDOFValue*, which controls the desired value for a single degree of freedom, to *CreateRecoil*, which changes a character's pose, warps the transition function, increases his tension and increases his tempo.

Each property accepts a small set of parameters. For instance, "look-at" takes a location and a weighting factor that indicates how directly the character should look at the spot. Properties also define rules for combining with other properties of the same type, allowing them to override or blend with previously applied properties. This is specified with a parameter accompanying an edit.

Properties are grouped into three categories. *Base properties* directly modify a low level attribute in the BRep, such as the duration of a transition. *Composite properties* generally modify higher level aspects of motion and do this by making calls to various base and other composite properties. *Generative properties* add additional actions to the movement script. For example, they can add idle motions, nervous ticks or hand flourishes at the end of a motion.

Properties can be applied at any level in the action hierarchy and cascade down to all the levels below. This reduces the amount of work required to specify the desired properties for a movement. Properties specified at lower levels take precedence over properties specified higher up in the hierarchy.

24.3.3 Modules

Some motions require a global view that cannot be isolated in a single property. This is due the need for effective code reuse, avoiding repeated implementations in multiple properties, and combining constraints from multiple, separately specified properties. We introduce *modules* that implement larger units of functionality and are parameterized by properties. Two of these will be discussed below: the *body shape solver* is used to calculate poses, and the *time planner* is used to arbitrate between time constraints.

24.4 Workflow

The animator's workflow follows an iterative specify-review-refine process:

1. Define new actions if needed.
2. Add actions to the performance script.
3. Specify/modify the character sketch.
4. Add/modify motion edits by applying additional properties.
5. Generate an animation.
6. Review, returning to earlier steps until satisfied.

Actions are defined in files and normally include an initial set of properties to give the action form. A library of actions has been built and can be extended as needed (Step 1). The script has multiple tracks, each containing a time-ordered list of actions (Step 2). For example, a script might specify a shrug, followed by a grand sweeping gesture to the right accompanied by a change in gaze direction. The character sketch specifies global edits, so is normally applied first (Step 3). An edit is simply a property along with a specific set of parameters and a label specifying the actions to which an edit is applied. Animator edits further refine the motion (Step 4). The process of generating the animation (Step 5) will be described in Section 24.11 below. After reviewing the generated sequence, the animator can go back to any step earlier in the process, make changes and regenerate the animation (Step 6). This process is repeated until convergence. Generally, the animator will start by making large changes via the character sketch in an exploratory phase and then make smaller changes via edits in a refinement phase.

24.5 Arts Literature

Most of the detailed discussion of the arts literature will be delayed to the sections below so that artistic ideas can be presented next to their technical implementation. In this section we will comment on our sources and approach and present some over-arching ideas about movement that do not fit into the more specific sections below.

Our study of the performance literature drew on three main sources: works on actor training and theory, traditional animation, and performance theory. The texts on actor training included work by developers of influential acting theories, such as Stanislavski [29–31] and Grotowski [10], as well as practical training texts, such as Alberts [1]. The traditional animation research leaned heavily on the Disney school, as represented by Thomas and Johnston [34] and Lasseter [15]. The movement theorists studied include Laban [14], Delsarte [28], and the founders of theatre anthropology, Barba and Saverese [4].

Our approach to the literature was not to rely on a single theorist, but to look for common properties presented by multiple researchers. We synthesized a list of these and used them as the basis for our computational tools. The properties were grouped into three categories: *shape*, or a pose at a particular instance in time; *transitions*, or transient properties related to how a character moves from one pose to the next; and *timing*, or properties related to the time structure of the movement. This categorization helped facilitate computational implementation, as different constructs are used for each of the categories. Examples of each category and corresponding implementations are presented below.

There are several lessons from the literature that do not fit into any of these categories, but shape our understanding of expressive movement. The first is simply that performance movement is different from daily-life movement. Barba [3] argues that performance movement is based on *excess* [3] in order to create a heightened impact with the audience. Clear communication is a fundamental requirement of performance movement. For this reason, performance draws on the twin principles of simplification [4, 15, 16, 19, 34] and exaggeration [3, 15, 34]. The amount of movement you show an audience is reduced and simplified in order to ensure that they notice the intended movements. These movements are then exaggerated to make them still easier for an audience to read.

24.6 Shape

24.6.1 Performance Theory

Numerous parts of the body have specific expressive uses. For instance, Delsarte [28] refers to the shoulder, elbow and wrist joints as thermometers because he feels they indicate *how much* rather than *what kind* of expression [28]. Raised shoulders act to strengthen any action; the more they are raised, the stronger the action. He argues no intense emotion is possible without the elevation of the shoulders, or forward contraction in the case of fear or backward pull to show aggression or defiance. The collar bones also play an important role in opening and closing the chest. Meanwhile, "The elbow approaches the body by reason of humility, and moves outward, away from the body, to express pride, arrogance, assertion of the will." [28, p. 41]

The shape of the torso is of significant expressive importance and has received limited attention in the computer graphics field. Laban suggests that there are three principal components of trunk movement: rotational movement about the length of the spine; "pincer-like" curling from one or both ends of the trunk and "bulge-like" shifting of the central area of the trunk out of its regular position [14]. The "S" curve or "Beauty Line" in the coronal plane involves the legs, torso and neck in making a large S curve with the entire body. It is a key pose in Indian dance, ancient Greek sculpture – the Venus de Milo offers a clear example (Figure 24.2) – and was taken up by Florentine sculptors in the 14th century [4].

Posture is one of the clearest indicators of both a character's overall personality and emotional state. Alberts suggests that posture is a combination of two components: the level of tension displayed in the body and the overall body position (e.g., standing, leaning, kneeling, lying down) [1]. Alberts proposes the following posture scale: hunched, stooped, slumped, drooped, slouched, sagging, tired, relaxed, straight, upright, uptight, erect and over-erect (at attention).

The term *recoil* is employed in two different ways. One, referring to an anticipatory effect, will be discussed in the Section 24.7. The other will be dealt with here. Delsarte's Law of Reaction [28, 33] suggests that the body recoils not in preparation for an action but in reaction to either an emotional stimulus or the climax of an emotional state. Any object that surprises us

Fig. 24.2. Venus de Milo.

will make the body react in recoil; the degree of recoil is related to the degree of emotion caused by the object [28]. Consider, for example, the reaction to a snake that has suddenly appeared, as opposed to seeing a cute puppy. In general, leaning one's torso away from an object indicates repulsion while leaning it toward an object indicates attraction [28].

Balance adjustments are fundamental to expressive movement. Indeed, Barba claims that the "dance of balance" is revealed in the fundamental principles of all performance forms [4]. Being near the edge of balance or slightly off balance greatly heightens the excitement of a performer's movements [14, 32]. When we stand in daily life, we are never still, but rather are constantly making small adjustments, shifting our weight to the toes, heels, right side, left side, etc. These movements should be modelled and amplified by the performer [4].

Extent or extension refers to how far an action or gesture takes place from a character's body. It can be thought of as how much space a character is using while completing an action. Full extension of arms held straight out would constitute maximal extension, while pulling the hands against the torso with elbows held tightly to the side would be minimal extension. Laban refers to the area around a person's body as the *kinesphere* and defines three regions within it [32]. The near region is anything within about ten inches of the character's body and the area for personal, intimate or perhaps nervous actions. The middle area is about two feet from the person's body and this is where daily activities take place, such as shaking hands. The area of far extent has the person extended to full reach. It is used for dramatic, extreme movements.

Amplitude is a concept related to extent that can also be quite powerful. One of Grotowski's exercises asks an actor to take a large action and perform it repeatedly, each time reducing the amplitude while trying to maintain the same energy [32]. The actor goes to 50%, then 40%, eventually working down to 3% and 2%. The result is a movement that is very small and subtle, but very energized and alive. Exaggeration works in the opposite direction, increasing the amplitude of a motion.

24.6.2 Implementation

As individual tools for shape modelling in the system are adapted from the performance literature, several aesthetic and practical constraints arise that must be accommodated when solving for a character's pose. For instance, we might want the character to assume a particular posture, with a particular balance shift, while also touching an object and keeping his feet planted on the ground. Many of these issues must be solved together, and different aesthetic aspects of movement will relate to different constraints. For this reason, we placed the logic for calculating pose within a module called a *body shape solver* that handles these issues collectively. The behaviour of this module is in turn controlled by individual properties that relate to specific aspects of pose.

The body shape solver combines balance adjustment, world space constraints on the character's wrists and feet, and soft aesthetic constraints in determining poses for a standing character. It uses a combination of feedback-based balance control and analytic and optimization based inverse kinematics (IK). In mime, the body is divided into four sections: the head, the torso, the arms and the legs [16]. We follow this organization when designing our IK routines because it allows us to construct handles for our shape solver that provide direct adjustment for each of these expressively different parts of the body. For instance, the approach allowed us to include direct control over a character's pelvic twist.

Balance control was included due to the expressive importance of balance shifts and is based on a simple feedback mechanism [23, 37]. A forward or lateral shift of the character's desired balance point can be specified by the animator. The character's actual centre of mass (COM) is projected onto the ground plane and an error term measures the distance between the actual and desired COM projections. This error is weighted and used to adjust the angles of one of the ankles by the feedback controller, reducing the error. The remaining angles in the lower body are solved using IK [23].

Analytic IK routines were derived for the lower body chain that goes from one foot to the other, the arms, and the gaze direction of the head. An optimization based routine is used to determine the final torso pose. This allows the solver to trade off reach constraints, which request a certain distance between the character's shoulder(s) and a world space location(s), against aesthetic constraints on the curvature of the spine [23].

Many of the aesthetically important aspects of body shape discussed above are of a low level, relating to movements of particular parts of the body, such as raising the shoulders or curving the spine in a particular way. These low-level aspects of body-shape were used to design the parameterization of pose space embedded in the body-shape solver. For example, the solver provides direct control over the swivel angle of the arm (the rotation around the axis extending from the wrist to the shoulder). This allows control over the distance the elbow lies from the chest. A reduced DOF parameterization of the spine was developed based on the key torso movements. The final spine shape is defined by three amplitudes: one for the transverse twist, or rotation along the axis of the spine; one for the sagittal curve, or forward and backward deformation; and one for the coronal curve, or sideways deformation. The transverse amplitude is evenly distributed between the spinal joints. Each of the sagittal and coronal amplitudes are multiplied by weighting factors to determine the value of each spinal DOF. These weighting factors induce one of a "small C", "large C", "small S", "large S", or no deformation in the spine as visualized for the coronal plane in Figure 24.3.

In a similar manner, the collar bones can be set either to align with or oppose each other and can be moved up and down, or forward and back. Taken together, the spine and collar bone movements span the important torso deformations identified in the arts literature and do so with a small set

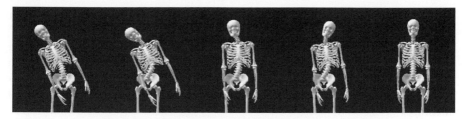

Fig. 24.3. Emphasized for illustrative purposes, the five coronal shape classes for the spine are shown from left to right: large C, small C, large S, small S, straight.

of parameters, providing efficient control. Each of these low-level parameters can be directly controlled by invoking the appropriate property.

Other aspects of shape are of a higher level, relating to a character's overall pose. Examples include recoil and postural erectness. These changes could be achieved by tediously adjusting low-level parameters. To allow more direct control, we introduce a new form of property called a *shape set* [23]. Shape sets encapsulate a particular type of higher level pose deformation and provide direct and simple control through a small set of parameters. As with all properties, shape sets consist of a short piece of code. Different poses can be specified as a vector of low-level parameters and shape sets define interpolation functions between a small set of such vectors. Alberts' posture range, discussed above, is defined in one shape set with a single parameter that specifies any location on the range between "over-erect" and "hunched". Another shape set implements "recoil" in the "Law of Reaction" sense of the term. It takes as parameters a world-space location and an intensity scalar. Invoking the shape set will cause the character to recoil back from the given location. Changing the intensity value changes how much the character recoils. Different recoil shape sets implement different forms of recoil.

Extent and *amplitude*, are completely defined within properties. The extent property allows the location of the wrists to be scaled up and down relative to the the centre of the character's body. The amplitude edit increases or decreases the angular range spanned by the DOFs during a movement sequence. Implementation details can be found in [21].

24.7 Transition

24.7.1 Performance Theory

Transition refers to how a character moves from pose to pose, along with other transitory effects. Disney animators found it effective to have the bulk of film footage near extreme poses (keyframes) and less footage in between in order to emphasize these poses [15, 34]. They referred to this as "slow in, slow out", indicating that when the animation is shown, most of the time is spent on the main poses. Interpolating splines are used to generate this result in computer animation [15], and this effect is known as an "ease-in, ease-out" curve. Often splines are parameterized to go beyond this, such as with tension, continuity and bias splines [13]. This reflects the importance of having some motions start slowly and end quickly and other motions do the opposite. We refer to this profile of the speed of a movement over its duration as the *motion envelope*.

Laban's Effort analysis describes transitory aspects of movement using the parameters flow, weight, space and time [14, 32]. Each parameter can either be indulged in or resisted. The work of Chi et al. [6] shows that much of this variation can be captured by properly parameterizing the transition curves that define the motion envelope.

The motion envelope is not necessarily spatially bound by the end points of the motion. For instance, *recoil* is one of the most frequently cited movement properties [3, 8, 14, 15, 28, 33, 34]. As a transition property, recoil involves first making a movement in the opposite direction of the intended movement, followed by the intended movement itself [8].[1] Thus, recoil serves to underscore and accentuate a movement [8]. It is one aspect of the more general concept of *anticipation* important in traditional animation [15, 34] and is described by Barba as *The Negation Principle* because by moving in a direction opposite to the action one is effectively negating the action before it is performed [3]. This creates a void in which the contrasting action can exist. In addition to recoil, some movements will exceed their end point before returning to it. This is referred to as *overshoot*.

The interplay of tension and relaxation is another widely cited movement property [3, 7, 14, 16, 28] and the other main transition property we consider. Tension and relaxation naturally interleave: there must first be relaxation in order for there to be tension and tension is followed again by relaxation. There is a consequent ebb and flow of energy that accompanies changes in tension [28]. This interplay between tension and relaxation is one way to create opposition and build interest in movement [3, 16, 28]. A rise in tension can serve to accent a movement [14]. Conversely, in his actor training, Stanislavski stresses the importance of an actor being relaxed and avoiding tension in his body [19, 29, 30]. Stiff arms and legs give the body a wooden quality, looking like a mannequin: "What emotions can a stick reflect?" [29, p. 102].

24.7.2 Implementation

In choosing an interpolation function to include in each Transition Element, it was necessary to span the full range of motion envelope variations described above. A cubic Hermite embedded in space and time was selected. We parameterize the curve on $[0, 1]$ to interpolate $(0, 0)$ and $(1, 1)$, yielding the following simplified version of the cubic Hermite curve:

---

[1] When used in terms of shape, recoil can refer to simply pulling back from an object.

$$H(u) = -2u^3 + 3u^2 + (u^3 - 2u^2 + u)R_0 + (u^3 - u^2)R_1,$$

where R_0 and R_1 are the start and end tangents. We thus represent a point on our time and position curve as the tuple (t, p), defined by:

$$t(u) = H_t(u) \tag{24.1}$$
$$p(u) = H_p(u) \tag{24.2}$$

where H_t is a Hermite curve embedded in time, H_p is a Hermite curve embedded in space and $u \in [0, 1]$ is a particular parameter value. Given a desired value of t, we can calculate the corresponding u and from that determine the value of p in the normalized $[0, 1]$ range. This value is then mapped to the actual range of the DOF for the movement to determine the desired position of the DOF. A transition function is thus defined by four samples: the initial and final tangents of p and of t, affording us a compact way to span the range of desired motion envelopes.

It is the importance of tension and relaxation that motivates our use of dynamic simulation. Humans increase joint tension by co-contraction of agonist–antagonist muscle pairs located on either side of the joint. Motivated by this observation, we place an antagonistic actuator at each rotational DOF in the skeleton [20]. The antagonistic actuator is a reformulation of a proportional derivative controller and consists of two angular springs arranged in opposition and an angular damper in parallel:

$$\tau = k_L(\theta_L - \theta) + k_H(\theta_H - \theta) - k_d\dot{\theta}, \tag{24.3}$$

where τ is the torque generated, θ is the current angle of the DOF and $\dot{\theta}$ is its current velocity. θ_L and θ_H are the low (L) and high (H) spring set points which serve as joint limits, k_L and k_H are the corresponding spring gains, and k_d is the gain on the damping term. The tension T, or stiffness of the joint, is taken as the sum of the two spring gains: $T = k_L + k_H$. The motion of the character is determined by forward time integration of Newton's equations using the accelerations generated by the torque each actuator produces.

Properties are used to set and vary the tension and damping values for each DOF during each transition. Equilibrium point control is used to achieve transitions. The angle of any DOF, at least at steady state, will be at the point at which all the forces and torques acting on the limb are balanced (summing to zero). Equilibrium point control generates limb movement by adjusting the stiffness of the two spring gains in order to vary the equilibrium point. This equilibrium point over time is called a *virtual trajectory* and may not precisely correspond to the actual trajectory of the limb. After merging all the tension properties, the system has a target starting and ending tension for each DOF in a transition. At the start of the transition, spring gains are calculated to achieve the desired starting and ending DOF values with the specified tension and stores these values in the BRep. The system then uses the transition functions to create virtual trajectories that interpolate these gain values, and updates the gains at each time step.

Tension variation affords important expressive nuances. It varies how a character will react to external forces. Stiff characters will behave more like rigid bodies, while loose characters will tend to "flop". Changing tension during a movement also provides a way to warp the motion envelope. Finally, adjusting tension allows control over secondary motion and overshoot. If a character brings his arm down to his side with a decrease in tension, the arm will swing slightly, adding natural pendular movement. When a character ends a movement in low tension, he will normally overshoot the end target before returning to it.

24.8 Timing

24.8.1 Performance Theory

The two main components of timing are *tempo* and *rhythm* [1, 14, 16, 19, 30]. Rhythm refers to the cadence of a set of movements. For instance, the pattern could be long, long, short, repeat. It deals with the patterning of full beats, eighth beats, etc. Tempo refers to the speed at which motions are completed; the speed of the beat. Tempo is independent of rhythm, in that a given rhythm could be performed with a fast or slow tempo. Tempo changes can hasten or draw out the action [30]. *Syncopation* can be used to put different accents on beats.

Stanislavski [19, 30] argues that each character has his/her own tempo-rhythm and it is the work of the actor to find it. The psychology literature also suggests that people have a characteristic tempo [9]. If a character is taking a strong decisive action, there will be only one tempo-rhythm, but a tortured soul like Hamlet will show several different tempo-rhythms at once [30]. Different tempo-rhythms can generate moods ranging from excitement to melancholy [30].

Other terms used to define timing include *duration* of an action [1] and *speed*, or the rate at which movements follow one another [14]. Clearly both are related to tempo and rhythm.

Another important property, *succession*, deals with how a movement passes or propagates through the body. Rarely will every limb involved in a motion start and stop at the same time. Delsarte [28] defined two types of succession: true or normal successions and reverse successions. In a *normal* succession, a movement starts at the base of a character's torso and spreads out to the extremities. In a *reverse* succession, the movement starts at the extremities and moves inward to the centre of the character. The concept of successions is also present in the work of numerous other researchers [10, 15, 30, 34].

24.8.2 Implementation

Outside of a dance setting, rhythm is a difficult concept for actors to develop a feel for, let alone to represent computationally. The French acting teacher

LeCoq writes: "Tempo can be defined, while rhythm is difficult to grasp. ... Rhythm is at the root of everything, like a mystery." [17, p. 32] We do not explicitly represent rhythm in our system. Instead, each transition element (*TElement*) contains two pieces of time data: a duration and a hold time. The duration is the amount of time it takes the character to move from the previous pose to the desired pose. The hold time is how long the DOF should be held at the desired value before the next transition begins. Properties can be used to set, scale and average each of these quantities. Performing scaling operations on the two quantities together changes the tempo of the movement. Everything is made faster or slowed down. Performing different scaling operations on the hold times to the durations, however, generates a change in rhythm. This is a provisional interpretation of rhythm, as more concrete definitions of the term are lacking. It is, however, useful in practice as will be seen with the "old man" sketch below.

Successions are implemented as a standalone property [21] and break up the time alignment of a pose. For a forward succession, a time delay is added for each joint, moving from the base of the spine out to the extremities. This is achieved by shifting the associated transition elements in a stair case manner.

Time planning benefits from a global view. Properties operating on individual actions may request synchronization constraints, scaling of TElement duration and hold times, and shifting of TElements that cascade through the timeline. A *Time Planner* implements a timeline semantics that works directly on the BRep as a post-process, making use of tag data added to the TElements by properties [22]. The time planner accommodates the various types of edits and implements elastic behaviour for the timeline where no TElements can overlap and there can be no gaps in the timeline. All time effects are ultimately achieved by adjusting the three time properties in the TElements: start time, duration and hold time.

24.9 Exploration and Refinement

The power of having a language to discuss movement becomes evident when we use that language to write novel descriptions of how a particular character moves, and then generate animation based on these. This is the fundamental idea behind a *character sketch*. The system allows an animator to write a short description of the style of a particular character and then uses that description to make whole scale changes to the specified animation script.

In this section, we will use a very simple animation to show the power of character sketches and how both exploration and refinement are supported. The animation is defined by three pairs of reach and look-at constraints and a neutral posture. The first constraint is placed far in front of the character, and causes him to reach for an object while looking at it. The second constraint is close to his chest and causes him to bring the object close to inspect it. The third constraint is again further in front of his body, but he now looks at the audience, giving the effect of him showing the object to the audience.[2]

We first apply an "old man" sketch to this motion that contains a stereotyped description of an old man's movement. This is one way the system supports exploration – the animator can very quickly try a completely different take on the motion sequence. The edits cause the character to move more slowly, take more time to prepare his actions, reduce movement to a more limited joint range and make less extended movements. Using a different scaling on the hold and duration times changes both the character's rhythm and tempo. The transition functions have been flattened to give a steadier, more linear pace to his movements. Some high frequency shaking was also added to the forearms, depicting what might occur with age or the onset of Parkinson's disease.

The character sketch for the old man is shown in Table 24.2. Each row defines an edit, the first term is the property to apply, the curly braces specify the recipient action (* indicates a global edit) and the rest of the entries are parameters. The specifier {* * * 16} associated with the SetDOFAngle edit specifies that the edit should be applied to DOF 16 in any pose, in any cycle, in any action. DOF 16 is the axial-angle of the neck. It should be noted that in a production system such information would be entered through a user friendly GUI in which the joint to control would be selected on the skeleton. The prototype system described here relies on script files.

The default posture for the old man is shown in Table 24.3. He has a bad hunch in his back, his collar bones are curled inward and down, his knees are

Table 24.2. The *old man* character sketch.

```
#Scale duration (increase it)
SetDuration {*} REL 1.8
#Scale hold time (increase it)
SetHoldTime {*} REL 3
#adjust shape settings
SetCharShapeParam {*} 1 load oldMan.shp
#flatten the transition functions
SetTransitionFunction {*} AVG 0 0 0 0
#reduce extent
VaryExtent {*} 0.8
#add a normal succession
VarySuccession {*} normal 0.01
#reduce neck y rotations
SetDOFAngle {* * * 16} REL 0.5
#add some shake to the left and right forearms
SetShake {* * * 24|33} ABS 0.017 7
```

---

[2] This sequence can be viewed online at
http://www.dgp.toronto.edu/people/neff/sketch.mp4

Table 24.3. The default posture for the *old man* sketch.

```
#add a slight bend to the arms
RArmLen .9              LArmLen .9
#create a hunch in the spine
SagClass SmallC
SagCentre 1.6
#create a forward and downward hunch with the collar bones
ColYCentre 1
ColZCentre -2.6
#arms should hang down when not in use
ArmsVertical true
#add some bend to the knees
rKnee .5               lKnee .5
```

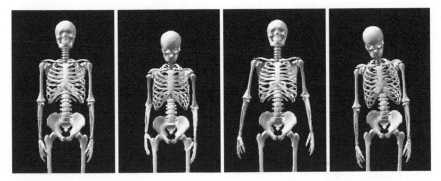

Fig. 24.4. Default posture for different character sketches. From the left, the first figure is the default posture, the second is the *old man* posture, the third is the *energetic* posture and the last is the *dejected posture*.

slightly bent and his arms are bent as well. This default posture is shown in the second frame of Figure 24.4. A GUI for pose control is provided and all amplitudes in the default postures were quickly determined using the feedback provided by this GUI.

In a second example, we demonstrate a refinement process by applying a series of edits to the initial animation. First, we add "beauty-line" posture changes to each movement, then we slow the timing, finally we apply a succession to increase the sense of flow. The resulting animation has a more languid, feminine look. Such edits can also be applied on top of previous character sketch edits.

Indeed, the refinement edits and the character sketch both use the same language of properties and so can be directly combined. To illustrate this, we apply an energetic character sketch (Table 24.4) to the more feminine, languid version of the animation we have just created. This sketch is largely the opposite of the old man sketch. The duration of movements is decreased and the hold time is almost eliminated. Extent is increased so the character

Table 24.4. *Energetic* character sketch.

```
#increase the speed of movements
SetDuration {*} REL .8
#almost eliminate the hold time between movements
SetHoldTime {*} REL .02
#specify a default posture
SetCharShapeParam {*} 1 load energetic.shp
#warp motion envelope to start more quickly
#this could also be done with a tension edit
SetTransitionFunction {*} AVG 0 0 3 0
#increase the extent of actions
VaryExtent {*} 1.3
```

Table 24.5. Base posture for the *energetic* sketch.

```
#slightly shorten the arms
RArmLen .95            LArmLen .95
#keep the arms out from the character's side (slight asymmetry)
#angle of arm relative to side
RArmAngle 10
LArmAngle 6.5
#arch backwards in the sagittal plane
SagClass SmallS
SagCentre 1.4
#raise the shoulders and pull them back
ColYCentre -.9
ColZCentre 1
#arms should hang down when not in use
ArmsVertical true
```

takes up more space with its movements. Transition functions are warped so that motions start more quickly. The result is a much perkier version of the motion.

Table 24.5 shows the default posture that is used with the energetic sketch. The character thrusts its chest out and pulls its shoulders up and back. The arms are held out from his side so the character can indulge in the use of space. Arms are bent slightly to suggest some tension in the elbows. Note: this blends with the previously specified beauty-line posture.

Finally, we apply a dejected character sketch (Table 24.6) to the original motion. *Dejected* is an interesting example as it shares much in common with the old man, and yet it should still make a distinct impression. The dejected sketch does feature a smaller extent edit than the old man sketch and joint ranges are not restricted. Similarly, both sketches slow the timing, but the exact value is different. The most significant difference is in how the motion envelope is warped. The old man sketch contains an edit that generates slightly

Table 24.6. *Dejected* character sketch.

```
#increase the duration
SetDuration {*} REL 1.3
#increase the hold time by a factor of 5
SetHoldTime {*} REL 5
#set the default posture
SetCharShapeParam {*} 1 load dejected.shp
#apply a slight forward succession
VarySuccession {*} normal 0.1
#reduce the extent of actions
VaryExtent {*} 0.9
#decrease shoulder tension during each movement for the shoulders
SetTensionValues {* * * 20-22, 29-31} ABS 400 100
#scale damping to match tension
SetDamping {* * * 20-22, 29-31} ABS 10
#similar edits are applied to the rest of the torso and arm joints.
```

Table 24.7. Base posture for the *dejected* sketch.

```
#arch the character to the right in the coronal plane
CorClass LargeC
CorCentre .4
#arch the character forward in the sagittal plane
SagClass LargeC
SagCentre 1.
#adjust the arms so they hang close to the character's side given
       the torso position
#angle of arm relative to side
RArmAngle 4
LArmAngle -3
#hunch the shoulders down
ColZCentre -1
#arms should hang down when not in use
ArmsVertical true
```

flattened kinematic transition functions. The dejected sketch instead uses tension reduction to warp the motion envelope. For each transition, the tension will start high and end quite low. This causes the motion to start slowly and speed up towards the end. The low final tension also leads to overshoot effects as the arm will waver in its final position or sway at the character's side. This looseness indicates a sense of indifference that is consistent with dejection. The tension decrease also gives the sense that the character hurls himself into each motion, but loses energy part way in.

The default posture for the dejected sketch is shown in Table 24.7. Like the old man, the collars are dropped, but they are not curled forward. Also a larger curve for the spine is used to suggest less of a hunch and more of a sense of weight. While the old man is fairly symmetrical, the dejected posture is assymetric to help generate a sense of unease.

24.10 Property Integration

Once the animator has completed the motion specification process via the various input channels, there are several active sources of information in the system that will all affect how the final animation is generated. These instructions must be combined in a consistent and predictable way in order to generate the final motion specification, as shown in Figure 24.1. This is the job of the *Movement Property Integrator*, or *MPI* for short, which must map all the higher level concepts into the base representation which is used to produce the animation. Before describing this process, we will first formalize the concept of the base representation and script as this will be necessary for the discussion below.

24.10.1 Formalization of the Base Representation and Script

The BRep is a random access data store that provides all the information necessary for either a kinematic or dynamic simulator to generate a final animation. Recall from its description in Section 24.3 that a BRep contains a track for every DOF in the skeleton and additional *signal parameter* tracks that specify control data for real time animation functions, such as varying the character's balance point or amount of pelvic twist. Each track is time ordered. A track T can be populated with either *Transition Elements* or *Control Parameters*:

$$T = \langle a_0, a_1, a_2, \cdots, a_k \rangle, a_i \in \{ControlParam, TransitionElement\}.$$

We define $\mathbb{T}$ to be the set of all such sequences T. TransitionElements are set by the planning process, described below (Section 24.10.2). They describe how a DOF should change over a short period in time and are used for planned motion. ControlParams are valid for one time step and are set by reactive controllers during the motion generation process to effect real-time adjustments (Section 24.11). They can override the values of a previously defined TransitionElement for a time step. Both elements provide the necessary data for either a dynamic or kinematic simulator to advance the state of the character.

Any instance B of a *base representation* is an m-tuple of tracks containing at minimum one track for each DOF. Thus $B \in \mathbb{T}^m$, or

$$B = (T_1, \cdots, T_m).$$

The index i identifies the DOF in the character's state to which track T_i is bound. Index values above the number of DOFs in the character state correspond to signal parameter tracks. We define the space of all base representations $\mathbb{B}$ to be all such B.

The BRep is incrementally and iteratively constructed during a motion specification process. This creates a sequence of refinements of valid BReps

$$\langle B^0, B^1, \cdots, B^n \rangle,$$

where $B^i \in \mathbb{B}$ and i indicates the ith iteration on the BRep. Any such B^i may well be executable by the simulator, but the final element B^n should be seen as a "converged" base representation. The initial base representation, B^0, is the null-operation, and is defined to be an m-tuple of empty tracks. With each iteration, more information is added or existing information is modified in the representation.

The Base Representation supports a set of commands and query operations which can be accessed by the properties. It is by using these operations that the properties ultimately achieve their functionality. The main commands and queries are summarized in the Appendix. Some commands simply add tags to transition elements that do not affect the final animation, but can be queried by other processes. For example, a tag might be added to increase the start time of subsequent actions, but the final change would be made by the Time Planner.

An animator does not interact with the BRep, and in general, need not be aware of it. The animator specifies the actions to be performed by a character in a script S (Section 24.4) containing a time-ordered series of actions that may overlap. The script can be thought of as a high level list of actions performed, whereas the BRep is similar to a keyframe system. More formally, a script consists of an unordered set of tracks R which each contain a time ordered sequence of actions:

$$R = \langle a_i \rangle, a_i \in Action \wedge StopTime(a_i) \leq StartTime(a_{i+1}),$$

$$S = \{R_1, \cdots, R_n\}, n \geq 1.$$

Normally, n is about 5. Actions (see Section 24.3.1) provide a rough description of a movement that is modified through the application of properties. The animator defines an initial script which outlines the animation sequence and can directly edit it during any iteration of the animation. It can also be modified by *generator properties*. Character sketches and animator edits are applied to change the form of the motion.

24.10.2 Generating a Motion Plan

The Movement Property Integrator is responsible for mapping the various forms of input into an executable motion plan, stored in the BRep. The stages involved in this process are shown in Figure 24.5 and described below. The script is first refined, all active properties in the system are then resolved and

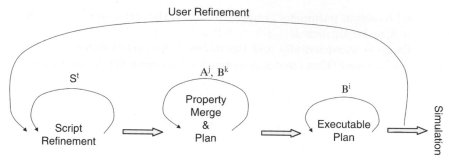

Fig. 24.5. Generation of an executable movement plan, showing the three stages: script refinement, property mapping and BRep refinement.

applied, and then the BRep goes through a final refinement stage. In a single iteration, the system transforms the user's input to an animation sequence. The system *always* generates an animation.

The MPI operates by applying properties, triggering solvers and running filters. Most of the "knowledge" of how to modify movement is contained in these entities and the MPI's job is rather one of coordination and arbitration. Multiple properties may attempt to modify the same underlying data (e.g., a DOF angle at a particular time) and the MPI determines how to correctly merge them. It must also order all operations so that data created by one process will be in place before it is needed by another. All this must be done in a manner that is consistent and predictable to ensure that high level animator adjustments have the expected result.

Due to the complex interactions among primitives and the current state of the motion plan, operators are required that clarify how high level instructions are decomposed into an executable base representation. The MPI makes use of four classes of operators for: script refinement ($\mathbf{Q}$), attaching properties to actions (α), executing properties ($\mathbf{M}$), and refining the BRep ($\mathbf{F}$).

Algebraically, any operator $\mathbf{Q}$ is a mapping from the script, together with information about how the script is to be modified, to a script. More formally, the operator $\mathbf{Q} : \mathbb{S} \times \mathbb{P}_G \to \mathbb{S}$, where $\mathbb{P}_G$ is the set of generator properties. The other operators can be defined in a similar manner. The operator α attaches properties to actions in order to apply animator edits and the character sketch. It can be represented as $\alpha : \mathbb{A} \times \mathbb{P} \to \mathbb{A}$ where $\mathbb{P}$ is the set of movement properties and $\mathbb{A}$ is the set of actions that are modified by these properties. The M operators execute the properties to update the BRep: $\mathbf{M} : \mathbb{B} \times \mathbb{A} \times \mathbb{P} \to \mathbb{B}$. Finally, the F operators also update the BRep, but take only the BRep as input: $\mathbf{F} : \mathbb{B} \to \mathbb{B}$. The MPI thus functions by composing a sequence of mapping operations: $\langle Q_1, \ldots, Q_k, \alpha_1, \cdots, \alpha_l, M_1, \cdots, M_m, F_1, \ldots, F_n \rangle$ where the comma is taken to mean the composition of functions. It is the MPI's job to specify, order and trigger these mapping operations. The ordering described here, and implemented in the current system, is designed to minimize conflicts between the various properties and to ensure that properties are applied ahead of when they are used. For instance, it will be seen below that signal parameter properties are applied before shape calculator properties because they will be needed by the shape calculator. The construction of an animation is essentially iterative.

In building the motion plan, the MPI begins by applying all the generator properties contained in the character sketch or animator edits. These properties act to update the script S^i by adding new actions to it, for instance, adding nervous hand twitches to a motion sequence:

$$MPI : S^i \rightarrow S^{i+1} .$$

The operator Q is then quite simply the application of a generator property: $\langle Q_i \rangle = \langle P_j \rangle$.

Once the generator properties have been dealt with, the remaining edits in the character sketch and animator edit lists act to attach properties to actions in the script. The MPI performs this binding through the α operators:

$$\alpha : S^i(\mathbf{A}^j), CharacterSketch, AnimatorEdits \rightarrow S^i(\mathbf{A}^{j+1}) ,$$

where $\mathbf{A}^j$ is the initial set of actions and $\mathbf{A}^{j+1}$ is the evolving set of actions as they are augmented with further properties. For example, during this stage, a default posture, reach constraints, weight shifts and posture edits might all be attached to an action.

Note that the ordering in which properties are attached does not matter because each property contains a priority tag which indicates the source of the property. Before properties are merged, they are sorted based on these tags.

Once the script has been completed and all properties have been attached, the MPI uses this information to develop the executable Base Representation. This is done by performing a sequence of mapping operations: $\mathbf{M} = \langle M_1, M_2, \ldots, M_n \rangle$, where once again, the operator has the form $\mathbf{M} : \mathbb{B} \times \mathbb{A} \times \mathbb{P} \rightarrow \mathbb{B}$. The ordering of these operators is based on the type of property being applied as some property types need to be applied ahead of others. The current mapping order is as follows:

1. Generate signal parameters (M_1).
2. Apply shape solver properties (M_2).
3. Apply other shape properties (M_3).
4. Apply timing properties (M_4).
5. Apply transition properties (M_5).
6. Apply properties that need to query the BRep (M_6).

Each operator normally consists of a *merge* phase followed by an *apply* phase. Composite properties are handled somewhat differently. The merge phase is the same as for other properties, but the apply phase acts by attaching a new set of low-level properties to the actions rather than writing directly into the BRep. Composite properties are processed first so that the low-level properties they generate can be merged with other active properties.

The effect of the six operators can be more precisely specified. M_1, M_3, M_4 and M_5 all take the basic form:

$$M(B^i, A, P_A) = B^{i+1},$$

where M is the mapping operator, A is an action included in the script and P_A is a set of properties of a given type associated with A. These mappings take all of the properties of a particular type that are attached to the specified action and map them to the BRep, generating its next iteration. In these mappings the properties are blind: i.e., they have no knowledge of the data that may already be stored in the BRep. The mapping operations are repeated for every action in the script. Some timing and transition properties will write tags into Transition Elements to request effects such as synchronization. Once all such properties have been applied, the timing and transition planners must be run in order to generate the result.

The shape solver operator M_2 is slightly more complicated and involves multiple steps. The first operator to be applied has the form

$$M(D^0, A, P_A) = D^1,$$

where D is a data store associated with the shape solver for the current action. All shape solver properties associated with the action write parameters into D. The shape solver is then invoked, which solves for a pose and writes the result back into the action as a set of SetDOFValue properties. This can be represented in operator form as:

$$\alpha(P_{ss}, A^i) = A^{i+1}$$

where P_{ss} are the properties determined by the shape solver. The final step is to merge and apply these properties using the M_3 operator:

$$M(B^i, A^{i+1}, P_A) = B^{i+1}.$$

The M_6 operator has the form

$$M(B^i, A, P_A(B^i)) = B^{i+1}.$$

The distinction here is that these properties can have a dependency on the existing iteration of the BRep whereas in previous M operators, they could not. Properties can query the BRep.

Merge: Before a property writes its data into the Base Representation, all properties of a given type acting on a specific action level are *merged*. Actions in general may have multiple properties of a given type attached, coming from different input sources: the initial action description, the character sketch and animator edits. It is important that all the applied properties can potentially affect the final motion, rather than one property replacing all earlier properties of a given type. This is what, for instance, allows the character sketch to *adjust* an existing movement sequence, rather than *define* it and greatly increases the flexibility and power of these edits. The "merge" resolution process effectively leaves a single property of each type attached to an action level.

The merge process involves three steps. First, properties are pushed down to the lowest level of the action hierarchy at which they can act. This ensures that all properties of a given type will be at the same level in the hierarchy when the merge is performed. Second, properties are sorted based on priority level. Priority level is determined based on the property source (default order: action description first, then character sketch, animator edits are last) and the level in the hierarchy the property is defined at (higher levels being used first). Finally, the sorted property list is merged. Prototype merge functions are available which support merges that overwrite (ABSolute), scale by a factor (RELative), average (AVeraGe) or add (ADD) the individual property parameters. Properties can also define their own merge semantics. This allows the property designer to decide how a property can best be merged. As an example, the *SetDuration* property supports absolute and relative duration specifications, the latter acting to scale a lower priority absolute duration.

Apply: Once the properties have been merged, the application process is quite straightforward. The one property of each type that is still active at an action node has its *apply* method called and simply writes its data either into the BRep or a data store associated with a solver. Once data has been committed to the BRep, it can be queried by future properties and filters.

BRep Refinement: Once the action-based properties have been applied, the BRep can be further refined through filtering or post-processing. These operators act in a similar manner, where for a filter F,

$$F(B^i) = B^{i+1}.$$

Unlike the M operators, these processes make no reference to the script, working solely from the information contained in the BRep. An example of such a filter is the Time Planner described previously which enforces the timeline semantics. Filters are a powerful notion as they admit the full extent of signal processing to be incorporated into the framework.

24.10.3 Property Implementation

As can be seen from the preceding discussion, properties must regulate both how they are to be combined with other properties and how they will update the base representation. In order to do this, they must define the data outlined in Table 24.8 and also implement the functions described in Table 24.9. If a property is applied to multiple actions, it will be replicated and a separate instance of the property will be applied to each action.

Table 24.8. Properties must include the above data. Many of these data items are used in property merging.

Data	Description
Combination type	Specifies how the property should be combined with other properties of the same type. (e.g., ABS, AVG, ADD and REL, as per the main text)
Category	Specifies which aspect of movement the property affects. Categories include: shape solver, shape, timing, transition, postburn in, generator, applicator, reactive controller and other
Priority level	Determines the order properties are merged based on the source of the property and the level in the action hierarchy at which it is applied
Minimum property level	This indicates the lowest level in the action hierarchy at which this property makes sense (e.g. cycle or DOFEntry)
bCombinable	Flag that states whether this property should be merged with other properties of the same type or should just be directly applied
Parameters	Most properties will define their own set of parameter values

Table 24.9. Properties must implement these three functions. Properties that are not combinable can implement stubs for GetParamString and Combine.

Function	Description
GetParamString	A property must be able to return its paramater values as a string. The strings can be passed between properties of the same type, which can decode them and use them in merge operations
Combine	Properties define how they can be merged, including what merge types they will support. This is done by writing a *combine* function
Apply	Every property must define this function. It is called to execute the property. This triggers the property to generate its output

24.11 Movement Generation

Once the BRep has been finalized, it must be executed to generate an animation. This process, for one time step, is shown in Figure 24.6. *Reactive controllers* allow the character to adjust its behaviour based on its state at the beginning of the time step. This is particularly important in dynamic simulation, where the effect of an action is not completely known ahead of time and adjustments may need to be made in order to ensure that a motion completes successfully. The main reactive controller in our system is used for balance control.

Reactive controllers receive the system state, the current BRep, and sensor information as input. *State* is a vector of position and velocity values for every DOF in the character. Reactive controllers can then update the BRep for the

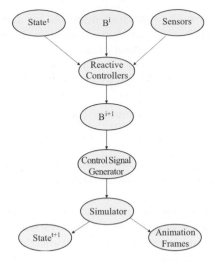

Fig. 24.6. Data flow for one time step of simulation.

current time step to attempt to better achieve the requested motion. They do this by writing Control Params into the BRep. As an example, the balance controller adjusts the desired DOF values for the charcter's lower body at each time step in order to achieve the balance adjustments, knee bends and pelvic twists specified in the signal parameter tracks of the BRep.

Once all the reactive controllers have made their adjustments, the *Control Signal Generator* takes the DOF tracks of the updated BRep as input and produces the control information required by the current simulator. For the kinematic simulator, this information consists of the value of each DOF. The dynamic simulator requires a torque for each DOF. The torque values are generated by either a proportional derivative or an antagonistic actuator positioned at each DOF and the Control Signal Generator provides the appropriate gain and set point data.

The simulator takes the input from the control signal generator and advances the simulation Δt seconds, updating the character state and producing animation frames. The dynamic simulator is based on code generated by the commercial software SD/Fast [11]. The kinematic simulator implements a standard articulated body hierarchy. The entire software system is built on top of the DANCE framework [27].

24.12 Discussion and Conclusions

We believe the incorporation of familiar concepts from the performing arts results in an appealing animation workflow that will attract a new and wider community to engage in the creation of expressive human animation. The project is not without considerable challenges, but we have presented an extensible system that permits the expressive concepts to be mapped into actions and properties in a manner that admits both a precise semantics and a way to resolve multiple properties into an overall motion plan.

Much more work is required to adapt a wider array of motion primitives on a more diverse set of motions. Better solutions to the balance problem are likely the main requisite for extending the approach to a wider array of physically simulated motions. In extending the expressive range of the system, the performance literature remains a rich resource that is far from exhausted. There are numerous concepts, such as the path a motion takes in space, that could be easily incorporated into the system, and many more that ellude easy computational definition, such as an actor's ability to react to another actor. The scaling properties of the motion property integrator have yet to be explored fully in light of the extensibility of the system. As the "language" of properties is increased and computational definitions developed, it is our hope that this work might contribute back to the performance community, serving as a test bed for our understanding of movement.

Theory-based approaches such as described here and data based approaches, as discussed in much of the rest of this volume, need not be antithetical. Indeed, data and learning based analyses could be used to generate models of specific movement properties. Similarly, the arts literature can provide guidance when pursuing data based approaches as to which aspects of movement are worth modelling. Hopefully this work will lead to the creation of more fine-scale data based models of specific aspects of movement, informed by the arts literature. Such approaches offer the potential to combine good animator control, as demonstrated here, with the detail of data based models. It should be possible to combine both data-derived properties and hand designed properties within our system.

Appendix

The following tables outline the major commands and queries that are used in writing the movement properties.

Table 24.10. Structural commands acting on the Base Representation

Command	Description
SetControlParams	Sets a control param that is valid for a single time step
AddTransitionElement	Creates and adds a transition element to a given track of the BRep at a given time
InsertPose	Inserts a new pose into the midst of a current action
RemoveTransitionElement	Removes a transition element from the BRep

Table 24.11. BRep commands that update animation attributes.

Command	Description
SetStartTime	Sets the time the TElement becomes active
SetStopTime	Sets the time the TElement ceases to be active
ShiftTimes	Shifts the whole transition element (start and stop times)
SetDuration	Sets the duration of a transition from the initial value of the DOF to the desired value
SetHold	Specifies how long the DOF value should be held for
SetDamping	Sets the damping value used during the transition
Set{Start|End}Tension	Sets the tension values used in dynamic simulation
SetDOFValue	Sets the desired DOF value for the end of a transition
Set{Start|Stop}Value	Used for control parameter transition curves, specifying both the initial and final value of the DOF
SetTransitionTangents	Set the four tangent values that define a cubic Hermite transition curve
ApplyOffsetCurve	Specifies an offset curve that is added to the transition curve

Table 24.12. Commands for adding meta data to TElements.

Command	Description
SetActionLabel	Sets the label that is used to identify the TElement. The label links it back to the generating action
SetTag	Allows a generic tag to be added to a given TElement. The tag has a name and a parameter

Table 24.13. Queries supported by the BRep.

Query	Description
GetTElementsWithName	Returns all the transition elements that match the given label. It can get a specific TElement, all the TElements for a given pose etc.
GetDOFValue	Returns the desired value of the Transition Element
GetStartTime	Returns the start time of a specified Transition Element
GetStopTime	Returns the stop time of a specified Transition Element
GetDuration	Returns the duration of a specified Transition Element
GetTag	Returns the tag data given a tag name

References

1. Alberts D. *The Expressive Body: Physical Characterization for the Actor.* Heinemann, Portsmouth, N.H., 1997.
2. Amaya K., Bruderlin A. and Calvert T. Emotion from motion. *Graphics Interface '96*, pp. 222–229, 1996.

3. Barba E. Dilated body. In E. Barba and N. Savarese, editors, *A Dictionary of Theatre Anthropology: The Secret Art of The Performer.* Routledge, London, 1991.
4. Barba E. and Savarese N. *A Dictionary of Theatre Anthropology: The Secret Art of The Performer.* Routledge, London, 1991.
5. Brand M. and Hertzmann A. Style machines. In *Proc. SIGGRAPH 2000*, pp. 183–192, 2000.
6. Chi D.M., Costa M., Zhao L. and Badler N.I. The EMOTE model for effort and shape. In *Proc. SIGGRAPH 2000*, pp. 173–182, 2000.
7. Dorcy J. *The Mime.* Robert Speller, 1961. Translated by Robert Speeler, Jr. and Pierre de Fontnouvelle.
8. Eisenstein S. On recoil movement. In A. Law and M. Gordon, editors, *Meyerhold, Eisenstein and Biomechanics: Actor Training in Revolutionary Russia.* McFarland and Company, Jefferson, North Carolina, 1996.
9. Gallaher P.E. Individual differences in nonverbal behavior: Dimensions of style. *Journal of Personality and Social Psychology*, 63(1):133–145, 1992.
10. Grotowski J. *Towards a Poor Theatre.* Odin Teatret, 1968. Edited by Eugenio Barba.
11. Hollars M.G., Rosenthal D.E. and Sherman M.A. *SD/FAST User's Manual.* Symbolic Dynamics Inc., 1994.
12. Hsu E., Pulli K. and Popović J. Style translation for human motion. *ACM Transactions on Graphics*, 24(3):1082–1089, Aug. 2005.
13. Kochanek D.H.U. and Bartels R.H. Interpolating splines with local tension, continuity and bias control. *Computer Graphics (Proc. SIGGRAPH 84)*, 18(3):33–41, 1984. Held in Minneapolis, Minnesota.
14. Laban R. *The Mastery of Movement.* Northcote House, London, fourth edition, 1988. Revised by Lisa Ullman.
15. Lasseter J. Principles of traditional animation applied to 3D computer animation. *Proc. SIGGRAPH 87*, 21(4):35–44, 1987.
16. Lawson J. *Mime: The Theory and Practice of Expressive Gesture With a Description of its Historical Development.* Sir Isaac Pitma and Son, London, 1957. Drawings by Peter Revitt.
17. LeCoq J. *The Moving Body: Teaching Creative Theatre.* Theatre Arts Books, 2001. with Jean-Gabriel Carasso and Jean-Claude Lallias. Translated by David Bradby.
18. Liu C.K., Hertzmann A. and Popović Z. Learning physics-based motion style with nonlinear inverse optimization. *ACM Transactions on Graphics*, 24(3):1071–1081, Aug. 2005.
19. Moore S. *The Stanislavski System: The Professional Training of an Actor.* Penguin Books, 1984.
20. Neff M. and Fiume E. Modelling tension and relaxation for computer animation. In *Proc. ACM SIGGRAPH Symposium on Computer Animation 2002*, pp. 81–88, 2002.
21. Neff M. and Fiume E. Aesthetic edits for character animation. In *Proc. ACM SIGGRAPH/Eurographics Symposium on Computer Animation 2003*, pp. 239–244, 2003.
22. Neff M. and Fiume E. AER: Aesthetic Exploration and Refinement for expressive character animation. In *Proc. ACM SIGGRAPH/Eurographics Symposium on Computer Animation 2005*, 2005.

23. Neff M. and Fiume E. Methods for exploring expressive stance. *Graphical Models*, 68(2):133–157, March 2006.
24. Perlin K. Real time responsive animation with personality. *IEEE Transactions on Visualization and Computer Graphics*, 1(1):5–15, 1995.
25. Pullen K. and Bregler C. Motion capture assisted animation: Texturing and synthesis. *ACM Transactions on Graphics*, 21(3):501–508, 2002.
26. Rose C., Cohen M.F. and Bodenheimer B. Verbs and adverbs: Multidimensional motion interpolation. *IEEE Computer Graphics and Applications*, 18(5):32–40, 1998.
27. Shapiro A., Faloutsos P. and Ng-Thow-Hing V. Dynamic animation and control environment. *Graphics Interface '05*, pp. 61–70, 2005.
28. Shawn T. *Every Little Movement: A Book about Francois Delsarte*. Dance Horizons, New York, second revised edition, 1963.
29. Stanislavski C. *An Actor Prepares*. Theatre Arts, 1936. Translated by Elizabeth Reynolds Hapgood.
30. Stanislavski C. *Building a Character*. Theatre Arts Books, 1949. Translated by Elizabeth Reynolds Hapgood.
31. Stanislavski C. *Creating a Role*. Theatre Arts Books, 1961. Translated by Elizabeth Reynolds Hapgood.
32. Tarver J. and Bligh K. The physical art of the performer, 1999. Workshop: A 30 hr. intensive introduction to Laban and Grotowski held at the Nightwood Theatre studio, Toronto.
33. Taylor G. Francois Delsarte: A codification of nineteenth-century acting. *Theatre Research International*, 24(1):71–81, 1999.
34. Thomas F. and Johnston O. *The Illusion of Life: Disney Animation*. Abbeville Press, New York, 1981.
35. Unuma M., Anjyo K. and Takeuchi R. Fourier principles for emotion-based human figure animation. *Proc. SIGGRAPH 95*, pp. 91–96, 1995.
36. Wiley D. and Hahn J. Interpolation synthesis of articulated figure motion. *IEEE Computer Graphic and Applications*, 17(6):39–45, 1997.
37. Wooten W.L. *Simulation of Leaping, Tumbling, Landing and Balancing Humans*. Ph.D. dissertation., Georgia Institute of Technology, 1998.

Index

Computational Imaging and Vision

1. B.M. ter Haar Romeny (ed.): *Geometry-Driven Diffusion in Computer Vision*. 1994
    ISBN 0-7923-3087-0
2. J. Serra and P. Soille (eds.): *Mathematical Morphology and Its Applications to Image Processing*. 1994   ISBN 0-7923-3093-5
3. Y. Bizais, C. Barillot, and R. Di Paola (eds.): *Information Processing in Medical Imaging*. 1995   ISBN 0-7923-3593-7
4. P. Grangeat and J.-L. Amans (eds.): *Three-Dimensional Image Reconstruction in Radiology and Nuclear Medicine*. 1996   ISBN 0-7923-4129-5
5. P. Maragos, R.W. Schafer and M.A. Butt (eds.): *Mathematical Morphology and Its Applications to Image and Signal Processing*. 1996   ISBN 0-7923-9733-9
6. G. Xu and Z. Zhang: *Epipolar Geometry in Stereo, Motion and Object Recognition. A Unified Approach*. 1996   ISBN 0-7923-4199-6
7. D. Eberly: *Ridges in Image and Data Analysis*. 1996   ISBN 0-7923-4268-2
8. J. Sporring, M. Nielsen, L. Florack and P. Johansen (eds.): *Gaussian Scale-Space Theory*. 1997   ISBN 0-7923-4561-4
9. M. Shah and R. Jain (eds.): *Motion-Based Recognition*. 1997   ISBN 0-7923-4618-1
10. L. Florack: *Image Structure*. 1997   ISBN 0-7923-4808-7
11. L.J. Latecki: *Discrete Representation of Spatial Objects in Computer Vision*. 1998
    ISBN 0-7923-4912-1
12. H.J.A.M. Heijmans and J.B.T.M. Roerdink (eds.): *Mathematical Morphology and its Applications to Image and Signal Processing*. 1998   ISBN 0-7923-5133-9
13. N. Karssemeijer, M. Thijssen, J. Hendriks and L. van Erning (eds.): *Digital Mammography*. 1998   ISBN 0-7923-5274-2
14. R. Highnam and M. Brady: *Mammographic Image Analysis*. 1999
    ISBN 0-7923-5620-9
15. I. Amidror: *The Theory of the Moiré Phenomenon*. 2000   ISBN 0-7923-5949-6;
    Pb: ISBN 0-7923-5950-x
16. G.L. Gimel'farb: *Image Textures and Gibbs Random Fields*. 1999 ISBN 0-7923-5961
17. R. Klette, H.S. Stiehl, M.A. Viergever and K.L. Vincken (eds.): *Performance Characterization in Computer Vision*. 2000   ISBN 0-7923-6374-4
18. J. Goutsias, L. Vincent and D.S. Bloomberg (eds.): *Mathematical Morphology and Its Applications to Image and Signal Processing*. 2000   ISBN 0-7923-7862-8
19. A.A. Petrosian and F.G. Meyer (eds.): *Wavelets in Signal and Image Analysis. From Theory to Practice*. 2001   ISBN 1-4020-0053-7
20. A. Jaklič, A. Leonardis and F. Solina: *Segmentation and Recovery of Superquadrics*. 2000   ISBN 0-7923-6601-8
21. K. Rohr: *Landmark-Based Image Analysis. Using Geometric and Intensity Models*. 2001   ISBN 0-7923-6751-0
22. R.C. Veltkamp, H. Burkhardt and H.-P. Kriegel (eds.): *State-of-the-Art in Content-Based Image and Video Retrieval*. 2001   ISBN 1-4020-0109-6
23. A.A. Amini and J.L. Prince (eds.): *Measurement of Cardiac Deformations from MRI: Physical and Mathematical Models*. 2001   ISBN 1-4020-0222-X

Computational Imaging and Vision